Discrete Simulation and Animation for Mining Engineers

Discrete Simulation and Animation for Mining Engineers

John R. Sturgul

CRC Press
Taylor & Francis Group
Boca Raton London New York

CRC Press is an imprint of the
Taylor & Francis Group, an **informa** business

CRC Press
Taylor & Francis Group
6000 Broken Sound Parkway NW, Suite 300
Boca Raton, FL 33487-2742

First issued in paperback 2017

ISBN-13: 978-1-4822-5441-9 (hbk)
ISBN-13: 978-1-138-74882-8 (pbk)

Visit the Taylor & Francis Web site at
http://www.taylorandfrancis.com

and the CRC Press Web site at
http://www.crcpress.com

Contents

Preface

A bucket-wheel excavator is the largest equipment in the world.

Mining simulation and animation can be a fun course! Ask any student who has taken a well-presented course on adding animations to the systems just simulated. It is one thing to write a simulation program that will create 100 different coloured stars but quite another to have these stars appear on the screen at random with various colours and make them blink, move, and then disappear. This course will teach you both how to do simulations using the powerful GPSS/H® simulation language and, along with it, how to make your simulation *come alive* with an animation software application known as PROOF®.

A simulation study represents an actual system as closely as possible, and the animation allows the user to *see* his or her system in motion. PROOF animation is one of the best such animations on the market and easily follows from GPSS/H. It can, of course, be used with other simulation languages (and non-simulation languages for that matter).

This course is the result of many years of teaching mining students from widely diverse backgrounds. It is also the evolution of notes used for short courses in many countries, such as Argentina, Australia, Brazil, Bulgaria, Canada, Chile, Peru, Poland, China, Romania, Russia, South Africa, Spain, Tanzania, and the United States. More importantly, it is the result of modelling actual mines and mining operations in the above countries plus others as well, such as Papua New Guinea, Borneo, Namibia, Ghana, and even a few others.

It is necessary to have both the student versions of PROOF and GPSS/H installed. These can be downloaded free of cost from www.wolverinesoftware.com.

All of the examples presented here assume that you have installed both Student GPSS/H and Student PROOF.

There will be quite a few files from which to learn. Those with Student PROOF will be in *pairs*: *.LAY and *.ATF, where the "*" could be any legal name such as MYFIRST.LAY and MYFIRST.ATF. Those with Student GPSS/H will be of form *.GPS. For example, a program might be named MYPROG.GPS and the animation pairs MYPROG.LAY and MYPROG.ATF.

If you already have other versions of the software, such as PROOF Professional® or GPSS/H® Professional, this is fine. You will need to know only in which directory on your computer each file is located and how to run the programs and the animation.

With every chapter, there are examples that should be run as exercises. In some cases, I debated whether to include the listing of the GPSS/H program when the actual program is given in a file. I decided to include certain listings where the student can benefit from important parts of it. In many cases, the student will be instructed to rerun the program with designated lines changed in order to compare results.

Comments, suggestions, and perhaps even some mining examples that the reader has created that might be added to these exercises are solicited. Send them to:john.sturgul@adelaide.edu.au.

The form of this course is a bit different from that of other courses. The normal way of teaching this subject is to cover first the simulation language and then, once the students are writing their simulation programs, to gradually introduce animation. Having taught simulation and animation for many years, I have noticed that the students really come alive when they start doing the animations. No doubt this is because of the opportunity for each to be his or her own artist. It is easy to be creative and make fancy animations, especially using the software presented here.

This course starts with animation. The first six chapters refer only to animation. Each topic or point to be considered will have an animation (or animations) to show the students what is being done. The simulation language is slowly introduced and, as each new topic is studied, there will be an animation to illustrate this topic. Thus, the student will always be guided to view an animation to *see* what is being taught.

Special thanks must go to James O. Henriksen, president of Wolverine Software, for his assistance in this book. Jim developed GPSS/H while a graduate student at the University of Michigan and continued its many morphs through the years. Dr. Thomas Schriber of the University of Michigan was also very helpful with his many suggestions and for his kindness in allowing me to use several of the examples from his classic textbooks on the GPSS and GPSS/H languages. It was his textbooks that got me started in this field, for which I am eternally thankful.

For the student: I envy you in a way. You are about to embark on a journey that will give you one of the strongest tools that exists for modelling the real world. It is rare that one can learn only one subject and then immediately be able to feel that he or she can solve real-life problems. Once you finish this course and work through the many examples and exercises, you will have a tool that will provide you such skills.

I tell my students during the first lecture that they will learn many subjects during their study to become mining engineers. These include subjects such as mine ventilation, rock mechanics, open-pit slope design, and underground mine design. All are important and necessary in any mining curriculum. However, no one is an expert without many years of actual practice. On the other hand, once a student learns mine simulation and animation, he or she can immediately put this training to use on day 1.

John R. Sturgul
University of Adelaide

Additional material is available from the CRC Web site: https://www.crcpress.com/product/isbn/9781482254419

Introduction

Simulation Models in General

The GPSS/H (General Purpose Simulation System/H) computer programming language is a special language that is used primarily to simulate what can be classified as *discrete systems*. A discrete system is one where, at any given instant in time, a *countable* number of things can take place. For example, students working in a computer room with multiple PCs comprise such a system; other examples are a truck being loaded, a truck dumping, or a shovel starting to load a truck. At any one instant in time, the number of things taking place can be counted. Many engineering systems are ideal examples of discrete systems. However, air flowing through a mine, water flowing through a pipe, and the deformation of a chair under the load of a heavy person are not examples of discrete systems.

Considering only discrete systems may seem restrictive, but it is not as restrictive as may appear at first reading. Nearly all of the problems one encounters in the study of queuing theory can be represented by discrete systems. Textbooks on the subject of operations research have numerous examples that can be considered as discrete systems. Some systems that may not appear to be discrete easily can be approximated by discrete systems. These might include cars and trucks travelling from one point to another. When travelling, their motion is certainly continuous but, for the purpose of studying traffic flow, their motion can be considered using discrete simulation. Material on a conveyor belt also can be considered as being part of a discrete system. We will study these and other examples in this book.

Some Examples of Discrete Systems

There are many examples of discrete systems. In general, whenever there is a queuing situation or a potential one, the system may be considered as a discrete system. A few such systems are as follows:

People coming to a news stand with a single worker: If the worker is busy, people wait until it is their turn.

People entering a bank with multiple tellers: The customers may either form individual queues for each teller or wait in a single queue (known as a *quickline*).

Trucks working at a construction site where a single shovel loads each truck: The trucks travel to a dump area where they dump and then return to the shovel. This is an example of a *cyclic* queue. The elements of the system, in this case, the trucks, do not leave the system.

The *barbershop* problem, where people come from a large population and there is a single barber: If the barber is busy, the customers will wait in the chairs provided. If all of the chairs are taken, customers will go away and, perhaps, return later or possibly go to another barber.

Ships entering a harbour with multiple berths: The ships need to be towed into a berth with one or more tug boats.

Telephone calls arriving at a central switchboard where they need to be routed to the correct extension.

Television sets on a conveyor belt arriving at an inspection station: If the set fails inspection, it may be sent back for adjustment or, in the worst case, be discarded.

Trucks breaking down and having to come to the repair shop for repairs.

There are many situations in a mine that give rise to discrete simulation. The basic operation of a mine itself can be considered such a system. Reliability of the equipment and inventory problems are also examples of discrete systems. Following the flow of the ore from the mine to the mill, to the loading docks and then off to the markets is also an example.

A complete treatment of simulation theory is beyond the scope of this book. However, an understanding of how simulation models are constructed and what they tell us is not too difficult.

Consider a simplified version of a bank with customers arriving and tellers giving service. Ignore the fact that the bank may offer multiple services such as insurance and new accounts. In this simplified bank, customers arrive only to visit one of the multiple tellers. All the possible events that take place in this bank are discrete events or can be considered as such. Possible events might be customers arriving, customers joining a queue if all the tellers are busy, customers going to a teller who is free, and customers leaving the bank when finished. Perhaps some of the customers will leave the queue if the waiting time is too long and go to another store or stores and return later. In most cases, we shall be modelling systems that will involve some queuing, for example, when all the tellers are busy in the bank, all the petrol pumps in a petrol station are being used, and all the checkout counters at the grocery store are in use.

The classic problem of trucks working in a surface mine site with a single shovel is one that has quite a history. Civil and mining engineers have been trying to solve this problem using classical means for many years. In this case, there are n trucks at the site and a single shovel that can load only one truck at a time. The loaded trucks travel to a single dump area where they dump their loads and return to the shovel. Since the shovel can load only one truck at a time, an arriving truck sometimes must wait in a queue until the shovel is free. The *problem* is to determine the production of the system as the number of trucks increases. For constant load, dump, and haul times, the solution is trivial, but for stochastic times, the problem can be solved for the general case only by simulation. In fact, it is easily solved in only a few lines of computer code using the software presented here (Figure I.1).

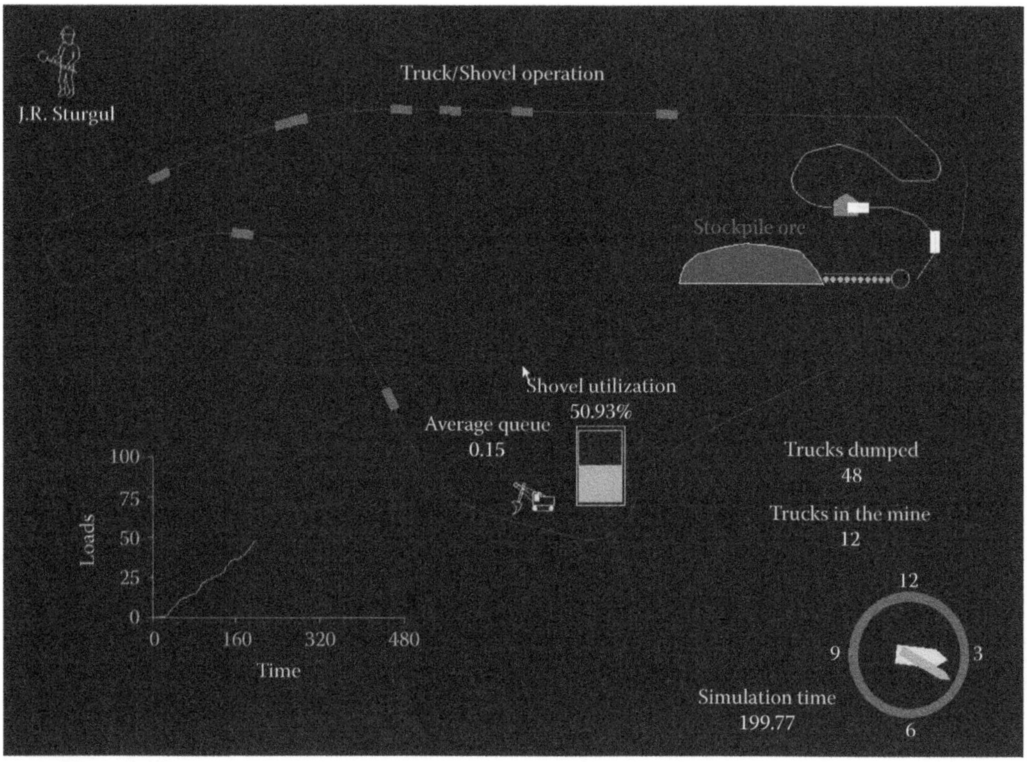

FIGURE I.1
Animation for a basic mine design problem.

The GPSS/H simulation language is excellent for simulating systems that have this type of queuing. As we shall see, it is very easy to model a great variety of very complicated systems using GPSS/H.

What Will Be Modelled?

The models we shall be studying will be primarily ones that mining engineers will encounter. However, the models might represent a bank working over a period of many months; an assembly plant that manufactures television sets; a barbershop where customers can obtain haircuts, shampoos, and manicures; or even a person doing the Saturday morning shopping. In some cases, the model may be only a small part of a large system, such as the tool crib in a large factory or the repair shop for small items.

It is important to understand that a simulation model does not in itself solve a problem. Instead, it tells the modeller how the system under study will work given certain parameters. For example, a simulation model might be constructed for a system that uses 10 trucks. The model might then be rerun using 11 trucks, then 12 trucks, and so on. It is the modeller who takes the results of the simulations and uses them to draw a conclusion. For example, the model might be a simple manufacturing system. Assuming the model is correct, the modeller might run it with four workers and determine, using the results of the simulation, that this number of workers might result in a profit of $500 per day. With 5 workers, the profit might be $526.79 per day and with 6 or more workers, the profit might be less than this figure. Hence, the conclusion to be drawn is that the system works best with 5 workers. This conclusion implicitly assumes that the model used for the simulation and all input data are correct. If the workers suddenly were given a pay raise, the model results may have to be re-evaluated to see if the same conclusion holds.

Thus, the simulation models we construct will not in themselves *solve* any problem directly but will provide information about how the system is working and, then, how it will work with certain selected parameters changed. Suppose a company has its own fleet of cars for its salesmen to use. If the cars need any service, whether it is of a routine nature or major repairs, it is done by one of two mechanics. The company is concerned that the mechanics are not able to keep up with the repairs and wonders whether it would be worth its while to hire another mechanic. Before the simulation model can be constructed, the company must define the problem to be solved in greater detail than has been given here. The following information is also needed:

1. The frequency of service and the distribution of times for the particular service.
2. What it would cost to hire a third mechanic, as well as the cost in lost sales when a car is not available.
3. When the managers bring their cars in for minor service, these cars have priority over those that are in for major service. This means that these cars are put in the front of the queue of any other cars that are waiting for service.
4. When the company owner brings his car in for service, it is given a special status and this car is immediately worked on. Thus, even if both mechanics are busy, one will put aside the car he is working on and start the repairs on the owner's car.

The information obtained from the simulation model might include the following:

How the system presently works. Obviously, the computer model has to accurately reflect the system as it is working before any reliability can be associated with the results from the changed model.

How the system will work with three mechanics.

Another Type of Problem to Be Studied and the Method of Solution

Consider the following system consisting of a five-person barbershop with customers arriving to have their hair cut. The shop operates 8 hours a day with no time off for coffee breaks or even for lunch. There are five chairs for the haircuts and only one chair is provided for the customers to wait if all five barbers are busy. Assuming that people do not like to stand around waiting, the system can only hold six customers. If a seventh customer arrives and finds the shop full, he will always leave. All five barbers are identical, so it can be assumed that they work at the same rate and the customers have no preference for any barber. The customers do not arrive at regular intervals, and the haircuts are not given at exact times but vary according to known statistical distributions (Figure I.2).

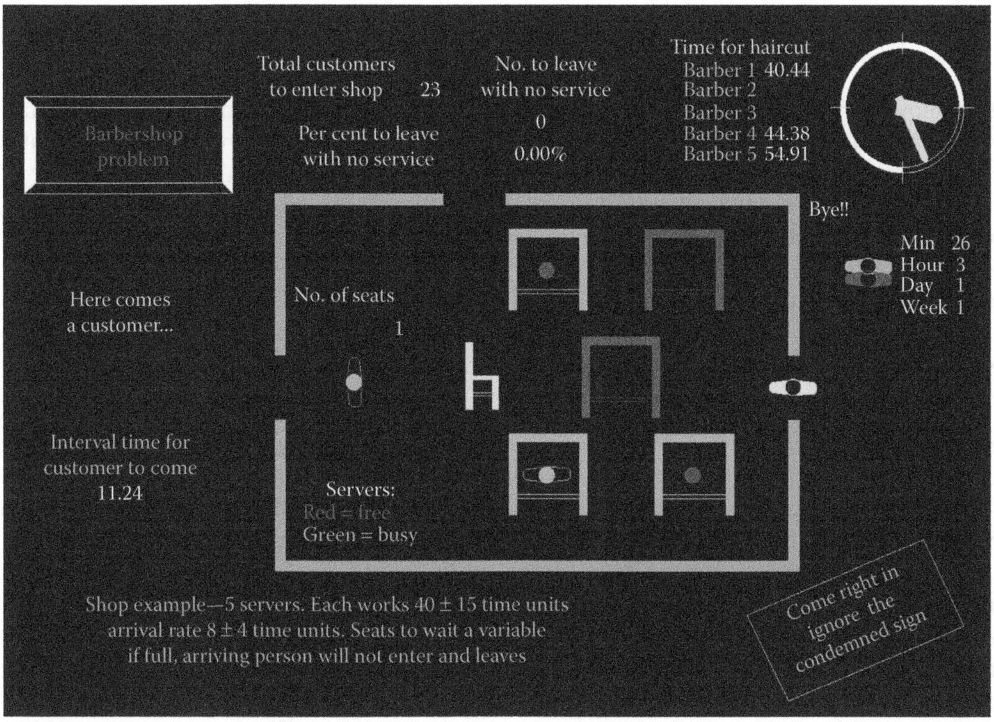

FIGURE I.2
Animation of barbershop problem—customers are represented by elliptical shapes.

The owner of the barbershop would like to study the shop to see whether it would be profitable to add another barber or simply add another chair for customers to wait. Perhaps it would be possible to purchase new equipment or provide training, so that the barbers can work even faster. Would the extra haircuts justify the expense of this equipment? GPSS/H will assist in building a model:

1. Predict how the system as outlined above works. The model will only be as good as the input data and the assumptions given above.
2. Once the model represents the barbershop as it presently is working, it can be modified to predict how it will work under different conditions.

This demonstrates another reason why GPSS/H is such a powerful problem-solving tool. As we will discuss, changes in GPSS/H programs often are made by changing only a few lines of code.

The fact that GPSS/H programs can so easily be changed to reflect the what-if?-*type questions a person may want to pose makes it an ideal language to use for simulation studies.*

Once the modeller is satisfied that the original model is correct, the simulation can be redone, but this time with six barbers. Alternately, the model can be run for more seats for customers. Finally, the model can be run for different combinations of haircut speeds for the barbers.

Using the cost data for the various combinations of barbers, lost customers, profit per haircut, and so on, the modeller can determine the economics of the system and make the correct choice.

A Simulation Model

The following example will illustrate a situation of a simulation model with constant arrival rates and constant service rates. Suppose a tool crib has 1 attendant to serve a large group of machinists. These machinists come for a single tool at a uniform rate of 1 every 5 minutes. It takes exactly 6 minutes to obtain the tool. Machinists earn $20 per hour and the tool crib attendant earns $16 per hour. The factory works an 8-hour shift but stops for a 1-hour lunch break. The crib is closed for lunch and at the end of the 8-hour shift. In order to simplify the calculations, if a mechanic is waiting for a tool either at the lunch break or at the end of the day, he or she will wait and be served. The tool crib operator does not receive extra pay for working overtime but the mechanics can include the time waiting in their actual working time. Should the company hire another tool crib attendant?

Solution

While this is a very simple situation and one that would rarely be encountered in practice, it will prove instructive to learn what simulation models can tell us. The problem will be solved first for one tool crib attendant, then two, and then for three.

The machinists arrive every 5 minutes, so there will 12 per hour arriving. In a 4-hour time period, 48 will arrive. The first arrives 5 minutes after the tool crib opens. There will be no wait for him. The second person arrives 5 minutes later and will experience a 1 minute wait until the attendant is free. Similarly, the third person has a 2 minute wait, and so on, up to the 48th person, who has a 47 minute wait. In a 4-hour period, there will be a total waiting time for the machinists of $1 + 2 + 3 + \cdots + 47$, or 1128 minutes at $20 per hour. This represents a loss of $(1128/60) \times \$20$ or $376. For the two 4-hour periods in a day, this represents a loss of $752. If two (or more) tool crib attendants are working, there will never be a wait for a free attendant (only the 6 minute wait for the tool). Table I.1 summarizes the results from considering the case of three attendants.

Clearly, it is advantageous to hire one additional attendant. The result of this simple model may or may not be useful to a company, depending on the original hypotheses, which were as follows:

Having an arrival rate of one machinist every 5 minutes. In practice, the arrival rate will be at random times. There may be an average rate of so many per hour but, in general, the arrival time for a particular machinist will be random.

Service rate of the tool crib attendants is constant. Here, too, in practice, the service rate will normally be random.

The tool crib closed every 4 hours. Anyone waiting for service immediately left when the 4 hours was up. In practice, the machinist about to be served still will obtain service.

TABLE I.1

Tool Crib Simulation

Number of attendants	1	2	3
Number of machinists who arrive	48	48	48
Total time waiting for attendant	1128	0	0
Cost of lost time per 4 hours	$376	0	0
Pay to attendant per 4 hours	$64	$128	$192
Total cost per 8 hours	$880	$128	$192

The length of the queue tended to grow to an eventual size of 47. This is not realistic. If the queue is too long, an arriving machinist will tend to leave and come back later when the line is shorter.

Each arriving machinist wanted only one tool. In practice, the number of tools needed may be 2, 3, or more.

It will be shown that, using the GPSS/H language, a model can be constructed easily to include all of the above possible changes to the original assumptions. Problems such as this are quickly and easily solved with GPSS/H.

Some Comments on Queuing Theory

The example of the queuing problem for the tool crib has an exact solution. There are very few such exact solutions available for most real-life problems. For example, systems involving cyclic queues and/or those involving queues for a finite population for the general cases of random arrival or service times have no solutions. Cyclic queues are those where the system under study has elements that do not leave, such as ships sailing from port to port. Here the ships might be loaded, haul, unload, perhaps load again, and travel to another port. Whenever there is a finite population, as soon as one element is doing a particular thing, the statistical distribution governing rates will change. Thus, if a company has a fleet of 10 cars to be studied, if 2 are being serviced, the probability of another car coming for service is no longer the same as when all 10 were up and running.

In general, in order to study complex systems where queuing takes place, it is necessary to build computer simulation models. It will be shown that GPSS/H is an ideal computer programming language to model such systems. In fact, one of the features of GPSS/H is that, as one learns the language, one automatically learns how to build complex simulation models. Before we actually start learning the software for simulation and animation, it might be instructive to review a few basic concepts from queuing theory. These have to do with the possible arrival distributions, service distributions, number of servers either in series or in parallel, the population size, and the queue discipline. Table I.2 gives the possibilities to consider.

TABLE I.2

Possibilities for Queuing System

1. Population
    - Infinite
    - Finite
2. Arrival time distribution
    - Constant
    - Poisson
    - Erlang
    - Uniform
    - Arbitrary
    - Normal

(*Continued*)

TABLE I.2 (*Continued*)

Possibilities for Queuing System

3. Service time distribution
   - Exponential
   - Erlang
   - Constant
   - Uniform
   - Arbitrary
   - Changing with the time of the day or queue length
4. Service facility
   - Single
   - More than one in parallel
   - More than one in series
   - Variable number, both in parallel and in series
5. Queue discipline
   - FIFO (first in—first out)
   - Random
   - LIFO (last in—first out)
   - Priority of one type of customer over another for the position in the queue
   - Ability of one customer to pre-empt another one being served
   - Priority of shortest or longest service time being served first or last
   - Balking (customer refuses to join if queue too long)
   - Switching from queue to queue
   - Leaving (customers will leave if waiting too long)
   - Being a member of more than one queue (a person can be in a shopping centre and take a number for meat service at the same time as he or she is waiting for his or her vegetable number to be called).

The following may come as a surprise to the person who has not formally studied queuing theory: it is *not* possible to obtain exact solutions to most of the above situations (although a lot of very fine mathematicians have tried). However, several problems do have solutions and these can be found in textbooks on operations research or queuing theory, such as the one by Phillips et al. (1976). As one learns how to construct simulation models, it is instructive to compare the results from the simulation model with what one expects to obtain from an exact solution.

Simulation versus Mathematical Solution

To illustrate a comparison of a simulation model with one that has an exact solution, consider the case of a store where customers arrive on the average of 24 per hour. The interarrival times are given by sampling from the Poisson distribution. Thus, after each arrival, the next arrival time is obtained by sampling from the Poisson distribution with the same mean as before. The single clerk in the store can handle a customer on the average of 1 every

2 minutes. The distribution for this service is exponential. Service is first-come, first-served. The customers do not mind waiting if there is a queue. It is desired to simulate the store for 50 days or 10 weeks of operation, where a day is 8 hours in duration and the store operates continuously. Compare the results with those obtained by an exact solution.

Solution

The problem will be recognized as a standard one that is discussed in any text on queuing theory. The exact mathematical solution is available and equations can be found for determining the probability of the clerk being idle, the probability of any number of customers being in the store, the expected number of customers in the store, the average time for a customer to wait in the queue, to be in the store, and so on.

GPSS/H is used to solve such problems. The simulation model used Monte Carlo simulation. This technique uses a random number generator to simulate both arrival times and service times. The simulation starts at simulated time $t = 0$ and runs until the program reaches a point in simulated time that the programmer feels is enough to yield correct results. First, a basic time unit needs to be selected. This is normally taken as the smallest time as given by the statement of the problem. For the example here, a time unit of 1 minute is selected. Thus, the customers will arrive on the average of every 2.5 time units. The clerk can handle a customer every 2 time units. The simulation is then done for times of 8, 50, 100, 200, and 400 hours, and so on. These have to be converted to minutes since the basic time unit is a minute.

Since the exact solution assumes steady-state conditions, the simulation is run for 4 hours (240 time units) and then is stopped. All relevant statistics, except for the customers in the system, are discarded. Then, the simulation is restarted and run for the desired simulated time. A selected portion of the output from this program for the simulated time of 400 hours is as follows:

Customers serviced	9605
Percent time clerk busy	801
Average number of customers in system	4.122
Average time in system	10.2 minutes

The theoretical values can be found by use of simple formulas that can be found in any book on operations research or queuing theory.

Customers served	9600 (customers arrive on average of 24/hour for 400)
Percent time clerks busy	800 (this is 24/30)
Average number of customers in system	4
Average time in system	10 minutes

As can be seen, the above results compare quite favourably with those obtained by the simulation. It is important in both interpreting and using the results of a simulation that the simulation has been allowed to run for a long enough period or the results may not be accurate. In performing a simulation, one would like to obtain results that can be reproduced nearly identically if other simulations are done with different random numbers. There is no set answer to the question of how many simulations are enough, as the proper number of time units to simulate for is a function of several variables. One is the nature of the simulation, that is, is the population infinite or finite? In the case just considered of

TABLE I.3

Results of Re-Running Simulation for Store

Simulate Time (hours)	Customers Served per Hour	Percent Time Clerk Busy	Avg. No. of Customers in System	Avg. Time Customers in System
8	26.9	92.2	4.34	9.58
50	23.6	81.7	4.105	7.87
100	24.0	82.6	4.061	10.63
200	23.9	79.7	4.810	11.52
400	24.2	80.3	4.122	10.2
Theoretical	24.0	80.0	4.00	10.2

an infinite population and Poisson arrival times with exponential service times, a large number of simulations have to be performed. In the case of a system where the parameters being simulated cycle through the system (such as workers in a factory) not quite so many simulations may be needed. The nature of the queue and the service facilities are also important. In addition, if the statistical distributions are relatively uniform, such as a normal distribution with small standard deviation, the simulations tend to achieve a level of stability rapidly. This last result is important (and comforting) for the person doing simulations who has a lot of data that is normally distributed. This is often the case for working times in a factory, truck haulage rates along a road, manufacturing times, and so on. If the statistical distributions are non-symmetric, the number of simulations required can be high. This will be demonstrated by means of an example later. First, let us again consider the example just solved.

Suppose, however, that, for the simple queuing system just studied, the simulation was done for less than 400 hours. What would the results have been? The answer depends, in part, on the sequence of random numbers. But it is instructive to redo the simulation for less than 400 hours and examine the results. Table I.3 summarizes the results from these different simulations.

As can be seen, the results for simulating for 8 hours are quite different from the theoretical ones. Simulating for 200 hours yields results that come close to the theoretical ones except for the average number of customers in the system. After 400 hours, the simulated results are quite close to the expected ones. If this problem was for a real store, the simulation may well have been run for an even longer time.

Simulation with Non-Symmetrical Distributions

Whenever the statistical distributions are non-symmetric (the Poisson is non-symmetric, the normal or Gaussian distribution is symmetrical), the number of simulations may have to be very large. This is easy to understand, since it is desirable to model a system over every possible situation and, in theory, repeat the simulation until the various parameters being studied do not change. To illustrate this concept of non-symmetric distribution, consider a simple example. Suppose a person is modelling his behaviour on a day-to-day basis, weekends not included. Each day this person stops at the local casino and bets $2 on number 7 on a roulette wheel. He makes only this bet and, whether he wins or loses, will leave. The probability of winning is 1/38 (the wheel has numbers running from 1 to 36 as

TABLE I.4

Results of Studying Simulation of Roulette Wheel

No. of days	First Set Random Nos.	Second Set Random Nos.	Third Set Random Nos.	
	Obs. Wins	Obs. Wins	Obs. Wins	Exp. Wins
380	8(−20%)	13(+30.0%)	5(−50%)	10
3,800	95(−5.0%)	95(−5.0%)	95(−5%)	100
38,000	1,019(+1.9%)	1,019(+1.9%)	1,017(+1.7%)	1,000
380,000	10,025 (+2%)	10,029(+.2%)	10,021(+.21%)	10,000

well as a zero (0), and double zero (00). How many simulated days are needed to produce satisfactory results for the simulation? Certainly not 38, as the expected number of wins is only 1. How about 380 or 3800? To study this, a short GPSS/H program was written. The simulation was performed for 380; 3,800; 38,000; and, finally, 380,000 days. It was then run for three different sequences of random numbers. Table I.4 summarizes the results of these three simulations.

As can be seen, the observed number of wins versus the expected number of wins start to approach each other only after a very large number of simulations. In fact, the results for 380 simulations give results that vary from the expected number of wins by as much as 50%. Thus, for situations such as the above, one must always be aware that a very large number of simulations may be needed (there's no guarantee that even 380,000 is enough, depending on the problem!). Fortunately, for most situations, the simulation can be performed successfully with a reasonable number of simulations.

Why Do a Simulation?

It always comes as a surprise to mining students to learn that it is rarely possible to obtain exact solutions to any but the most elementary problems that lead to queuing situations. We all experience queuing daily whenever we enter a bank that has many customers and we have to wait for a teller or shop in a large grocery store and wait in the checkout line, and so on. One would think that such problems can be solved quite easily, but this is not the case. Although queuing theory has been studied by mathematicians for many years, very few complex problems have been solved.

A computer simulation can rapidly and accurately solve most elementary and complex situations where one encounters queuing situations. Many mining engineers still do design work using average values, such as the average time a truck takes to be loaded, the average time to travel from point A to point B, and the average time to dump. It will not take long for the student to realize, however, that this type of design leads to incorrect results. For example, if a different person enters a barbershop with a single barber for a haircut exactly every 5 minutes and the time to give each person a haircut is also exactly 5 minutes, then the system is *perfect* in the sense that the barber will always be busy and no person will ever have to wait for the barber. However, if the interarrival time for each customer is obtained by sampling from the Poisson (exponential distribution) with a mean of 5 minutes and the time for a haircut is also exponential with a mean of 5 minutes, then the expected queue of waiting customers becomes infinite! This is one of the few queuing problems that have an exact solution. The complete results of this simple example can be found in any textbook on operation research or queuing theory.

What Is Meant by a *Solution* to a Simulation Problem?

When we obtain a solution to a simulation problem, it is important to understand just what is meant by our solution. A simulation model does *not* solve a problem but tells us how a system will operate under a given set of parameters. For example, the model might tell you that, if you have 9 workers on an assembly line, your profit would be $457. Adding a 10th employee and doing another simulation might then tell you that the new profit will be $505. Doing a further simulation for 11 workers might tell you that the new profit will be $480. Thus, we conclude that the optimum number of employees is 10.

Another Example of a Simulation Model

Consider a simple example of a single shovel loading trucks at a mine site. This shovel can only load a single truck at a time. After each truck is loaded, it travels to a dump area where it dumps the load and then returns to the shovel. If the shovel is free (no other truck is being loaded), it immediately begins to be loaded again. If not, it waits in a queue until the shovel is free.

Assume that you are going to study this system (the trucks and shovel and the travel paths make up the system). You are told by the engineer in charge who has studied the shovel and the haulage routes that the shovel can load a truck in *exactly* 5 minutes (this is a slow shovel, but don't worry about this for the present). It takes *exactly* 8 minutes to drive to the dump, *exactly* 2 minutes to dump, and *exactly* 6 minutes for a truck to return to the shovel.

If you had a single truck in the system, it would load in 5 minutes and then take 16 minutes to return to the shovel. The shovel would be busy for every 5 minutes out of 21 minutes or 23.8% of the time. There would be a load dumped every 21 minutes or approximately 3 per hour. In an 8-hour shift, you would expect that there would be slightly less than 24 loads dumped (this construction site does not allow the workers any breaks).

If you added another truck to the system, you would expect the production to double and the shovel to be twice as busy. Adding a third truck would likewise increase production, and the shovel would be busy about 72% of the time. What will happen when you add a fourth truck? The answer is that the system will experience no problems yet as, at any one time, there can only be one truck being loaded (this takes 5 minutes) and the other three can be travelling to or from the dump. It is only when you have five trucks working that you start to have queuing problems. However, production will increase by having five trucks as compared to having four. The question is how much will the increase be and will this be worth it? To answer this last question, additional data is needed. Let us assume that each load carried by the trucks somehow represents a contribution to profit of $45 and that each truck costs $225 per day to run.

Table I.5 gives the results of a computer simulation for the above problem. The simulation was run for 10 days with each day being an 8-hour shift. At the start of the simulation,

TABLE I.5

Results of Computer Simulation

N	Loads	Shovel Util.	Avg. Que.	Profit per Day
1	229	0.239	0.000	$805
2	458	0.477	0.000	$1611
3	686	0.715	0.003	$2412
4	914	0.952	0.006	$3213
5	959	1.000	0.807	$3190
6	959	1.000	1.807	$2965

all of the trucks were at the shovel and the trucks worked for 10 days straight. The results of the simulation are as above.

It is easy to see that the number of trucks to have for optimum profit is four. Note that adding a fifth truck will increase production but will not result in a greater profit.

A Change to the Problem

The problem just completed assumed that all of the times used in the model were constant. This is certainly not the case in real life. Things do not happen in exact times. The time to load a truck will depend on several parameters; for example, the time to drive to the dump will not be constant. Let us assume that you did some time studies at the construction site and found that the time to load a truck took an average of 5 minutes but the statistical distribution that best describes it is the exponential distribution. The travel times and the dumping times are best described as coming from normal distributions as follows:

Travel to dump	Mean of 8 minutes	Standard deviation of 1.5 minutes
Dump	Mean of 2 minutes	Standard deviation of .3
Return to shovel	Mean of 6 minutes	Standard deviation of 1.15 minutes

As can be seen, the mean times have remained the same.

The computer program was modified to allow for these changes and run for 100 days; the reason for this is that the exponential distribution was used—whenever this is used, the number of simulations or the simulated time is substantially increased (more on this later). Table I.6 gives the results of the simulations.

Notice that the number of loads keeps increasing as the number of trucks is increased until 10 trucks are working. Also note that the profit is a maximum for 6 trucks. This profit is $2676, which is considerably less than the previous profit of $3219. Thus, the number of trucks needed is 50% more than before and the optimum profit is 21% less. Also note that it really would not make much difference if 5 trucks were used rather than 6. Do you see what happens when too many trucks are in the system? Notice that the average queue length for 9 trucks is 4.84 and, for each additional truck, the average queue will increase by 1. This means that each additional truck will, in effect, add 1 to the average queue. No increase in production will result.

TABLE I.6

Results of Simulation

N	Loads	Util. Shovel	Avg. Queue	Profit
1	2281	0.241	0.00	$801
2	4355	0.445	0.10	$1509
3	6056	0.627	0.35	$2050
4	7515	0.768	0.73	$2481
5	8285	0.876	1.36	$2603
6	8948	0.936	2.08	$2676
7	9243	0.974	2.94	$2584
8	9434	0.987	3.87	$2405
9	9423	0.997	4.84	$2215
10	9547	0.999	5.82	$2046
11	9550	1.000	6.82	$1820
12	9523	1.000	7.82	$1585

Although this seems like a simple model, this was studied by numerous investigators in the 1960s and 1970s to determine the optimum number of trucks to have for construction projects. Numerous lengthy reports and papers were written on this problem alone.

Some comments on the GPSS/H language follow. The software that will be used for the animation is called PROOF. This was developed by the same company that makes GPSS/H, so it is only logical that it was selected as the software to be used for the animation. However, it is a superb tool for use in making animations, as shall be shown. It is easy to use and can provide extreme detail to any system that is being modelled.

Why Use the GPSS/H Language?

All of the examples in this course are solved using the GPSS/H simulation language. Since this is probably not the first computer language taught in university courses, it is appropriate to learn why GPSS/H was selected. In fact, there are or were multiple versions of the original GPSS available, and the one used here is GPSS/H, which is designed specifically for the personal computer. This introduction is intended to answer questions about the GPSS/H language, which will be done in a question and answer format.

What Is GPSS/H?

GPSS/H is both a computer language and a computer program. It was designed for studying systems represented by a series of discrete events. A discrete system is one where a countable number of events can occur at any one time. These discrete events might be trucks being loaded, ships entering a harbour, people entering a bank, cars travelling on a road, parts on a conveyor belt, and so on. GPSS/H is a high level, non-procedural language.

For a fascinating look at the development of computer simulation with interviews of some of the pioneers in the field, go to http://d.lib.ncsu.edu/computer-simulation/.

Where Did GPSS/H Come From?

GPSS originally was developed by Geoffrey Gordon for IBM in the early 1960s and released to the public around 1963 or 1964, so it has been around for quite some time. However, it is a dynamic language in that new versions are introduced every 3 or 4 years. It is now a multi-vendor language and various versions are available. It is widely used on both mainframes and PCs. By 1972, there were at least 10 versions of GPSS available, but few of these have survived. Greenberg (1972) presents details about these early versions of GPSS and traces their histories. GPSS/H is the most recent version of GPSS. This was developed by James O. Henriksen, who is the president of Wolverine Software, the vendors of GPSS/H.

What about Modern Versions of GPSS/H?

Since GPSS has been around for quite some time, it is natural that people who were introduced to it at an early stage of its development may not be aware of how it has changed. An excellent summary of recent developments is given by Schriber (1988). Schriber lists versions of GPSS such as GPSS/H, GPSSV, GPSS/PC, GPSSR/PC, and GPSS/VX and some relevant information regarding each. It is possible to add animation to the results of a simulation and to view the simulation in *cartoon* fashion. In fact, it is possible to find certain functions performed by GPSS embedded in other languages, such as GPSS-Fortran, APL-Fortran, and PL/1-GPSS. Schriber's work cites references for these languages.

What Is a Non-Procedural Language?

A non-procedural language is one that anticipates what the programmer is attempting to do and allows the computer code to be very short. Often the programming code for a non-procedural language appears very similar to the problem it has been designed to solve. For example, sorting an array of data using a procedural language is done in one of several ways. One way is to find the smallest (or largest) element, place this at the front of the numbers, and then sort through the remaining numbers for the next smallest (or largest), find it and place this second, and so on, until the array is sorted. This involves numerous comparisons of data. A non-procedural language that is used for handling databases where it is common to sort data may have a single command, namely, SORT, to do this. In simulation studies, one often encounters queues. To model a queue in GPSS/H to gather certain statistics, the single line of code (known as a *block*) might look as simple as the following:

```
QUEUE  DUMP
```

No code for output is needed: GPSS/H will automatically gather relevant statistics and output them when the program is finished. Of course, it is possible to customize the output and have it sent either to the screen or to an output file. This will be done in this book, but the original versions of GPSS did not include this facility. Today, thanks to the introduction of PROOF animation, the results of the simulation are not only placed in a data file but also used to run the animation of the system being modelled.

People seeing a GPSS/H program for the first time tend to remark, Is that all that there is to it? As we shall see, this is one of the remarkable features of the language. Since it was designed specifically for solving certain problems, it is, indeed, quite compact.

In the study of queuing theory, one soon encounters a system having a single server for people arriving at random times from an infinite population. One case of this is known as the M/M/1 (the first M stands for Markov to indicate that the arrival rates are Poisson, the second M is for an exponential server, and the 1 indicates a single server). This is modelled in GPSS/H by writing only seven (!) lines of code, which are known as *programming blocks.* The equivalent Fortran program would take many hundreds of lines of code. In addition, to make changes in a GPSS/H program to answer the what-if? questions often takes only a few lines of code. If a bank is being studied with eight cashiers working, the relevant line of code may be as follows:

```
STORAGE  S(CASHIER),8
```

To study the same system with nine cashiers may involve changing the above line to

```
STORAGE  S(CASHIER),9
```

Is GPSS/H Hard to Learn?

GPSS/H is not any more difficult to learn than any other programming language. Most people find it easier to learn than traditional engineering languages such as Fortran, Basic, C++, or Pascal. After about 30 or 40 hours of instruction, most students find that they can proceed on their own with writing practical simulation programs. Since it is a very popular simulation language, numerous short courses are offered throughout the world.

Will Knowledge of Other Languages, Such as Fortran, Help in the Learning of GPSS/H?

Not really. The logic behind GPSS/H is so different that knowledge of other procedural languages may even be a hindrance. Of course, knowledge of any other simulation language is a different matter.

What about Using Other Simulation Languages?

Other simulation languages exist that are quite good for solving simulation problems. Some of these are ARENA®, SIMCRIPT II.5 (MODSIM)®, SLAM®, SIMPLE+®, Extend®, ProModel®, POSES++®, AutoMod®, and FlexSim®. The solutions obtained by investigators using these languages may well be as accurate as those obtained by GPSS/H. However, GPSS/H was selected as the language for this book for the following reasons:

1. It is continually being upgraded.
2. It is widely available.
3. It is written in machine language and, therefore, is inherently very fast.
4. It can solve a wide variety of problems rapidly and accurately. These problems come from many sectors such as manufacturing, engineering, business, and science.
5. It has withstood the test of time, having been introduced by IBM in 1961. Other simulation languages have fallen by the wayside.
6. It is extremely flexible and can solve a wide variety of problems that not all other simulation languages can solve. It has been used for the simulation of some of the largest models ever done. Programs with over 100,000 (!) lines of code have been constructed. As you will learn, most GPSS/H programs are surprisingly short, so programs of this size represent models of gigantic and complex systems.
7. It has a proven track record for modelling mining operations. In short, it works!

Will GPSS/H Replace Languages Such as Fortran, Pascal, C++, or Visual Basic?

No. There are a large number of problems that should and will always be solved using traditional computer languages (this does not exclude the possibility of having packages available to solve specific engineering problems based on traditional languages). Learning GPSS/H enhances a person's computer skills rather than replacing any. In this regard, it can be looked upon as adding a computer skill such as word processing or learning how to construct a spreadsheet. Knowledge of these does not replace programming skills using traditional procedural languages.

How about a Comparison between Fortran and GPSS/H for a Simulation Study?

Below is a rough comparison. The actual values will depend on the particular problem. This might be for the simulation study of the trucks in a mine with two shovels.

Comparison of GPSS/H with Fortran

	GPSS/H	Fortran
Time required to write program	2–3 days	Many months
Execution time	<1 minute CPU	3–4 hours CPU
Ease of changing program	Trivial	Up to a week
Lines of computer code	500–600	20,000–50,000
User friendly?	Yes	No
Animation?	Available	Not standard

But Aren't There Simulation Packages on the Market?

Simulation packages are available on the market but they tend to be very expensive (some cost around $100,000), to be hard to change, to be not user friendly, and to take a great deal of time to run. In addition, they are not as flexible as a program that is specifically constructed for a particular model.

But Why GPSS/H?

Of all the versions of GPSS ever developed, GPSS/H is by far the most advanced. It contains features that other versions of GPSS do not have. However, learning GPSS/H does not exclude the use of other versions. Switching or moving from one version to another is not at all difficult. However, many of the features of GPSS/H will not work on other versions.

A Bit about PROOF

PROOF® also comes from Wolverine Software. It is one of the highest quality animation software programs available for any simulation model (it may well be the very best). Since it does not have predefined icons or layouts, each animation needs to be drawn separately (but, once drawn, one can cut and paste to other animations). Thus, it may take a bit longer to use in the beginning than some of the other animation software applications, but the quality is extremely high. The detail on the animation will depend on the artistic talents of the person making the animation.

The student will soon find that making subsequent animation layouts can be done quite rapidly. Students who have access to PROOF Professional can import other layout files such as the ones obtained from software like AUTOCAD®. Students who may have had an introduction to the use of AUTOCAD or similar animation software will find many similarities between PROOF and what you have already learned.

Although this book is concerned with GPSS/H and PROOF, it should be noted that PROOF can be used with other software. It is not directly connected with GPSS/H or any other simulation language for that matter. In this book, it will only be used with GPSS/H.

A Brief History of Mine Simulation

Mining engineers have been interested in using computers to build simulation models of mining operations ever since the computer was introduced and accepted into their industry. A brief history of the development of such models in the mining industry is presented here. Many of the early works on the subject are very difficult to find, such as the ones from smaller conferences. For example, several of the computers in mining conferences held in Australia are often neglected by researchers in other countries. Some of the proceedings of the early APCOM (Application of Computer and Operations Research in Mining) conferences are difficult to find since they were distributed only to the people who attended. The proceedings from these early conferences were often just bound together, whereas, today, they are published in book form, normally through a professional society.

The first example of a mine simulation was a work by Rist (1961), which is a most important contribution. The simulation was done using a special language, Symbolic Programming System Language, that is no longer available. Rist's problem was taken from an actual underground molybdenum mine, where his model was used to determine the optimum number of trains to have on a haulage level. Loaded trains had to queue at a portal and wait until the single track was clear as well as wait until the crusher area was free (only two trains could be at or waiting near the crusher). Other restrictions were imposed on the system to make it as lifelike as possible.

An interesting early work was by Falkie and Mitchell (1963), who studied the complex underground rail haulage for a coal mine in Pennsylvania. This work laid the groundwork for how Monte Carlo methods are incorporated into stochastic simulation models.

The work by Harvey (1964) was the first application of the GPSS language for a mining operation, and this was done at the same mine, namely, the molybdenum mine of Climax Molybdenum Company, where Rist's work was done. Unfortunately, these two important works did not have the effect on other researchers who, at the same time, were using traditional languages for their simulation work. Further work at the same mine was also reported by Achttien and Stine later the same year (1964).

Morgan and Peterson (1968) presented a detailed investigation on how to set up a stochastic simulation of a surface mining operation. This work discussed how one can go from a working mine with known production and travel and loading and dumping times and predict production for some future time. The simulation model was then used to determine the correct fleet size for both optimum production and minimum cost per ton.

There were several simulation languages available in the 1960s and early 1970s, such as SNOBOL, GPSS, GASP, SIMSCRIPT, and SIMULA. The potential value of using one of these special languages for mine system simulation was pointed out by Bauer and Calder in 1973 at the APCOM in South Africa. Their conclusion was both interesting and prophetic:

> [A]s such it (*the model as obtained from a simulation language*) should play an essential part in every open pit planning operation. (emphasis added) (p. 276)

Unfortunately, this important advice seems to have gone unheeded for many years.

An excellent source of references for simulation models is the proceedings from the APCOM conferences. The first of these was held in 1961 at the University of Arizona in Tucson, Arizona. There have now been 36 such conferences held in many different countries with similar conferences now held in Canada and Australia. APCOM 36 was held in Brazil; APCOM 37 held in Alaska.

The mining engineer should be aware of the classic work done in attempting to obtain exact solutions to mining examples. Even though few examples of systems with stochastic behaviour have such solutions, some important work has been done in mining. The classic work by Koenigsberg (1958) presents the first solution in closed form of a cyclic queue. It is of interest to note that, to validate his results, he used several underground coal mines in Illinois. In 1982, Koenigsberg wrote a paper that any researcher looking into the development of cyclic queues would consider as *essential*. The title of the work is "25 Years of Cyclic Queues: A Review." Koenigsberg lists 66 references of contributions to the field of cyclic queues. Some of the works in this section were annotated if it was felt that they would be of sufficient interest to the mining researcher. Other works, such as the early ones on some theoretical aspect of queuing theory, are given for information only. Other works have to do with the development of the use of queuing theory, some with potential mining applications. The early work for the use of cyclic queues for excavation by Douglass (1964) and O'Shea et al. (1964) had a direct bearing on mining applications. Other works include the work by Maher and Cabrera (1975), which solved a multi-stage cyclic queue problem and that of Griffis (1968), who solved the simple truck–shovel cycle problem for construction applications. These last two works were published in journals of interest to civil engineers.

There have been some attempts to apply queuing theory results to mining operations. The work by Luo and Lin (1988), for example, was an attempt to use the results of queuing theory for a mining example. The work by Barnes, King, and Johnson (1977) was an application of queuing theory to a mining problem. It used discrete simulation as a check on the mathematically obtained equations using basic queuing theory.

Several works by James O. Henriksen are worthy of note, since he was the developer of GPSS/H. This is the first of the general-purpose simulation languages and was introduced in the early 1960s by Geoffrey Gordon, who was working for IBM. Gordon had no idea that his language—intended only to model simple manufacturing systems—would grow into its present form. Some mining engineers may have been exposed to one of these early versions that grew out of the early versions of GPSS and not realized the great changes that have been made over the years. Several myths relating to GPSS/H have arisen that other languages such as SLAM and SIMAN seem to have avoided. In his 1983 work, Henriksen exploded these myths.

Dr. Thomas J. Schriber, who is at the University of Michigan, wrote the main textbook on GPSS and has done considerable work in various aspects of simulation. His 1974 textbook is a classic and is known amongst simulators as simply *The Big Red*. References to this book as well as ones on other languages are given here.

Mine Simulation Methods

There are four main approaches to doing a simulation for a mining operation. All have been used by researchers. These can be summarized as follows:

- Formulate equations relating to the model using queuing theory and solve the equations exactly.
- Write a computer program in a traditional language such as Fortran or Pascal to simulate the problem. When this has been done, the program is nearly always based on a clock that starts at zero and is incremented a small amount.

- Use a general-purpose package. Usually these have been developed by workers using techniques from one of the above languages.
- Use a special language designed for simulation studies. The main languages for this are GPSS (including GPSS/V, GPSS 360, GPSS/PC, and GPSS/H), SLAM (including SLAM II, SLAMSYSTEM, and GASP), and SIMAN (including ARENA).

There have been a few applications of SIMCRIPT II, but this language is no longer supported, having been replaced by MODSIM. Two examples of SIMULA have been found, but both are by the same person. Each of the above techniques will be examined next.

Exact Solutions

The number of exact solutions available for even the simplest queuing theory example are very small. As noted previously, the first solution to a cyclic queuing example was provided by Koenigsberg in 1958. A few other solved examples of cyclic queues are in the literature. For example, Koenigsberg in 1976 solved the problem where the entities doing the cycling can enter into more than one server. He considered ships on the ocean travelling from port to port. At each port there can be more than one server (dock). The ships load or unload at each port and then travel to the next port. After solving the problems mathematically, the solution was checked using a computer simulation model.

Aside from the unwieldy mathematical analysis and scarcity of known solutions, there is another major drawback in using the exact solution that many mining researchers seem to have overlooked. The few problems where solutions are available generally make the assumption that both arrival and service times are exponential. In the case of the 1958 study by Koenigsberg, he solved the problems of different work crews mining coal in an underground room and pillar mine. The time to do each separate task (drilling, blasting, loading, and timbering) was exponential and the various tasks must be done in sequence. When one crew was done with its task, it immediately went to the next face, which took no time. This is generally the case for room and pillar mines. If the face was clear (meaning that the crew that works before it was done), the new crew began to work. The travel time can be incorporated into the work time since the faces are so close together. However, the example of Maher and Cabrera for multi-stage cyclic queues has the units travelling from one service to another in a cyclic manner with both the travel times and service times being exponential.

In an actual mining operation, the times to load a truck are often exponential (or, at least, skewed to the left) but the travel times are certainly not. Dumping times also are not exponential. If anything, they are best represented by a normal distribution (often without the two tails). This means that any results from applying exact solutions need to be looked at very carefully. In fact, one exercise that students who are learning a simulation language for mining applications often do is to consider a simple cyclic queue for a mining operation with n trucks and only one shovel. The students are asked to make plots of production versus number of trucks under the assumption that the various times are from the common statistical distributions such as

1. Constant, that is, everything deterministic
2. Uniform, that is, $A \pm B$

3. Normal or Gaussian
4. Exponential
5. Erlang

A study of these five plots can be very revealing in that, when the exponential distribution is used, the production results are substantially below those obtained from using symmetrical distributions. This exercise indicates how an assumption such as exponential travel times can significantly change the results. Since it is impossible to find mathematical solutions to all but the simplest queuing models, there seems little point in the mining engineer looking for practical applications in this area.

Clock Increment Models (Mostly Fortran)

During the 1960s, other investigators started building computer simulation models of mining operations. However, the computer language used was primarily Fortran. This was a logical choice since mining engineers were taught this language, and it was (and is) universally used. Programming was much slower using punch cards, and a considerable amount of time and effort was spent in writing and debugging the programs. Few mining engineers were concerned with learning a second or third language.

One particular problem in mining that concerned engineers was to construct models of conveyor belts, especially for underground coal mines. Sanford (1965) appears to be the first to undertake this problem and simulated a conveyor belt system for his MS thesis using Fortran. A large Fortran program, BELTSIM (1968), was developed at the Virginia Polytechnic Institute. An application of BELTSIM to an Australian mine was described by Talbot (1977). Hancock and Lyons (1984) describe a package known as SIMBELT4 and reviewed the work done by the National Coal Board in England. Their package is designed to predict the rate of flow at various points in a belt system and to estimate the ore in each storage bunker.

Suboleski and Lucas (1969) discuss a Fortran program (SIMULATOR1) that would simulate room and pillar mining operations. O'Neil and Manula (1967) used a simulation model for materials handling in an open-pit mine, and Venkataramani and Manula (1970) were able to simulate truck haulage in an open pit. Waring and Calder (1965) discussed a simulation package developed for a particular mine in Canada. Madge (1964) was able to simulate truck movement in an open-pit mining operation also in Canada. The outline of a large package capable of simulating a wide variety of operations was described by Manula and Rivell (1974).

One of the best early examples of a computer simulation model for a working mine was by Cross and Williamson (1969). This model had to do with a working open-pit copper mine in the southwest of the United States. The mine was a truck-and-shovel operation with five shovels loading trucks that dumped at the ore crusher, the waste pile, or a leach area. Initially, the trucks were dedicated to a particular shovel. A simulation model was then constructed to determine if a dispatcher could be used to route the trucks to different shovels so as to minimize queuing time. The model assumed that all of the times were deterministic and was able to indicate that a dispatcher would indeed improve operations. The program was several thousand lines of code and took considerable time to write.

One theme that all of these approaches have in common is that the models were very difficult to construct and took considerable time to write, debug, and verify. Little was said about the problems in changing the code to answer typical *what-if?* questions. Programs consisting of 50,000 lines of code were common in such models. The programs often took several man-years to write. Many of these studies were done by students for their MS or PhD research. Execution times of up to 4 hours on a mainframe were common.

Why are these programs so bulky and time consuming both in development and execution time? The reason is found in examining how these clock-increment models work. The program starts with an imaginary internal clock set to time 0. This clock moves forward a small time increment. Next, tests are made to see which, if any, of the many possible events that might occur in the system have indeed occurred. If, for example, there are 100 possible events that could happen in the system, 100 tests must be performed for every clock advance. If any of the events happens, the code associated with it needs to be executed. If no event takes place, a great deal of testing is needlessly done. Consider the case where the clock is going from one time to another 8 minutes later. Suppose that the clock increment is 1 second, and there are 100 possible events that can happen. Thus, there needs to be $60 \times 100 \times 8$ or 4800 tests done in the interval. If no event took place, these tests still have to be done. Over the course of a year's simulated time, the number of such null tests becomes astronomically high. And, what about writing the code? Since the program needs to jump from one section to another when an event takes place, this can be a programming nightmare. Writing the code for even simple mine simulations can require several hundred lines even if one can use multiple subroutines.

Even though many of these clock advance models were widely advertised and distributed, it is doubtful if they served their purpose. The last two such models for a mining operation were both published at around the same time. Both had to do with the simulation of conveyor belt systems. These were by Tan and Ramani (1988) and by Thompson and Adler (1988). Thompson and Adler replied to comments made by Sturgul (1988) by giving their reasons for rejecting GPSS and not selecting SLAM or SIMAN. The Fortran program they developed took about 9 months to code and consisted of 6000 cards. It took several hours to run on a PC. The interested reader is referred to Henriksen (1983b) and his section "Myths Surrounding GPSS." The widespread availability of simulation languages has done away with such models.

Event-Based Simulation Models

The first general-purpose simulation language was GPSS, which was developed by Geoffrey Gordon of IBM in 1961. This language has gone through many changes and revisions since its inception. Some of the versions are GPSS II, GPSS360, GPSS/V, GPSS/VS, GPSS/R, GPSS/PC, GPSS-10, micro GPSS, and GPSS/H. SIMSCRIPT was developed at the RAND Corporation in the early 1960s and SIMSCRIPT II.5 was introduced in 1972. SIMULA dates from the 1960s. SLAM, which was the successor to GASP, was introduced in 1979 and GPSS/H in the same year. SIMAN was introduced a few years later. There are other simulation languages such as Q-GERT and SIMPLE+ for discrete system simulation. However, only GPSS, SLAM, SIMAN, SIMSCRIPT, and SIMULA seem to have found favour with mining researchers. Of these GPSS/H, SLAM, and SIMAN are by far the most popular. There are also special languages for continuous system simulations such as

DYNAMO, CSMP II, ACSL, and EASY 5x. These are designed to solve systems where the various entities are controlled by differential equations (a parachutist jumping from a moving airplane, for example). Each of the different languages for discrete system simulation works in a different way. However, they all have one thing in common: they all are event oriented. A good summary of the different ways each of these languages works is given by Law and Kelton (1982). They also give a few remarks on the comparison of simulation languages with general-purpose languages. Their conclusion after contrasting Fortran-based models and those obtained using special languages is not surprising: "[W]e believe, in general, that a modeler would be prudent to give serious consideration to the use of a simulation language" (p. xiii).

The textbooks by Schriber (1974, 1991) are excellent sources for information on GPSS and GPSS/H.

Special Languages in Mining

As indicated, the first application of a special language was by Rist in 1961 with Harvey's use of GPSS in 1964. BETHBELT-1 was done for the Bethlehem Steel Corporation by Newhart (1983) using GASP/V. However, there were not many other applications of these languages for a decade until Wilke (1970) used GPSS to model underground train haulage for three coal mines. In an important work, Bauer and Calder (1973) pointed out the advantages of using GPSS for open-pit operations, especially to simulate load–haul–dump circuits. Brake and Chatterjee (1981) used SIMULA to study the effect of dispatching for a large coal mine in Australia. Steiker (1982) reviewed various simulation models and used GPSS for the simulation of an underground train haulage system. Interestingly, this was also for the same mine that Rist and Harvey did their work. Wilson (1984) gave an interesting background to the decision to use GPSS for the simulation of the surface rail ore transport for platinum mines in Zambia, Africa. Basically, a local university reviewed their problem and suggested that they learn and use GPSS. Students hired during the summer were given the responsibility of gathering the relevant statistics while the model was written and refined.

The mid-to-late 1980s saw the use of special languages in mining increase from a trickle to a steady flow. The work by Borkovic (1982) is an excellent example of using GPSS for mine planning for many years in the future. Borkovic used the Panguna mine of CRA on the island of Bougainville for his study. This MS work has been overlooked for many years. The advantages of using GPSS/H for mining applications were pointed out by Sturgul (1987). A more complicated model of a surface mine based on the work of Cross and Williamson (1969) and using stochastic times for the various operations is given by Yi and Sturgul (1987) to demonstrate the power of the GPSS/H language. Their program runs to less than 100 lines of code and took less than 1 hour to write. SIMAN was used by Tan and Ramani (1988) to study belt networks. The MSc work of Kolonja and Mutmansky (1993) at Penn State is an example of using SIMAN for studying the various dispatch criteria for open-pit mines. This is a very detailed and important contribution. (This thesis was presented at the SME annual meeting in Denver, Colorado, February 1994.)

The GPSS/H language has been used for a wide variety of mining applications and some of these are given by Sturgul and Harrison (1987). How the personal computer and GPSS/H can be used for mining operations is given by Sturgul and Singhal (1988).

Other examples are by Harrison and Sturgul (1988). An example of how to use simulation to determine the optimum location of in-pit crushers is given by Sturgul (1987). A state-of-the-art paper on mine system simulation was presented by Sturgul (1992) at an APCOM in Tucson, Arizona.

Another major change in using simulation models has been the animation that is now possible. Every simulation language now has animation. For example, SIMAN has CINEMA, SLAM has SLAMSYSTEM, and GPSS/H uses PROOF. PROOF can, in fact, be used with any ASCII file and is not confined to GPSS/H. The animation allows the user to view the simulation in *cartoon* fashion. Not only is this an excellent way to present the results of the simulation, but it helps to verify that the simulation is indeed correct. The work by Payne et al. (1994) in modelling the tar sands operations features animation. The study by Sturgul and Jacobsen (1994) for the model of the proposed surface gold mine in the South Pacific is the first example of the use of a simulation model to be used for the design of a new mine from scratch.

References*

Abed, S. Y., Barta, T. A., and McRoberts, K. L., A qualitative comparison of three simulation languages: GPSS/H, SLAM, SIMSCRIPT, *Computers & Industrial Engineering*, vol 9, 35–43, 1985.

Achttien, D. B. and Stine, R. H., Computer simulation of a haulage system, *Mining Congress Journal*, vol 50, 41–46, 1964.

Barnes, R. J., King, M. S., and Johnson, T. B., Probability techniques for analyzing open pit production systems, *Proceedings of the APCOM*, pp. 462–476, 1977.

Bauer, A. and Calder, P., Planning open pit mining operations using simulation, *Proceedings of the 10th APCOM*, South Africa, 1973.

Borkovic, A., *Computer Simulation of Truck and Conveyor Haulage at Bougainville Copper ltd's Panguna Mine*, MS thesis, University of New South Wales, Kensington, Australia, 1982.

Brake, D. J. and Chatterjee, P. K., Evaluation of truck dispatching and simulation methods in large-scale open-pit operations, *Proceedings of the 16th APCOM*, Tucson, AZ, pp. 375–383, 1979.

Cross, B. K. and Williamson, G. B., *Digital Simulation of an Open-Pit Truck Haulage System*, APCOM, Salt Lake City, UT, pp. 385–400, 1969, published by SME of AIME.

Douglass, J., *Prediction of Shovel-Truck Production: A Reconciliation of Computer and Conventual Estimates*, Technical Report #37, Bureau of Public Roads, 1964.

Falkie, T. V. and Mitchell, D. R., Probability simulation for mine haulage systems, *Transactions of the Society of Mining Engineering*, vol 226, 467–473, 1963.

Greenberg, S., *GPSS Primer*, Wiley-Interscience, New York, 1972.

Griffis, M. J., Optimizing haul fleet size using queueing theory, *Journal of the Construction Division*, vol 94, 75–87, 1968.

Griffis, M. J. and Cabrera, J. G., A multi-stage cyclic queueing model, *International Journal of Production Research*, vol 13, no 3, 255–264, 1968.

Hancock, W. E. and Lyons, D. K. G., Operational research in the planning of underground transport, APCOM, London, pp. 389–399, 1984.

* For more early publications on the subject of queuing theory, see Sturgul's articles "History of Mine System Simulation," "Annotated Bibliography of Mine System Simulation," and "Mine Simulation," *Proceedings of Internet Symposium*, 1996, published by Balkema, Rotterdam, the Netherlands.

Harrison, J. and Sturgul, J. R., GPSS computer simulation of proposed underground rear dump truck ore haulage system at Z C mines, Broken Hill, *Proceedings of the AusIMM Illawara Branch, 21st Century Higher Production Coal Mining Systems—Their Implications*, Wollongong, Australia, April 1988.

Harvey, P. R., Analysis of production capabilities, *Quarterly of the Colorado School of Mines*, vol 59, 713–726, 1964.

Henriksen, J. O., State-of-the-art GPSS, *Proceedings of the 1983 Summer Computer Simulation Conference*. The Society for Computer Simulation, San Diego, CA, pp. 918–913, 1983a.

Henriksen, J. O., State-of-the art GPSS, *Proceedings of the 1983 Summer Simulation Conference*. The Society for Computer Simulation, Vancouver, Canada, July, 1983b. Several of the other papers referenced by Henriksen can be obtained from Wolverine Software (www.wolverinesoftware.com).

Koenigsberg, E., Cyclic queues: Queue theory applied to mining, *Operational Research Quarterly*, vol 9, no 1, 22–35, 1958.

Koenigsberg, E., Twenty five years of cyclic queues and closed queue networks: A review, *Journal of the Operational Research Society*, vol 33, 605–619, 1982.

Koenigsberg, E. and Lam, R. C., Cyclic queue models of fleet operations, *Operational Research*, vol 24, no 3, 516–529, 1976.

Kolonja, B. and Mutmansky, J., Simulation analysis of dispatching strategies for improving production of truck haulage systems, *Paper Presented at SME Annual Meeting*, Reno, NV, February 1993.

Law, A. and Kelton, W. D., *Simulation Modeling and Analysis*, McGraw-Hill, New York, 1982.

Luo, Z. and Lin, Q., Erlangian cyclic queueing model for shovel-truck haulage system, *Mine Planning and Equipment Selection*, R. Singhal, ed., Balkema, Rotterdam, the Netherlands, pp. 423–427, 1988.

Madge, D. N., Simulation of truck movement in an open pit mining operation, *Canadian Operational Research Society*, 32–40, 1964.

Maher, M. J. and Cabrera, J. G., A multi-stage cyclic queuing model, *International Journal of Production Research*, vol 13, no 3, 255–264, 1975.

Manula, C. B. and Rivell, R., A master design simulator, *Proceedings of the 12th APCOM*, Colorado School of Mines, Golden, CO, 1974.

Morgan, W. C. and Peterson, L. L., Determining shovel-truck productivity, *Mining Engineering*, vol 20, 76–78, 1968.

Newhart, D. D., *BETHBELT-1, A Belt Haulage Simulator for Coal Mine Planning*, Research Report File 1720-2, Bethlehem Steel Corporation, Research Department, Bethlehem, PA, 1983.

O'Neil, T. J. and Manula, C. B., Computer simulation of materials handling in open pit mines, *Transactions of the AIME*, vol 1, 137–146, 1967.

O'Shea, A., Slutkin, G. N., and Shaffer, L. R., *An Application of the Theory of Queues to the Forecasting of Shovel-Truck Fleet Productions*, Technical Report for NSF, Grant No. G-23700, Department of Civil Engineering, University of Illinois, Urbana, IL, 1964.

Payne, F. R., Heinz, D. A., and Ellis, D. W. H., Mine design using simulation modeling at syncrude, *Paper Presented at the CIM Annual Meeting*, Toronto, Ontario, Canada, 1994.

Phillips, D. T., Ravindran, A., and Solberg, J. J., *Operations Research*, John Wiley & Sons, New York, Chapter 7, 1976.

Rist, K., The solution of a transportation problem by use of a Monte Carlo technique, *Mining World*, November. This paper was also presented and published in the Proceedings of the 1st APCOM held at the University of Arizona in Tucson, pp. L2-1–L2-15, 1961.

Sanford, R. L., Stochastic simulation of a belt conveyor system, *APCOM Proceedings*, Tucson, AZ, March, pp. D1–D18, 1965.

Schriber, T., *Introduction to Simulation Using GPSS/H*, John Wiley & Sons, New York, 1991.

Schriber, T., Perspectives on simulation using GPSS, *Proceedings of the Winter Simulation Conference*, M. Abrams, ed., The Society for Computer Simulation, San Diego, CA, 1988.

Schriber, T., *Simulation Using GPSS*, John Wiley & Sons, New York, 1974.

Steiker, A. B., Simulation of an underground haulage system, *Proceedings of the 17th APCOM*, Denver, CO, pp. 599–613, 1982.

Sturgul, J. R., Advances in simulation models of mining operations, *Proceedings of the 23rd APCOM*, Tucson, AZ, April 1992.

Sturgul, J. R., Discussion: New simulator for designing belt system capacities in underground coal mines, by S. D. Thompson and L. Adler, *Mining Engineering*, 1124–1126, 1988.

Sturgul, J. R., How to determine the optimum location of in-pit moveable crushers, *International Journal of Mining and Geological Engineering*, vol 5, 143–148, 1987.

Sturgul, J. R., Simulating mining engineering problems using the GPSS computer language, *Bulletin Proceedings of the AusIMM*, vol 292, no 4, 1987.

Sturgul, J. R. and Harrison, J., Computer simulation studies to assist in mine equipment selection, *AusIMM*, Equipment in the Mineral Industry, Kalgoorlie, WA, pp. 155–159, 1987.

Sturgul, J. R. and Jacobsen, W. L., A simulation model for testing a proposed mining operation: Phase 1, *Paper Presented at 3rd International Symposium on Mine Planning and Equipment Selection*, Istanbul, Turkey, October 1994.

Sturgul, J. R. and Singhal, R., Using the personal computer to simulate mining operations, *First Canadian Conference on Computer Applications in the Mineral Industry*, Laval University, Quebec City, Balkema Publishing Company, Rotterdam, the Netherlands, March 1988.

Suboleski, S. C. and Lucas, J. R., Simulation of room and pillar face mining system, APCOM, Salt Lake City, UT, pp. 373–384, 1969, published by SME of AIME.

Talbot, K., Simulation of conveyor belt networks on coal mines, *Proceedings of the 15th APCOM*, Brisbane, Australia, July, pp. 297–304, 1977.

Tan, S. and Ramani, R. V., Continuous materials handling simulation: An application to belt networks in mining operations, *Transactions of the AIME*, vol 286, pp. 1803–1836, presented at SME meeting Phoenix, AZ, January 1988.

Thompson, S. D. and Adler, L., New simulator for designing belt system capabilities in underground coal mines, *Mining Engineering*, pp. 271–274. This paper was also presented at the Annual SME Meeting, April 1988.

Venkataramani, R. and Manula, C. B., Computer simulation of bucket wheel excavators, *Transactions of SME*, vol 247, 274–280, 1970.

Waring, R. H. and Calder, P. N., The carol mining simulator, *APCOM 1965*, University of Arizona, Tucson, AZ, 1965.

Wilke, F. L., Simulation studied computer controlled traffic underground in large coal mines, Decision Making in the Mineral Industry, 9th International Symposium, Can IMM, vol 12, 344–351, 1970.

Wilson, J. W., Simulation of ore transport on surface rail at Impala Platinum Ltd, *APCOM*, London, pp. 411–418, 1984.

Yi, R. and Sturgul, J. R., A comparison between Fortran and GPSS for mine simulation, *Wuhan Mining Journal*, 1987, in Chinese.

1

What Is PROOF Animation?

A portion of the Chuquicamata Mine in Chile. This mine has been in operation as a major open-pit mine since 1910. It has most probably been worked for centuries before this.

1.1 PROOF Animation

1.1.1 Getting Started

Proof Animation and GPSS/H, products of Wolverine Software, are needed for Chapters 1 and 2 of this book. You can download and install the software from http://www.wolverinesoftware.com. On the main page of this website, click on the Downloads link. On the Downloads page, click on the link supplied for downloading the Student/Demo versions of Wolverine Software's products. Follow the installation instructions carefully. Unless you override the installation procedure, all software will be installed in a folder named C:\wolverine. The examples for this book will be installed in a folder named C:\wolverine\SturgulExamples. Installing the software will create a Wolverine folder on your desktop. Inside this folder you will find a variety of Getting Started icons that will be of help.

1.1.2 Practice with PROOF Animation

Start by clicking on the Student P5 icon in the Wolverine desktop folder. Once Student P5 (SP5) comes up, click on the File menu item, click on Open Layout + Trace, navigate to the C:\Wolverine\SturgulExamples folder, and select the file named PAPA.LAY. Then a portion of the screen will look as shown in Figure 1.1.

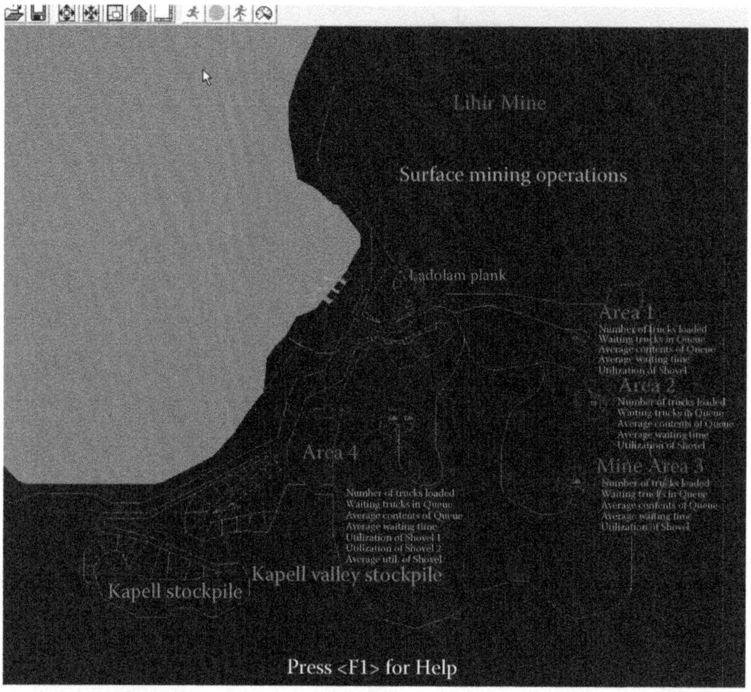

FIGURE 1.1
Portion of screen when running animation PAPA.LAY. This is the Lihir Mine, which is located on an island in Papua New Guinea. This is the first mine to have been developed from the start using a simulation and animation model. (Panagiotou, G. N., and Sturgul, J. R., Mine simulation. *Proceedings of the 1st International Symposium on Mine Simulation via the Internet,* Balkema Publishing Company, Rotterdam, the Netherlands, May 1997.)

Start the animation by clicking on the green *running man* on the SP5 toolbar. (Note that explanations of icons are displayed when the mouse hovers over them.) This icon is found in the top menu bar. The animation given by PAPA.LAY is of the Lihir Mine.* This is a surface gold mine located on an island northeast of Papua New Guinea. The mine started operations around 1994. The animation model is for when the mine was just starting to strip waste. The ore body is a caldera (collapsed volcano). Trucks are loaded by the shovels and travel along paths to the dock where they load the barges.

The subdirectory STURGULEXAMPLES will have different animations of systems. Some of the animations are simple, illustrating one aspect of the whole system. For example, MILL just shows a turning ball mill. BELT is an animation of a simple conveyor belt system.

The example DUMP10 has a long history. For many years, engineers have been trying to model this simple-appearing example of a single shovel and multiple trucks without success. They tried finding exact solutions but, if there is uncertainty in the various times for loading, hauling, dumping, and returning, the problem is impossible to solve except for a few cases, such as when all times are exponential. This is rarely the case for travel times. Using GPSS/H,

* Panagiotou, G. N., and Sturgul, J. R., Mine simulation. *Proceedings of the 1st International Symposium on Mine Simulation via the Internet*, Balkema Publishing Company, Rotterdam, the Netherlands, May 1997. This book contains several papers of interest to the mining engineer. They include a 'History of Discrete Mine System Simulation' and 'Annotated Bibliography of Mine System Simulation', both by J. R. Sturgul. In particular, the paper, 'Simulation and Animation Come of Age in Mining', Sturgul, J. R., *Eng. & Mining J.*, October 2003, feature article and cover devoted to article. pp. 38–42, includes work on the Lihir Mine.

one will learn both how to model this example and how to do the animation you have just seen. Rather than spend many months on this example, this system can be modelled in a matter of minutes, not months.

You may want to examine some of the other animations at your leisure.

Before we start learning the GPSS/H simulation language, we will first examine how the animation works. This is called PROOF animation and provides a very high-quality animation. It is easy to learn, probably because the student can *see* everything that is happening. If you draw a line, there it is right in front of you on the screen. It is also a fun topic, as you will soon discover.

No simulation project can be considered *complete* until one has added the animation. This is essential, as this is the *proof* that your simulation model is correct. Dr. Alan Pritsker, a famous simulation expert, once said words to the effect, "Every good simulation contains a surprise." While this is a bit of a simplification, there is a lot of truth to this. When one learns how to simulate complex situations, many times the results will not be what one would expect using simple averages. For example, when one studies queuing theory, one of the first examples one encounters is that of a single server and arrivals coming from an infinite population. If the service rate is exponential and the arrival rate is Poisson (also exponential), this system has an exact mathematical solution. Using these statistical distributions, suppose that the arrival rate has a mean time between arrivals of 5 minutes and that the service time also has a mean time of 5 minutes. One (somehow) obtains the solution, perhaps by looking it up in one of the books on queuing theory. One of the characteristics of the system is the expected length of the queue over time. If λ is the mean arrival rate and μ the mean service rate, then the expected queue length is as follows:

$$\text{Queue} = \frac{\lambda}{(\lambda - \mu)}$$

Thus, since λ and μ are equal, you have division by zero and, as a result, the expected queue length will be *infinite*. If one has never seen this result, it might be quite a surprise. This is a correct result and can be found in any textbook on either operations research or queuing theory. It serves to illustrate why one should *not* do design work using average values.

This is one of the main reasons for *always* including animation when simulating any system. The computer program merely cranks out numbers, but the animation shows the results. There will be other results from seemingly simple examples where the results will be a bit surprising, as you will learn as you progress through the book.

PROOF animation allows you to present your simulation in an animated or *cartoon* manner. This animation guarantees that the logic of your simulation is correct. One can *see* on the screen just what is happening during the running of the simulation.

The animations that you will create for your simulations can be as simple or complex as your artistic ability will allow. As one is learning how to create these animations, it is probably best at first not to be too fancy but to try to draw simple objects and layouts. Later, as one develops proficiency, one can increase the complexity of the animations, especially when one starts to make animations to show other people.

1.1.3 Learning PROOF

To begin, start by opening up PROOF animation. This can be done either from the icon on your screen or from the command prompt. It should be located in the subdirectory C:\SP5. If not, you should be told where it has been loaded. It may be in the subdirectory C:\PROOF> or it may be on your desktop in Windows.

Once PROOF is started, there will be a blank screen with a menu bar at the top.

The categories that are highlighted are the ones you can click on. They are as follows:

There is nothing to do at this time with File or any of the other icons except for Mode. We want to practice the drawing of a layout, so we first click on Mode and see a pull-down menu as follows:

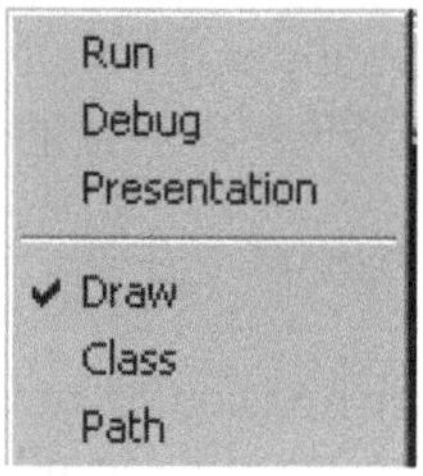

We have nothing to Run, Debug, or any Presentation to view. Thus, we will be concerned only with Draw, Class, and Path at the present time. Before we can do anything with PROOF, we need to have a layout.

The *layout* for an animation consists of the following three things:

1. The actual *layout* shows all the items in the animation that do not move but remain static.* This is made from the Draw option. Think of a bank and the layout of the floor, the cashiers, and so on. This layout might be what one would see in a plan view of the bank.
2. The Classes will be used to make the objects that will move in the animation. These might be the trucks in the mine. There might be multiple objects that are created from a single class. For example, in the mine, the Class might be a truck. In the animation, the objects that are created from this Class would be the different trucks.
3. The Paths are the paths on which the cars and customers move. These are created from the Path mode.

* There is one exception to this that will be introduced later. This is called a *layout object* that can move, change colours, change to another object, and so on.

We will look first at the Draw option, as this is the necessary first step for your animations. You actually can have an animation with a blank layout, so that there might be only objects moving across the screen but, for the most part, there will be a layout on which objects will move. In most cases, you will want to put your name or a title of the animation on the screen as part of the layout.

1.1.4 Making the *Static* Layout

Click on the Draw option and you will see a new menu bar below the top one. You also will notice that the screen has changed. This will be discussed shortly. The menu bar will appear at first to have a bewildering array of options. These are all necessary as there are many things you can do in making up a layout. The options are as follows, going from left to right:

Each of these icons will be considered separately. PROOF is not difficult to use. Don't let the many icons shown above intimidate you (Figure 1.2).

FIGURE 1.2
Screenshot of animation given by program CHAP1A.LAY.

	Open a new layout trace file
	Save current layout file
	Zoom in (magnify)
	Zoom out (shrink)
	Zoom in using a draggable box view
	Select home view
	Show/hide scroll bars
	Snap to option

(Continued)

Trim unwanted portion of lines/arcs

Capture elements in boxes, transfer them

Undo the previous operation

Draw one or more fillets

Draw a line

Draw a polyline

Draw an arc

Draw static text

Draw a message (dynamically alterable text)

Draw a bar graph

Draw a line plot

Place a layout object

Draw, edit, or recolour an area

Insert a bitmap picture

As mentioned, you will also notice that the screen has changed from the solid black to a grid showing the grid lines and some of the grid values (in units of 10). The grid goes from (0, 0) to (80, 58) but you cannot see these corner points. Actually, what you are looking at is a small window into a very much larger grid. We shall see how to move this window, shrink it, expand it, and so on. Most of the animations we will be doing in learning PROOF will fit nicely on the grid you now see. However, it will be instructive to learn just how large this grid actually can be. Many large animations will make use of the fact that this grid can be much larger than what you see on the screen.

Next, we are going to make a somewhat fanciful layout. Click on File and Open Layout [not: Open Layout + Trace (ATF)]! Click on CHAP1A.LAY and examine what the layout is. It looks like something out of an abstract painting. Figure 1.3 is this layout.

You will shortly make one that is similar to this, although it is highly unlikely that yours will be identical to the one that you will see now on the screen.

After you have examined this for a while, you are ready to draw a similar layout. Click on File and Close to return to the original drawing screen. Then click on Mode and Draw. You will see a different screen with more icons. Each will be explained in subsequent chapters. For now, we only want to use the icons to make a layout similar to the one in Figure 1.2. If you position the mouse over each icon, a summary of what each does appears on the screen.

This is the screen where you will be creating the layouts. It is shown with a grid to assist you. The grid can be expanded when needed. You are actually drawing on a tiny portion of a very large pallet. For most of our learning work, the grid you see on the default screen will be sufficient.

The first thing we are going to do is draw a line, so click on the Line icon. This is the icon:

You will see a pull-down menu appear in the centre of your screen. The line is to go from (20, 30) to (60, 30), so first click on the dark blue line at the top of the menu and drag it aside.

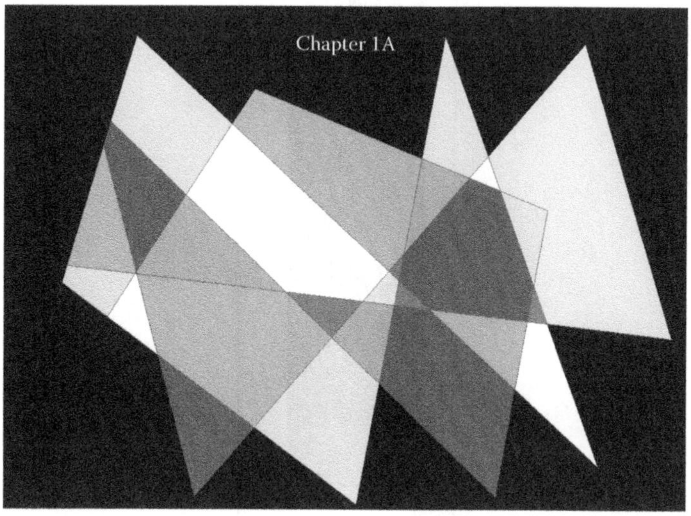

FIGURE 1.3
Menu portion of the screen in Draw mode.

It can be placed anywhere on the animation, so that it does not interfere with what we are doing. Now, click on the point (20, 30) and drag the line to the point (60, 30). You will notice how easy this is to do as the line seems to *snap to* the coordinate. We will learn more about this feature shortly. The colour of the line is RED or F2. Let us examine the menu again. It looks as follows:

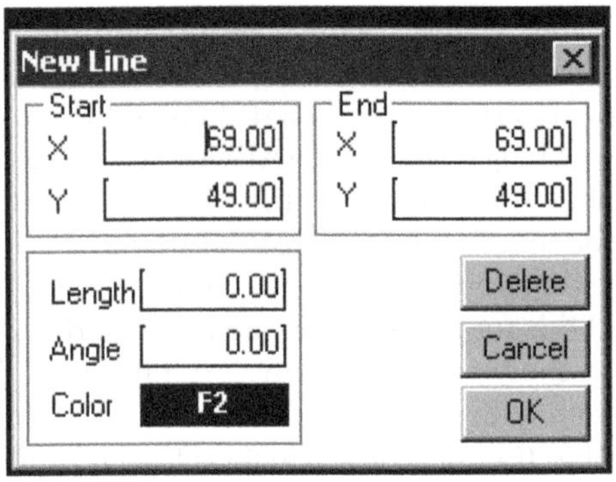

It would have been possible to draw the line only from the menu by clicking on the appropriate boxes and keying in the values. However, it is much easier to do this, as we have just done. There will be cases for complex layouts where you might want to type in the coordinates of the start and end points.

Click on the colour RED (F2). You will see another menu on the screen. This shows the 32 colours you can use for PROOF. Each colour has a code starting with the letter F and then a number. Later, we will use these in our animations to either set a colour of an object or change a colour.

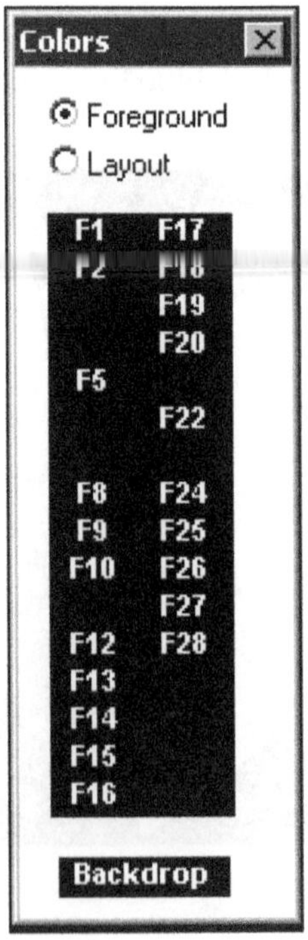

The *colour* backdrop is in the menu. This means that the colour will *not* show on the animation layout. Instead, it will be the same colour as the background and not be visible. This is useful if you want to use the line as a path in the layout where objects will move and the paths are not shown. For example, a plane flying across the screen will move on one of these *invisible* paths.

If you click on Backdrop, the colour of the red line becomes a light blue. This is only for your benefit, so that you can see it when you are making and/or changing the layout. Change the colour back to RED again.

The Backdrop is normally going to be black, the same as the background of the layout. However, you may wish to have another softer colour for the background of your animations. For example, some of the grey colours make a nice backdrop. You can change this background colour by clicking on Setup → Colors. You will see the pull-down menu with three *handles*, which can be adjusted to change the background color once you click on Backdrop. You will see how the backdrop colour is made up by moving the three-colour handles up and down. You may wish to experiment with these handles by moving them up and down to obtain other possible background colours. If you do this—and you may well want to later as you gain proficiency in PROOF animation—you will need to be careful to remember that, when you learn topics such as messages, which will put numbers or text on the screen, when one message is replaced by another, you will want the *erase* colour to be the background colour also. Otherwise, the new message will appear to override the old one. This point will also be discussed in Chapter 6 when the topic of messages is discussed.

Now, draw two more lines that are vertical but at an angle, so that they intersect the original line. Make sure that each is in a different colour. This should be very easy to do.

The icon next to the line is for drawing what are called *polylines*.

These are lines with different segments, not necessarily in the same direction. Click on this icon and you will first see a menu:

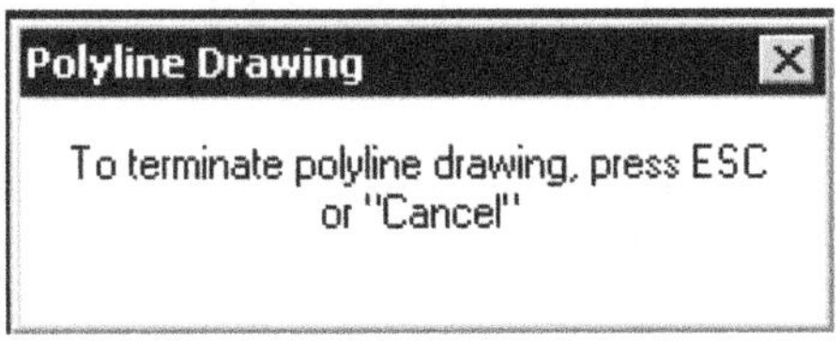

This is the new menu that comes up (the first one goes away).

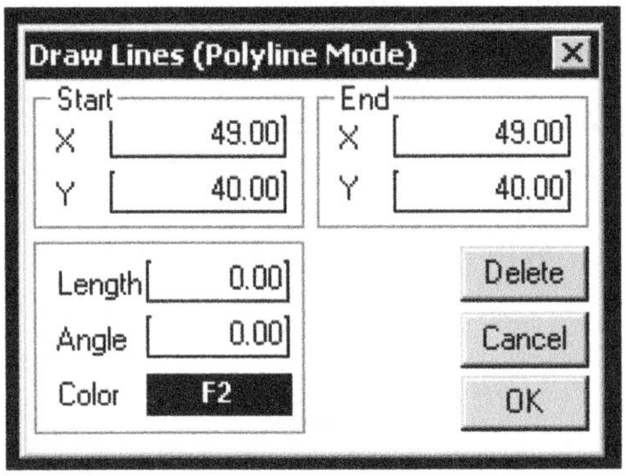

This menu is quite similar to the one you saw when you drew a line. You may wish to move this menu off to the side while you experiment with polylines. Move the cursor to any place on the screen, press the mouse, and drag the cursor to another location. Left click and continue to drag the line to another location at an angle to the first line. Do this for about four or five other segments. Make a few other polylines of different colours on the screen.

If you are like other students, you now have a screen that looks like some sort of abstract artwork with lines and shapes going every other way.

Let's fill in a few of the areas that are fully surrounded by lines. At the far right of the menu bar is an icon of a paint can that is pouring out. This is the *fill* icon.

This icon allows you to fill in areas on the layout. It will also allow you to remove fill from a filled area. Click on the icon, and you will see the following menu:

Move the menu out of the way and click on any enclosed area. The area is now filled with the red colour. You can change the colour and fill some more areas and, eventually, you should have a layout worthy of display in any museum of modern art.

If you happen to be filling an area and the whole background takes on the colour of the fill, you tried to fill an area that was not entirely closed. It may well have appeared to be closed, but the filled screen tells you otherwise. You need to click on Done and re-open the menu. Click on the same fill icon and click on Remove fill; then click anywhere in the background colour.

Your design probably has a few lines that stick out of the main layout. This often happens in making a layout where you are going to use the fill option. To remove these extraneous lines and have a layout that has no lines sticking out of the main layout, we will use the *trim* option. This is the icon that is grey with the large T:

Click on this and the new menu appears:

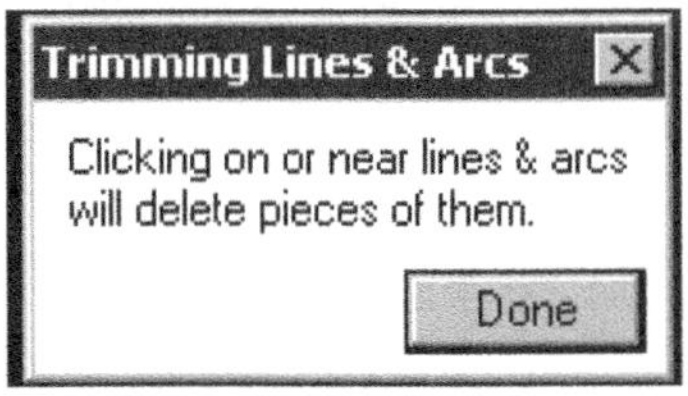

Go to each line segment that sticks out on your layout and click on it. It will disappear. This is easy and quickly done. Often, if you are drawing an unusual shape that is to be filled, you will deliberately have lines that cross to ensure that the lines that eventually make up the perimeter of the area to be filled are connected. You then can click to remove the extra lines.

When you are done with your work, it is a good idea to save it. Click on File and you will see a large pull-down menu.

The following are the only options we care about at present:

- Save Layout
- Save Layout As

The extension of each layout must be .LAY. If you do not give the layout such an extension, PROOF will do so automatically. Click on either of the two options and give your layout a name such as MYART or some name that you will easily remember.

Congratulations! You now have a layout artwork no doubt worthy of display in most museums of modern art. Your layout should look something like Figure 1.3.

1.2 Further Work

Examine more of the animations that are included and practice running these multiple times. You also can practice navigating the animations using the various icons to zoom in or out. You can speed up or slow down the animations by clicking on the clock icon and changing the speed.

1.3 Exercises

The files referred to in these exercises are in the subdirectory STURGULEXAMPLES.

1.1. Make a layout that will show on the screen as a rectangle. It should be filled with the colour green. The figure will have corners: (10, 10), (40, 10), (40, 40), and (10, 40). Notice that, even though the figure would normally represent a square, because of the way PROOF is configured, the shape on the screen is a rectangle.

1.2. Make a layout that is a five-sided star by making it as a series of different lines that will intersect. Fill the star with colours you like depending on how you drew it.

1.3. Experiment with different backgrounds for your next layout. Click on Setup and Colors. Click on Color from the dropdown menu and select Backdrop. You will see a menu on the centre of the screen. From this you can move the three-colour handles up and/or down to change the background. Take the layout CHAP1A as an example and change the background to all light grey. An example of this is given by the layout CHAP1B.LAY.

The following exercises are designed to give practice in running animations. The student is encouraged to run the animations and use the various PROOF icons presented in this chapter to learn the animation options. These files are all in the subdirectory STURGULEXAMPLES.

1.4. There is a layout file called MINER.LAY. Examine this and draw a similar one.

1.5. A classic problem in queuing theory is a single worker and customers coming for service. There are many variations of this. One such example is given by the file CLASSIC, which is an example of a barbershop. Customers arrive for a haircut and wait if the barber is busy.

1.6. There is an animation called BARGE. This shows barges coming to a dock and being loaded by a FEL. Run the animation.

1.7. There is an animation called CARWASH. This is an animation of a car wash. Cars come in and move on an assembly line where they are washed and then dried. Run this animation.

1.8. There is an animation called TRAFFIC. This shows traffic flow on six streets with many intersections. Run this animation. Simulation is very useful in studying such systems.

1.9. There is an animation called LEMMINGS, which shows lemmings jumping off a cliff. Run this animation.

1.10. There are two animations, BANKA and BANKB. BANKA has the animation of a bank where arriving customers use the first available teller. If all tellers are busy, arriving customers wait in a single line, called a *quickline*. BANKB has arriving customers wait in individual queues if all tellers are busy. No *queue jumping* is allowed once an arriving customer selects a teller. Such models are a classic example for simulation.

1.11. There is a layout given by CHAP1C.LAY. This is an example of your name in lights. To have the program run with your name, you need to run the program that created the file to run the animation. This is done by keying in

GPSSH CHAP1 <cr>

from the subdirectory where you have installed Student GPSS/H. When the program runs, you will be prompted for your name (only your first name, please). After the program runs, a file CHAP1A.ATF will be created. Copy this to where you have Student PROOF (SP5). Now, when you open CHAP1C, the program will run showing your name in lights.

2

Drawing a Layout

St. Barbara, patron saint of miners.

2.1 Making a CLASS

In this chapter, we will complete our first actual animation. We also will learn some of the basic PROOF commands such as `CREATE` and one form of the `SET` command.

Open up PROOF animation and click on Mode Draw. Draw a line from (20, 30) to (60, 30). Let this line remain the default colour RED.

The animation to be created will show a small, white square moving on this line from left to right in 30 time units. The only thing that will be shown in our layout is this one line. Thus, we are now done with the layout portion of the animation.

To make the object, click on Mode and then Class. You will see the following pull-down menu choice.

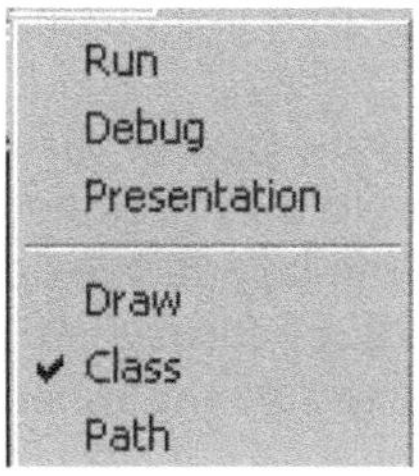

The top menu bar shows new icons. One is as follows:

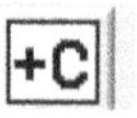

This says that you can now add a new CLASS. You can either click on this or click on Class. If you do the latter, you will see the drop-down menu:

The Select option is grayed out, so we click on New and see the following box:

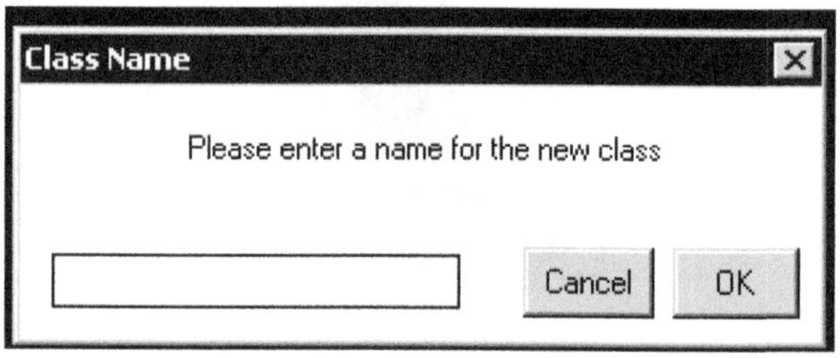

Each CLASS needs a name. *The name is case sensitive so be very careful that you remember what you named it!* For example, CAR, Car, and car are different names. This is the source of many errors for beginning PROOF users. A suggestion is that you always use all capital or small letters, so that you can always be sure of what you named the CLASS. In the examples here, only capital letters will be used. It is also possible to use several pieces of text for each CLASS:

NEW TRUCK
OLD MACHINE
Large truck for hauling
Small truck for men

It is suggested that you use short names until you start making larger animations, where the use of mnemonics will be more meaningful. If you are going to draw a CLASS that is a square, call it SQUARE; a CLASS that is a circle, CIRCLE or BALL; and one that is a triangle, TRI or TRIANGLE.

Notice that the drawing pallet now is greatly shrunk. It goes from around (–4, –2) to (4, 2.5). This is because the objects are to move on the layout and you will, quite naturally, want them to be much smaller than the layout. We don't want to draw a car that is larger than the road or the garage! You can, of course, shrink or expand the CLASS drawing palette, but you are normally going to use the default palette. Most of the time, this is sufficient.

Once you give your new CLASS a name, there is also a pull-down menu that looks as follows (what each option does will be explained later):

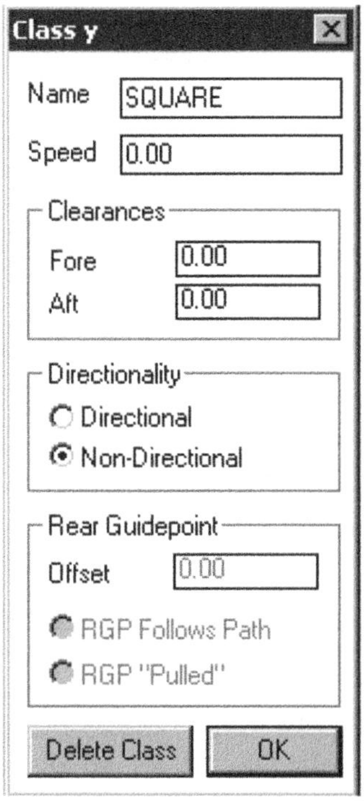

For now, we will use the default values as given in the menu.

The name of our CLASS is to be SQUARE. It could be anything else, but this mnemonic is easy to remember. It is from the CLASSes that one makes objects that move on the screen. We will see how PROOF does this shortly. One can use the same CLASS to make any number of different objects. It is the objects that will move on the screen.

The point (0, 0) is called the *hot point. This is very important.* When you place an object *on* a path or at a position in the layout, it is the hot point that goes on the end of a path or at the point in the layout. For example, suppose your object is called a *car* and that it is simply a filled rectangle. The hot point for a car is located in the middle of the rectangle in the centre. When it is placed on a path, the car will appear to be half way above the path and half way below. It will also stick out from the path as the hot point will be on the end of the path. If the hot point is at the bottom of the car rectangle in the centre, the car will appear to be placed on top of the path with the front half on the path and the bottom half off. If the hot point is at the bottom front of the object, the car will be just to the left of the path, that is, the hot point is placed on the front of the path. In fact, if the CLASS was drawn with the hot point not even on or in the car, the car would appear to be in the air above the path or possibly below the path.

Our animation is going to have a square that moves on a path, so draw a white square with sides 2 units having the hot point in the bottom centre. It will not appear to be a square as the height and width of the drawing layout are not the same. However, each side should be 2 units as measured on the layout. It is necessary to fill this in with the colour WHITE. The following is the icon to do this:

When this is clicked on, the following pull-down menu appears:

If the colour is not WHITE (F3), click on Color and select the white colour. Then position the cursor inside the square and click Done. The square is now white. There are other options associated with this icon, which will be explained later.

The square will look something like the one shown in Figure 2.1.

To accept the default value, click on OK. The square disappears from view.

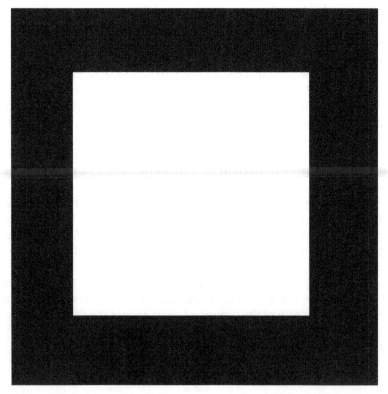

FIGURE 2.1
Class SQUARE.

2.2 Making a Path

We next need to make the path on which the SQUARE will move. Click on Mode → Path and you will notice that the Path on the top menu bar becomes bold. There is also a new icon that is highlighted:

The layout also changes to show the original layout with the layout lines in grey. If we had put more than just the one line in the layout, all of them would be grey.

You can either click on this icon for a new path or click on the Path icon on the top menu bar. If you select the latter, the menu appears:

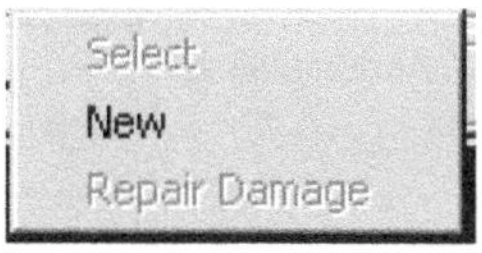

Since we do not have any paths, only New is highlighted. Click on New and a dialog box similar to the one for CLASSes appears:

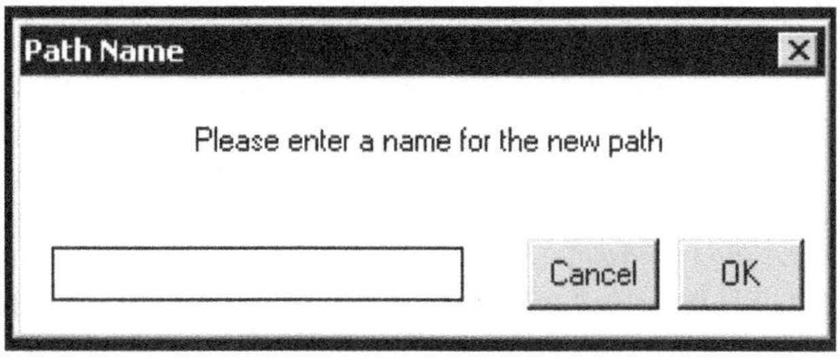

Again, be careful, as the name of the path is case sensitive. Key in P1 for the name of the path (or some similar mnemonic that you will not forget—it is a good idea when creating many paths that you write each down for reference later). You will notice that the original layout will appear on the screen. Each line, line segment, or whatever you have drawn in the layout is shown in light blue. It is possible to make a path out of just about anything in the layout.

Notice that the cursor is now a large arrow. Move this to the right of the line segment and click on it. It will turn red and the label P1 will appear at the left of the diagram. There should be an arrowhead on the right of the new path. This indicates that objects placed on it will move from left to right. You may have to move the new pull-down menu to see the path you have just created.

In the event that your path goes from the right to the left, you need to click on the third icon from the right on the top menu bar to reverse the motion.

Be very careful to use only a single click when you click to make a path. It is possible to have two paths using the same line segment, one going from right to left and one going from left to right. If you double click (as is easy to do), you may have a path that goes both ways on the segment when you only wanted a single path. *This is a very easy error to make.* Click on the OK and see the path turn blue.

You will also notice a pull-down menu to the left of the layout as shown below.

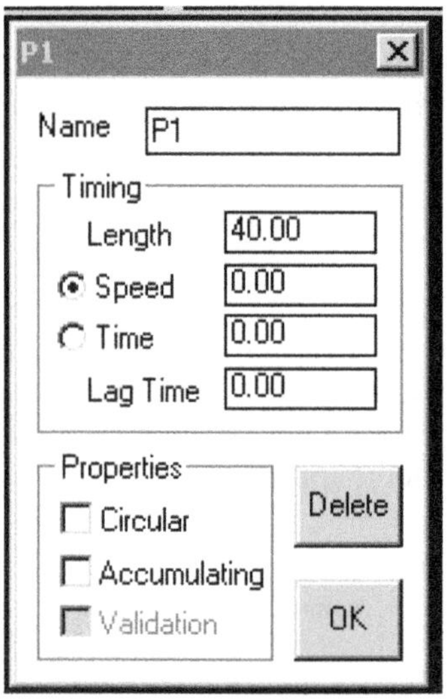

The only item of interest in the menu for now is that the length of the path is 40.00. This is correct, as we can see from inspection. If the length were to be 80.00, we would have seen two arrowheads on the path as this would mean that we had double clicked on the line segment in error. We would have to either delete the path or use the icon second from the right to delete the incorrect segment. Paths are easy to make, but great care and caution must be exercised when making them, so that one does not accidentally click on a path more than once. If so, the path becomes a *double* (or even a triple or more path!). An object will travel from one end to the other and then suddenly go back and redo the same travelling. It is always a good idea to check the length of the path just created to see if it is correct. If the length is wrong, it is usually easier to simply delete the path just created and go back and make a new one.

We are now done with the layout and can save the file. Call it CHAP2A. Click on File and the Save Layout (or Save Layout As). You will be prompted to type in a name for the layout. *It is not necessary to key in the extension.* It will automatically be saved with the extension .LAY, so it will be CHAP2A.LAY.

2.3 Making the .ATF File

We are now ready to make the second file that is necessary to run the animation. This is called the *automatic trace file (ATF).* This file *drives* the animation. It must have the same name as the .LAY file and have the extension .ATF. This file consists of a series of very English-like commands that are separated by time increments that PROOF uses to move the objects, write messages, change colours, and so on in the layout. PROOF uses a *look-ahead* technique as it takes the current program time and then looks to the next time increment and does everything up to that time increment. For example, if you had the following as a part of a PROOF file:

```
TIME 40.0000
.......
.......
(Proof commands)
.......
TIME 45.4567
```

PROOF would be at time 40.000 and look ahead to the next time increment, which is 45.4567. It would do the English-like commands in between the two time increments and then go to time 45.4567 and look to the next time increment. Notice that four decimal spaces are used for the time increment. It is suggested that you always use the same number of decimals throughout the file and use at least four digits with even six recommended for large animations.

Since you cannot go backwards in time in PROOF and PROOF uses integer truncation, if you do not have enough decimals, it may be possible to generate a time increment that is not more than the last one. Thus, the two different times of 4.56123 and 4.56534 would be taken as both 4.56 if the field for the decimals is only two digits. This would lead to an error in running the animation.

There are two basic ways to make the .ATF file. The first is to make it line by line by hand. The second, and much more convenient, way is to let the GPSS/H program make the files. Since .ATF files can be many thousands of lines in length, this is the only logical way to create these files for most animations. However, we will make a few .ATF files by hand to demonstrate them and study how they work.

2.4 The `CREATE` Command

The first of the English-like commands to be encountered is necessary for any animation. One has to use the `CLASS`es we have created to make an object that will move in the animation. The command for this is as follows:

```
CREATE (class name) (object name or number)
```

Notice that the `CREATE` is written starting in position 1. While this is not necessary, it is the format that will be followed here. Some examples of `CREATE` are as follows:

```
CREATE SQUARE 1
CREATE SQUARE S1
CREATE SQUARE SQUARE1
CREATE SQUARE BOX
Create SQUARE SQ5
```

The `CREATE` command is not case sensitive, whereas both `CLASS` and `SQUARE` commands are. However, all PROOF commands will be done in capital letters.

Thus, in our example, if we had written:

```
CREATE Square S1
```

we would obtain an error message, as there is no Square in the CLASSes, only a SQUARE.

This is an easy command to remember. After you `CREATE` an object from a CLASS, you need to do something with it in the layout. You will normally place it on a PATH or at a point (or, in the rare case, in another object).

2.5 The `PLACE` Command

The forms of this command are also easy to remember. Some of them are listed below:

```
PLACE (object ID) on (path name)
PLACE (object ID) on (path name) at (units from end point)
PLACE (object ID) at (coordinates)
PLACE (object ID) in (object ID)
```

The first of these will place the object on the path with its hot point on the end point of the path. The next will place the object on the path but so many units from the end point. The third will place the object at the coordinates specified but the coordinates are given without a comma and parenthesis. For example,

```
PLACE S1 at 35 30
```

would place the object S1 on the layout at coordinates (35, 30). However,

```
PLACE S1 at 35, 30
```

and

```
PLACE S1 at (35, 30)
```

would both give error messages.

It is beneficial to mention at this point that PROOF has an excellent Help at the top of the menu bar. You might want to check it now to see still another option with the `PLACE` command.

You may recall that, in the menus that appeared when we were making the CLASSes, there was a place for the SPEED and the PATHs had places for both SPEED and TIME. If we had input values for SPEED on the CLASS menu, this is the speed at which the object would travel on the path. However, if we had input values for SPEED on the PATH menu, this is the speed at which the object would travel on the path; also if we had input a value for the `TIME`, this is the time it would be on the path. If we had input values for all of these, the order of priority would be: lowest is SPEED of object, next is SPEED of path, and highest is `TIME` on path. However, the most common command is as follows:

```
SET (object ID) TRAVEL (time)
```

If the next time increment is the same as the time specified in the `SET TRAVEL`, the object will travel to the end of the path in exactly the same time as this time. For example, consider the following code:

```
TIME 0.0000
CREATE SQUARE S1
```

```
PLACE S1 ON P1
SET S1 TRAVEL 30.0000
TIME 30.0000
END
```

The `END` is needed for every PROOF.ATF file. This trace file will create the object S1 at time 0.0000, place it on the path P1, and set its travel time to 30.0000 units. The next time increment is 30.0000, so PROOF takes the object and moves it to the end of the path in *exactly* 30.0000 time units. It will move smoothly on the path.

Did you notice that `TIME` is given with four decimal points? This is not needed here, but, as mentioned, it is good practice to use at least four decimal places (and sometimes six). The reason is that, in more complex examples, one has to be careful not to have subsequent times that have been rounded off and might give negative time increases. If the animation is correctly created, a negative time increase cannot happen if you have enough decimal spaces.

If you had the same `SET  TRAVEL` but the next time increment was `TIME` 40.0000, the object would arrive at the end of the path at time 30.000 and sit there for another 10 time units. If you had the `TIME` increment at time 20, the object would still be travelling on the path when the next PROOF commands would be executed.

Remember that only the CLASS and the PATH are case sensitive, so this program could have been written as follows:

```
time 0.0000
create SQUARE s1
place s1 on P1
set s1 travel 30.0000
time 30.0000
end
```

In this book, only capital letters will be used, but this is not necessary as long as you remember what you named the various entities.

PROOF uses 1 second of actual viewing animation time to equal 6 animation time units. Thus, one will see the animation on the screen for 5 seconds of viewing time (30 animation time units = 5 actual viewing seconds).

Now, we will write this code and see if, indeed, this is what happens. You can use any editor to create the file as long as it is saved as an ASCII file. Probably the easiest to use is the old DOS editor, which is given by command prompt from Windows. (In some setups, you may have to click on Start and Run and key in CMD.)

Once this file is created and saved as CHAP2A.ATF, we are ready to view our first animation. The file name is CHAP2A.ATF and must be in the same subdirectory where you have the CHAP2A.LAY. Normally, this is just where you have PROOF.

Open up PROOF if it is not already open and click on File. Click on Open Layout + Trace (.ATF). When you click on this, you will see the file CHAP2A.LAY. You will not see if there is a file CHAP2A.ATF. If there is no corresponding .ATF file, PROOF will let you know.

Click on CHAP2A.LAY and open it. You will see the layout you made previously.

To run the animation, you can either click on Run and Go or just the running man on the menu bar.

FIGURE 2.2
Animation of square on path.

Assuming you have no errors in your .ATF file, you will see the SQUARE on the path begin to move on the path from left to right. You will also notice a timer in the upper right-hand side of the screen. This is moving as the SQUARE moves. When the timer hits 30.00, two things happen. First, the SQUARE is at the end of the path and then the animation is over.

The animation looks as shown in Figure 2.2. To run the animation again, click on the icon with the runner jumping the hurdle.

This brings up a menu that asks for the time to jump to. Input 0 and OK, and you can rerun the animation as many times as you like. If you key in any other number, the animation will restart from this time.

Notice the *clock* on the menu bar.

Click on this and the following menu appears.

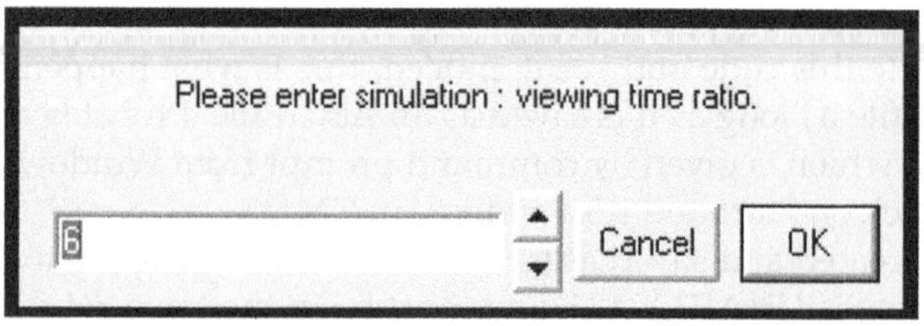

You will see the default time of 6. This is the viewing time ratio. This means that you will see six animation seconds in one second of actual time on the screen. This can be changed to allow the animation to take longer (lower the number) time or speed up (increase the number). This can be done by either keying a new viewing time ratio or using the up or down arrows.

If you want to check your animation, the file CHAP2A.LAY is an animation that is identical to what yours should be.

Change the clock setting to 1. Run the program again. Notice how slowly the square seems to move. How long does it take to travel on the path in real seconds?

Change the clock setting to 12. Notice how much faster it runs.
Go back to the program and change it as follows:

```
TIME 0.0000
CREATE SQUARE S1
PLACE S1 on P1
SET S1 TRAVEL 30.0000
TIME 40.0000
END
```

Run the animation and you will see that the SQUARE reaches the end of the path 10 time units before the animation ends.

Changing the time to TIME 20.000 will result in the simulation being over before the SQUARE reaches the end of the path.

2.6 The MOVE Command

The MOVE command is not used as often as the PLACE ON command is as, usually, you want to have movement on a path. However, the MOVE command can be used to move an object from one point to another. The coordinates of the place where the object is to move to need to be given, as well as the duration of the MOVE. The general form is as follows:

```
MOVE (object) (time duration) (coordinates of new location)
```

For example, suppose there was an object, BALL1, located at position (10, 10). To move it to position (20, 50) in 25 time units, the command would be as follows:

```
MOVE BALL1 25 20 50
```

The object, BALL1, would move from (10, 10) to (20, 50) in 25 time units. The motion would be smooth.

Example 2.1: An Animation of a Moving Square

Make a new layout that is blank. Go to CLASS and make a square called SQUARE. The hot point is to be at the bottom middle of the line. Save the layout as MYCHAP2B.

Make the .ATF file MYCHAP2B.ATF as follows:

```
TIME 0.0000
CREATE SQUARE S1
PLACE S1 at 20 20
MOVE S1 30 50 50
TIME 30
END
```

When your animation runs, you will notice that the object, S1, is moved from its initial position to the position (60, 30) in 30 time units.

You may want to examine the file CHAP2B. This is the stored version of the file you just created. Figure 2.3 shows what the animation will look like.

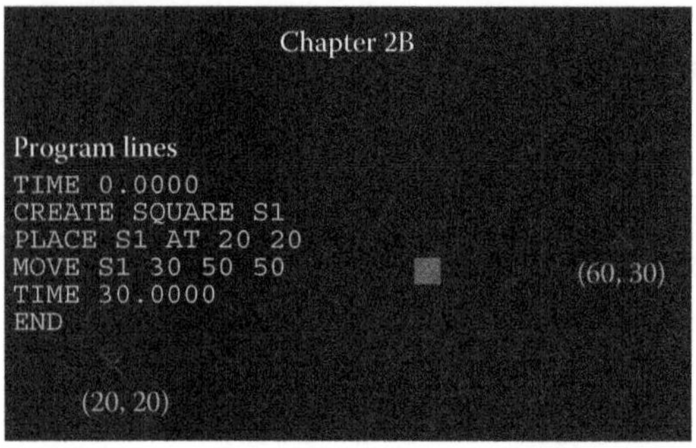

FIGURE 2.3
Screenshot from program CHAP2B.

2.7 Study of TIME and SPEED

Open up the file CHAP2C. This is basically the same as CHAP2A but not with a path and the object, SQUARE, is given a speed of 5.0000. This can be seen by looking at MODE—CLASS—SELECT—SQUARE. As you run the animation and examine the .ATF file, you will note that there is no command such as `SET S1 TRAVEL 30.0000` as there was in CHAP2A. Instead, the object will move with a speed of 5.0000 and no further travel time or travel speed is needed.

Run the animation and you will see that the object, S1, reaches the end of the path at time 6.00. This is because the path length is 30 and the speed is 5.0000. You might want to experiment with different speeds of the object such as 30, 10, 2, and 1. Figure 2.4 gives a screenshot from this animation.

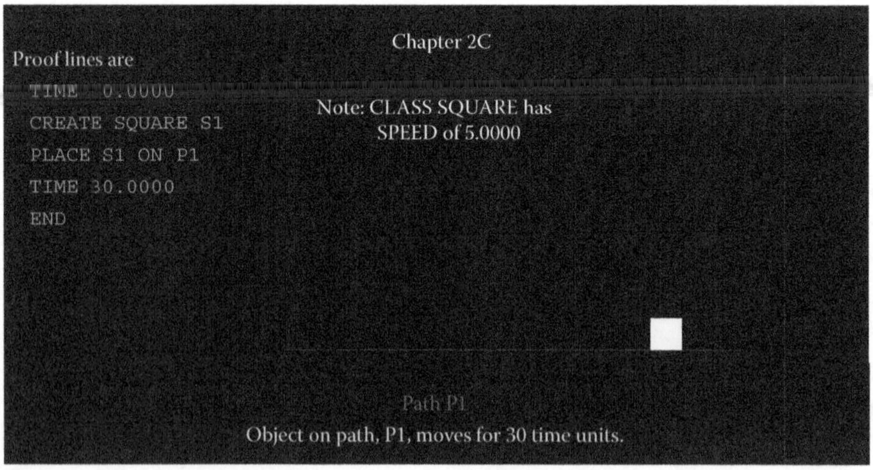

FIGURE 2.4
Screenshot from program CHAP1C.

2.8 Exercises

2.1. For this exercise, you will need three CLASSes. Call them STAR1, STAR2, and STAR3. Each is in a different colour (later we will see how to do this with only one CLASS). This is the whole layout. Call this MOVE1. Write the .ATF file called MOVE1 to create a different star every 5 time units, place it at (0, 0) and have it immediately move to (80, 58) in 10 time units. You will start with STAR1 appearing at time 0, STAR2 appearing at time 5, and then STAR3 appearing at time 10. The animation continues for 25 time units.

2.2. You may have noticed that the stars in Exercise 2.1 do not move exactly on the actual diagonal across the screen. Copy the file MOVE1.LAY to a new file MOVE2. LAY. Try to guess the correct coordinates. For example, you might try something like (0, −5) and (80, 50). Now, copy the MOVE1.ATF file to MOVE2.ATF. Modify this so that the stars start and end at the new coordinates. Run the animation MOVE2. Adjust the coordinates so that the stars move from the bottom left corner of the screen to the top right hand corner. Once you are satisfied that the stars are moving correctly, add code to the program, so that new stars move across the screen starting from the lower right corner and ending at the upper left corner. These are to move starting at the same time as the previous stars. Thus, you will see two identical stars appearing on the screen, each moving on different diagonals of the screen.

2.3. Have a layout consisting of four horizontal lines. They should be evenly spaced on the screen. There is a single CLASS called SQUARE. This CLASS is a square, side 2 and hot point at the mid-point of the bottom. Each of the four lines is to represent a path. The top path goes from left to right, the second path from right to left, and so on. At time 0.0000, a SQUARE travels from the start of the path to the end in 25.0000 time units. The paths and layout can be any colour you choose.

2.4. Make the .LAY and .ATF files to have the animation that will consist of a straight green line going from (20, 20) to (20, 70). Have a white SQUARE go from the bottom to the top of this in 40 time units. Do not worry that it does not move as you may want it to. The green line should be centred on the screen.

2.5. Make the .LAY and .ATF files for the following animation. This consists of two lines that are diagonal lines on the screen. One goes from (0, 0) to (80, 58) and the other from (80, 0) to (0, 58). These lines can be the same colour. Make a single CLASS, a square with each side 2 units, hot point at the bottom centre. Have two squares both be created at time 0 and travel along the two lines in 50 time units.

2.6. Make the following animation: The layout is a single CLASS, a star. The star is to be initially placed at point (20, 20) in the layout. It is to move immediately to position (10, 30) in 10 time units. At time 10, you will create another star and place this at position (40, 20). This will move to (10, 20) in 10 time units. The whole simulation is to last for 25 time units.

2.7. Make a circular path. Run the animation given by C:\PATH. This is found in the sub-directory STURGULEXAMPLES. This is an example of circular paths. When you examine the layout, you will notice that there are two paths, both circular. When an object comes to the end of a circular path and the time is not ended, it goes back to the front of the path and repeats the motion *even if the path is disjointed.* You are to create a similar animation. The layout consists of a square and a straight line. Make these both circular paths. P1 is the path along the square and P2 is the path along the line. Draw a CLASS such as a square or circle to move on one of the paths. The animation file will be something like the following:

```
Time 0.0000
Create CIRCLE C1
Place C1 on P1
Set C11 travel 12
Create CIRCLE C2
Place C2 on P2
Set C2 travel 8
Time 200
END
```

2.8. Run the animation called BOX that is in STURGULEXAMPLES. Notice that the animation consists of two bouncing balls in a box. The balls will bounce until the animation is over. Draw a similar animation. Your box can be of any shape such as a rectangle. Draw the single CLASS, which is the ball that will bounce on the inside of the shape you have drawn. If the radius is 1, the paths must start one unit away from the walls else the bouncing ball will appear to enter the wall. You want the animation to show the balls bouncing once they hit the walls. If the paths are too far away from the wall, the balls will bounce before hitting the wall. In the animation, BOX, the balls have radius 1 and the paths are always 1 unit away from the wall. Make the paths circular. A circular path is one where the object on them will pop back to the front of it if it is to continue to move on the path. For example, if you told an object to move on a circular path for 10 time units and the time to do this is 100 time units, the object will make 10 complete *revolutions* on the path. If the path is a circle, this would correspond to 10 complete revolutions.

3

More on Layouts I

Beach sand mining using a dredge, New South Wales, Australia.

3.1 More on Layouts

We will continue to learn more about making layouts and what many of the icons and options are. Make a layout similar to the sketch given in Figure 3.1. This is going to be a path on which a SQUARE will travel. It will start at (20, 20), go to (40, 40), and then to (60, 20) in 80 time units. Open PROOF and click on Draw. The new menu bar is as shown below:

Some of these icons have been used previously but were not covered in full detail. Each will now be covered in this and subsequent chapters.

Making the lines in the layout is rapidly done. If you want to have the coordinates on the sketch also, this is called *drawing static text*. This is alphanumeric text that does not move or change during the animation. The icon for this is as shown below:

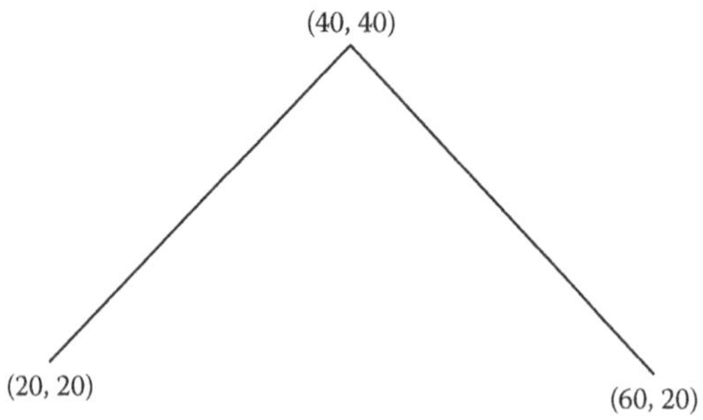

FIGURE 3.1
Simple path.

When you click on this, you will see the following menu:

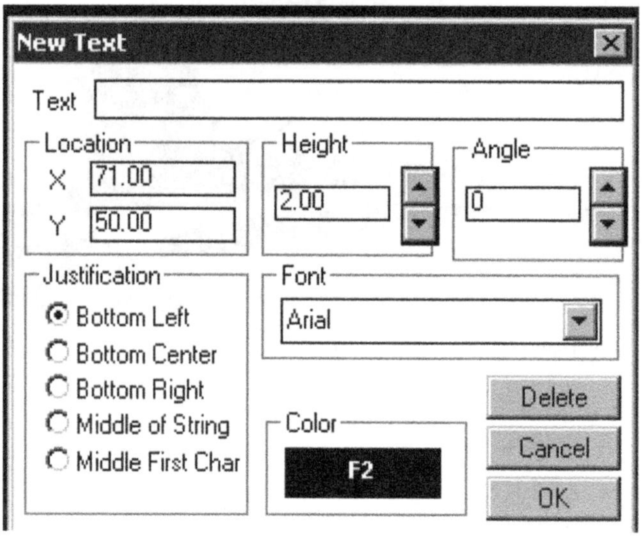

The text to be put on the screen needs to be typed on the line in the dialog box labelled *Text*. However, you should first take the cursor where you want the text to be placed on the screen. Click and then type your text. In this case, you would type (20, 20). Hit enter and you will see the *text* placed on the screen where the cursor was. If you do not like the location of the text, you can click on the option *bottom left* (see the dialog for the various options of where you want to click) and drag the text to where you want it. You can change the colour of the text by clicking on the RED colour and changing it. You can change the font using the many options in the dialog box. You can also change the height of the text as well as the angle, if you want the text to be slanted. All of these options are easy to do.

This is how one puts text on a layout that is easy to edit even at a later time if so desired. When you have put the coordinates on the layout, you may decide to add a bit more text. Eventually, you can save the file as CHAP3A.LAY

We are now ready for making the CLASS and PATHs. First, we shall make the CLASS. Click on Mode and Class. Then select Class again from the top menu bar. Select New Class and there will be a new menu on the centre of the screen. This menu will say *Please enter a name for the new class*. Give it the name SQUARE and make a white square with sides of 2 units. Make sure that the hot point is in the bottom edge at the middle of it.

This is the same as you did before in Chapter 2. When you are done, click OK on the drop-down menu. The SQUARE disappears but has been saved. It is not necessary at this time to select any other options.

We next need to make the path. Click on Mode and Path; name it P1. The path will consist of two segments. You will see this when you make it. When you have finished, click on OK from the pull-down menu. Do not select any other options. Save the layout as MYCHAP3A.

That is all that is to be done now for the animation. We are ready to make the .ATF file. This is as follows:

```
TIME 0.0000
CREATE SQUARE S1
PLACE S1 ON P1
SET S1 TRAVEL 80.0000
TIME 80.0000
END
```

Save this as MYCHAP3A.ATF. Since we now have both the .LAY and .ATF files, we are now ready to run the animation.

If all has gone well, you can open PROOF and run the animation. You will see the white square start at the beginning of the path and move in a *stiff* motion with the hot point always on the path.

If your animation is not as you thought it would be, run the stored animation CHAP3A. Check your animation and redo it until it is correct. The animation will appear similar to what is shown in Figure 3.2.

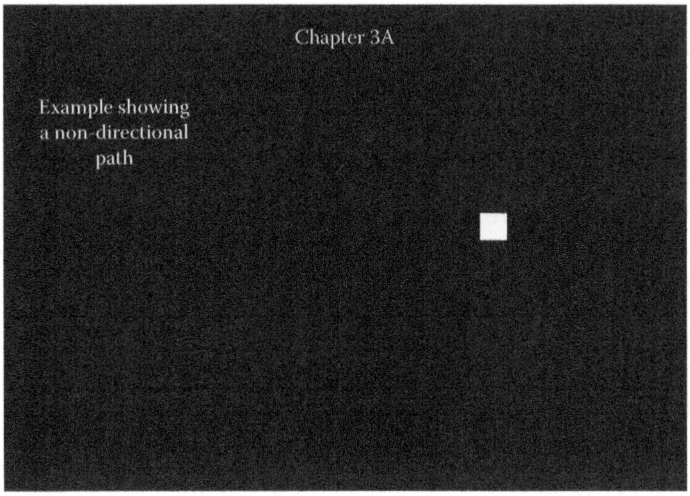

FIGURE 3.2
Screenshot of animation from program CHAP3A.

Notice how the object is placed on the path and how it moves. This is probably not the way you would like to have the motion on the path take place. Let us go back to the menu that was on the screen when we created the CLASS.

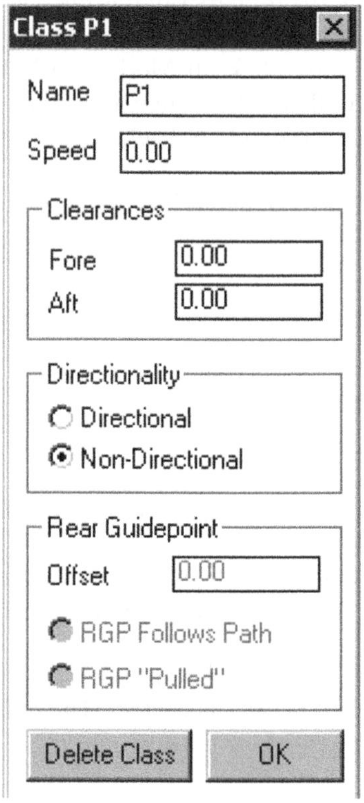

Notice the heading Directionality

Underneath this are the two options:

- Directional
- Non-Directional

Non-Directional means what we have just seen, that the hot point will remain firmly on the path and the object will move this way. If we had selected Directional, the object would move with its side along the path, that is, in a much more natural motion. Hence, go back and change the directionality by clicking on it. Save the layout and then run the animation again. Notice how the object now moves on the path. This is a much *smoother* animation to watch. If the object were to represent a car, the car would travel in a directional motion rather than in a non-directional one as we would expect.

The file CHAP3B shows the same layout and object, but now the object has been changed to be directional. Figure 3.2 shows what the animation looks like when the object is now directional. Most simulation models in mining will have the objects as directional (Figure 3.3).

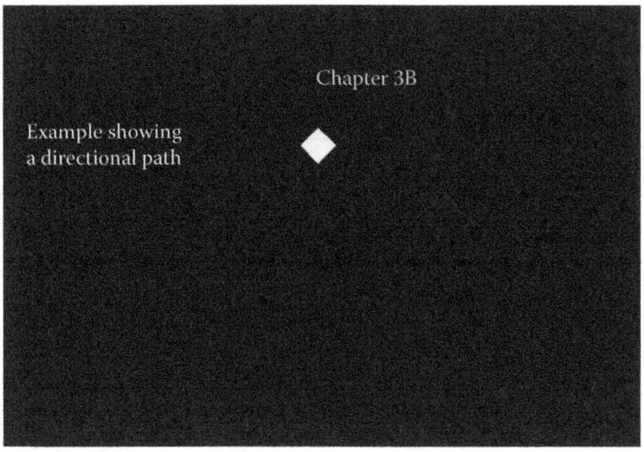

FIGURE 3.3
Screenshot of a directional object.

3.2 Making Curves

PROOF can make curves but it is not a straightforward, simple process. Rather, it is one that takes a bit of practice to accomplish. However, like the rest of PROOF, once this has been practiced a few times, it becomes quite easy. Go to Draw mode and draw two lines that come together at an angle. This angle can be any that you select up to 180°, for example, look at Figure 3.4.

Your sample lines should be approximately the same length. Now, click on the icon that looks like the following. This is for making curves or *fillets*.

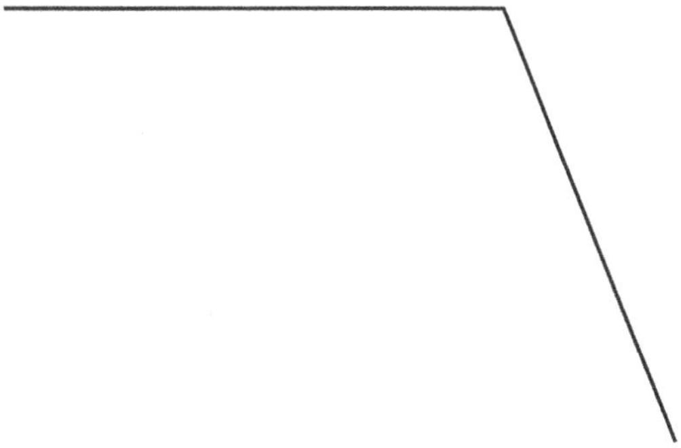

FIGURE 3.4
Sample lines where a curve is to be drawn.

When you click on this, you see the following pull-down menu:

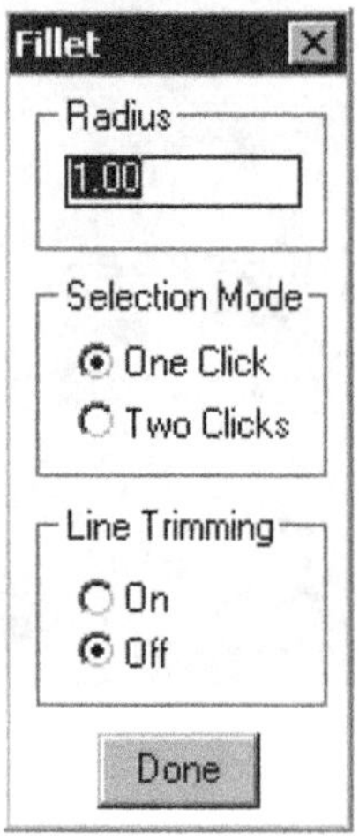

Notice that, by default, the radius of the fillet is 1.00. This may have to be changed if the lines are too long or the angle too acute. To visualize how a fillet works, imagine that a circle with radius 1.00 is to be placed in between the angle.

You will notice that you can have *One Click* or *Two Clicks*. What is going to happen here is that you will click on the inside of the angle near one of the lines. With the One Click option, there will need to be only one click of the mouse, and PROOF will find the closest line making up the angle and assume that this is what you meant. With the Two Clicks option, you need to click on both lines. It is strongly recommended that you choose the Two Clicks option, so that you will have to click on both lines (actually, you click just inside the lines).

The trim option will automatically trim the fillet of the lines that are outside of it. You can use this option or trim the extra lines yourself. It is recommended that, until you acquire fluency in using fillets, you do not use the trim option.

At this time, you should be aware of the *undo* option. This will come in very handy. This icon is as shown below:

Clicking on this icon will undo the last step you did in the animation. This is especially useful if you need to increase the radius of the fillet or perhaps you have trimmed the wrong line or segment.

Example 3.1: Making Fillets

Practice making several fillets for different angles. For example, consider a layout such as that shown in Figure 3.5. Smooth these out so that they make continuous curved paths using fillets.

You may wish to examine file CHAP3C. This shows lines similar to the ones shown in Figure 3.5 and how they might look with fillets to round the intersections (Figure 3.6).

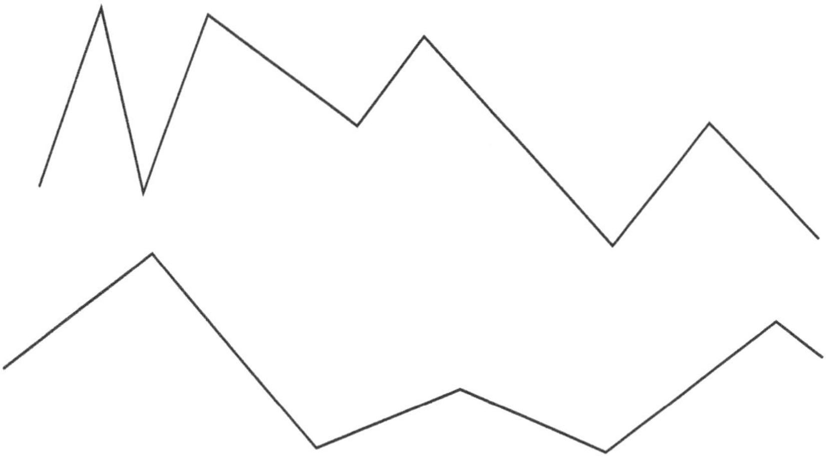

FIGURE 3.5
Possible layout elements for learning fillets.

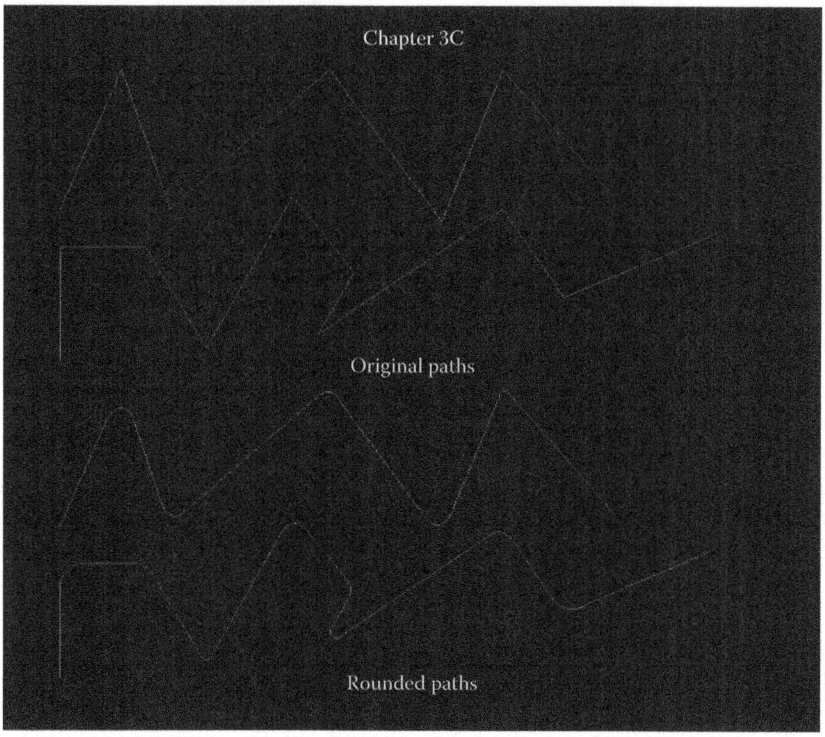

FIGURE 3.6
Rounding of paths.

3.3 Making Circles

There are often times when you will want to add a circular object to your layout or use circles in your paths. The icon for this option is shown below:

This icon is used for making arcs. When you click on this, you will see another dialog appear on the screen, which is as follows:

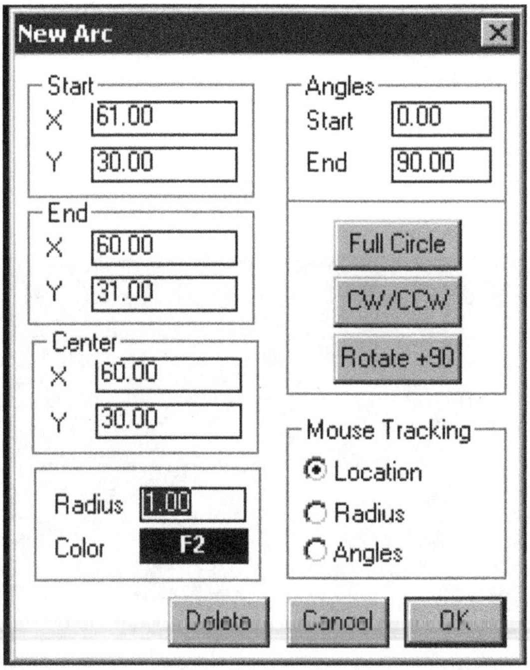

There are quite a few options with this. Notice that there will be a small arc on the screen that goes from 0° to 90°. You can move this wherever you want by clicking on the centre with your mouse. The options to use with this dialog box are very easy to understand. For example, if you want the circle to have a radius of 5.00, simply change the Radius option from the default of 1.00 to 5.00. If you want a full circle, click on the Full Circle box. You can change the angle of the arc by inputting different values. You can also rotate the arc by 90°. The box CW/CCW stands for *Clock Wise* and *Counter Clock Wise*. The Mouse Tracking option has to do with how you want to move the arc. In most cases, the default of Location (i.e., using the centre) will be sufficient. You should practice changing this option to see how tracking with the mouse also changes.

3.4 Moving the Grid Coordinates

It is possible to navigate the default drawing pallet by several means. You can make it larger (smaller) by a certain percentage or move the pallet in the X and Y direction yourself. The first of these icons to move the pallet in the X or Y direction is shown below:

Click on this and you will see the right side of the layout drawing become a blank line about a quarter inch wide with a small square in its centre. The same holds for the bottom of the layout. If you click on either of these squares, you can drag your drawing up or down or to the left or right. When you release the mouse button, the pallet is at different coordinates and the squares return to the centre of the lines.

If you want to make your drawing pallet larger or smaller, you use either of the icons given below:

Clicking on the first brings up the following pull-down menu:

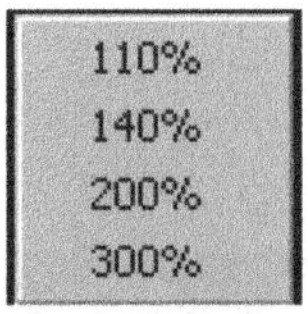

Clicking on any of these four options will make the pallet *smaller* as it zooms in by this amount. You can experiment with this to see how the drawing grid changes. The second icon gives the following pull-down menu:

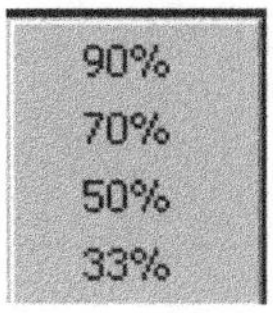

Clicking on any of these options—which are the inverse of the ones on the previous menu—will make the pallet *larger.* As you practice using these features, you will see how the drawing pallet changes.

Example 3.2: Practice Using Different Icons

Try using each of the icons to navigate your pallet. In particular, have your pallet set up so that the following is the grid approximately at the coordinates given. It will not be possible to have the grid exactly as suggested.

a. The bottom corners are approximately from (0, 0) to (280, 0).
b. The centre is (0, 0) and the lower left corner is (–15, –20) and the upper right corner is at (20, 10.5).
c. The lower left corner is at (–100, –100) and the upper right corner is at (300, 160).

Example 3.3: Drawing Circles

Draw a series of six concentric circles with centre at (30, 30). The first will have radius 1, the second radius 3, the third radius 6, and so on, up to 18. Make each of these a circular path, the innermost going clockwise and the next going counterclockwise, the next clockwise, and so on. Next make a single CLASS, a solid red circular BALL with radius 1. When all of the paths are done, click on path 6 (the outermost path). Give this a speed of 4 in the dialog box on the left. You are going to write the .ATF file with a different BALL on each of the concentric circles. Call the objects created from the class BALLS: B1, B2, B3 … B6. You will want each ball on each path to take the same amount of time to make a complete revolution. When you give the path a speed of 4, you will notice that the *time* box is filled in as taking 28.27. This is the length of the path, 113.10 divided by 4. In order for the ball on path 5 (P5) to make a complete revolution in this same time, its speed should be 3.333 (94.25 divided by 28.27). Similarly, the speeds for paths 4, 3, 2, and 1 need to be as follows: 2.667, 2.00, 1.33, and 0.667, respectively.

In case it is needed, the animation is given by CHAP3D. This shows the balls all travelling along each concentric path at the same time. You will notice that since the paths all start at the same angle (zero, if measured from the x-axis), they are all horizontal at 180° and 360° (Figure 3.7).

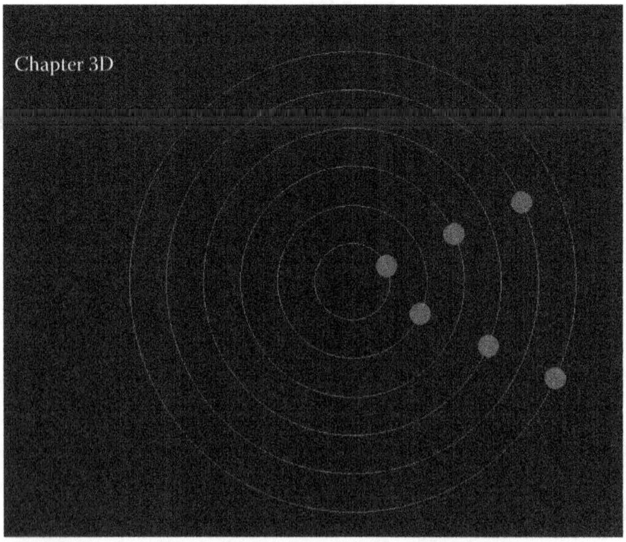

FIGURE 3.7
Screenshot of balls moving on concentric circles—this is program CHAP3D.GPS.

You may wish to change the speeds of the objects, so that each is different from the animation you are viewing. Some of the paths should be very fast and others very slow.

You have noticed that by changing the speed on each path, the TIME automatically was filled in. This is the time it will take for an object to travel from the start of the path to the end. Since the paths were all circular, the objects kept travelling as long as the simulation took place.

You can also change the time on a path rather than the speed. Try doing this with the same example.

3.5 Changing the Colours of Objects

The PROOF command to change the colour of an object is as follows:

```
SET (object ID) color (name or Fn)
```

When you clicked on the colour while making a line or polyline, you would have noticed that each colour has a name. If you specify a colour for an object, it will change to that colour. Alternately, you can use the name of the colour as shown in the table below:

BLUE	F1
RED	F2
WHITE	F3
YELLOW	F4
PINK	F5
TAN (but *not* ORANGE)	F6
GREEN	F7
BACKDROP	Bac

However, for the rest of the colours, you would need to use the specifications F8 through F24. The various layout colours that vary from shades of grey are from L1 through L8. Most programmers tend to use the F numbers for the colours. These are easier to use (especially later when you will be generating animation files using GPSS/H programs).

Some examples are as follows:

```
SET B1 COLOR GREEN
SET BOX4 COLOR F5
SET MAN COLOR BACKDROP
SET TT56 COLOR BLUE
SET HILLS COLOR F4
```

Examine the file CHAP3E.ATF. This is essentially the same as the animation used with CHAP3D, as the layout is the same. However, now lines of code have been added to change the colours of the various objects. Run the animation given by CHAP3E (Figure 3.8).

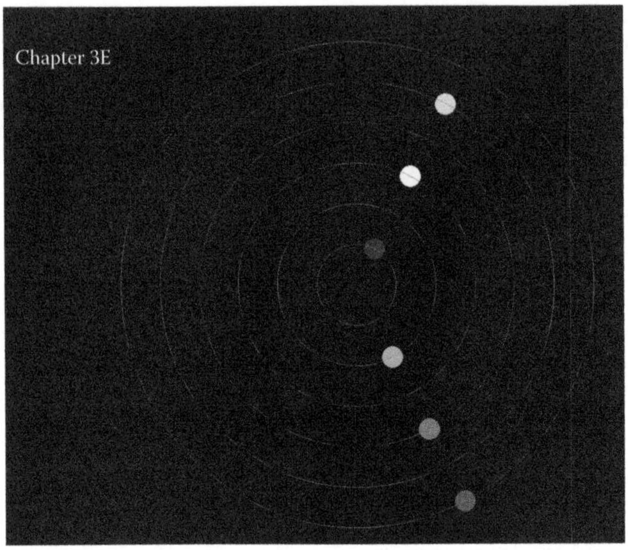

FIGURE 3.8
Animation given by changing colours of objects in the previous animation.

3.6 Exercises

3.1. This exercise is an example taken from the PROOF manual. It is a nice layout to practice on. At first, this will take some time to draw exactly as given. However, with a bit of practice, it should be quite easy and could be rapidly drawn. Notice the coordinates of the drawing. Also note the radii of the fillets. It is suggested that you practice this example until you can draw it in a few minutes (Figure 3.9).

3.2. Examine the animation given by CHAP3F. This shows a single ball that travels on a path from left to right but does a loop-de-loop as it nears the centre of the screen. Make a similar animation. An exercise such as this needs to be done with care. Be careful when you make your paths, so that you do not accidentally click on the same path two times. This is very easy to do. The animation given by CHAP3F uses three paths. If you happen to have an error in making the paths, you will see it in the animation.

3.3. Examine the animation given by CHAP3G. This is an example of a *circular path.* When the path is created, you can click the *circular* option on the pull-down menu. A circular path is one where the object will reach the end of the path and then start over at the beginning. In this case, the circular path is a closed path. In the next exercise, you will see what will happen if the path is not closed. Of course, in order for the path to show its circular nature, the time on it must be more than the time to make one loop. Change the path from circular to noncircular to see the difference.

3.4. You were cautioned to be careful about making paths so as not to inadvertently double click on them. However, it is possible to have a path be on the same line more than once. For example, suppose a car is to travel from left to right on a path and then return from right to left on the same path. Examine the animation

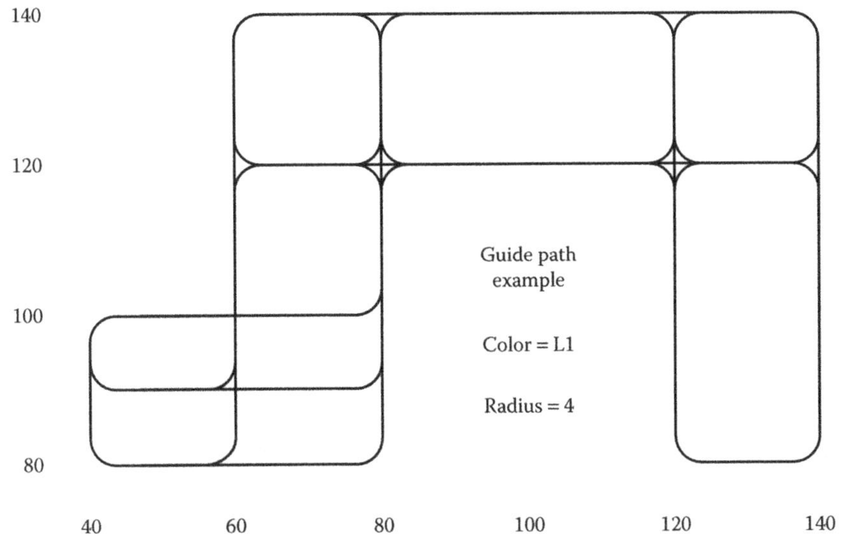

FIGURE 3.9
Example of a layout taken from the PROOF manual.

given by CHAP3H. This shows an object that goes first from left to right, and then returns to the left. About half way back to the left, it reverses and returns to the right side of the path where it *disappears*. The PROOF command DESTROY is used to do this. This is an easy command to understand. When an object reaches this command, it is removed from the animation. Make a similar animation.

3.5. It is possible to make many objects from the same CLASS, although each will have the same form. For example, one could have the following code:

```
TIME 0.0000
CREATE BALL B1
PLACE B1 ON P1
CREATE BALL TINY
PLACE TINY ON P2
SET B1 TRAVEL 20.0000
SET TINY TRAVEL 25.0000
TIME 25.0000
DESTROY B1
DESTROY TINY
END
```

Examine the animation given by CHAP3I, which might go with the above code. Here there are two paths: P1 and P2. The single CLASS is BALL and the objects are B1 and TINY. You will notice that B1 reaches the end of path P1 before object TINY. It sits at the end of the path for 5 time units and, then, when object TINY reaches the end of path P2, it is destroyed along with object TINY. The animation is then over. Make a similar animation but with three paths and three classes. Have the objects travel on the paths at different times, so that you can see the effect of this.

3.6. Draw your layout so that it is two concentric circles, each colour grey (F26, for example). The inner circle will have a radius of 8 and the outer a radius of 16. The single class will be a ball, with radius 1 and colour white. The inner circle will be a circular path going clockwise and the outer circle going counterclockwise. Write the .ATF file to have three objects equally spaced circling around on the inner circle and four objects also equally spaced circling on the outer circle. You can use any speed that you wish for the paths. In case you need to view this animation, it is given by CHAP3J. If you examine this animation layout, you will see that the TIME for path P1 is given as 30. This means that objects are created at time 0, 5, 10, 15, 20, and 25. If one were to be created at time 30, it would not be shown on the animation, as it would be placed on top of the object that was created at time 0. Similarly, the objects created for path P2 are created at time 0, 8, 16, …, 56. There will be eight of these.

Suppose there were 7 objects on the inner path and 14 objects on the outer path. What would be the time to create each of the objects?

3.7. Draw the layout that has a large green border in the shape of a rectangle with corners at (24, 28), (54, 28), (24, 32), and (54, 32). The sides of the border are to be 1 unit thick. It should fill most of the screen and be reasonably centred. The CLASS is called PERSON and is to be a shape as shown in Figure 3.10.

The colour of the PERSON is to be white (F2). Its radius is 1 and its hot point is at the bottom centre.

The .ATF file is as follows:

```
TIME 0
TIME 10.0000
CREATE PERSON 1
PLACE 1 AT 30 25
TIME 20
CREATE PERSON 2
SET 2 COLOR F2
PLACE 2 AT 40 25
TIME 30
CREATE PERSON 3
PLACE 3 AT 50 25
SET 3 COLOR F4
TIME 50.0000
END
```

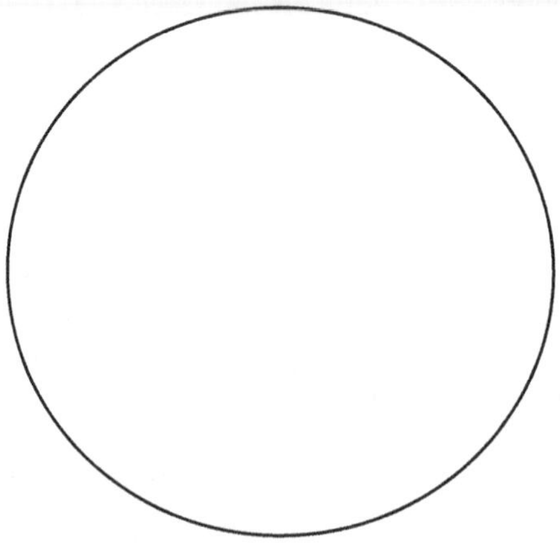

FIGURE 3.10
CLASS PERSON.

To centre the layout, you need to do the following:

Click on the zoom box and centre it over the layout. Move the sides of the zoom box until the rectangle is approximately centred. Click on *OK*

Click on View → Save view as (home). Then click on FILE and save the file.

Run the animations and explain what is happening at each time step.

3.8. Use the same layout as for Exercise 3.7. The .ATF file will be as follows:

```
TIME 0.0000
CREATE PERSON 10
PLACE 10 AT 40 22
SET 10 COLOR RED
TIME 10.0000
CREATE PERSON 1
PLACE 1 AT 30 25
SET 1 COLOR GREEN
TIME 15
CREATE PERSON 2
PLACE 2 AT 40 25
SET 2 COLOR F11
TIME 20.0000
CREATE PERSON 3
PLACE 3 AT 50 25
SET 3 COLOR YELLOW
TIME 50
END
```

Run the animation and explain what happens at each time step.

3.9. Examine the animation CHAP3K. This has several *bouncing* balls and two circular paths. Draw a similar layout and write the appropriate .ATF file.

3.10. There is a layout file named CHAP3L. This file consists of the following:

a. The layout consists of two lines: the first goes from A to B, and the second from C to D.

b. Two CLASSes: CAR1 and CAR2. CAR1 is red and CAR2 is green. Both are otherwise identical. The fore and aft clearances are both 1.25 units. They are both *directional*.

c. Two PATHs: P1 and P2. P1 goes from A to B and P2 goes from C to D. Both are accumulating (see the layout).

You are to examine each of the following .ATF files and state what each animation will do *before* creating the file and running the animation. After you study the files and determine what you would expect the animations to look like, you should create the .ATF file and run the animation to see how it actually runs.

```
1.TIME 0.0000
  CREATE CAR1 C1
  PLACE C1 ON P1
  SET C1 TRAVEL 30
  TIME 30.0000
  END
```

2.
```
TIME 0.0000
CREATE CAR1 C1
PLACE C1 ON P1
CREATE CAR2 C2
PLACE C2 ON P2
SET C1 TRAVEL 40
SET C2 TRAVEL 30
TIME 40.0000
END
```

3.
```
TIME 0.0000
CREATE CAR1 C100
CREATE CAR2 C200
PLACE C100 ON P2
PLACE C200 ON P1
SET C100 TRAVEL 25
SET C200 TRAVEL 25
TIME 25
PLACE C100 ON P1
PLACE C200 ON P2
SET C100 TRAVEL 30
SET C200 TRAVEL 30
TIME 55.0000
END
```

4.
```
TIME 0.0000
CREATE CAR1 C1
PLACE C1 ON P1
SET C1 TRAVEL 40
TIME 10.0000
CREATE CAR1 C2
PLACE C2 ON P1
SET C2 TRAVEL 40
TIME 20.0000
CREATE CAR1 C3
PLACE C3 on P1
SET C3 TRAVEL 40
TIME 60.0000
END
```

5.
```
TIME 0.0000
CREATE CAR1 C1
PLACE C1 ON P2
SET C1 TRAVEL 30
TIME 10.0000
CREATE CAR2 C2
PLACE C2 ON P1
SET C2 TRAVEL 40
TIME 30.0000
PLACE C1 ON P1
SET C1 TRAVEL 30
TIME 50.0000
PLACE C2 ON P2
```

```
   SET C2 TRAVEL 30
   TIME 80.0000
   END

6. TIME 0.0000
   CREATE CAR1 C1
   CREATE CAR2 C2
   PLACE C1 ON P1
   PLACE C2 ON P2
   SET C1 TRAVEL 40
   SET C2 TRAVEL 40
   TIME 20.000
   SET C1 COLOR F3
   SET C2 COLOR F2
   TIME 40.000
   DESTROY C1
   DESTROY C2
   END
```

4

More on Layouts II

South African carving of a miner.

4.1 More on Paths

This chapter will expand our knowledge of paths and help us learn what various options are available. When objects reach the end of a path, they will do one of several things:

- Remain there until some later time
- Be destroyed (taken out of the animation)
- Be placed on another path
- MOVE to a point

In actual simulations, we know that objects may reach a certain point and wait there for a time before they move to another path. This might be the case for cars coming to a stoplight that is red. Trucks arriving at a shovel to be loaded when another truck is being loaded is another example. PROOF handles this situation by a two-part operation. First, it gives the various CLASSes clearances that can be in front of the object (*fore clearance*) and/or at the back of the object (*aft clearance*). These clearances are distances whereby another object cannot impinge on it. *Both of these clearances are positive numbers.* For example, a CLASS may be 1 unit in length with the hot point in the middle (.50 unit from the front and back). If there is no clearance, another object from this CLASS will totally override it if both come to the end of the same path and you will see only the second object. If the fore clearance is .50 unit and the aft clearance is also .50 unit, a second object will come to rest exactly at the back of the first object. This will become clear by running the animation CHAP4A. A screenshot when the animation is finished is given in Figure 4.1.

Here you have as the object a square with side 1 unit in length. The fore clearance is .75 unit and the aft clearance is also .75 unit. The animation shows one object created at time 0 and placed on the path, P1. It will travel to the end of the path in 60 time units. At time 30, a second object is placed on path P1 and will also be directed to travel to the end of the path. Notice how the objects rest at the end of the path when the animation is over. The program should be rerun with the clearances removed to see what happens.

We shall better understand the concept of clearances from the next example. Here you will construct an animation to illustrate the principles involved in constructing such animations with objects accumulating on paths.

Open PROOF and draw the three lines for the layout as you see in Figure 4.2. Call this layout CHAP4B.

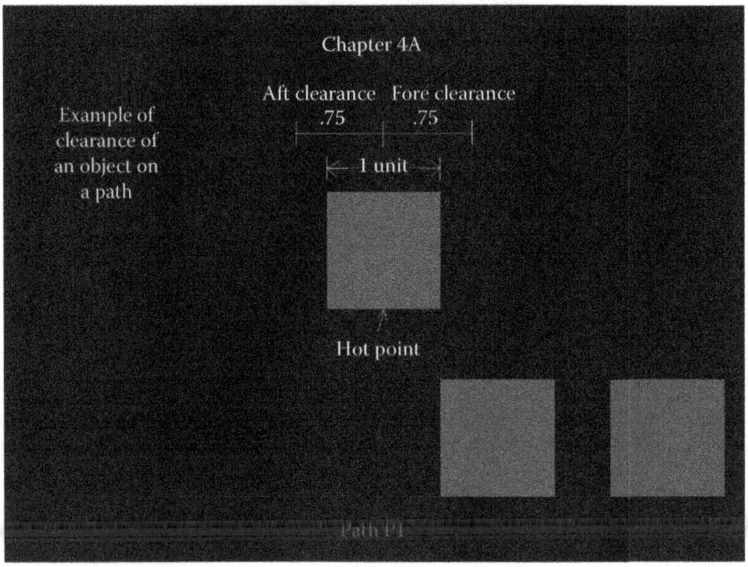

FIGURE 4.1
Screenshot of animation illustrating the concept of clearance.

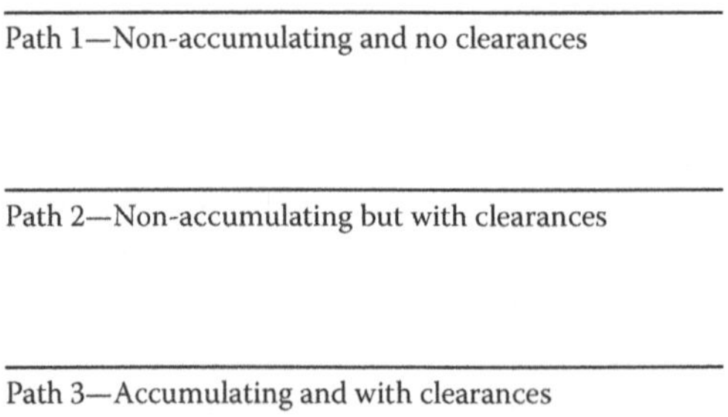

FIGURE 4.2
Paths needed for animation.

The lines will go from (10, 50) to (70, 50), from (10, 30) to (70, 30), and from (10, 10) to (70, 10). Thus, each will be 60 units in length and horizontal. They are to be in the colour red.

There will be two CLASSes. One will be BALL1. This is a white ball with radius 2 and hot point at the bottom, centre of the ball. Thus, when placed on a path, it will travel on top of the path. The other will be BALL2. It will be identical to BALL1, except for what is called *Clearances*. First, make BALL1. The pull-down menu on the left of the screen is as follows:

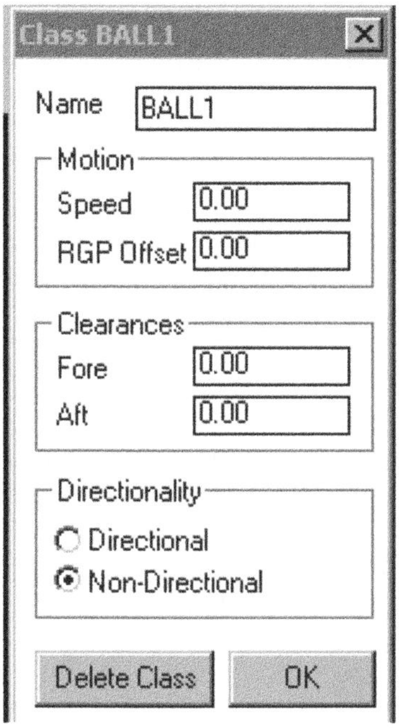

This is the default menu. Notice that all the options are initially taken as zero. For BALL1, we shall keep it this way. Since the objects made from BALL1 will be travelling horizontally on the paths and are circles, the directionality is not important. It might be important for other shapes of the CLASS and different paths. It most certainly would be important if the path were not horizontal. This path will be a non-accumulating path, which it is by default.

The second CLASS will be called BALL2. It will be identical to BALL1. However, look at where the front and rear of it are positioned on the grid and the options under Clearances in the pull-down menu. Under Clearances, we see Fore and Aft. These represent the distances objects will come to rest in relation to another object's hot point when there is more than one object at an end of the same path. However, giving objects clearances is not enough. There is one more thing that is necessary. The path must be an *accumulating* path. The clearances also must be larger than the size of the objects that come together at the end of a path. For example, if the fore clearance of one object was only 1, a second object with no fore clearance would still impinge on this object at the end of a path, although it would not fully overlap the object, as we shall see. For BALL2, make the clearances: Fore = 2.5 and Aft = 2.5. Notice that the aft clearance is a positive number even though the class is numerically listed as going from –2 to +2. This is always the case. *Do not input a negative number for the aft clearance.*

Now, go to the paths. The pull-down menu is as follows:

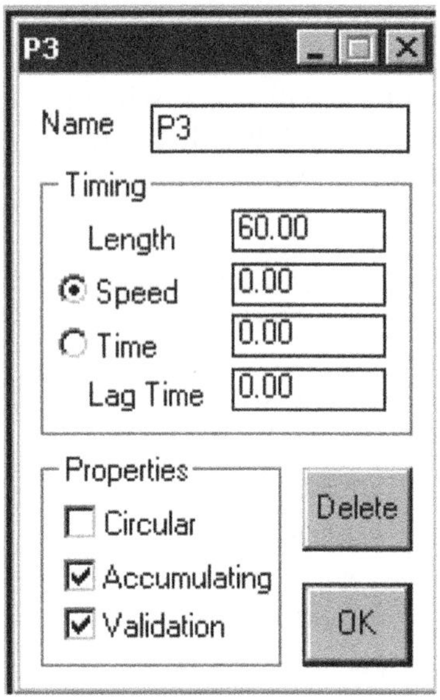

This is for path P3. Notice that, now, under the category of Properties, the option Accumulating is ticked. The option Validation is also ticked, but this is not important for our exercise.

We are now ready to make the .ATF file. Save your layout as CHAP4B.LAY.

The .ATF file to drive the animation is as follows:

```
TIME 0.0000
CREATE BALL1 B1
CREATE BALL2 B2
CREATE BALL2 B3
PLACE B1 ON P1
PLACE B2 ON P2
PLACE B3 ON P3
SET B1 TRAVEL 60.0000
SET B2 TRAVEL 60.0000
SET B3 TRAVEL 60.0000
TIME 30.0000
CREATE BALL1 B4
CREATE BALL2 B5
CREATE BALL2 B6
PLACE B4 ON P1
PLACE B5 ON P2
PLACE B6 ON P3
SET B4 TRAVEL 60.0000
SET B5 TRAVEL 60.0000
SET B6 TRAVEL 60.0000
TIME 130
END
```

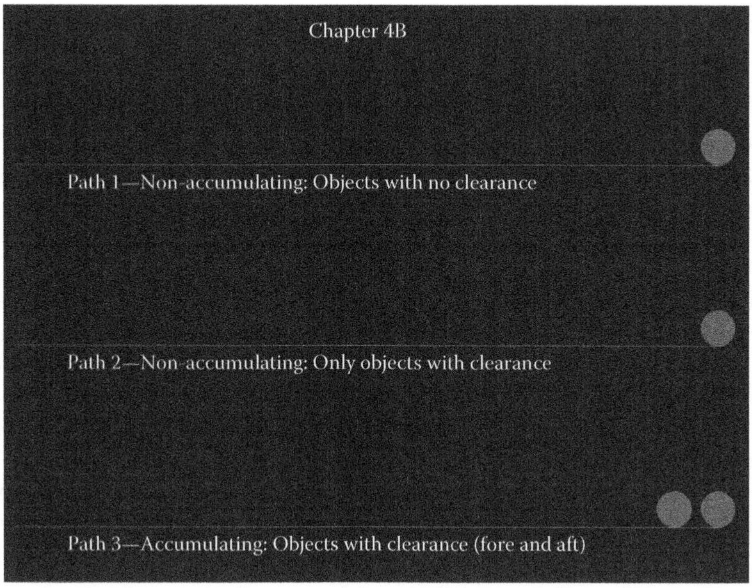

FIGURE 4.3
Screenshot from CHAP4B.LAY. There actually are six BALLs on the screen but two are covered by other BALLs.

You may run this or try to write your own program first. Figure 4.3 illustrates the situation just before the end of the animation. The BALLs on the two upper paths will override each other, whereas the BALLs on the bottom path will come to rest separated by a space.

The simulation will last for 130 time units. At time 0.0000, three objects are created: one from the CLASS BALL1 and two from the CLASS BALL2. They are placed on the three paths as the .ATF file indicates. The paths are named P1, P2, and P3. Notice the ability to use the same CLASS to create three different objects. Each of the three balls is given a time of 60.000 units to travel from the start to the end of the path.

At time 30.000, another three new objects are created from CLASSes BALL1 and BALL2. These are named B4, B5, and B6. The first three balls will be at the mid-point of their respective paths when these new objects are created. When time 60.000 comes for the first three balls, they are all at the end of the three paths. They will come to rest there. In the case of path P1, at time 90, the second object B4 arrives at the end of the path. It will impinge on the first object since the path is non-accumulating and the object does not have any clearances. The animation will show only one object at the end of the path.

For path P2, the same thing happens, since the path is non-accumulating. It makes no difference what the clearances are for the object.

For the path P3, the two objects will not impinge but will be as shown since the path is accumulating *and* the object has clearances. This is most probably what we would want to happen for most of our animations. Hence, when making CLASSes, it is important to add appropriate clearances.

In modelling mining systems, whenever there can be a queue, *the paths must be accumulating and the objects that move on them (normally trucks) need appropriate clearances.*

4.2 Non-Closed Circular Paths

Circular paths are ones where the object(s) keeps on moving from the start to the end point and then back to the start point and then again moves to the end point, and so on. The start and end points need not be connected as this example will illustrate. In this case, the object will *jump* from the end of the path back to the start. Consider the layout as shown in Figure 4.4.

Go to PROOF and draw a similar layout. Colour it RED. This can be done using a single polyline. Make the drawing almost large enough to fill the screen. The total length of the polygon will be around 140 units. The CLASS will consist of a single circle, called BALL, with hot point at (0, 0) filled with RED colour. It will have radius 2. Now make the whole layout a single path, called PATH1. This path will start at the beginning where you started to make the figure. As shown in Figure 4.2, click on *Circular* for the path. Note its length. This is the complete layout, so save it as CHAP4C.

Write the .ATF file to have the BALL travel from the beginning of the path to the end and continue to rerun the same path for a long time. The following file will work for the .ATF file:

```
TIME 0.0000
CREATE BALL B1
PLACE B1 on P1
SET B1 travel 40.0000
TIME 1000.0000
END
```

Run the animation. Notice that the BALL, B1, will continue to run on the path. When it comes to the end of the path, it *jumps* back to the beginning and retraces its route. This will continue until time 1000.0000 (assuming you want to watch the screen for this length of time). The animation for this example is stored as CHAP4C.LAY. A screenshot is given by Figure 4.5.

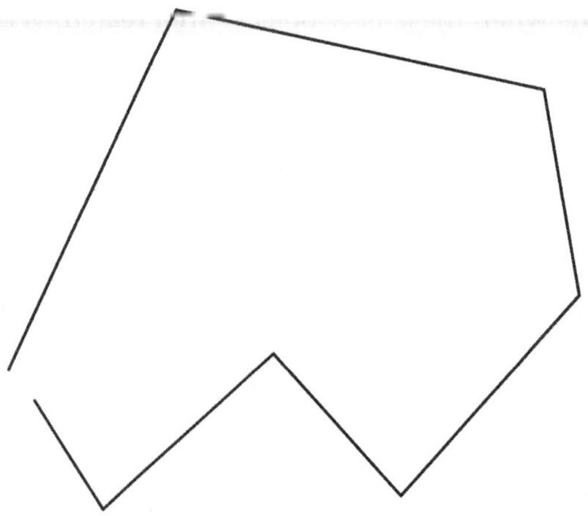

FIGURE 4.4
Example of a *circular* path in PROOF.

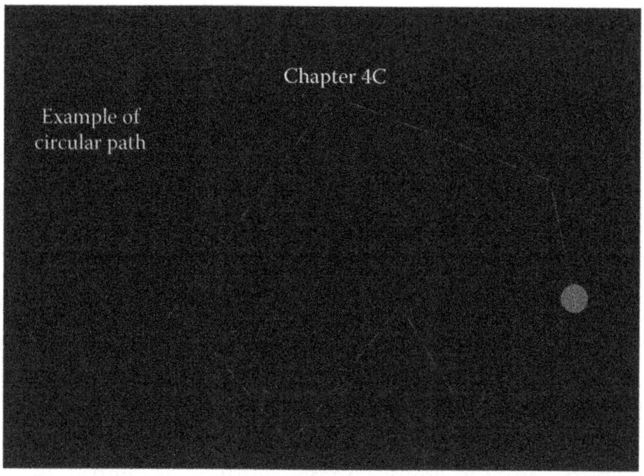

FIGURE 4.5
Screenshot of circular path.

4.3 The *Snap to* Option

PROOF has several options that aid in drawing the layouts. These are ones that allow lines or polylines to snap to other already drawn parts of the layout or to start drawing lines. For example, to draw a straight line from (10, 10) to (34, 50), you will use the *snap to* grid option. This icon is as follows:

When you click on this, you will see the following pull-down menu:

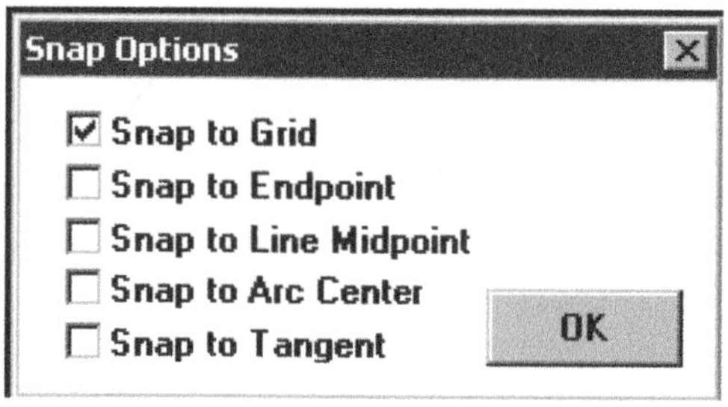

Notice that the *Snap to Grid* option is the one that is given by default. This means that the line or polyline that you are drawing will snap to the nearest grid point. This is why so many of our layouts were so easy to draw in previous examples. All you had to do was make sure that the line or polyline you were drawing was near the grid point that was 1 unit away from the side of the rectangle. The other options are important to keep in mind as they are useful when drawing other types of layouts. For example, if you have a line that is not on the grid, anytime you want to add another line to it, the *Snap to Endpoint* option will be used. In fact, if you try to connect the lines without this option, you will not be able to do so as the new line will always snap to the nearest grid point. The other options are self-explanatory.

If you wanted to draw a line from the point (10.5, 10.5) to (34.25, 50.4), and did not want to change the grid, you would need to turn the *snap to* option off. In this case, you would most probably want to draw the line as close as you could with your mouse but then type in the correct coordinates into the pull-down menu dialog.

You will not make circular paths for too many examples, but they can be very interesting to watch. For example, consider the layout in Figure 4.6.

This is stored as CHAP4D. You may wish to run it before attempting to make your own animation.

The layout consists of either a box or a rectangle. On the inside is the *circular* path coloured backdrop. An object (or objects, if you want) will travel on the path and appear to be bouncing off the sides of the box. The speed of the object will depend on what speed you select for it to travel in the .ATF file. There is still another way to have the animation speed up that we have not used yet. This is to click on the CLOCK icon on the top menu bar. This icon was discussed in Chapter 2.

Draw the layout. This is quickly done. The path will be as shown. Be careful *not* to have the path hit the sides of the box! If you have the path hit the sides, the object will appear to not bounce off the sides but to go through the sides.

Make the single CLASS a circle with hot point at (0, 0) and with radius 1. It can be any colour you select. The path is as given. It can be called P1. The .ATF file is as follows*:

```
TIME 0.0000
CREATE CIRCLE B1
PLACE B1 on P1
SET B1 TRAVEL 20.0000
TIME 1000.0000
END
```

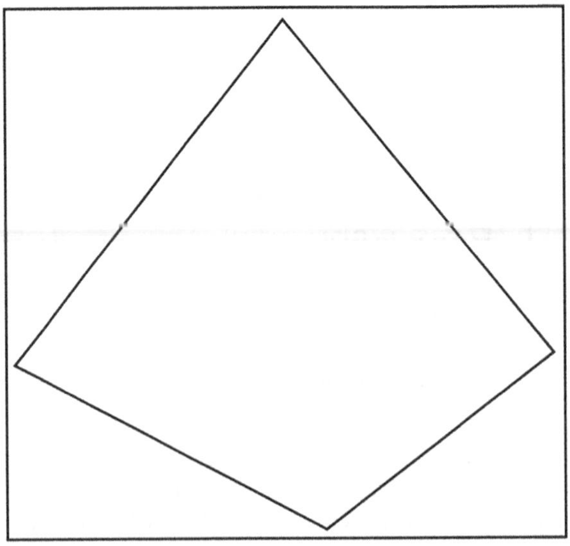

FIGURE 4.6
Layout for *bouncing ball* animation, CHAP4D. The paths are shown here on the figure but actually will be hidden on the animation (backdrop).

* The SET command will be covered in Chapter 5.

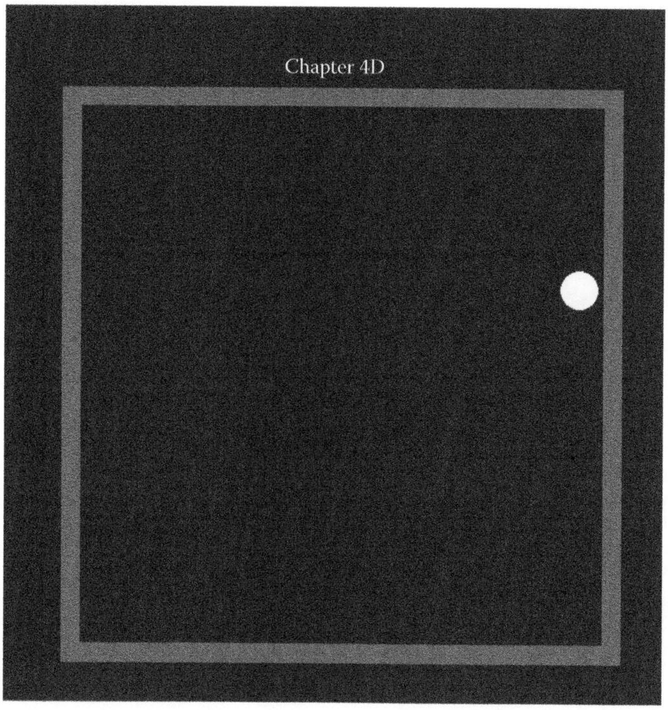

FIGURE 4.7
Bouncing ball animation.

The ball will appear to bounce off the sides of the rectangle or square as it moves on the invisible path, as shown in Figure 4.7.

4.4 The DESTROY Command in PROOF

To remove an object from the viewing screen, the DESTROY command is used. This is very easy to remember. For example, see the following code:

```
TIME 0.0000
CREATE BALL B1
PLACE B1 on P1
SET B1 TRAVEL 20.0000
TIME 20.000
DESTROY B1
END
```

will create an object, B1, from the CLASS BALL and place it on the path P1. It will travel from the beginning of the path to the end of the path in 20 time units. At time 20, it will disappear from the screen.

Example 4.1: A Blinking Star

The next example will involve a star that blinks on the screen. This is one of the rare examples where no visible layout is needed. The animation is given by CHAP4E. Open this and run it.

Make your own similar animation as follows. From Draw mode, click on CLASS and draw a star. This can be any colour that you want. It can also be multi-coloured. There is no path. The star is to be in the centre of the screen and will blink every 5 time units. Call this layout CHAP4E.

The .ATF file is as follows:

```
TIME 0.0000
CREATE STAR S1
PLACE S1 at 40 29
TIME 5.0000
DESTROY S1
TIME 7.0000
CREATE STAR S1
PLACE S1 AT 40 29
TIME 12.0000
DESTROY S1
TIME 14.0000
CREATE STAR S1
PLACE S1 AT 40 29
TIME 19.0000
DESTROY S1
TIME 21.0000
CREATE STAR S1
PLACE S1 AT 40 29
TIME 26.0000
DESTROY S1
TIME 28.0000
END
```

There is another way to do the same animation. You may have thought of it. Rather than destroy and re-create the STAR, you could simply use the command:

```
SET S1 COLOR BACKDROP
```

4.5 Exercises

4.1. Modify the layout for the blinking star example (CHAP4E.LAY) to make use of the command:

```
SET S1 COLOR BACKDROP
```

This will change an object's colour. By setting the colour to BACKDROP, it remains on the screen but is invisible.

4.2. Run animation CHAP4A but now with the aft clearance of the CLASS set to 0. What happens when the two objects reach the end of the path?

4.3. Make an animation that has a star in the centre of the screen. The star will do the following:

a. Remain in view for 6 time units
b. Disappear for 2 time units
c. Reappear for 6 time units
d. Disappear again for 2 time units
e. Reappear for 6 time units

The animation can run for 40 time units. The star can *disappear* by changing its colour to BACKDROP as was done in Exercise 4.1.

4.4. Make an animation that will have four circles, one approximately at each corner of the grid. Have them *blink* one at a time every 2 time units.

4.5. Make an animation that has a single class that is a circle named BALL that has a radius 2 and is red. The animation will have six of these circles all placed at (−60, −40) at time 0. They all move at once in 10 time units: one moves to (50, 40), one to (30, 40), one to (10, 35), one to (20, 20), and one to (30, 10). They are then all removed from the animation.

4.6. Make an animation that has eight balls that will travel in a clockwise direction on a rectangle that is in the centre of the screen. The balls will be all different colours and evenly spaced on the rectangle. The balls can be of any size and the rectangle also of any size.

Hint: The path will be a circular one with time 8. The balls can be coloured, F1, F2, F3, and so on, up to F8, and spaced 1 time unit apart. So as to not have the rectangle the same colour as one of the balls, it can be colour F26. A portion of the .ATF file might be as follows:

```
TIME 0.0000
CREATE BALL B1
PLACE B1 ON P1
SET B1 COLOR F1
TIME 1.0000
CREATE BALL B2
PLACE B2 ON P1
SET B2 COLOR F2
TIME 2.0000
```

4.7. Suppose that the animation in Exercise 4.6 was changed as follows:

The time for the path was now 1.00. What happens to the animation?

4.8. Examine the layout shown in Figure 4.8.

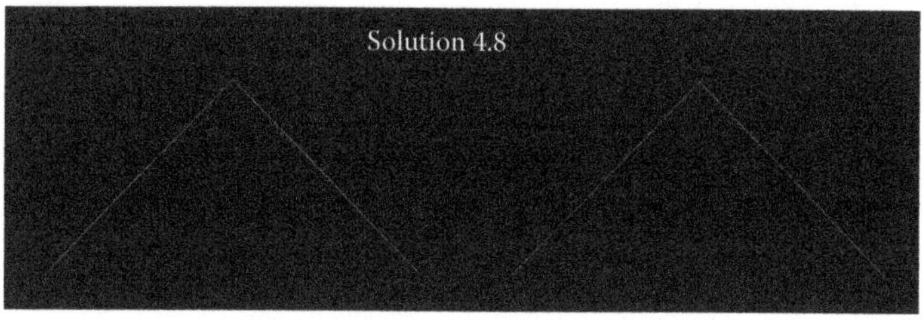

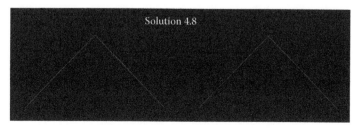

FIGURE 4.8
Layout for Exercise 4.8.

The line on the left goes from (−40, 0) to (−20, 20) and to (0, 0). The line on the right goes from (10, 0) to (20, 20) and to (50, 0).

Draw a single path to include both lines. PROOF will prompt you and ask if you want a path that is not connected. Make this path circular with a time of 20. The single CLASS will be a BALL with radius 1. Make the .ATF file to have the BALL travel on the circular path.

5

Some PROOF Icons

Decommissioned shovel at Chuquicamata Mine, Chile.

5.1 The SET Command and More PROOF Icons

We will continue to explore more of the PROOF icons. We will go over a few of the options associated with each and will introduce the few remaining options later as they are needed when we learn the GPSS/H code. There is an excellent help facility with PROOF and the student is urged to consult this if any question should arise about a PROOF command. This is found in the top menu bar and comes up when you click on the Help tab.

Some PROOF options are rarely used, such as changing the default settings on the PROOF animation screen, and will not be considered here. These are covered in the PROOF manual from Wolverine Software.

To illustrate more commands, let us use a simple animation model. Go to PROOF and create the layout, which is shown in Figure 5.1. This is a square with sides of length 10 and the corners at (30, 20), (50, 20), (50, 40), and (30, 40).

There will be four CLASSes.

- A solid BALL of radius 2
- A solid SQUARE with sides 2 units and hot point in the lower centre
- A solid TRIANGLE with base 2 and hot point in the middle of its bottom side
- A solid, inverted TRIANGLE with base 2 and its hot point at the middle of its base

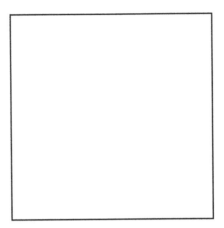

FIGURE 5.1
Basic layout of a square for the animation program CHAP5A.

The above CLASSes can be any colour you choose.

As the BALL travels on each edge, it will change CLASS until it makes a full circuit. When it is back where it started from, it will revert to its original CLASS.

The animation to do this is given as CHAP5A. After you view this, you should attempt to write your own .ATF file, as well as draw your own animation layout. Yours can be called something such as MYCH5A (Figure 5.2).

The .ATF file will be one that has a single object, the BALL, travelling from the lower corner around the edge of the square. As it goes around, it will change to three different CLASSes using the following command:

```
SET (object ID) CLASS (different class)
```

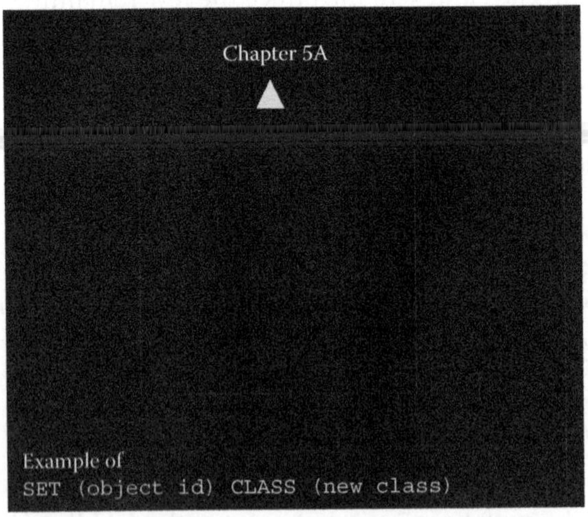

FIGURE 5.2
Animation of the program CHAP5A—object is shown at the top of the square.

5.2 Other Forms of the `SET` Command

There are other forms of the `SET` command that we will encounter. In Chapter 3, we encountered the `SET (object ID) COLOR (new color or color ID)`. To illustrate again how this form of the `SET` command works, let us make some changes to the animation given by CHAP5A. The original code was as follows:

```
TIME 0.0000
CREATE BALL B1
PLACE B1 ON P1
SET B1 TRAVEL 80.0000
TIME 20.0000
SET B1 CLASS SQUARE
TIME 40.0000
SET B1 CLASS TRI1
TIME 60.0000
SET B1 CLASS TRI2
TIME 80.0000
SET B1 CLASS BALL
TIME 120.0000
END
```

The objects take 80 time units to travel the edges of the square. As they come to each corner after 20 time units, they change CLASSes. The .ATF file for CHAP5B is as follows:

```
TIME 0.0000
CREATE BALL B1
PLACE B1 ON P1
SET B1 TRAVEL 80.0000
TIME 10.0000
SET B1 COLOR TAN
TIME 20.0000
SET B1 CLASS SQUARE
TIME 30.0000
SET B1 COLOR GREEN
TIME 40.0000
SET B1 CLASS TRI1
TIME 50.0000
SET B1 COLOR F3
TIME 60.0000
SET B1 CLASS TRI2
TIME 70.0000
SET B1 COLOR WHITE
TIME 80.0000
SET B1 CLASS BALL
TIME 120.0000
END
```

An example is as shown in Figure 5.3. Notice that the colours are now changed, as each object is halfway along each edge of the square.

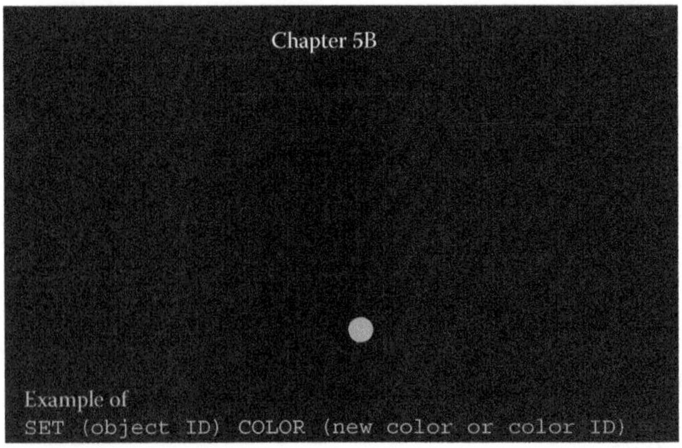

FIGURE 5.3
Screenshot of the animation program CHAP5B.

5.3 SET CLASS

Let us first examine the forms of the SET command that change a whole CLASS at one time.

```
SET (CLASS) CLEARANCE (fore) (aft)
```

Normally, one gives clearances to an object when it is created. It is possible to give a clearance to an object during an animation. If one had the command:

```
SET CLASS SQUARE CLEARANCE 3 2
```

The fore clearance would become 3 and the aft clearance would be changed to 2. This might be done for a portion of an animation where the programmer would want objects stacked upon a path closer than on a different path.

```
SET CLASS (class name) DIRECTIONAL (or NONDIRECTIONAL)
```

would cause a CLASS to change the directional (or non-directional) property of a CLASS. For example,

```
SET CLASS TRUCK DIRECTIONAL
SET CLASS CAR NONDIRECTIONAL
```

This change will only affect objects created *after* the command, not for objects already in the animation.

CHAP5C illustrates this. This is an example taken from two files in Chapter 3, namely, CHAP3A and CHAP3B. There the directionality was different for each animation. Now, the directionality will change when a new object is created. The animation shows that an object is created at time 0. It is placed on path P1 for 60 time units. At time 30, a second object is created and the CLASS is then set to be directional. Notice that the object that is at the top of the path will not have its directionality changed, only new objects.

```
SET CLASS (class name) SPEED (value)
```

This will change the speed of objects that will be created from a CLASS to the new value. Existing objects are not affected (Figure 5.4).

Run animation CHAP5D. This shows two circles that objects will move on. These are identical circles and each is a circular path. At time 0, an object is created from the CLASS CIRCLE. This has a speed of 1. It is placed on the circle at the left of the animation and will rotate counterclockwise for the duration of the animation. At time 20, a second object is created from the same CLASS and placed on the second circle. This new object is given a speed of 4. Notice how much faster this rotates with the different speed. The object created at time 0 will not have its speed changed (Figure 5.5).

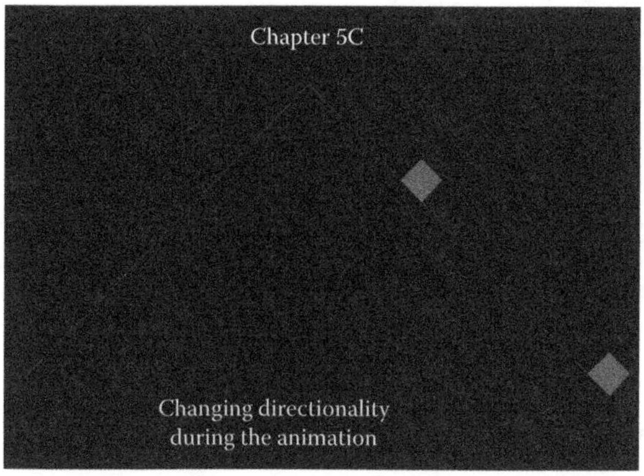

FIGURE 5.4
Changing directionality during the animation.

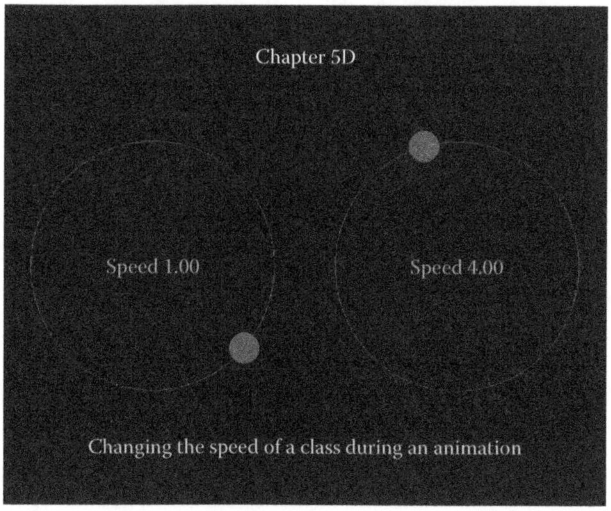

FIGURE 5.5
Changing the speed of an object during the animation.

5.4 SET Object

The commands to change a property of a CLASS are similar to those to change a whole CLASS and, in fact, are more commonly used. There are more of these than for the SET CLASS. For example, consider the following command:

```
SET SQUARE COLOR RED
```

This will *not* work if SQUARE is a CLASS as the SET command will only change the colour of an *object*.

The various commands are as follows:

```
SET (Object ID) CLEARANCE (fore) (aft)
SET (Object ID) DIRECTIONAL (NONDIRECTIONAL)
SET (Object ID) SPEED (new value)
SET (Object ID) TRAVEL (travel time)
SET (Path name) SPEED (new speed)
SET (Path name) TRAVEL (new travel time for path)
```

5.5 More PROOF Icons: Navigating the Top Menu Bar

5.5.1 The File Tab

You can use one of the icons to save your file. Click on the File tab on the menu bar. The pull-down menu is as shown below:

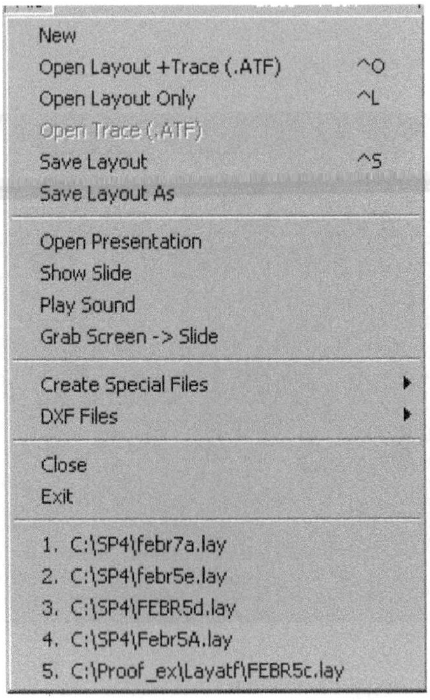

This menu is easy to navigate. *New* is when you want to create a new layout. If you have a .LAY and .ATF already made, you open them with *Open Layout + Trace (.ATF)*. When you click on this option, you will only see the files in the subdirectory you select that have the .LAY extension.

Open Layout Only opens a layout file. You would do this to edit the layout.

Open Trace (.ATF) will not be used. It is available only when there is an error in the animation file (.ATF). If there is an error, PROOF will prompt you with the line in the .ATF file where the error is located. This makes it easy to correct the error. However, only the first error will be noted by PROOF. As you become more proficient in writing and creating .ATF files, there will (should?) not be many errors to correct.

Save Layout is used whenever you create or make changes to the layout file. The *Save Layout As* is for saving the layout as a different name or for the first time you make a layout.

The next three commands, *Open Presentation, Show Slide,* and *Play Sound* will not be used. Information about them can be found in the PROOF manual from Wolverine.

It is possible to make a slide from any animation. This is done by the *Grab Screen → Slide* command. When you click on this, you can capture the screen image of the animation and send it to the desired format such as .PCX or .BMP file. This file then can be imported into other documents such as Word files. It also is possible to convert this to another format assuming that the user has the necessary software. The slide can be with the borders (the top line menu) or without it. It needs to be given a name.

Create Special Files is used mainly to convert .ATF files to those with extensions .PTF. These are animation files that run without the Wolverine PROOF security key. There is no need to create these with SP5. *DXF Files* can be used with the student version of PROOF. These can save a great deal of time in drawing complex layouts.

Close and *Exit* are the same as any other file. *Close* closes the current file and keeps you in PROOF, but *Exit* leaves PROOF. You may have noticed that, when you open File, the bottom of the pull-down menu contains the past five animations that were opened. Simply clicking on any of these brings it up. This can save a bit of time when editing animations.

5.5.2 The Mode Tab

Now, click on Mode.

The pull-down menu that appears is as shown below:

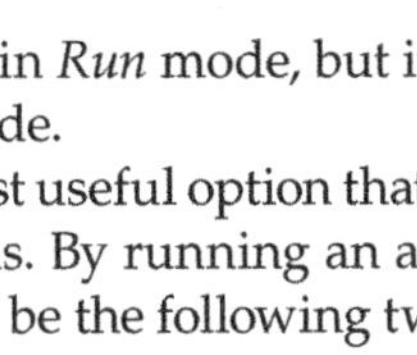

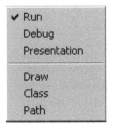

Run does nothing if we are already in *Run* mode, but it will put you back in the *Run* mode if you are in *Draw, Class,* or *Path* mode.

Debug is an interesting and the most useful option that can be very handy if there are errors in the animation that are not obvious. By running an animation using this option, you will see a change in the menu. There will be the following two icons that are added to the screen:

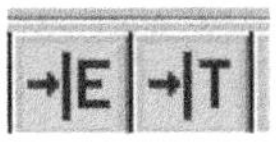

The first of these is for running the animation in *event mode* and the second for *in time mode*. If you select event mode, the animation will take place but only move according to the various events that are in the .ATF file. You will have to click on the icon after the animation goes from one event to the next. For our example, if you reopen the file, CHAP5A, this is what you will see after each click of the mouse:

```
Trace Line 1: TIME 0.0000
Trace Line 2: CREATE BALL B1
Trace Line 3: PLACE B1 ON P1 (you now see the B1 in the lower left hand
corner)
Trace Line 4: set B1 TRAVEL 80.0000
Trace Line 5: TIME 40.0000 (B1 moves to the upper right hand corner)
Trace Line 6: SET B1 CLASS SQUARE
...
...
```

Thus, at each step of the animation trace file, you can *see* what is happening in the model. This assists in debugging animations. However, many animation files are quite long and this method can take a great deal of time.

Running the animation in *Debug* mode and clicking on the time icon gives the following:

```
Trace Line 5: TIME 20.0000 (B1 ball moves)
Trace Line 7: TIME 40.0000 (object changes CLASS and moves)
Trace Line 9: TIME 60.0000 (object changes CLASS and moves)
Trace Line 11 TIME 80.0000 (object changes CLASS and moves)
Trace Line 13 TIME 120.0000
```

We will not worry about the Presentation mode at this time.

5.5.3 The View Tab

The next tab moving across the top menu bar is the View tab. Clicking on it brings up the pull-down menu:

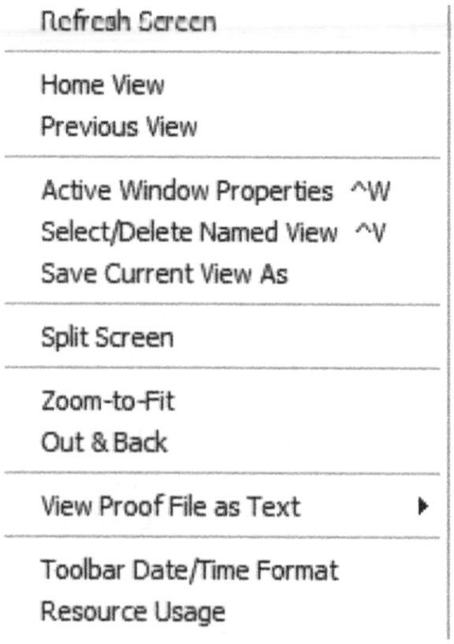

The *Refresh Screen* option is used when the screen does not display the animation correctly. At times, depending on the video card in the monitor, you *may* see a small box appear on an animation that does not belong there. Click on this option to make a fresh screen. The *Home View* takes the animation back to the original view that was shown when the layout was opened. The *Previous View* is obvious. For large animations, there might be many views to be considered.

The next of these options is *Active Window Properties ^W*. This shows the various default parameters that are used in the window you see for your animation. This looks as follows:

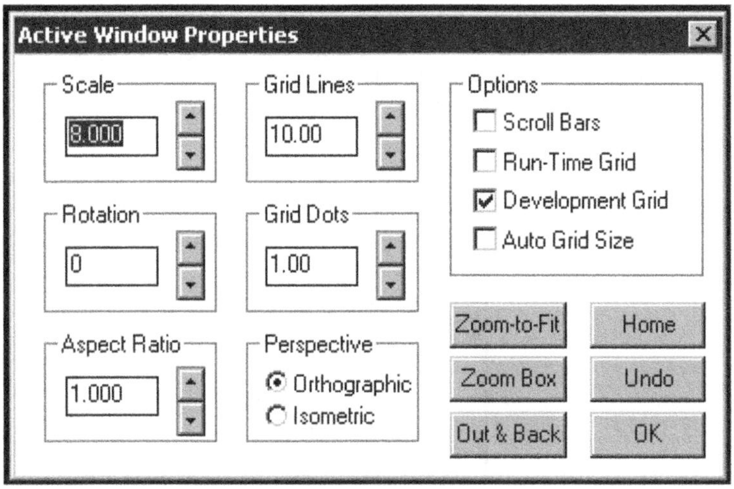

These default values are rarely changed. They specify that the grid lines are to be on the screen every 10 units and that the grid dots are 1.00 units apart. The rest of the options are clear and not to be changed. For example, the Perspective is either Orthographic (as you have been seeing it) or Isometric. Switching to isometric gives the drawing a 3D effect but will distort the view. It is not recommended that this option be used.

Select/Delete Named View ^V is something you will do a lot of when you have large animations. It is possible to capture the view you have on the screen and save it as a named view. Then, one can go to the *Select/Delete Named View ^V* and recall the view. We will give several examples of this later. At present, we only have the home view.

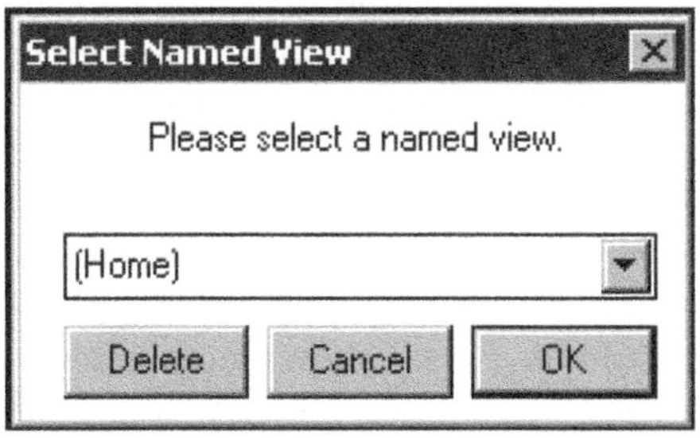

If we had more than just the (Home) view, we could click on the down arrow and select which of the ones we wanted to view. This will become apparent when we have large layouts and want to look at just a portion of the layout. We need to learn a bit more about PROOF before we can do much with this option, so we will put it aside for a short time.

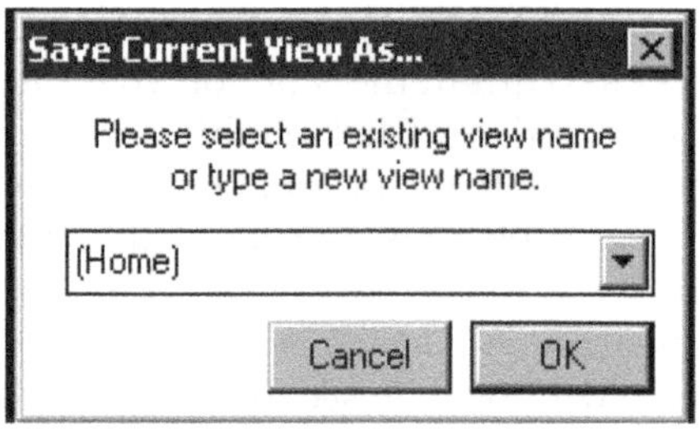

This simply allows the user to input a name (or a string of names) for the current view that is on the screen. One also can change the (Home) view using this option. For example, if the speed of the animation is given as the default of 6, and in the current view it is only 1.5, changing the (Home) view will result in any further viewing of the animation to have a speed of 1.5 by default.

The *Split Screen* is nice to use if you have a large monitor and are making a presentation of your animation. It allows you to split the screen either horizontally or vertically (or both directions) and view different aspects of the animation simultaneously. The menu for this option is as follows:

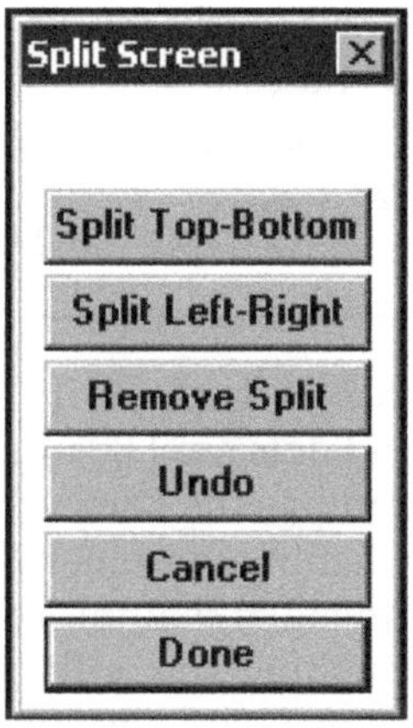

Clicking on either *Split Top-Bottom* or *Split Left-Right* will split the screen with double lines. One then can view part of the animation in each of the new screens. The main problem is that many current monitors are not large enough for complex animations and splitting the screen tends to clutter up the animation. To remove the split screen, click on the *Remove Split*. Then click just inside the double line on the split screen.

In general, it is better to have different saved views for each *View*. This way, each new *View* will occupy a new, full screen. We will not be using the *Split* option in any of our chapters.

Zoom-to-Fit takes all of the animation, even the parts not shown on the current view and puts them all on the current viewing screen. Often, you will make a layout that is much larger than the current view. You have been shown how to move the viewing area around. If you then use *Zoom-to-Fit*, you can view all of the layout at the same time. The *Out & Back* option is similar to the *Zoom-to-Fit* except that it allows you to select a new centre point for the layout. It is not used very often.

The *View Proof File as Text* option is rarely, if ever, used. If you need to view the layout file in text form, you can do so with any standard editor. This will not be used here.

The *Resource Usage* is another option that is rarely used. However, it does tell you a bit about your layout. Again, this will not be used here.

5.5.4 The Run Tab

The *Run* option brings up the following menu:

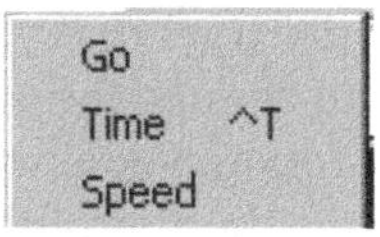

Go simply starts the animation. This is the same as clicking on the green runner icon in the menu bar. The *Time ^T* option allows you to jump to a time for the animation to begin and *Speed* has to do with the animation speed. PROOF uses 6 animation units to equal one second of viewing time. Thus, the default speed is set at 6. This can be changed to either speed up or slow down the animation. Both of these options can be done with other icons, as we shall shortly see.

5.5.5 The Setup Tab

The Setup tab has to do with the initial settings to view the animations and most probably will not be used very often. The pull-down menu is as follows:

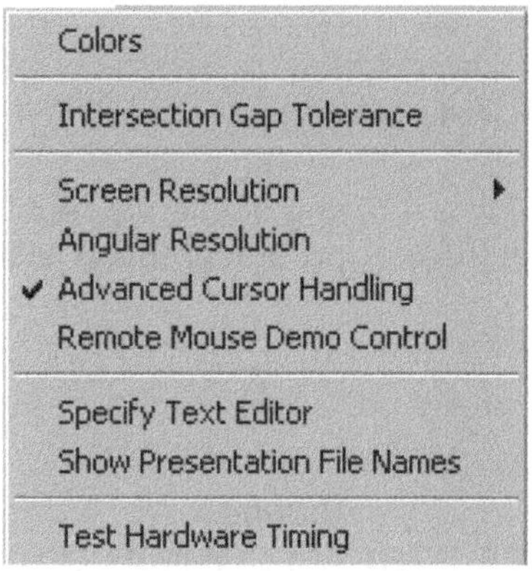

Clicking on *Colors* allows you to change the shading of any of the various colours should you wish to do so. The most common change is the backdrop colour. Initially, the screen uses tan for the background. However, clicking on this option brings the following menu up:

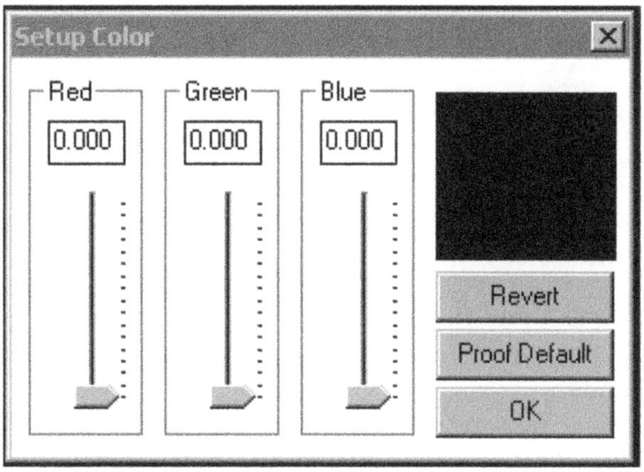

If you click on the *Color* tab, you will be shown a pallet of all of the colours that PROOF uses. You can click on any colour you want to change, and it will be brought up. As mentioned, the most common usage of this option is to change the backdrop colour from a harsh black to a more pleasing one. Other than that, it is rarely used.

The *Screen Resolution* tab allows you to change the screen viewing pixels. The choice will depend on the monitor you are using. You may wish to experiment with different settings to see which appeals to you best. Only the highlighted screen resolutions are allowed depending on the video monitor you are using.

The rest of the tabs in *Setup* are not used in this book. They are covered in the PROOF manual.

5.5.6 The Help Tab

The last tab is Help. Clicking on this gives the menu:

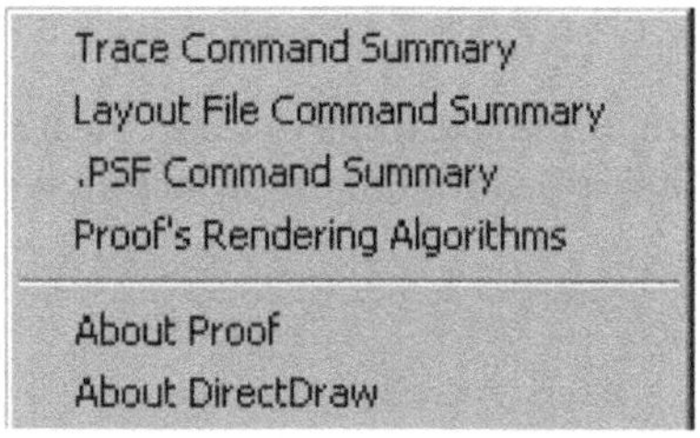

The *Trace Command Summary* gives a listing of every possible PROOF command. A brief explanation is given for each of these. A portion of this is as follows:

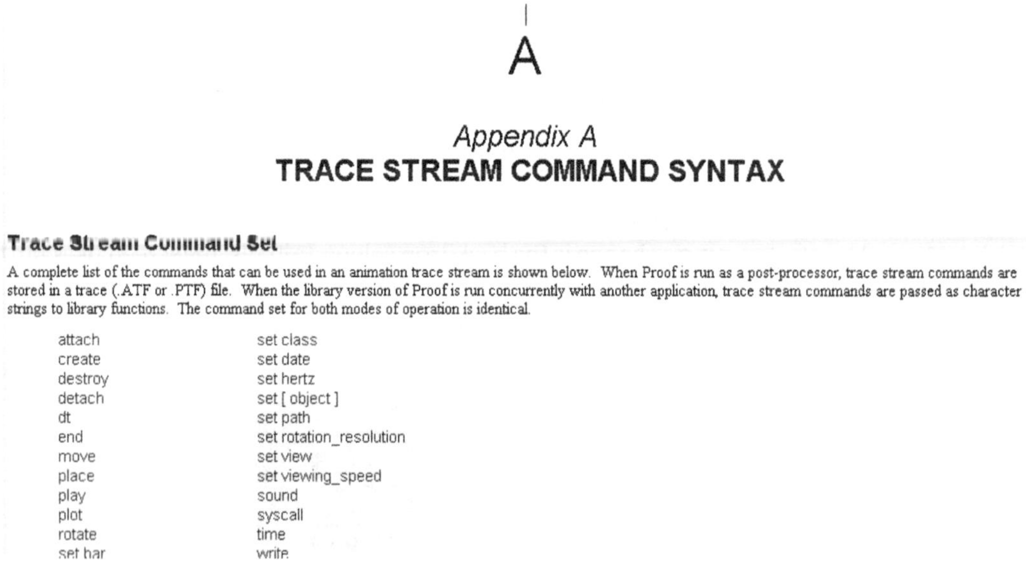

A

Appendix A

TRACE STREAM COMMAND SYNTAX

Trace Stream Command Set

A complete list of the commands that can be used in an animation trace stream is shown below. When Proof is run as a post-processor, trace stream commands are stored in a trace (.ATF or .PTF) file. When the library version of Proof is run concurrently with another application, trace stream commands are passed as character strings to library functions. The command set for both modes of operation is identical.

attach	set class
create	set date
destroy	set hertz
detach	set [object]
dt	set path
end	set rotation_resolution
move	set view
place	set viewing_speed
play	sound
plot	syscall
rotate	time
set bar	write

This is identical to what is found in the PROOF manual, Appendix A. This contains more commands, which will be covered later. This is an excellent quick reference to the various PROOF layout commands.

The *Layout File Command Summary* is identical to Appendix B in the PROOF manual. The screen will show the text for topics such as rules for names, layout file order requirement, and so on. Both this tab and the previous one will be referred to from time to time when developing animations. A screenshot of the tab looks as follows:

B

Appendix B

LAYOUT FILE SYNTAX

Proof Animation stores drawings and definitions in layout files. Every animation requires a layout file. Layout files have an open, ASCII format that is documented in this appendix. For most applications, layout files are written by Proof; however, some applications may generate their own layout files under program control. The use of an open, ASCII format makes it straightforward to develop such programs. It also makes it possible for human beings to read a layout file, although this is rarely necessary. This appendix is provided primarily for people who are interested in using programs other than Proof, such as model generators, to write layout files.

Layout command words, like all Proof commands, may be any mixture of upper and lower case letters. By convention, Proof capitalizes the first letter of command keywords when it writes layout files.

Some examples of layout file commands are given in this appendix, but the best source of examples is layout files that have been written by Proof Animation. When you are running an animation, you can click on View, View Proof File as Text, Layout File and select a layout file to be viewed. Although some examples take up more than one line due to lack of space on the page, please note that each layout command must be written on a single line in an actual layout file.

Rules for Names

Classes, Objects, Bars, Plots and Messages defined in a layout file, require specification of an alphanumeric name. Names are case-sensitive; i.e., "joe" is different from "Joe." Each name must start with an alphabetic character (A-Z or a-z). The remaining characters in a name can be letters, digits or underscore characters.

The "name spaces" for each of the five categories mentioned above are independent. So, for example, it is possible to have an Object named STATUS and a Bar named STATUS. This is acceptable, but not recommended.

A sixth type of name is the name of a View. The rules for View names are explained in the section on the "define view" command later in this appendix.

Unless the reader is an advanced PROOF programmer and wants to change the layout, this is rarely used and the layout commands are automatically placed into the .LAY file as the layout is drawn by the mouse. However, there are layout file commands that are also described here. It is not recommended to manually change the layout file, unless the user is highly proficient in PROOF. No reference to this option will be made in this book.

The .PSF command summary is Appendix C in the Wolverine PROOF Manual. This has to do with making presentation files. The various commands that are used in making up such files are covered there. A presentation can be a series of animations with sounds, slides, time delays, and so on. Presentation files can be made once you learn how to make animations.

PROOF's rendering algorithms is Appendix D in the Wolverine PROOF Manual. This will not be used here. The advanced PROOF user eventually will want to import files such as .DXF ones into a layout. For our work, all layouts will be created on screen. To use this option, you need to have PROOF Professional.

The last two items simply give the address of Wolverine Software as the version of Direct Draw you have.

5.6 Exercises

5.1. Draw a new layout. This will consist of a single, small triangle. Fill this with whatever colour was used to make the boundary of the triangle. Use the box edit to make another six of these. Then box edit across the screen and then six above it. Thus, you will have 48 triangles in all. Practice the use of the box edit until you can do this quite rapidly.

5.2. Draw a new layout. Make only a single CLASS, a ball with radius 2, fill it, and colour it white. Save this layout with any name you choose, such as MYCHAP5. Now, box edit the ball and click on COPY. The ball disappears. Close this layout and open a new one from the File menu. Click on Class → Class → New C. You will need to give this new CLASS a name. Now click on box edit and Paste. The saved boxed item from the previous layout is now a new CLASS in the new layout. *This copying using box edit can be a tremendous time saver for making future layouts.*

5.3. Make a layout that consists of two parallel lines. The first goes from A to B, where A is at (−40, 30) and B is at (40, 30), and the second from C to D, where C is at (40, 10) and D at (−40, 10). The upper line will be path P1 and goes from left to right; the other is P2 and goes from right to left. There will be two classes, TRUCKL and TRUCKU. TRUCKL represents loaded trucks and TRUCKU represents unloaded trucks. The unloaded truck, TRUCKU, will be a simple rectangle with base 2, hot point (0, 0). The loaded truck, TRUCKL, will be identical except that it will have an irregular shape on top to represent a load of ore. The animation will have the loaded truck travel on the path from A to B in 30 time units. It then becomes an unloaded truck and returns on path C to D in 15 time units.

Figure 5.6 shows a screenshot of the animation.

5.4. Modify the .ATF file for Exercise 5.3, so that the mining car now pauses for 5 time units at point B for dumping. While it is dumping, it will turn into a truck that is dumping and then returns to point D via the path from C to D. The dumping truck will be a new class, TRUCKD.

5.5. Make an animation that has a path named P1 going from (−40, 0) to (40, 30). This path should be colour F26. The only class will be TRUCK, which can be any simple icon for a truck. Write the .ATF file that has the following:

a. A truck is created at time 0 and then every 6 time units until five are created.
b. The trucks are to be put on path P1 for 25 time units.
c. The animation is to run until all the trucks are at the end of the path.

5.6. Add the code to Exercise 5.5, so that the trucks are now removed from the animation when they arrive at the end of the path P1.

5.7. Extend the path P1 used in Exercises 5.5 and 4.6 to go from (40, 30) to (70, 0). Call this path P2. Now have the trucks in Exercise 5.6 placed on this path for 15 time

FIGURE 5.6
Screenshot of a mining car.

units. When they come to the end of this new path, they are to be removed from the animation.

5.8. Make an animation that has two paths. The path PATH1 goes from (–50, 0) to (20, 0) and path PATH2 goes from (–30, –10) to (20, –10). Both paths are coloured F26. A single CLASS is TRUCK that can be any simple shape for a mine truck icon and any colour. The animation is to be as follows: the TRUCK is to be placed on path PATH1 starting at time 0 and travel for 30 time units to the end of the path. It then is immediately placed on the path PATH2 and travels to the end in 25 time units. It is then to be moved in 20 time units to position (30, 25). It sits there for 5 time units and then is placed at position (50, –10) for 5 time units until the animation is finished.

6

More PROOF Icons

The bucket wheel excavator is the largest machine in the world.

6.1 More PROOF Icons: Navigating the Lower Menu Bar

When PROOF is opened in Run mode, the following is the lower menu bar. Notice that several of the icons are not highlighted.

6.1.1 Open File Icon

The first icon on the left of the menu bar is shown below:

Opening this option is exactly the same as using the File and *Open Layout + Trace (.ATF)* option above it. This will allow the user to open any .LAY and .PTF file. The files are shown only as .LAY files. The menu that opens will allow you to either double click on the file to open or type in the name.

6.1.2 Save Icon

Next to open file icon is the save icon, which is shown below:

This icon, which looks like a PC floppy disk, not only will save your file but also will let you know how much of the Student PROOF you are using and the byte limit.

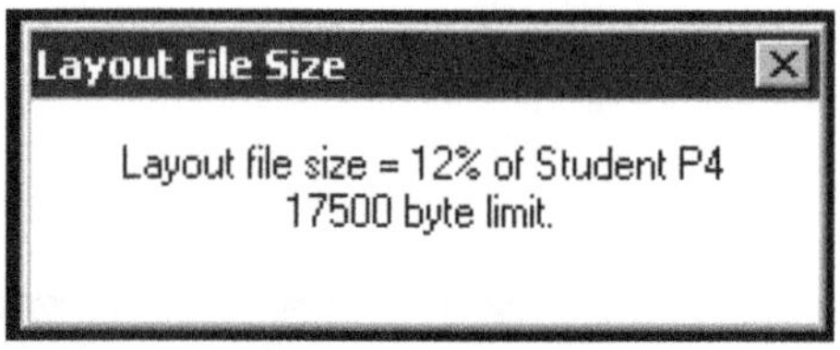

6.1.3 Zoom Icons

The next two icons are as shown below:

These allow you to zoom in or out of the animation or drawing palate. The one on the left is for zooming in (magnify), and the one on the right is for zooming out (shrink). When either is clicked on, there are four options to select for zooming. For example, clicking on the zoom in icon brings up the following menu:

The percentages are what zoom factor you want to select. The other zoom tab also gives four percentages, which are the inverse of the above.

6.1.4 Dragging the Zoom Icon

The next icon to consider is the following:

This is a very important option and one that we will use quite often. It is called a *draggable zoom box*. It allows you to move selected different portions of your layout and make them larger or small. It is similar to the two icons that increase or decrease the layout size, but this option allows you to select a portion of the animation to zoom in or out on. Clicking on this option gives the following dialog box:

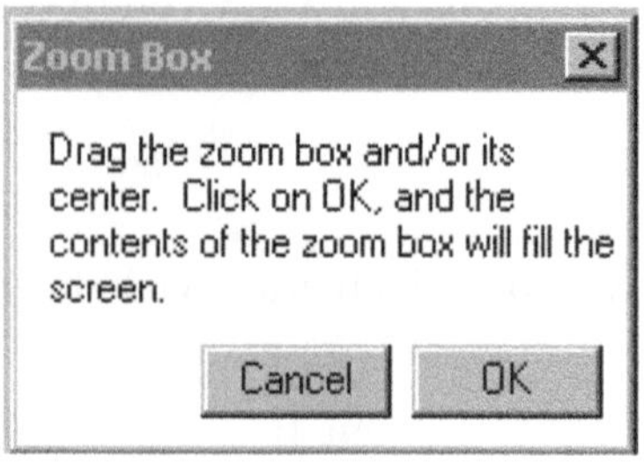

This menu gives the instructions for using the zoom box. When you click on this icon, you will notice a red box on the screen in the centre. It will have small cross hairs in the centre and lines radiating from the sides. For example, suppose we were viewing the animation given by DUMP10 as shown in Figure 6.1. When we click on the icon, the screen shows the following:

You now can manoeuvre this by clicking on the cross hairs and dragging the box to where you want the centre of the screen to be. Doing this and clicking on OK will change the screen view to where the cross hairs are at the centre of the screen. The box will be the sides of the new layout. Thus, the new screen might look as shown in Figure 6.2.

You also can click on any side of the box to drag and zoom in or out to increase/decrease the size of the layout. This option is especially nice to use when you need to zoom in on just a specific area of the layout. For example this can be used to see if the various lines actually do totally enclose an area. It is also very handy for trimming only the extra lines or arcs.

6.1.5 Home Icon

This icon simply takes you back to the (home) view.

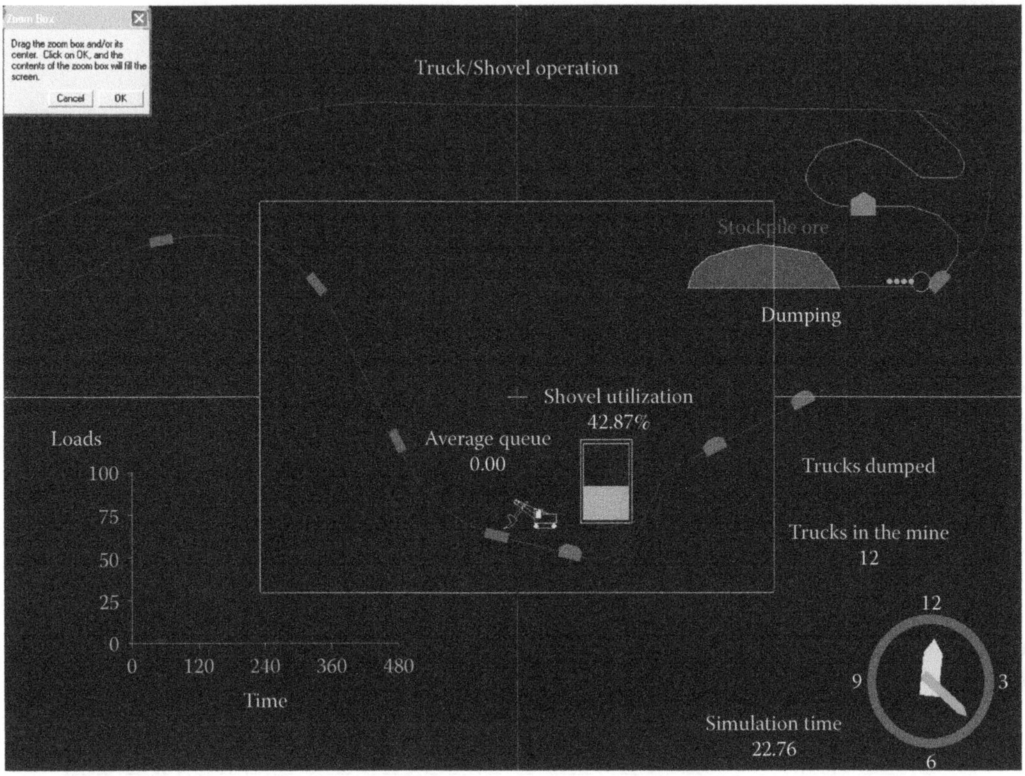

FIGURE 6.1
Screenshot after clicking on the draggable zoom box.

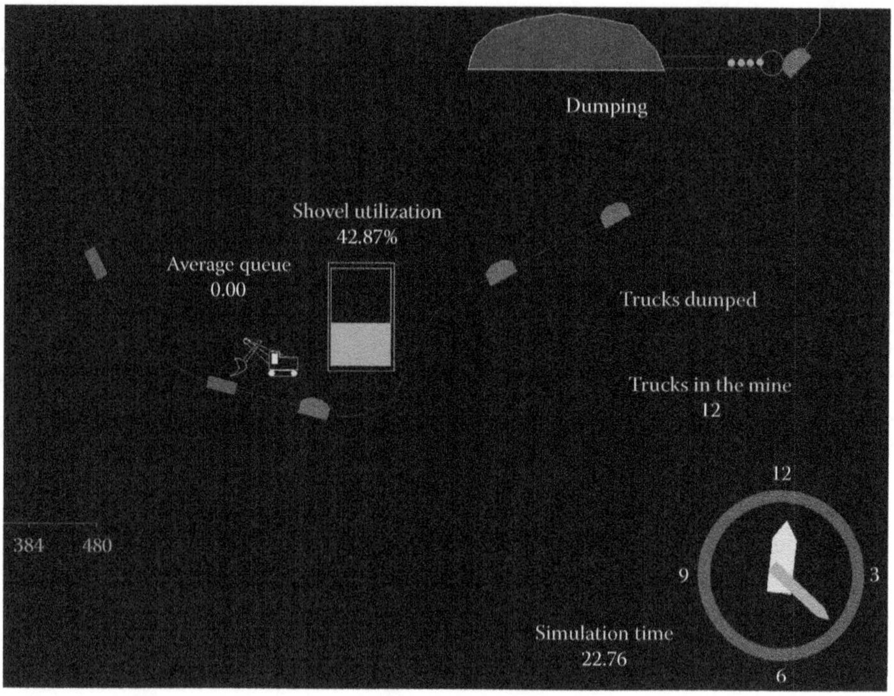

FIGURE 6.2
New screen after using draggable zoom box.

6.1.6 Scroll Bars Icon

The next icon is used to change the drawing palate by moving the view either up or down. The icon is as shown below:

When this icon is clicked on, the bottom and right sides of the screen have small margins added with a small square in the middle of each. By clicking on these, one can move the screen up or down or to the left or right.

The next icons to be discussed are ones that appear when you are in Draw mode. These are used to make the layouts and need to be examined closely.

When you click on *Mode* and select the *Draw* option from the pull-down menu bar, another set of icons appears in the menu bar. The screen also has a grid that is lined every 10 units. It appears as in Figure 6.3.

The corners are located at (10, –5), (90, –5), (90, 65), and (–10, 90). The grid is numbered every 10 units (the corner numbers are not shown on the screen) and there are grid lines every 10 units also. The new icons are as follows:

Let us examine these, as they will all play a role in creating the layout.

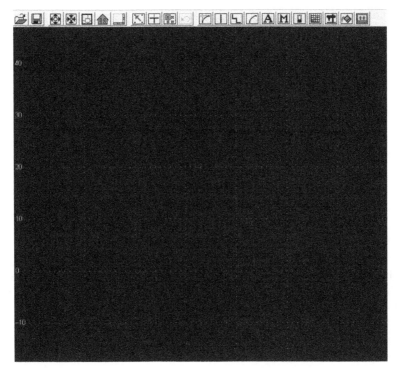

FIGURE 6.3
New screen for the Draw mode.

6.1.7 Snap to Icon

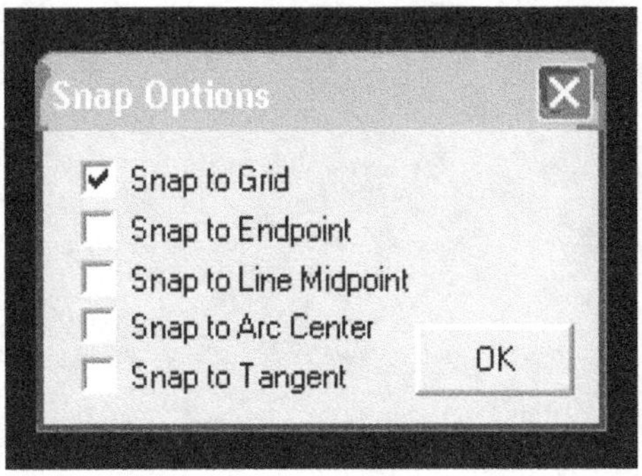

This is called the *snap to* option. The default option is shown by the check mark being next to Snap to Grid. When drawing a line, no matter where you start or finish, the line will automatically attach itself to the nearest grid point. By clicking on any of other options, you can choose to snap to what they are. Normally, one uses either this option or the Snap to Endpoint option. The other options are easy to understand. One often will switch back and forth among these options. If you remove all the options, then the line will move where the cursor places it.

6.1.8 Trim Icon

The next option is useful when you need to have lines end at exact places. For example, if a line is to be exactly inside a circle, it can be drawn so as to extend outside the circle and then trimmed off. When this option is clicked on the pull-down menu, it is as shown.

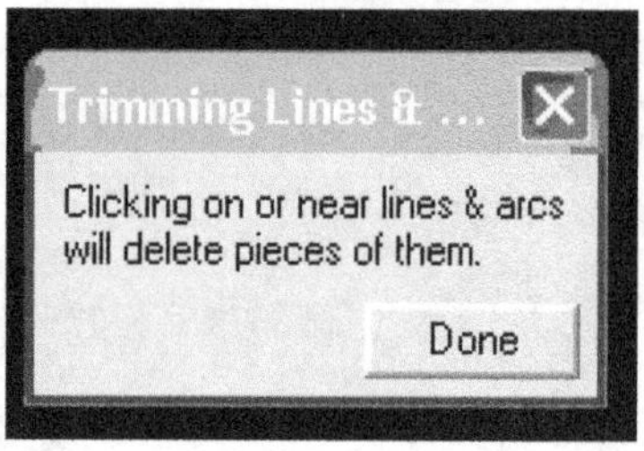

Figure 6.4 shows one application of the option. Suppose it is desired to draw a circle with four lines drawn as diameters. The lines are to be vertical, horizontal, at 45° to the horizontal, and at 135° to the horizontal. The vertical and horizontal lines are easy to draw but the ones at 45° and 135° can present a problem especially if the snap to option is on. One can easily draw the lines by drawing them as shown in the left-hand side of Figure 6.2 and then trim them using the trim option.

6.1.9 Undo Icon

The next icon, the undo icon, has already been covered in Chapter 3.

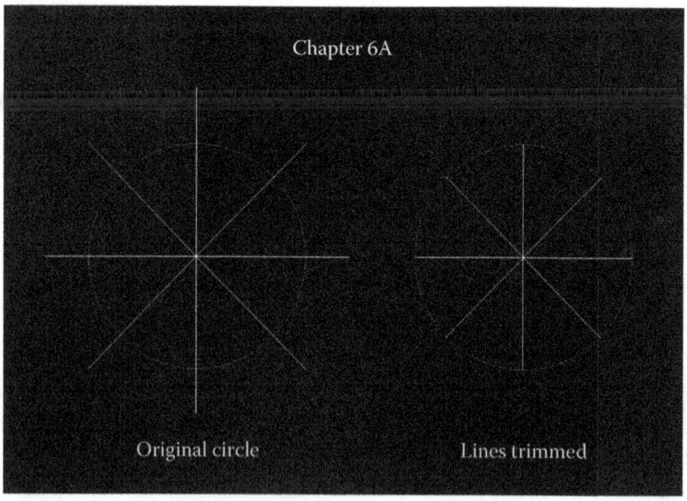

FIGURE 6.4
Layout showing the use of the trim option. The original circle has the lines drawn as shown and then with the ends trimmed.

6.1.10 Box Edit Icon

The box edit icon is one of the most useful icons in PROOF. It will be used a great deal in most layouts. The icon is as shown below:

This icon allows us to move parts of the layout, copy them to both the current layout or import them to other layouts, and also to increase or decrease the size of selected elements of the layout. It also can be used for CLASSes. Clicking on it gives the following dialog:

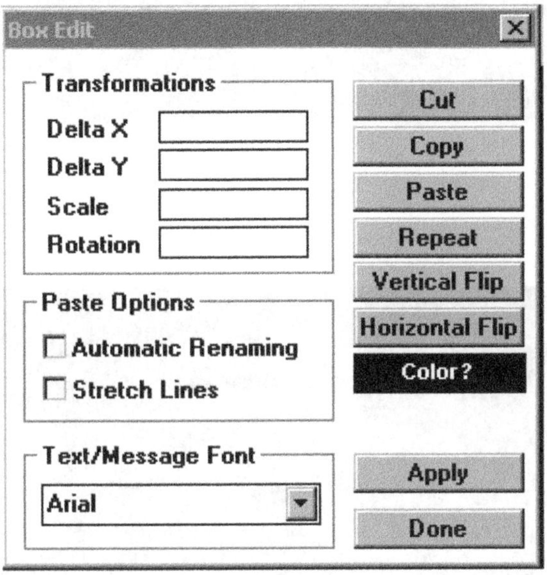

In order to use the box edit option, you need to *box* the elements first. These elements can be any portion of the layout, including text. Move the cursor to the upper left-hand side of the box and make sure that it is outside of the box (the box can actually be made from any point outside the elements). As you click and drag, a new box will appear that will enclose the original layout. Release the mouse and the elements inside the layout box will blink. This indicates that the elements inside the box have been selected. To move it to a new location, click on the cross hairs and drag to the new location. This is frequently done if you want to change the hot point of a CLASS. It will take a bit of practice to move objects to exactly where you want them, since the box normally will be a bit outside of the elements, and it is the whole box that is moved.

Let us examine the various options with the layout CHAP6A. Click on Mode and then *Draw* and then shrink the two circles by 50%. This is done by clicking on the icon and positioning the cursor to the upper left side of the area to be changed. Then, when the mouse drags the cursor, a red box is formed to include the two circles. Figure 6.5 shows this.

Next, one keys in .5 into the pull-down menu where it says *Scale* and the animation becomes as shown in Figure 6.6.

To make duplicates of the object, you need to click on *Copy* and then *Paste*. The boxed item or items will blink. You then drag the box (actually, in our case, the original box with the new box outside of it) to where you want the box to be copied and release the mouse.

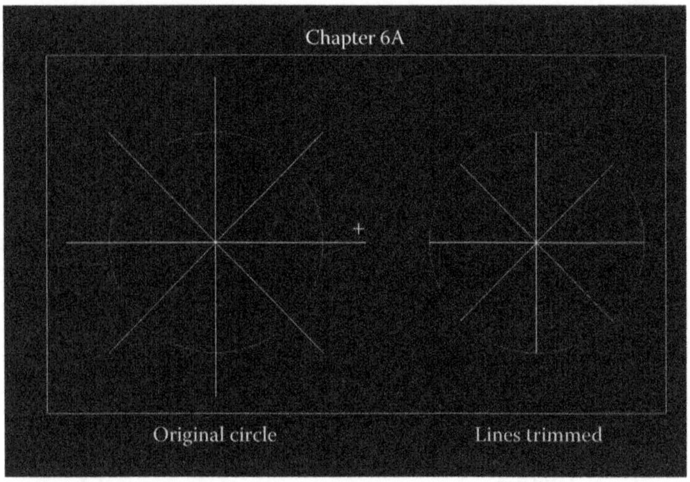

FIGURE 6.5
Boxed area.

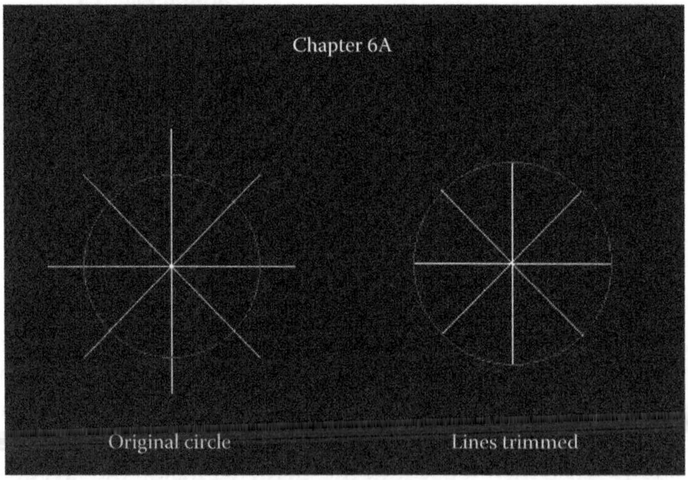

FIGURE 6.6
Application of box edit icon.

To make identical copies of the box located at the same distance away from the original box, click on *Repeat*. To make additional copies of the object but located where you wish, simply repeat the process. Do *not* just click on *Copy* and move the box, as this will only move the objects in the box to the new location.

You can also input an angle (in degrees) to rotate the object that is boxed.

The other options are the *Vertical Flip* and *Horizontal Flip*. These are designed to change the object's orientation. It is possible to use the *Cut* option to cut and then *Paste* the elements to another animation if so desired.

Once you box the elements, you can change the colour of all of them by clicking on the *Color?* and selecting a new colour.

It is possible to build a catalogue of different icons to use to import into a new layout. As can be appreciated, this icon can be very useful in rapidly making layouts.

Example 6.1: The Box Edit Icon

The box edit icon can do quite a few things. Start with the layout given by CHAP6B. This is given in Figure 6.7.

Click on the box edit icon and place the cursor in the upper left side of the objects, just above the square, triangle, and square on the upper left side of the screen. As you drag the cursor, a red box appears. Drag the cursor until the box on the screen includes both the squares and the triangle (Figure 6.8).

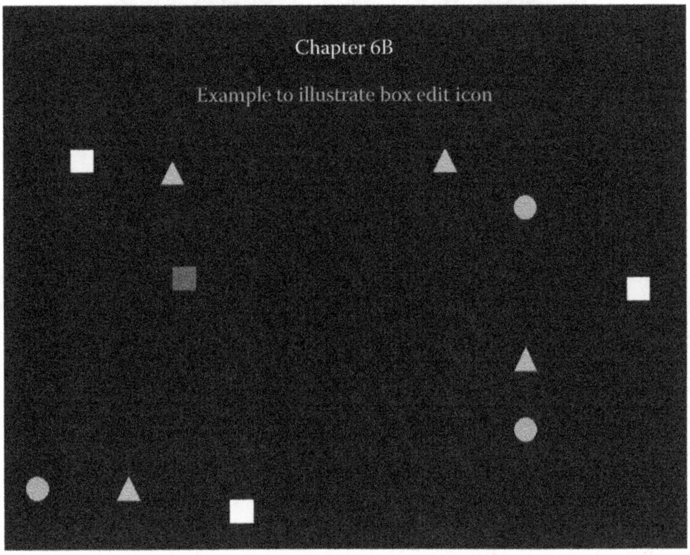

FIGURE 6.7
Screenshot of animation for CHAP6B—this will be used to illustrate the important box edit icon.

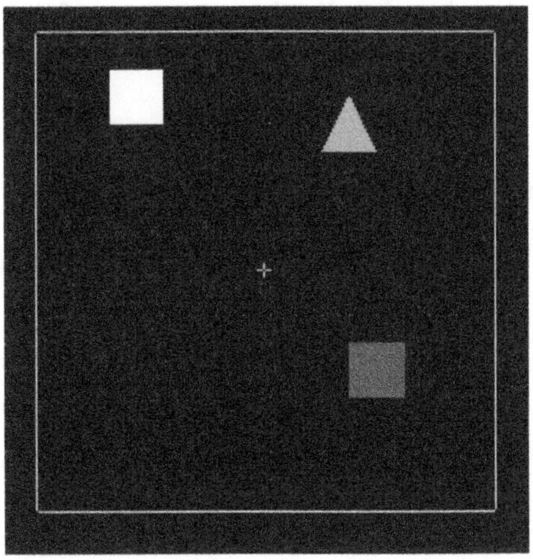

FIGURE 6.8
Portion of screen showing box formed around the squares and the triangle.

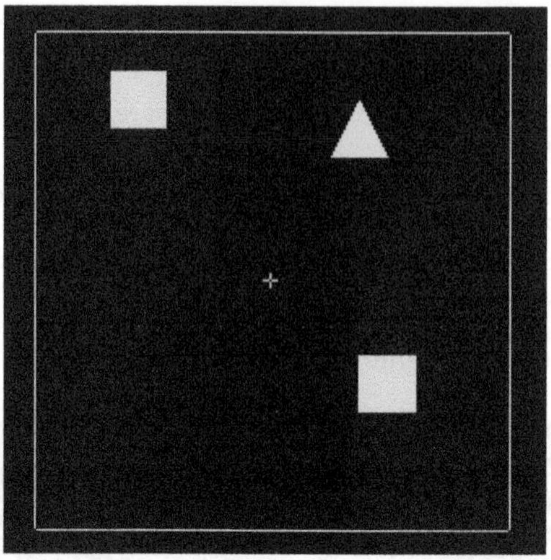

FIGURE 6.9
Portion of screen after objects in box have their colour changed to yellow.

One square is white, the triangle is green, and the other square is red. Now click on *Color?* in the dialog box. You can change the colour of the three objects by simply clicking on a new colour. Actually, it does not matter where you position your cursor when you use the box edit, just so long as the objects to be changed can be included in the box you are going to create (Figure 6.9).

You can now change the font to one of many different fonts using the Text/Message Font in the dialog box.

As mentioned, if you want to make an object larger, box edit it and change the scale by either clicking to increase or decrease it. Alternately, you can type in a number for the scale. This is especially useful when you are making a CLASS and it appears to be too small (or large) in the animation. You can simply box edit it and change the scale. *When you do this, be careful to note that the hot point may have moved.* However, you may have noticed the small cross hairs in the centre of the box when you box edit an object or objects. Putting the cursor on this and moving it will move all the objects in the box to a new location. Do this with a few of the objects on the screen. Similarly, the *Rotation* in the dialog box allows you to rotate the item(s) in the box to different angles.

6.1.11 Fillet and Text Icons

The next icons, the fillet and text icons, have been discussed in Chapter 3.

The text icon will now be covered in more detail than was done in Chapter 3. The pull-down menu when the icon is clicked is as follows:

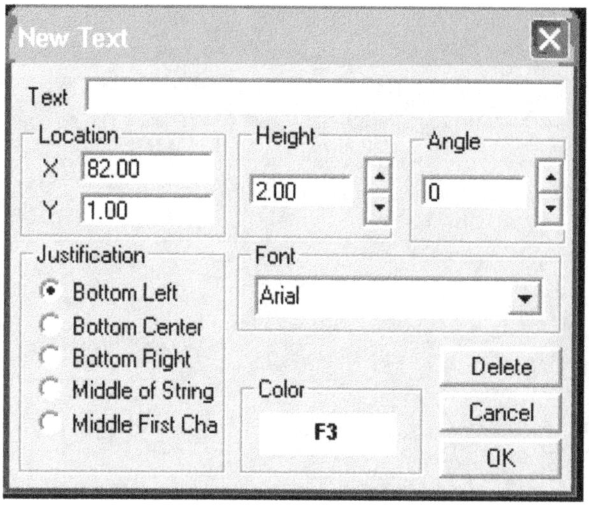

Notice all the options of this icon. The location of the text can be changed by either keying in different coordinates or moving the cursor on the screen. Under *Justification* are the standard placement options. By default, this is the *Bottom Left*. This means that the text can be dragged across the screen by clicking on the bottom left of the text string. This is one of five settings. This can be changed if so desired. In the example shown here, the cursor is at (82.00, 1.00). If you key in *HELLO, KIND PEOPLE* the text will be located with the *H* in HELLO at this point at the bottom left. If the justification is *Bottom Right,* the second *E* in PEOPLE will be at this point. The colour of the text is white (F2) and the font style is Arial. The text will be horizontal but can be changed by either inputting an angle or using the arrows to increase or decrease the angle.

Example 6.2: Different Text Messages

Examine the layout given by CHAP6C. This is shown in Figure 6.10. This shows some of the many possible text messages that might be on the screen. You should practice with the Text menu until you are satisfied that you understand it clearly. It is a rare animation that does not have text on it.

If you examine this in Draw mode, you will notice variations in *Justification* of the text.

6.1.12 Message Icon

A message is either text or numbers that will dynamically change when the animation takes place. For example, the number of loads from a shovel will be increased as mining takes place, the utilization of the loader changes, and the amount of ore dumped increases. The icon for a message is as follows:

When this is clicked on, the user is prompted to input a name for the message. This is case sensitive so M1 and m1 are different.

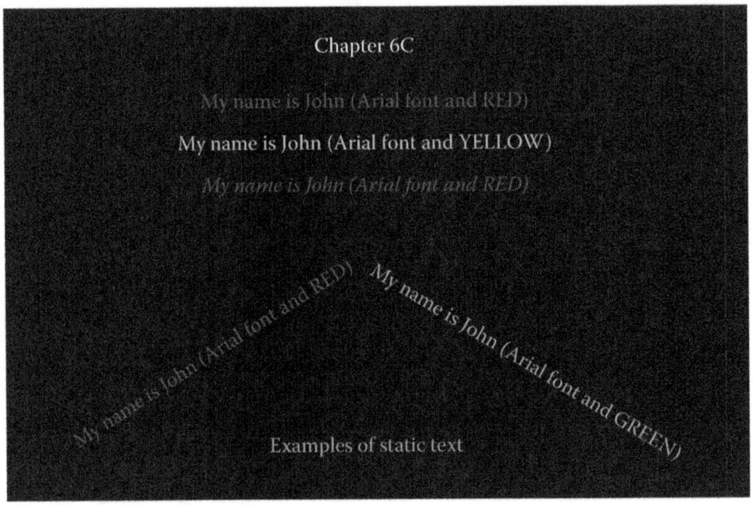

FIGURE 6.10
Examples of text.

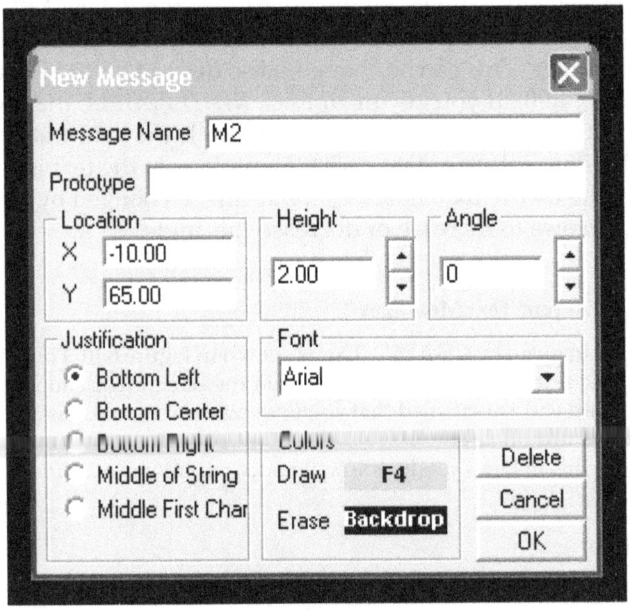

After a name is given for the message, the above dialog box opens. There are several options with which the user should be familiar. The Prototype is simply a sample value or text that will show where the message will appear on the screen and what it will look like. For example, if the message is how many tons of ore have been loaded, Prototype might be 1245.67. If the message is text, it might be something like:

```
Hello, John
```

The actual message will be whatever is in the .ATF file. The rest of the options are self-explanatory and should be studied and experimented with.

6.1.12.1 Putting a Message on the Screen

A message will be placed on the screen according to the options selected in the dialog box by the command:

```
WRITE (message name) (value or text)
```

Thus,

```
WRITE M1 1
```

will result in the value of 1 being placed on the screen where message M1 was created.

```
WRITE M123 35.45%
```

will result in message M123 having 35.45% being written on the screen.

```
WRITE Mess1 Hello, kind folks
```

Will result in the text *Hello, kind folks* being written on the screen.

Example 6.3: A Mine Crusher Animation

Write the layout and .ATF files to show how a mine crusher will show the number of loads dumped. For this example, assume that there is a load dumped every 5 time units. The animation is given by CHAP6D.LAY and a screenshot is in Figure 6.11.

Suppose that a layout has a single message, and that this is named M1 and Prototype is the word *hello*. Click the box edit icon and box the message. Click on Copy and Paste to make a new message. Move the cursor where you want the new message and this will bring up a dialog box saying Message M1 clone and the box will show M2 as the name for the new message. Every time you click on Paste and move the edit box, a new

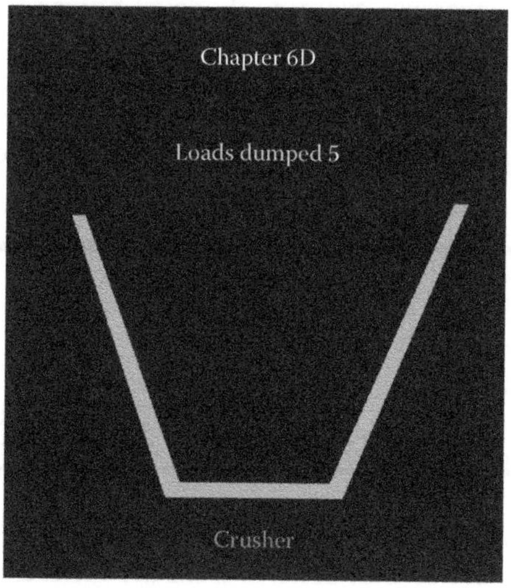

FIGURE 6.11
Screenshot of animation for CHAP6D.

message will be created. The name of the new message will increase by 1. This can be a real time saver in creating large layouts.

There is a layout file called ICON.LAY. This was created by Wolverine Software. Open this. Notice the many figures it has. Suppose you are making a layout and need a CLASS as a miner. Click on Mode and Class and then Class again. Then click on Select. Select the MINER. It may be too large for the screen, so minimize the figure. Box edit it and click on Copy. Then, open the layout you are working on and click on Class → Class → New and give it a name. Then click box edit and Paste. Your MINER will look as follows.

There is your miner as a new CLASS. *Having a file of these icons can greatly reduce the time it takes to make new layouts and CLASSes for animations.*

6.1.13 Layout Objects Icon

The icon given next allows you to create what are called *layout objects*. These can be very important for your layouts as you will see.

These are objects made from CLASSes that will be shown on the layout but which can move, change colour, and so on. For example, a boom on a quarry shovel will rotate as it loads a truck. You will want this boom shown as part of the layout when the shovel is not loading the truck. For this exercise, examine the layout given by CHAP6E.LAY. You will notice that it appears to be a normal layout with two white circles, three orange squares, and four green triangles. Except for the text, these are layout objects. These objects came from CLASSes called FIRST (the ball), SECOND (the box), and THIRD (the triangle). These were made by clicking on the icon given above and seeing the following menu on the screen:

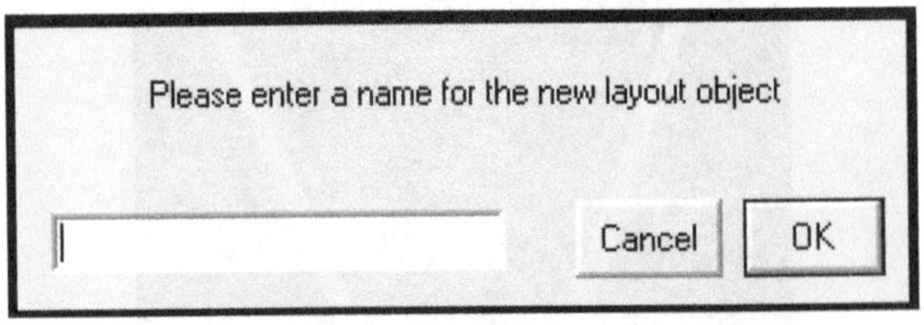

You need to give each object a separate name based on the CLASS it comes from. Be careful! The name is case sensitive so Car, CAR, and car are different. Give it a name and click on OK. It is a good idea to keep a list of names you have given to each object when you are going to have a lot of them. The next menu appears as shown below:

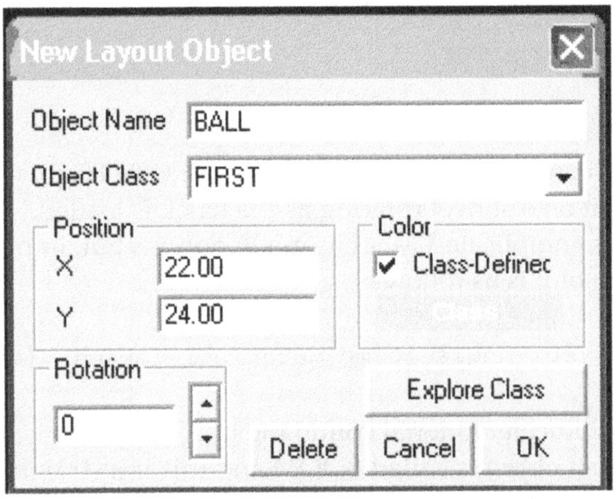

Here, the Object Class FIRST was selected and the name given to this is BALL. When you hit <enter> you can position the new layout anywhere you wish by dragging it with the mouse. Other new layout objects are similarly created.

If you want to create an object from the second CLASS (the box in the previous paragraph), you need to have SECOND in the menu, give the object a name, and position it as before. *Be careful that your cursor is on the viewing screen when you do this, so that you can see the object on the viewing screen.* When you click on OK, the new layout object appears on the screen.

Once you have these layout objects on the screen, they can be made to move, change colour, change size, and so on. Figure 6.12 shows the layout. It can be studied by clicking

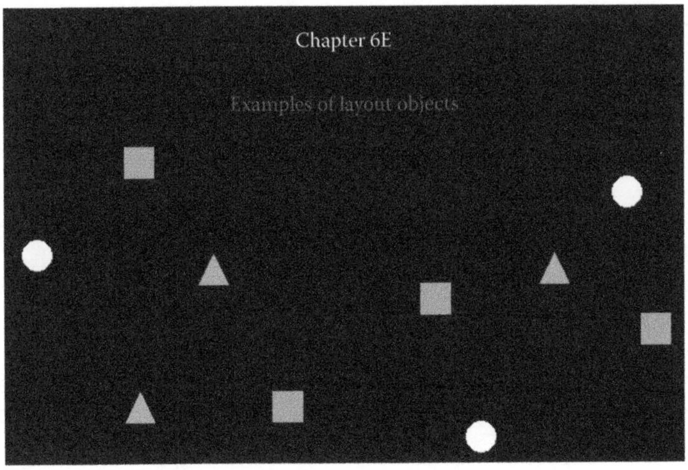

FIGURE 6.12
The circles, squares, and triangles are layout objects.

on Mode → *Draw* and placing the cursor at each object to see their names. Most mining animations will include numerous such layout objects.

There are several more icons to be covered and they will be covered in Chapter 30. But, first, there is another PROOF command to be presented.

6.1.13.1 The `ROTATE` Command

One of the most useful commands for layout objects is ROTATE. One common use in mining is to show the boom on a shovel rotating as the truck is loaded.

The command seems complicated as it has many options but, in practice, only a few are used. The general form of it is as follows:

```
ROTATE (object ID) [to] angle [time duration] [speed] [step degrees]
```

The square brackets [] indicate optional commands.

Assume that the layout object is called BOOM and this is a straight line, some examples of the ROTATE command are as follows:

```
ROTATE BOOM to 45
```

This rotates the layout object BOOM 45° counterclockwise from its initial position. If the initial position was horizontal, the boom rotates 45° counterclockwise. If the next command is `ROTATE BOOM to 30`, the boom will move to a position 30° counterclockwise *from its initial position*. The rotation is instantaneous. Rotation is about its hot point.

```
ROTATE BOOM -60
```

This rotates the layout object BOOM 60° clockwise from its initial position.

```
ROTATE BOOM to -110
```

The `BOOM` will rotate to 110° clockwise no matter what its initial position is.

```
ROTATE BOOM SPEED -90
```

This will rotate BOOM continually clockwise about its hot point. It will make a complete rotation every 4 animation time units.

Examine and run CHAP6F. Here you start with three layout objects, ARM1, ARM2, and ARM3. The CLASS is ARM.

As the animation takes place, one of the arms rotates according to the text that is written below it. You may wish to change the clock speed to assist in watching the movement of the arms (Figure 6.13).

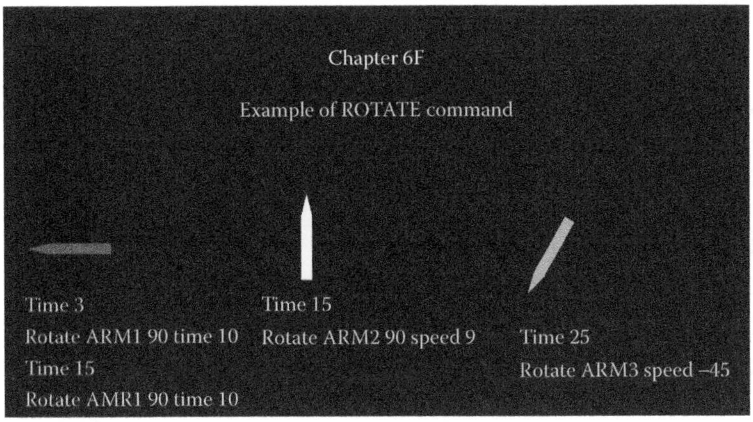

FIGURE 6.13
Portion of animation given by CHAP6F.LAY—examples of the ROTATE command.

6.2 Exercises

6.1. Make an animation having the layout objects:

BOX sides 5 units long and 1.5 units high

CAR length 5 units long and 2 units high

TRUCK length 4.7 and height 1.8 (approximately)

Have the hot point in the middle of all the CLASSes.

Position all other CLASSes horizontally at various locations on the layout (all are to be layout objects). The animation will do the following:

At time 1, the CAR rotates 70°

At time 2, the BOX rotates –50°

At time 3, the TRUCK rotates to 75°

At time 4, the CAR rotates 80°

At time 5, the BOX rotates –60°

At time 6, the TRUCK rotates to –75°

At time 7, the CAR rotates 50°

At time 8, the BOX rotates –50°

At time 9, the TRUCK rotates to 75°

At time 10, the animation is finished.

In order to slow your animation, the clock speed should be set to 1 time unit. Then click on View → Save → Save Current View As → Home. Save the layout, and now, when the layout is opened again, the clock speed will be 1 time unit.

6.2. Make four layout objects that all come from the CLASS ARM. ARM is a horizontal CLASS, shaped as a rectangle with length 8 by 1 but with the right side pointed. It can be any colour. The hot point is slightly to the left of the bottom centre. Call the four layout objects ARM1, ARM2, ARM3, and ARM4. They can be different colours if desired. Position them in the layout horizontally a bit apart from each other. All four layout objects are to rotate continuously as follows:

- ARM1 is to rotate with a speed of –90
- ARM2 is to rotate with a speed of +90
- ARM3 is to rotate with a speed of –45
- ARM4 is to rotate with a speed of +45

6.3. This is an important example of a model. By now you should have an idea of how to put an analogue clock on the screen for your animations. Look at the animation given by CHAP6G. This is just the clock of the animation time. Do you see how easy this is to make? The minute and hour hands are layout objects and the ROTATE command has to be selected to match the animation clock. Recall that 6 animation time units is 1 second of viewing time. Thus, the ROTATE commands are simply as follows:

```
ROTATE S1 speed -6 step 6
ROTATE HOUR speed -.5 step 6
```

The *step* commands specify the number of steps in each rotation of 6°. The use of the step overrides the PROOF default of 30°.

A screenshot of the animation is given in Figure 6.14.

Modify this to make it appear a bit more realistic. For example, you might want to add other clock numbers. Also, the hands are a bit rough.

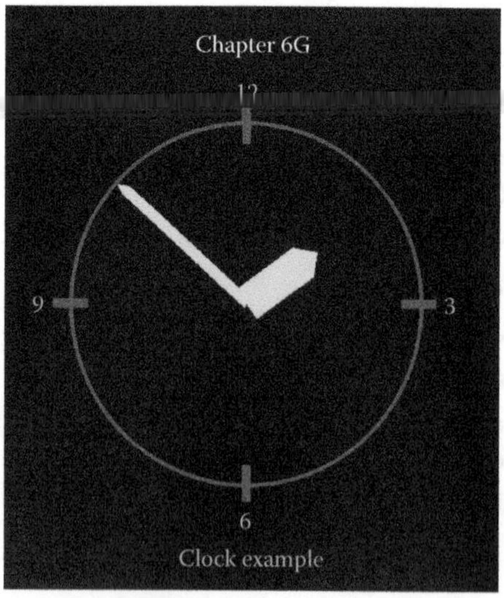

FIGURE 6.14
Screenshot of the clock model.

6.4. Make a layout with the following as layout CLASSes:

- Star
- Moon
- Sun
- Cloud

Write the .ATF file so that, after 5 time units, one of the layout objects *disappears*. After another 5 time units, another disappears, and so on, until the screen is blank except for the caption.

6.5. You can treat the layout objects as you would any other objects, such as changing their colours, moving them and, as we shall see, rotating them. For example, if you had a CLASS STAR, colour yellow, and the layout object was SS1, the following ATF command:

```
SET SS1 color F3
```

would change the color of SS1 to white.

For example, suppose your ATF file is as follows:

```
TIME 0.0000
TIME 5.0000
SET B1 COLOR F2
SET B2 COLOR F3
SET C1 COLOR F4
SET C2 COLOR F5
SET C3 COLOR F6
SET T1 COLOR F7
SET T2 COLOR F8
SET T3 COLOR F9
SET T4 COLOR F10
TIME 10.0000
SET B1 COLOR F11
SET B2 COLOR F12
SET C1 COLOR F13
SET C2 COLOR F14
SET C3 COLOR F15
SET T1 COLOR F16
SET T2 COLOR F17
SET T3 COLOR F18
SET T4 COLOR F19
TIME 15.0000
SET B1 COLOR F20
SET B2 COLOR F21
SET C1 COLOR F22
SET C2 COLOR F23
SET C3 COLOR F25
SET T1 COLOR F2
SET T2 COLOR F3
SET T3 COLOR F4
SET T4 COLOR F5
TIME 20.0000
SET B1 COLOR BACKDROP
```

```
SET C1 COLOR BACKDROP
SET T1 COLOR BACKDROP
END
```

The animation would show the objects in their original colour until 5 time units have passed. At this time, their colours change, then at time 15, they change again. At time 20, three objects disappear (their colours are now the same as the backdrop colour). Make an animation with three classes: BOX, CIRCLE, and TRIANGLE. From BOX, make the layout objects B1 and B2. From CIRCLE make the layout objects C1, C2, and C3 and from TRIANGLE make the layout objects T1, T2, T3, and T4. They can be of any size and positioned at various places on the screen. Run the animation.

6.6. In making layout objects, PROOF offers a time-saving device. Suppose you have a layout object B1. If you box edit it and Copy and Paste and then drag it to another position, you will be prompted to give it a name B2. The prompt will be *Object B1 clone name B2.* Click on OK and a new object will appear in the spot where you have moved the clone. Use the layout to make different clones of the various layout objects.

6.7. Make an animation that has four stars roughly at the four corners of the screen. These are to be layout objects S1, S2, S3, and S4. After 10 time units, star S1 will disappear; after another 10 time units, star S2 will also disappear; after another 10 units, star S3 will move in 5 time units diagonally across the screen and then disappear; finally, after another 10 time units, the last star will also move diagonally across the screen in 5 time units and disappear.

6.8. Make an animation that has a square box as its layout. The corners of the box are at the coordinates: (–20, –20), (–20, 20), (20, 20), and (20, –20). The box can be any colour. There will be four CLASSes. They are as follows:

Square box	BOX
Round ball	BALL
Triangle	TRI
Large box	BIGBOX

The first three should have a *size* of approximately 2 units with hot points in their centres. The BIGBOX should have side 4 units. They can be different colours. The single path will be the sides of the layout square going clockwise and starting at the lower left corner. This path should have a time of 20.

Write an .ATF file to have the animation as follows: At time 0, the BOX travels from the lower left corner of the box to the top left corner in 5 time units. It then becomes the BALL and travels to the top right corner in 5 time units where it becomes the TRI and travels to the lower right corner in 5 time units when it becomes the BIGBOX and travels to the lower left corner in 5 time units. After 30 time units, the BIGBOX disappears.

6.9. Refer to the layout used in Exercise 6.8. Write an .ATF file so that the animation is as follows: The BOX is created at time 0.0000 and placed on the square at the lower left-hand corner. It will travel around the square in 20 time units. At time 5.0000, the BALL is created and placed on the lower left-hand corner. It, too, will travel each side of the square in 20 time units. At time 10.0000, the TRI is created and placed on the square at the lower left-hand corner again travelling the square in 20 time units. Finally, at time 15.0000, the BIGBOX is created and placed at the lower left-hand corner to also travel the square in 20 time units. The animation will run for 50 time units and, at the end, all the four CLASSes will be on the square at the lower side. All the objects should be visible at the end of the animation.

NOTE: In order to have this animation correct, it is necessary to have the path accumulating and each of the CLASSes to have appropriately chosen clearances.

7

Introduction to the GPSS/H Simulation Language

The Wieliczka Salt Mine, Poland. The mine features an underground cathedral with numerous statues. All statues are carved from rock salt. The mine dates from the thirteenth century and is still active. It is a major tourist attraction and has a health spa.

7.1 Introduction to General Purpose Simulation System Version H

In the previous chapters, we have learned how to make both layouts (.LAY files) and the animation trace (.ATF) files that drive these layouts. However, for most mining animations, the .ATF files can be very large. Also, there may be many events taking place separated only by a very short time period. In this chapter, we will begin to learn how to construct programs that will both simulate a system and create .ATF files.

We will learn the GPSS/H* simulation language using animations to illustrate each of the many GPSS/H commands. This will be a bit awkward at first, as the actual programs we will run to create the .ATF files will be complete and contain code that we haven't learned yet.

A GPSS/H computer program consists of two types of commands. These are either *statements* or *blocks*.†

A statement can be considered as a *control* command. It will be executed only once (in general), such as before the main program for initializing variables or putting a line or lines of code on the screen to a file. Similarly, a statement might be executed after the program is run and also can display output on the screen or be written to a file.

* The code for GPSS/H will be shown in Courier 10 type to distinguish it from regular text.

† The work *Block* as referred to in the GPSS/H sense will be shown as capitalized.

There are many other statements that will be described as each is used. For example, one might have a program that first prints out on the screen what the program will do and then prompts the user to key in values to be used in the program. For example, when one runs the computer program CHAP7A, a portion of the screen will show the following:

```
TRUCKS COME FOR REFUELING
TRUCKS ARRIVE EVERY 10 MINUTES
IT TAKES 8 MINUTES FOR REFUELING
```

As we shall learn, the code to print these lines is done via statements and are normally executed only once before the actual lines of simulation code are executed.

7.2 Rules for Writing Code in GPSS/H

Remember that the original GPSS was introduced back in 1962 and has undergone multiple changes since then. We will be using the most advanced form of GPSS, known as GPSS/H.* The original GPSS was designed for use with punch cards, and the format was very rigid. For the sake of uniformity, we will follow a rigid format although it is possible to use *free format*. This will be discussed later in this chapter.

Imagine that you are writing a line of code and that you need to be aware of where each character you type is on that line. This is the way GPSS/H code used to have to be typed. Although today there is the free format, we will begin by using a fixed format for the sake of uniformity. There still remain some strict rules for where some of the code must be typed. If you write your code in this *fixed* format, you will not have any problems. Later, as you become more proficient in writing GPSS/H code, you can easily switch to free format.

A typical line of GPSS/H code can have up to four parts. These are the label, the operation, the operand, and a comment. The position of these is as follows:

```
123456789012345678901234567890
(label) (Operation)(Operand) Comment
```

The *1* above refers to position one on a line, the next *1* would be position 11, and so on.

Positions 1, 10, and 22 were blank in the original GPSS but this has been changed with GPSS/H. One caution is that you do *not* put the start of the operands *beyond* position 25. (This, too, can be overridden, but this will not be important until much later.†)

For fixed format, the (label) starts in position 2. (Position 1 is normally blank.) The (Operation) is in position 11; the (Operand[s]) is in position 23. Leaving at least one space after the operands, a comment may be inserted.

* The *H* in GPSS/H comes from IBM. When IBM introduced the 360 computer series, it used letters of the alphabet to specify the machine memory size. The position of the letter in the alphabet was the power of two of the memory size. Thus, H-level machines had 256K of memory, *H* being the eighth letter of the alphabet and $2^8 = 256$. Thanks to James Henriksen for this (personal communication).

† If one has the statement `OPERCOL n`, where `n` is a positive integer at the start of a program, the GPSS/H code operands can extend beyond position 25. Thus the statement `OPERCOL 30` would allow the operands to start on or before position 30.

Thus, typical lines of GPSS/H might be as follows:

```
UPTOP         QUEUE       WAIT       Truck comes to a shovel. Waits in a queue
NEXT          ADVANCE     10,4       Truck travels from a to B
              GENERATE    ,,,3       3 trucks are created
```

UPTOP is the label. The *U* in UPTOP is typed in position 2; the QUEUE is the operation and the *Q* in QUEUE is in position 11; WAIT is the operand and the *W* in WAIT is in position 23. The rest of the line is considered as a comment and is ignored by the compiler. The *T* in *Truck* can begin in any space as long as there is one space after the operand (in this case, the operand is WAIT). In the next line of code, the operands are 10 and 4. They are separated by a comma, but not a space. If the operands 10 and 4 had been typed with a space between the 10 and the 4 such as

```
ADVANCE                 10,(space) 4
```

a compiling error would occur. If it had been typed as:

```
ADVANCE                 10(space),4
```

There would be no compiling error as the operand would be taken by GPSS/H as simply the number 10 alone. The rest of the operand, 4, would be considered as a comment.

Errors of this sort that the compiler will not catch are the hardest errors to find in debugging a GPSS/H program. If one is careful to always line up the code using the old, strict format for writing the code in the spaces suggested, these errors are less likely to happen.

Even though the rules governing typing of code in these positions are no longer strict, it will be the policy in this book to type GPSS/H code in this manner.

The following is important:

Only capital letters were originally allowed in the GPSS/H code except for comments. Even though this restriction has been dropped, all GPSS/H will be written with only capital letters here. The (label) and comments are often missing from a line of code. In a few cases, even the operand may be missing. We shall see many examples of GPSS/H code as we continue in the remaining chapters of this book.

The last line of code has no label, the operation is `GENERATE`, the operands are ,,,3, and the comments follow. That line could also have been written as follows:

```
GENERATE ,,,3
```

Another way to make a comment in GPSS/H is to put an asterisk in position 1. For example,

* This is an example of a comment line. It is completely ignored during compiling.

* This is another comment line. It can start anywhere but the asterisk is *must.*

* should be in position 1.

A blank line between lines of code is ignored by the compiler. Thus, code such as

```
ADVANCE        56,2
TRANSFER       ,BACK
```

could have been written as follows:

```
ADVANCE         56,2
(blank line)
(blank line)
TRANSFER        ,BACK
```

7.3 Continuing a Line of Code

If a line of GPSS/H code is too long to fit on a line, it can be continued using the underscore (_). This is a holdover from the days when GPSS/H was confined to only 72 characters per line. However, the underscore still is used to continue a line of code and will come in handy in later chapters. Thus, care must be exercised in not using this for anything other than a continuation of a line of code. The GPSS/H compiler will take the underscore as a continuation.

7.4 Running a GPSS/H Program

The best way to learn GPSS/H is to run numerous programs each time you are introduced to a new topic. A GPSS/H program will have many different commands, so it will not be until a few more topics are introduced that you will be able to write programs yourself. However, it is possible to type up the programs and run them right away. For the present, do not worry about what all the different commands are—each will be explained in chapters yet to come.

7.5 What You Will Need

You will need to have Student GPSS/H loaded on your PC. In the example here, the assumption is that it is located either on the desktop of your PC or in the subdirectory, C:\GPSSH>.

You must know how to create and edit files. The creation of files can be done using the old DOS command prompt editor (probably the easiest), a word processing software such as Microsoft Word or Notepad, which can create an ASCII file, or some other editor. The creation and editing of files will not be covered here.

The first example we are going to simulate is one where trucks enter a refuelling depot every 10 minutes starting at time 10. It takes *exactly* 8 minutes to refuel a truck. Only one truck can be refuelled at a time. Obviously, this is an ideal situation that would not happen at a mine, as there is always randomness in real-life situations. However, it will introduce us to what the GPSS language is like and what the output from the program looks like. The example is to simulate the depot for half an hour, starting at $t = 0$ when there are no trucks in the depot. The timescale with customers arriving and leaving during the 30 minutes is as follows:

Time	Event
0.0000	Simulation starts
10.0000	First truck arrives
18.0000	First truck leaves
20.0000	Second truck arrives
28.0000	Second truck leaves
30.0000	Simulation ends

The refuelling shop will have nothing to do until $t = 10$ when the first truck arrives. The first truck finished refuelling at time 18.0000. It is easy for us to understand this system and be able to explain what is happening. Our first GPSS/H program will simulate the depot. Although it is not necessary to follow the format given below for this exercise, try to type the program exactly as given.

In the program for this model, there are no labels, so the Blocks or statements will begin in position 11. In some cases, there will be associated operands, and these will begin in position 23. The program will look as follows (Figure 7.1).

This program is stored as CHAP7A.GPS. However, you should attempt to write and save the program yourself to start learning how to develop such programs.

If you have never seen a GPSS/H program, this must look strange, but, like any programming language, this will become familiar to you with practice. Notice that there are no commands that correspond to input or output such as READ or WRITE. This is because you normally do not read data into a simulation program. What about output? Here is where GPSS/H is so helpful. Whenever certain Blocks appear in the program, there will automatically be output associated with that Block. If you are studying a queuing situation as happens in our truck depot, output automatically will be produced as will be the case for other Blocks to be discussed later. If you want to have customized output, this is possible but will not be covered until later.

```
          SIMULATE
*******************************
*    PROGRAM CHAP7A.GPS TO    *
*    SIMULATE TRUCKS          *
*    COMING TO A REFUELING    *
*    DEPOT                    *
*******************************
          PUTSTRING   ('  TRUCKS COME FOR REFUELING')
          PUTSTRING   ('  ')
          PUTSTRING   ('  TRUCKS ARRIVE EVERY 10 MINUTES')
          PUTSTRING   ('  IT TAKES 8 MINUTES FOR REFUELING')
          GENERATE    10      TRUCKS ARRIVE
          QUEUE       REFUEL  JOIN QUEUE
          SEIZE       DEPOT   IS DEPOT FREE?
          DEPART      REFUEL  YES, LEAVE QUEUE
          ADVANCE     8       REFUEL
          RELEASE     DEPOT   FREE DEPOT FOR NEXT TRUCK
          TERMINATE           TRUCK LEAVES
          GENERATE    30      TIMER TRANSACTION FOR 30 MINUTES
          TERMINATE   1
          START       1
          END
```

FIGURE 7.1
Listing of program CHAP7A.GPS.

All GPSS/H programs *must* have the extension .GPS. Suppose the name of this program is MYCHAP7A.GPS.

7.6 Running Programs Using the Command Prompt—Up to Windows 7

To run the program MYCHAP7A.GPS, you need to be in the GPSS/H subdirectory and then key in the following*:

```
GPSSH MYCHAP7A NOXREF NODICT <cr>
```

The extension .GPS is not needed in the file MYCHAP7A.GPS. The commands `NOXREF` and `NODICT` are also optional. They represent *no cross reference* and *no dictionary*. If they are omitted, there will be considerable output. However, when a program is first written and might have typing errors, it is helpful to have as much output as possible to assist in the debugging process. The programs written here will omit these commands to have maximum output to assist in debugging should programming errors occur.

If your program was written with no errors, you will see a screen as shown in Figure 7.2.

The return of the DOS prompt means that the program ran successfully. At the completion of the program, GPSS/H creates a list file that has the same name as the original file but now has the extension .LIS. To view the program, you need to examine this file. This can be done using the same text editor that was used to create the file (or, if you are in the command prompt mode, simply by typing TYPE CHAP7A.LIS | MORE).

Depending on the editor you are using, the top portion of the .LIS file will look as shown in Figure 7.3.

Let us now examine this file. It is going to look strange at first, but you will soon become accustomed to interpreting the results.

Scrolling the file leads to a screen that will be similar to the next one (Figure 7.4).

The first line of interest to us will be the following one:

```
Relative Clock: 30.0000 Absolute Clock: 30.0000
```

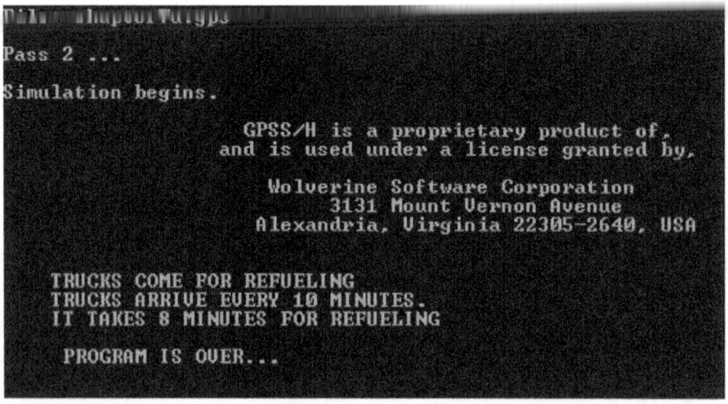

FIGURE 7.2
Screenshot after running program CHAP7A.GPS.

* If your system is running under Windows 7 or higher, skip to the next section.

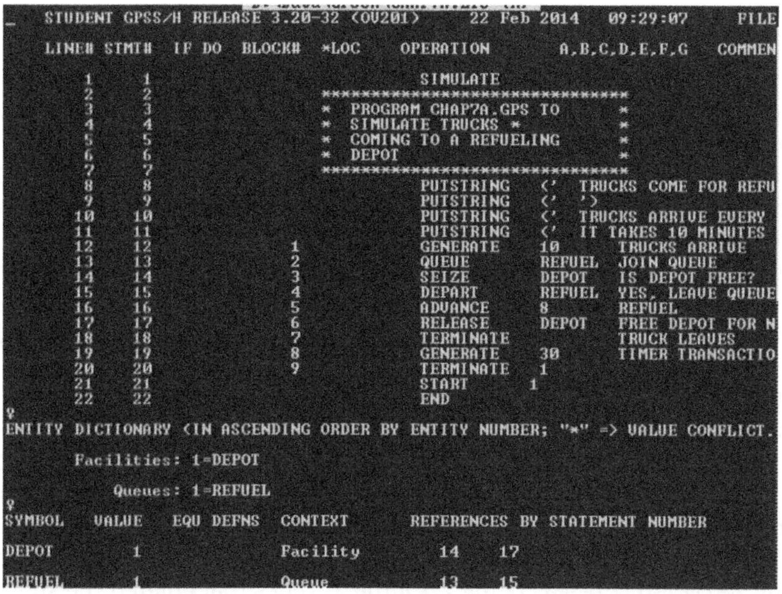

```
  STUDENT GPSS/H RELEASE 3.20-32 (OV201)     22 Feb 2014   09:29:07     FILE

    LINE# STMT#  IF DO  BLOCK#  *LOC     OPERATION       A,B,C,D,E,F,G   COMMEN

       1     1                            SIMULATE
       2     2                   ******************************
       3     3                   *  PROGRAM CHAP7A.GPS TO     *
       4     4                   *  SIMULATE TRUCKS *         *
       5     5                   *  COMING TO A REFUELING     *
       6     6                   *  DEPOT                     *
       7     7                   ******************************
       8     8                            PUTSTRING    ('  TRUCKS COME FOR REFU
       9     9                            PUTSTRING    ('  ')
      10    10                            PUTSTRING    ('  TRUCKS ARRIVE EVERY
      11    11                            PUTSTRING    ('  IT TAKES 10 MINUTES
      12    12              1             GENERATE     10      TRUCKS ARRIVE
      13    13              2             QUEUE        REFUEL  JOIN QUEUE
      14    14              3             SEIZE        DEPOT   IS DEPOT FREE?
      15    15              4             DEPART       REFUEL  YES, LEAVE QUEUE
      16    16              5             ADVANCE      8       REFUEL
      17    17              6             RELEASE      DEPOT   FREE DEPOT FOR N
      18    18              7             TERMINATE            TRUCK LEAVES
      19    19              8             GENERATE     30      TIMER TRANSACTIO
      20    20              9             TERMINATE    1
      21    21                            START      1
      22    22                            END
ENTITY DICTIONARY (IN ASCENDING ORDER BY ENTITY NUMBER; "*" => VALUE CONFLICT.

      Facilities: 1=DEPOT

          Queues: 1=REFUEL

SYMBOL     VALUE    EQU DEFNS   CONTEXT       REFERENCES BY STATEMENT NUMBER

DEPOT        1                  Facility          14     17

REFUEL       1                  Queue             13     15
```

FIGURE 7.3
Portion of output from program CHAP7A.GPS.

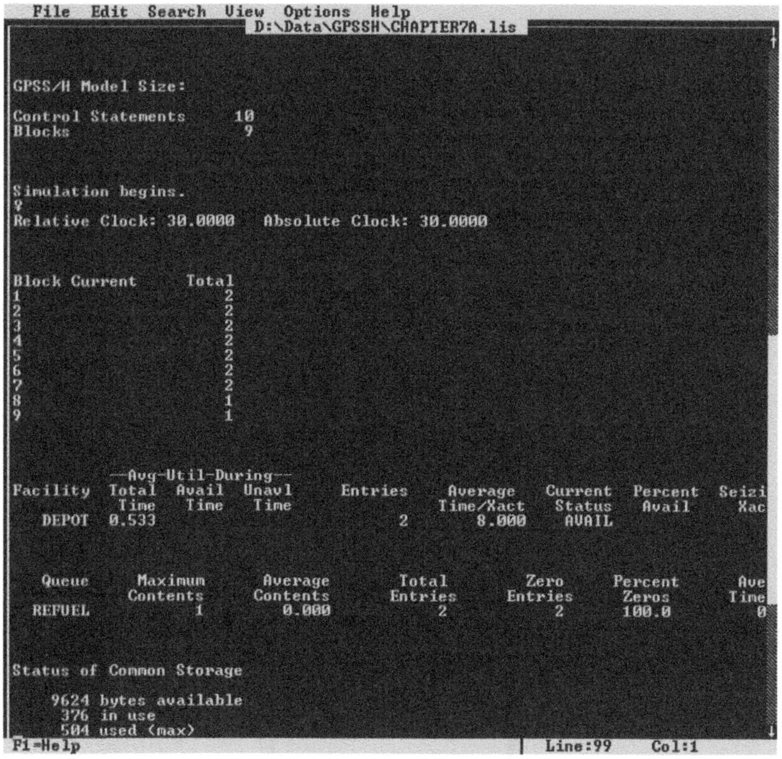

```
   File  Edit  Search  View  Options  Help
                              D:\Data\GPSSH\CHAPTER7A.lis

GPSS/H Model Size:

Control Statements      10
Blocks                   9


Simulation begins.

Relative Clock: 30.0000   Absolute Clock: 30.0000


Block Current      Total
1                      2
2                      2
3                      2
4                      2
5                      2
6                      2
7                      2
8                      1
9                      1


                 --Avg-Util-During--
Facility  Total  Avail  Unavl       Entries     Average   Current  Percent  Seizi
           Time   Time   Time                  Time/Xact   Status    Avail     Xac
   DEPOT  0.533                           2       8.000    AVAIL


    Queue     Maximum      Average       Total       Zero      Percent       Ave
             Contents     Contents      Entries     Entries      Zeros      Time
  REFUEL            1        0.000          2           2        100.0         0


Status of Common Storage

     9624 bytes available
      376 in use
      504 used (max)
F1=Help                                             | Line:99    Col:1
```

FIGURE 7.4
Portion of output file from program CHAP7A.GPS.

This indicates that the simulation went for 30 simulated time units. Notice that there are two imaginary clocks in GPSS/H. Both start at time $t = 0$. We shall learn how to run a simulation for a certain period, stop execution, and restart it with most, but not all, of the statistics set back to zero. When this is done, the absolute clock would keep going, but the relative clock is reset to zero each time the program is restarted.

The two Blocks that always have output associated with them are `QUEUE` and `SEIZE`. The lines of code associated with the `DEPOT` (called a *facility*) also give (among other things) the statistics of the utilization of the facility. This is the percentage of time that it was being used during all of the simulation. This is given as the number .533, which is interpreted as 53.3%. This means that the fuel depot was busy for this percentage of the time of the simulation (18/30). If the simulation was run for a much longer time, this figure would approach 80%. However, the fuel depot is free until the first truck arrives at time 10. It is then busy for 8 minutes. If the simulation were run for 1 hour, the utilization would be 66.67%; if run for 2 hours, it would be 73.33%; and so on.

Some of the statistics for the `QUEUE` Block are given next. Under `Maximum Contents`, we see that there was a maximum of one truck waiting. The `Total Entries` indicates that two trucks entered the queue. Even the first truck that entered the refuelling depot and went immediately to the refuelling area is counted. This truck is listed under the next heading `Zero Entries` as being 2. Of the 2 trucks that entered the shop, not 1 was delayed in the `QUEUE`, so the `Percent Zeros` is 100.0. There is more to the output file, but that will be covered later. For the present, we are learning how to construct and run simple GPSS/H programs.

7.7 Running Programs from Windows 7 and Higher

If you try to run GPSS/H programs from Windows 7 or higher, by just clicking on GPSSH.EXE, the program will run but all that you will see on the screen is a flash. The output will not be shown. Instead, one needs to use the command prompt as follows.

Click on the lower left button on the screen to bring up the menu that looks something like the following depending on your computer:

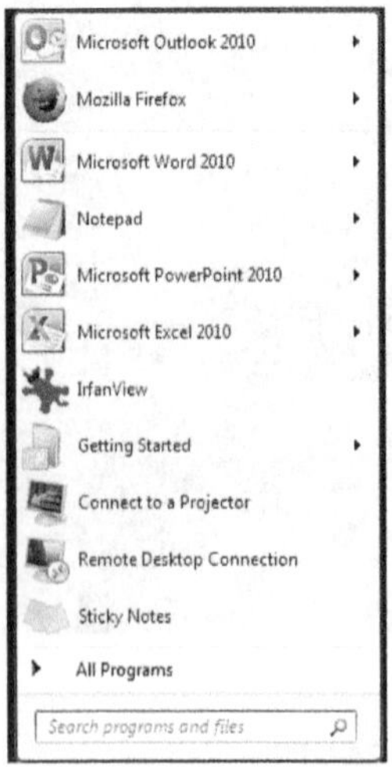

In the *search programs and file* box, key in the text CMD. This will bring up the following command prompt:

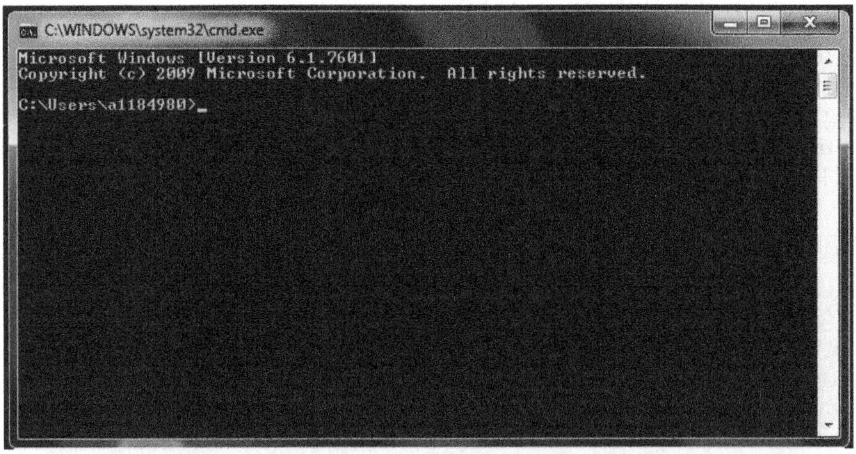

The screen view will depend on your system. For most computers, the default subdirectory for GPSS/H will be as follows: C:\WOLVERINE\GPSSH>. If this is the case, the next few lines can be omitted.

Assuming your subdirectory is not the Wolverine default one, you next need to switch to the subdirectory where GPSS/H is stored. For example, if the file GPSSH.EXE is in the subdirectory

U:\DATA\WOLV\GPSSH\

one would key in the following:

U: <enter>

This brings up the U: drive prompt:

U:\>

Then, key in CD DATA\WOLV\GPSSH <enter>

This will bring up a prompt on the screen as follows:

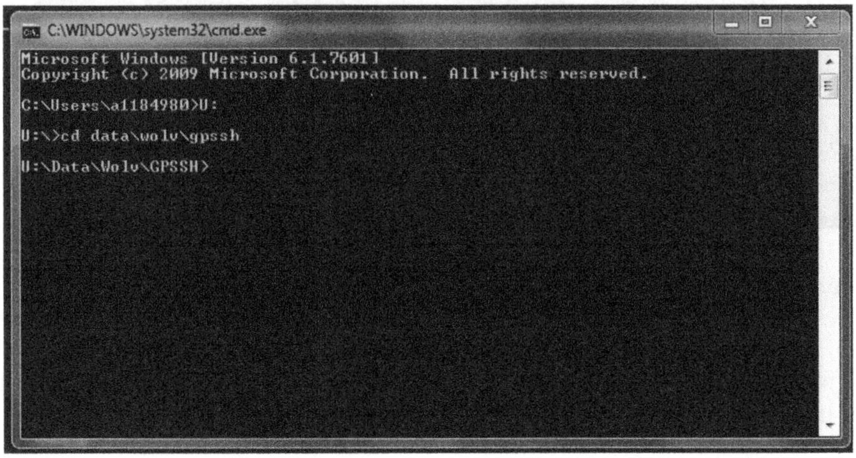

Again, the exact commands will depend on where GPSS/H is located on your particular computer.

Now, the command to run a GPSS/H program is as follows:

```
GPSS/H (name of program) <enter>
```

If the program to be run is not in the same subdirectory as where GPSSH.EXE is, one has to drag it using the mouse from where it is to the screen. For example, after the command GPSSH is keyed in, the screen will show the following:

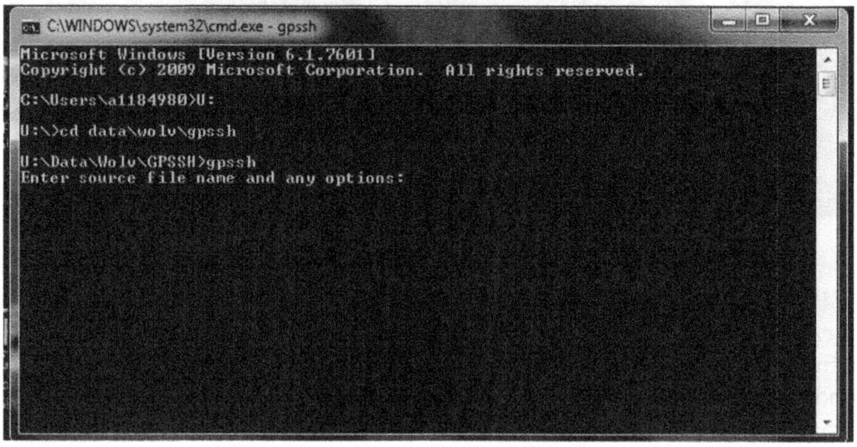

The file containing the GPSS/H program needs to be dragged to where it says

Enter source file name and any options:

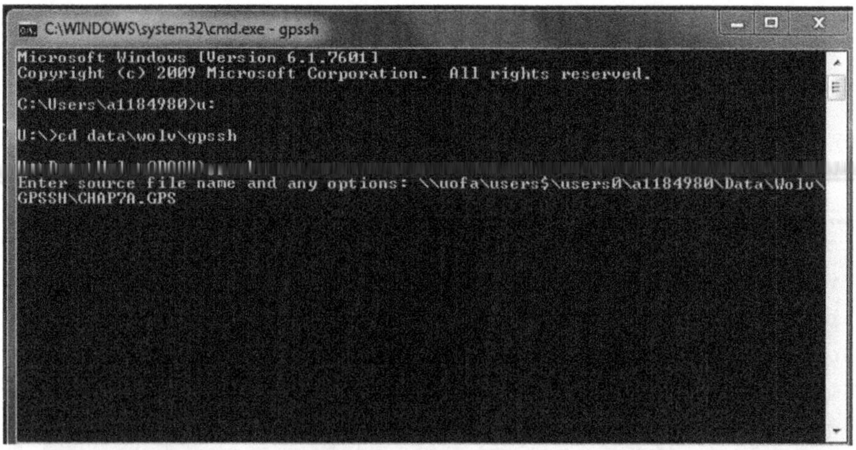

The cursor must be after the program CHAP7A.GPS and should be blinking. No spaces are to be in the path to the program. For example, if one had the subdirectory *u of a* instead of *uofa*, the program would not run. Now, when one hits the <enter> key the program will execute as was done previously. It may take a bit of practice to run the programs. It is suggested that the GPSS/H programs are stored in the same subdirectory as where GPSSH.EXE is.

7.8 Making Changes to a GPSS/H Program

The next example to be run shows how easy many GPSS/H programs are to change.

Suppose that the truck arrival times and service times were reversed, so that the trucks arrive every 8 minutes and the refuelling depot can refuel a truck in 10 minutes. Further assume, for the sake of the example, that the mine has an infinite supply of trucks. The modifications necessary to do this simulation are as follows:

Change the line

```
GENERATE 10
```

to

```
GENERATE 8
```

Change the line

```
ADVANCE 8
```

to

```
ADVANCE 10
```

Make these changes and see if you can interpret the results. This program is stored as CHAP7B.GPS. One thing to note from the file CHAP7B.LIS is that the utilization of the facility `DEPOT` is now .733 (Figure 7.5).

7.9 Giving Names in GPSS/H

Care must be taken when giving names to the labels and operands in GPSS/H statements. The label must be either a letter or a number, but not more than eight characters in length. It is possible to mix numbers and letters providing the first character is a letter. A few examples are

```
          SIMULATE
*******************************
*  PROGRAM CHAP7B.GPS         *
*  MODIFICATION OF CHAP7A.GPS *
*******************************
          GENERATE     8        TRUCKS ARRIVE
          QUEUE        REFUEL   JOIN QUEUE
          SEIZE        DEPOT    IS DEPOT FREE?
          DEPART       REFUEL   YES, LEAVE QUEUE
          ADVANCE      10       REFUEL
          RELEASE      DEPOT    FREE DEPOT FOR NEXT TRUCK
          TERMINATE             TRUCK LEAVES
          GENERATE     30       TIMER TRANSACTION FOR 30 MINUTES
          TERMINATE    1
          START        1
          END
```

FIGURE 7.5
Listing of program CHAP7B.GPS.

```
BIGTRUCK
JOEANNE
BILLYBUD
UPTOP
DOWN1
BACK1
UPTOP7
A123
```

but not

```
1JOE
B23$K
1
J&B
```

As in other languages, GPSS/H has reserved words that only can be used for their special meanings. Some of these are `END`, `START`, `DO`, `GOTO`, `IF`, `PRINT`, and `WRITE`. In general, do not use names that may be reserved words.

If you use a label for a Block that is never again referred to, there will be a warning message on the screen. For example, suppose the line of code in the program CHAP7A.GPS was

```
MYLABEL GENERATE 10
```

This would not result in a run time error but would cause a warning message to be printed. Do not be alarmed; this is just the way GPSS/H works. The warning message will look like the following:

*** WARNING: THE FOLLOWING ENTITIES HAVE BEEN DEFINED BUT NOT EXPLICITLY REFER
Blocks: MYLABEL

Once you give a Block a label, GPSS/H expects you to refer to it in the main program. It looks a bit messy but messages such as this can be a real help in debugging long programs.

7.10 Another Example

Even though we still have a way to go in writing our programs, we can use the program CHAP7A.GPS to learn a bit more about GPSS/H. From now on, we will speak of the *processor* when we refer to how the program is being executed. To simulate the same shop for the refuelling working a bit faster, simply change the line

```
ADVANCE 8
```

to

```
ADVANCE 5
```

Now, the trucks are being refuelled in 5 minutes.

To simulate for 2 hours, change the line

```
START 1
```

to

```
START 4
```

By now, you should be able to understand the results of these changes to the simulation.

7.11 Free Format

Since GPSS/H programs are now written on the screens of PCs, the restrictions that apply to fixed format have been relaxed. The way GPSS/H statements can now be typed is as follows:

1. Labels can begin in any position as long as the operand starts before position 26.
2. If no label is used, the operation portion can start in column 3 (or later).
3. Operands can start after just one space following the operation.*
4. Comments are placed in the program just as for fixed format.
5. Statements may be continued to another line as in fixed format.
6. No operand can start after column 25. (See footnote to rule 3.)

Thus, it is possible to return to our first program and type it as shown in Figure 7.6.

```
            SIMULATE
*******************************
*   PROGRAM CHAP7A.GPS TO      *
*   SIMULATE TRUCKS            *
*   COMING TO A REFUELING      *
*       DEPOT                  *
*******************************
 TRUCK     GENERATE      10       TRUCKS ARRIVE
             QUEUE     REFUEL   JOIN QUEUE
     SEIZE        DEPOT    IS DEPOT FREE?
            DEPART       REFUEL   YES, LEAVE QUEUE
         ADVANCE     8         REFUEL
            RELEASE      DEPOT    FREE DEPOT FOR NEXT TRUCK
        TERMINATE            TRUCK LEAVES
            GENERATE     30       TIMER TRANSACTION FOR 30 MINUTES
            TERMINATE    1
         START       1
            END
```

FIGURE 7.6
Listing of program in free format (not recommended).

* There is a statement, OPERCOL, that overwrites this. If you had OPERCOL 30, then the operands can start up to and including position 30. This is rarely used.

This would run without a compiling error (except for a warning message). There really is no advantage in writing programs as above. The programs presented in this book will all be written in fixed format for sake of uniformity. If you choose to write your programs in free format, it is perfectly acceptable.

7.12 The `SIMULATE` Statement

Every GPSS/H program must contain a `SIMULATE` statement. Although it need not be the first statement in the program, it usually is. Most programmers always start their programs with the `SIMULATE` statement, although it can be any place in the program. The general form of it is as follows:

```
SIMULATE      n
```

The operand `n` is optional. If used, it limits the amount of time in minutes that the program will run. This can be handy to use in the debugging stage to avoid infinite loops. For example,

```
SIMULATE      2.5
```

would limit the program to 2.5 minutes running time. It is possible to have the time limited to so many seconds if the letter S is placed after the `SIMULATE` operand:

```
SIMULATE      100S
```

would limit the execution time to 100 seconds.

There is a caution with this statement. Be careful that you do not put a comment after it. If you do, the comment must start after position 25.

7.13 The `END` Statement

GPSS/H programs must have an `END` statement. This is simply the last statement in the program and acts as a directive to the compiler. *Any lines of code or text after the END statement are ignored.* As we shall see, GPSS/H programs are often written in different segments. If a program needs debugging and only a portion of the whole program needs to be examined, one can sometimes move the working and debugged segments to after the `END` statement until they need to be inserted back in the main program.

Of all the many Blocks and statements encountered with GPSS/H, `SIMULATE` and `END` should never cause you any problems.

7.14 Further Example

Let us do the example of the refuelling shop again. Do not despair: soon we shall be doing other, more interesting problems. In the last example, you were told to run the program for 2 hours of simulated time by changing the `START 1` to `START 4`. The following program will simulate the refuelling depot for 10 trucks being refuelled. Write this program and run it.

```
SIMULATE
GENERATE      10
QUEUE         REFUEL
SEIZE         DEPOT
DEPART        REFUEL
ADVANCE       8
RELEASE       DEPOT
TERMINATE     1
START         10
END
```

After you run the program, you should examine the output to see the program results. After a few more exercises, you will understand the differences between the programs.

7.15 Exercises

7.1. Trucks in a mine pass point A every 5 minutes. They travel to point B in 8 minutes. A single inspector there takes 4 minutes to do the inspection. Only one truck can be inspected at a time. Next, the trucks are moved to a single repair bay in one minute. These repairs take 3 minutes to do. The trucks then continue back to the mine. Do a *hand simulation* for the first 20 minutes of operation starting at time 0.

Your result should show the event, the time it takes place, and a description of what is being done. The first part of your solution will look as follows:

Event	Time	Description
1	5	First truck comes to point A, next truck will come at time 10
2	5	First truck scheduled to arrive at B at time 13
3	10	Second truck comes to A, next truck will arrive at time 15
4	13	First truck arrives for inspection, will finish at time 14

NOTE: The listing of the GPSS/H program to do the simulation is given next. The program is also stored as CHAP7C.GPS (Figure 7.7).

Run the above program and interpret the results as much as possible from the .LIS file.

7.2. Run the program in Exercise 7.1 for 8 hours. In order to run the program for 8 hours (480 minutes), it is necessary to change the line of code:

```
GENERATE 20
```

to

```
GENERATE 480
```

```
          SIMULATE
***************************************
*   PROGRAM CHAP7C.GPS                *
*   TRUCKS IN A MINE BEING INSPECTED  *
*   TRUCKS ARRIVE AT POINT A          *
***************************************
          GENERATE    5      TRUCKS ARRIVE AT A  EVERY 5 MINUTES
          ADVANCE     8      TRAVEL TO POINT B
          QUEUE       WAIT1  JOIN FIRST QUEUE
          SEIZE       INSP   IS INSPECTOR FREE?
          DEPART      WAIT1  LEAVE FIRST QUEUE
          ADVANCE     4      TRUCKS INSPECTED
          RELEASE     INSP   FREE THE INSPECTOR
          ADVANCE     1      MOVE TO REPAIR BAY
          QUEUE       WAIT2  JOIN SECOND QUEUE
          SEIZE       REPAIR IS REPAIR BAY FREE?
          DEPART      WAIT2  LEAVE SECOND QUEUE
          ADVANCE     3      MAKE REPAIRS
          RELEASE     REPAIR FREE REPAIR BAY
          TERMINATE          RETURN TO MINE
          GENERATE    20     TIMER TRANSACTION
          TERMINATE   1
          START       1
          END
```

FIGURE 7.7
Listing of program CHAP7C.GPS.

Make this change and rerun the program.

7.3. Trucks in a mine are loaded by a single shovel and travel to A in a mine. At this junction, 65% of the trucks travel to the crusher; the rest travel to the waste dump. Only one truck can dump into the crusher at a time, but no such restriction applies to the waste dump. After dumping either ore or waste, all the trucks return to a single shovel. The system operates for a single shift of 8 hours of work.

What sort of information do you think this model might give you?

7.4. A shipping yard has only three docks for unloading and loading ships. There are three types of ships that come to this yard. These are types A, B, and C: 35% are of type A, 50% type B, and the rest are type C. All ships wait at the harbour entrance until they can be taken by a single tug to berth. If there is no dock available, the ship also must wait. The single tug is used for both berthing and unberthing. Write a flowchart for this system.

7.5. Widgets come along a conveyor belt at random times. They are first sent to a single forming machine and next to a moulding machine (only one machine available); then they go to one of two identical stamping machines (whichever is free). Finally, a single painting machine paints each. From there, a single worker inspects each; 5% are thrown away, 15% are sent back to the beginning for reworking, and the rest are sent (some place) to be boxed. Sketch this system and draw a flowchart. What information might be obtained from this? (In other words, why would someone want such a study?)

7.6. Suppose that a system consists of two identical workers who take 18 ± 5 minutes to work on a part. These parts come along every 10 ± 4 minutes starting at zero. The following data is from a GPSS/H program that gives the arrival time for each part and the time each will be worked on once a worker is free. This data is for the first 50 parts to come along and leave the system. Make a table that can be used as a detailed report for the system as it is working. Your report should show the following:

a. Utilization of workers as a per cent of total time
b. The maximum queue at any one time
c. The length of time that the simulation ran
d. The average number of parts in the queue at any one time

The data is as follows:

Part #	Arrival Time	Time to Finish Part	Part #	Arrival Time	Time to Finish Part
1	0.00	20.40	26	268.09	19.01
2	9.99	20.68	27	279.10	17.87
3	16.96	20.92	28	287.96	13.77
4	29.84	14.86	29	301.60	21.84
5	36.36	20.34	30	312.69	13.76
6	49.48	19.84	31	325.53	20.79
7	61.94	19.91	32	336.31	21.14
8	73.46	13.98	33	344.94	19.42
9	86.07	16.16	34	355.47	21.80
10	96.63	19.27	35	364.25	19.86
11	110.20	22.12	36	374.15	22.65
12	123.93	19.65	37	386.54	22.93
13	137.70	15.80	38	398.96	22.75
14	149.91	21.42	39	412.63	19.82
15	160.96	17.14	40	424.88	20.34
16	171.03	19.77	41	438.01	17.33
17	178.65	16.64	42	451.96	16.02
18	186.00	18.66	43	464.43	22.35
19	198.19	13.88	44	477.06	19.13
20	208.92	15.42	45	486.18	22.75
21	219.39	15.39	46	494.04	15.52
22	225.98	16.96	47	500.85	13.53
23	239.87	14.10	48	507.86	21.75
24	247.51	16.97	49	516.85	22.10
25	259.47	17.50	50	523.09	16.03

Your table should begin with headings such as the following:

Time	Event	Time to Work On	Number Being Fixed	Number in Queue	Time Finished
0	First part	0.00	1	0	20.48
9.99	Second part	20.68	2	0	30.61
16.96	Third part	20.92	2	1	41.40
...		...	...		...

```
          SIMULATE
***************************************
*    PROGRAM CHAP7D.GPS                *
*    PROGRAM TO SIMULATE EXERCISE 7.6  *
*    PARTS COME TO TWO WORKERS         *
***************************************
          STORAGE     S(WORKERS),2
          GENERATE    10,4       PARTS ARRIVE
          QUEUE       WAIT       JOIN QUEUE
          ENTER       WORKERS    IS EITHER WORKER FREE?
          DEPART      WAIT       LEAVE THE QUEUE
          ADVANCE     18,5       WORK ON A PART
          LEAVE       WORKERS    FREE THE WORKER
          TERMINATE   1          PARTS DONE
          START       1000
          END
```

FIGURE 7.8
Listing of program CHAP7D.GPS.

This will take some time to finish.

The actual results from running a GPSS/H program are given next. The program was run for 1000 parts being finished.

Utilization of both workers	90.3%
Maximum content of queue	3
Average queue content	0.265

The program ran for 100,025.72 time units. This is not unexpected. There were 10,000 parts to be worked on, and they arrived at an average of 10 time units. The mean working time is 18. This program is stored as CHAP7D.GPS. The GPSS/H program took only a second or less to run. The code is surprisingly short. Before too long, the meaning of each line of code will become clear. Figure 7.8 gives the program listing.

If the result of running the program for only 50 parts being finished is desired, the line of code that says

```
START    1000
```

needs to be changed to

```
START    50
```

8

The GENERATE Block and Transactions

Coal loading facility, El Diablo Mine near Maracaibo, Venezuela.

8.1 The GENERATE Block

The key to any simulation program is the GENERATE Block. This Block creates *transactions* that move through the program from Block to Block. These transactions can represent anything that you wish to model, such as ships entering a harbour, trucks in a mine, ore on a conveyor belt, cars on a highway, and people entering a bank. To give an idea of how a transaction is used, think of a truck arriving at a shovel where it is to be loaded with ore. The truck is the transaction. It will arrive at the shovel and then, once the shovel is free, it will spot and wait while the shovel loads it. The sequence might be as follows:

1. The shovel is free, so the truck spots and the loading begins.
2. The shovel is busy, so the truck waits until it is free.
3. Once loaded, the truck leaves the shovel, and another truck can commence loading.

We shall learn soon how to simulate the above situations with GPSS/H.

A system to be studied normally will involve the passage of time. An imaginary clock is used by the GPSS/H processor. This will be discussed next. GPSS/H programs belong to a class of computer programs that are called *non-procedural.* This means that they do not run sequentially from one line of code to the next, but their execution will depend on some other factor. Thus, a GPSS/H program may run according to the commands in the first five Blocks in sequence and then *jump* to the command on lines 21 through 25 and then back to line 6. What controls this sequence is an internal clock and what is called the *processor* that determines the execution of the program. This may seem complex at first, but remember that GPSS/H is designed to solve real-life problems. Imagine that you have all of the steps to model a single mine shovel written in a sequence. If you tried to explain how the shovel worked by referring to these steps, you sometimes would have to *jump* from one step to another that is not in sequence. Imagine that the steps are as follows:

1. Truck arrives at shovel
2. Truck checks to see if shovel is free
3. If free, truck spots
4. If the shovel is not free, truck will wait until it is free
5. A truck is done being loaded
6. Loaded truck leaves
7. Next truck spots

You will notice that if you tried to explain the working of the shovel during any time interval, the above sequence is not followed step by step. For example, suppose that it is time 100.000 and there is a truck being loaded and two trucks waiting. One of several things can happen as time passes:

1. The truck at the shovel may finish
2. Another truck may arrive at the shovel

If item 1 happens first, the program proceeds to step 3. If item 2 happens, the arriving truck proceeds to step 4. At this point, time may pass until something happens at the shovel. Now suppose that the truck being loaded is finished. Control passes to step 5. The finished truck will then leave the shovel (step 6) and frees the shovel to work on a waiting truck if there is one. If there is no waiting truck, the shovel is idle until one arrives.

8.2 The Internal GPSS/H Clock

GPSS/H uses an internal clock that starts at zero when the program begins execution. The processor moves the clock forward in time as the program is executed.

These time units can represent seconds, minutes, hours, one-tenth of a minute, and so on, depending on what the programmer chooses them to represent. The programmer selects the time unit appropriate to the problem being studied. In some cases, a time unit, called the *basic time unit*, of 1 second may be selected; in others, the basic time unit may be 1 hour. The clock in GPSS/H advances from event to event taking place in the simulation. For example, if some event is to take place at time $t = 345.765$ and the next one at $t = 420.511$, the internal clock will be advanced from where it is to time 345.765, execute the commands between time 345.765 and time 420.511, and then will advance to time 420.5.

If a shop is being analysed and time studies indicate that customers take 18 minutes to shop and 2 minutes to check out, the basic time unit may be taken as 1 minute. If the simulation is to run for 8 hours, then the 8 hours need to be converted to the basic time unit or 480 time units. This basic time unit is generally obvious from the statement of the problem being modelled. In most mining operations, the basic time unit is taken to be 1 minute.

8.3 Creating Transactions

Transactions are created by the `GENERATE` Block. Every GPSS/H program will have a `GENERATE` Block. This Block can have up to nine operands and can be quite involved. The simplest form of it is given by the following:

```
(label) GENERATE A
```

where A is either a positive integer or can be a variable (we will learn what forms these can have later). The operand A gives the times at which a transaction will be created.

The (label) is optional. We will learn when one is used. For the present, it is not needed.

For example,

```
(a) GENERATE 5
(b) GENERATE 100
```

In (a), a transaction is created every 5 time units, whereas in (b), a transaction is created every 100 time units. In both cases, as long as the simulation is taking place, *transactions will continue to be created every 5 or 100 time units.* What actually happens is that the first transaction to be created by a `GENERATE` Block is scheduled to leave at its first time during compiling. Once it leaves, the second transaction is then scheduled to leave. Thus, for the above two examples, during compiling, transactions are scheduled to enter the system at times 5 and 100. After a transaction enters the system at these times, another transaction is then scheduled to enter the system. In (a), the second transaction is scheduled to leave at time 10 and in (b), the second transaction is scheduled to leave at time 200.

An example of how the `GENERATE` Block creates transactions can be seen in the animation in Figure 8.1. This shows a miner being created every 12 time units and placed near the centre of the screen. After 9 time units pass, the miner is removed from the animation. This continues until 10 miners have been created and removed. The computer program to generate the .ATF file is found as CHAP8A.GPS.

An examination of this shows how the `GENERATE` Block creates a transaction every 12 time units. Nothing happens with the animation until time 12 as the `GENERATE` Block is as follows:

```
GENERATE      12
```

A screenshot of the animation is given in Figure 8.1.

A more fanciful example of the GENERATE Block is shown in the animation given by CHAP8B. This shows pieces of ore (transactions) that come along every 20 time units and fall into a pool of water. The program to make the animation file is CHAP8B.GPS. As you run this animation, you will note that in every 20 time units, pieces of ore (transactions) are created. The first piece of ore enters the system at time 20. It moves along a conveyor belt and then is dropped into a pond. We will shortly learn how this can be changed to have the first transaction enter the system at any time, such as time 0. The animation is shown in Figure 8.2.

FIGURE 8.1
Screenshot of animation as given by program CHAP8A.

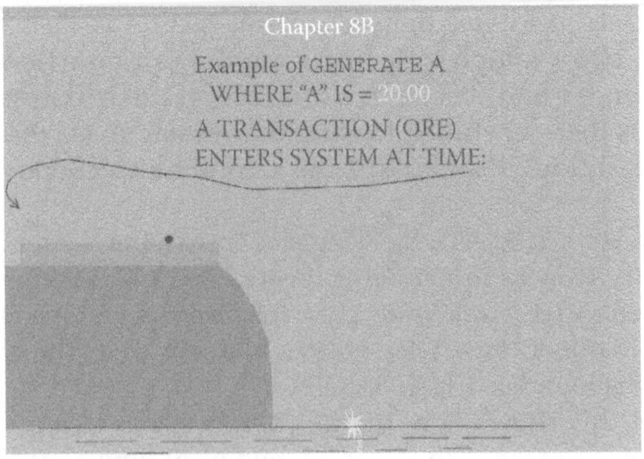

FIGURE 8.2
Animation as given by program CHAP8B.GPS.

In general, once the GPSS/H program is running, the processor will move a transaction from Block to Block in a sequential manner starting at time 0. It will move the transaction in this manner unless one of three things happens. These things will be introduced shortly.

During compiling, transactions are scheduled to enter the system depending on the GENERATE Blocks. For example, suppose a program has the following GENERATE Blocks:

```
GENERATE 12
...
...
GENERATE 6.5
...
...
GENERATE 480
```

Transactions will be scheduled to enter the system at times 12, 6.5, and 480. If these are the only GENERATE Blocks in the program, once the program starts execution, the first transaction to be moved is the one at time 6.5 even though the GENERATE Block may be located somewhere in the middle of the code.

In most systems to be studied, there is a degree of randomness involved. Trucks do not arrive at a shovel every 35 seconds, a crusher does not take exactly 1.2 minutes to crush a load of ore, and so on. Thus, it is necessary to have randomness as a part of the simulation. One of the features of GPSS/H is that it is so easy to incorporate randomness into the program. One way to have randomness in the generation of transactions is given next.

8.4 The B Operand

The second form of the GENERATE Block uses the B operand:

```
GENERATE            A,B
```

B can be a variable, but, for the present, it will be a positive integer. The above will generate a transaction over the interval A ± B, with each time having equal probability of happening. This is called the *uniform distribution* with mean A and spread of B.

```
GENERATE            8,2
```

means that a transaction is created by sampling from the distribution given by 8 ± 2 time units. This means that a transaction will be created at a time from 6 to 9.99999, each having an equal probability of occurrence. This interval is written as [6.0000, 10.0000). The brackets [a, b) indicate that the times go from 6.0000 to 9.999999.* This means that, if the internal clock is at t = 1013.000000, and the GPSS/H processor sets t = 6.000000 as the time before the next transaction is created, the next transaction will

* The number of decimal points used in GPSS/H depends on the computer you are using. For mining work, four decimal points will normally be sufficient. At times, six may be used.

enter the system at simulated time $t = 1019.000000$. We will learn later how to generate transactions that enter the system according to *any* statistical distribution.

The animation seen in CHAP8C is an example that illustrates how the `GENERATE A, B` Block works. This animation shows a ball that is created every 20 ± 8.25 time units. This ball moves across the screen from left to right in 30 time units and then is removed from the animation. The time that a ball is created is given on the screen. The time that it will be removed is also given. This is always the time of creation plus 30. You will be asked in one of the exercises to modify the program to change this for a random time. Figure 8.3 is a screenshot of the animation.

When you run this example, you will create an output file that gives the times that a ball is created and the time it will be destroyed. This file is called CHAP8C.OUT.

Output from this is shown in Figure 8.4 (What happened to BALL 2? This transaction was indeed created, but this was done during compiling and has to do with the second `GENERATE` Block.):

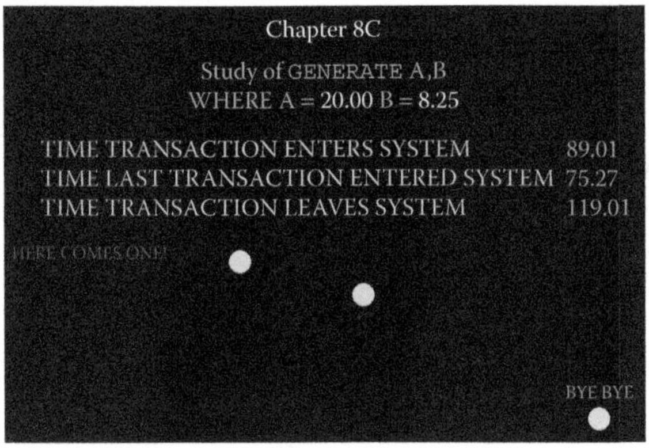

FIGURE 8.3
Screenshot of animation as given by program CHAP8C.

```
BALL NUMBER  1 IS CREATED AT TIME  16.21
BALL NUMBER  3 IS CREATED AT TIME  36.20
BALL  1 IS DESTROYED AT TIME  52.21
BALL NUMBER  4 IS CREATED AT TIME  60.16
BALL  3 IS DESTROYED AT TIME  72.20
BALL NUMBER  5 IS CREATED AT TIME  75.27
BALL NUMBER  6 IS CREATED AT TIME  89.01
BALL  4 IS DESTROYED AT TIME  96.16
BALL  5 IS DESTROYED AT TIME 111.27
BALL NUMBER  7 IS CREATED AT TIME 113.44
BALL  6 IS DESTROYED AT TIME 125.01
BALL NUMBER  8 IS CREATED AT TIME 132.46
BALL  7 IS DESTROYED AT TIME 149.44
BALL NUMBER  9 IS CREATED AT TIME 158.39
BALL  8 IS DESTROYED AT TIME 168.46
BALL NUMBER 10 IS CREATED AT TIME 183.21
BALL  9 IS DESTROYED AT TIME 194.39
BALL NUMBER 11 IS CREATED AT TIME 196.49
```

FIGURE 8.4
Output from CHAP8C.GPS.

Remember that the above times arise from sampling the distribution 20 ± 8.25. The output should be carefully studied to understand how the program CHAP8C.GPS works.

In using the GENERATE Block one must be careful to avoid a Block such as

```
GENERATE            10,12
```

as this would eventually lead to an attempt to generate a transaction at a negative time, which is not allowed. However, it is possible to have

```
GENERATE            10,10
```

8.5 A Word about Random Numbers

Although it is not of concern how the processor works, you might be curious how the processor might generate the various times using random numbers. The processor has a built-in random number generator that we will be referring to from time to time. Suppose you want to generate times from 10 to 24 with equal probability. Call these *X*. The random number will be called RN, where RN is a number from 0 to 1. Now consider the following formula:

$$X = 10 + \text{RN} * 14$$

Every time the random number is called up, a new value of *X* is obtained. As can be seen, a stream of times between 10 and 24 will be obtained. This is similar to the way that the GPSS/H processor works.

Before continuing with the GENERATE Block, let us do another exercise. By now, a GPSS/H program should be starting to make a bit of sense.

Example 8.1: Truck Arrives at an Inspection Station

A truck arrives at a single inspector's station every 15 ± 6 minutes. The inspector needs to check the tyres, and it takes him 14 ± 8 minutes to measure and record the tyre pressure. Simulate for 20 trucks having their tyres inspected. Assume that the inspector works continuously; that is, he does not leave until 20 trucks are finished. Determine how busy the inspector has been.

The program is CHAP8D.GPS. A listing of the program to this is shown in Figure 8.5.

The modified program to make the animation is stored as CHAP8E.GPS.* The animation shows the trucks arriving and then obtaining service.

* The observant student will notice that the loading times in this program have been changed from the original times. This is so that the animation will show the loaded trucks being placed on the travel paths, being delayed a short time and then the waiting trucks, if any, being positioned to load. Otherwise, any waiting trucks will immediately replace a loaded truck. In an actual mining situation, one would have different times for spotting after a loaded truck leaves the loader. This is just one more reason why animation is so important for a simulation model.

```
                SIMULATE
***************************************
*  PROGRAM  CHAP8D.GPS                *
*  EXAMPLE OF TIRE INSPECTOR IN A MINE *
***************************************
                GENERATE     15,6        TRUCKS ARRIVE
                QUEUE        WAIT        IS INSPECTOR FREE?
                SEIZE        INSPECT     YES, ENGAGE INSPECTOR
                DEPART       WAIT        LEAVE THE WAITING QUEUE
                ADVANCE      14,8        TIRE INSPECTION
                RELEASE      INSPECT     FREE THE INSPECTOR
                TERMINATE    1           TRUCK LEAVES
                START        20
                END
```

FIGURE 8.5
Program listing of CHAP8D.GPS.

8.6 Results of Simulation Programs

Do not be too concerned at this time that you do not understand all the code given above. In fact, even interpreting the results will appear strange at this time. However, if you successfully run the program and look at the list file created by GPSS/H, you will see the following as a part of the output (Figure 8.6):

```
BLOCK CURRENT      TOTAL
1                   21
2           1       21
3                   20
4                   20
5                   20
6                   20
7                   20
```

The above is interpreted as follows: The inspector worked until 20 trucks had been inspected. He was busy 94.0% of the time. This result is to be expected. Trucks arrive at

```
  --AVG-UTIL-DURING--
  FACILITY  TOTAL  AVAIL  UNAVL     ENTRIES    AVERAGE    CURRENT   PERCENT
  SEIZING  PREEMPTING
                 TIME    TIME    TIME              TIME/XACT    STATUS    AVAIL
  XACT        XACT
   INSPECT  0.926                        20        14.942      AVAIL

     QUEUE       MAXIMUM        AVERAGE        TOTAL          ZERO        PERCENT
  AVERAGE        $AVERAGE       QTABLE      CURRENT
                CONTENTS       CONTENTS        ENTRIES      ENTRIES        ZEROS
  TIME/UNIT     TIME/UNIT        NUMBER      CONTENTS
       WAIT              2        0.405            21             7         33.3
  6.219          9.329                          1
```

FIGURE 8.6
Portion of output from program CHAP8D.GPS.

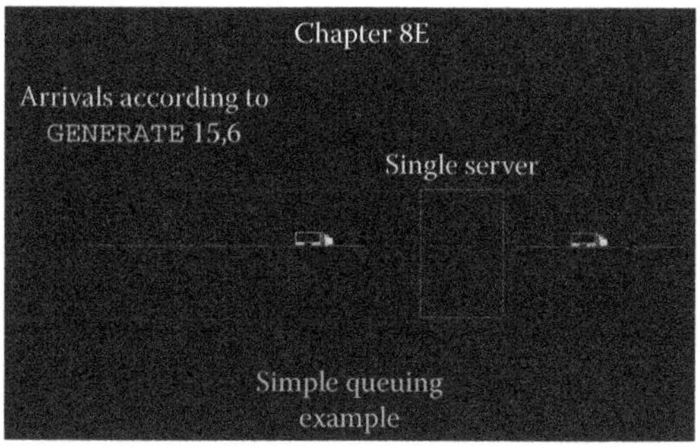

FIGURE 8.7
Screenshot of animation as given by program CHAP8E.

random but with an interarrival time of 15 minutes. The inspection takes an average of 14 minutes. Hence, in the long run, one would expect the inspector to be busy for 14/15 or 93.33% of the time.

When the 20th truck was done, the program was finished (Figure 8.7).

Example 8.2: Example of GENERATE A,B

To further illustrate the GENERATE Block, consider the program and animation given by CHAP8F. This can be used to study the GENERATE Block:

```
GENERATE           10,2
```

Here a transaction is generated according to the uniform distribution with a mean of 10 and a spread of 2 time units. Each transaction represents a truck. Each truck will appear on the screen for 6 time units. The animation shows the truck appearing on the screen, gives the time when the truck was created, and the time when the truck will leave the animation. The speed of the animation is set to 1 (normal is 6), so that the viewer can better observe the creation and removal of the truck. Notice that the first truck to be created arrives at time 9.08. In the next section, we learn how the first truck (transaction) can be made to arrive in the animation at any specified time. The program listing is given in Figure 8.8.

The animation of this example is given in Figure 8.9.

8.7 More General Cases of the GENERATE Block

There are other forms of the GENERATE Block. The ones we will use at present have the operands as shown below:

```
GENERATE           A,B,C,D,E
```

The C, D, and E operands do the following:

C is called the *offset time* for the first transaction. No transaction will enter the system until this time.

```
SIMULATE
******************************
*  PROGRAM CHAP8F.GPS         *
*  ANIMATION OF TRUCKS TO     *
*  APPEAR ON SCREEN FOR 6     *
*  TIME UNITS.                *
******************************
   ATF     FILEDEF     'CHAP8F.OUT'
           GENERATE    10,2  TRUCK APPEAR
           BPUTPIC     FILE=ATF,LINES=6,AC1,AC1,AC1,AC1+6,XID1,XID1
TIME *.****
WRITE M1 **.**
WRTIE M2 **.**
WRITE M3 **.**
CREATE TRUCK T*
PLACE T* AT 65 30
           ADVANCE     6      WAIT FOR 6 TIME UNITS
           BPUTPIC     FILE=ATF,LINES=5,AC1,XID1
TIME *.****
WRITE M1
WRITE M2
WRITE M3
DESTROY T*
           TERMINATE          TRUCK LEAVES
           GENERATE    200
           TERMINATE   1
           START       1
           PUTPIC     FILE=ATF,LINES=2,AC1
TIME *.****
END
            END
```

FIGURE 8.8
Listing of program CHAP8F.GPS.

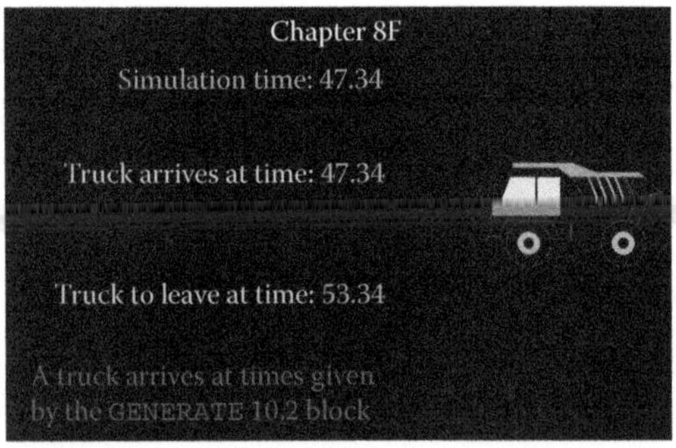

FIGURE 8.9
Screenshot of animation as given by CHAP8F.

Some examples are as follows:

```
GENERATE 1,,0
```

will schedule a transaction to enter the system at time 0. The next transaction will enter at time 1, the next at time 2, and so on.

```
GENERATE 10,5,100
```

will schedule transactions to enter the system according to the distribution 10 ± 5. The first one will enter the system at time 100.

D is the maximum number of transactions to be generated.

```
GENERATE            10,,,2
```

will generate transactions every 10 time units but only two will be generated. The first will enter the system at time 10 and second will enter at time 20.

E is called the *priority*. This can be very useful.

```
GENERATE            ,,,5,2
```

will generate five transactions at time 0, each will have a priority of 2.

When the E operand is used, the transactions are given a priority level specified by it. Often this operand is omitted and the priority level is, by default, 0. The priority levels are integers that can be from –2,147,483,632 to +2,147,483,632. In practice, only a few priorities are needed, and so one normally uses priorities such as 1, 2, 5, and 10.

If one transaction has a higher priority than another, it is placed ahead of it in queues as well as given preferential service in case of a time tie between transactions. In most queuing systems, the service criterion is known as *First in-First out*. This means that if the first person to arrive for service has to wait, he or she will be served before later arrivals. A typical example to illustrate the case of a time tie is the situation when a car arrives at a petrol station. Suppose that there is only room for six cars total and, if there are six cars at the station, an arriving car will leave. A time tie occurs when a car arrives at exactly the same time as one is through being serviced. The arriving one will not leave if it has a higher priority than the one that is about to leave after being serviced. This, of course, is what will happen in a real-life situation. We will examine the concept of priority later.

Several examples of the GENERATE Block are considered next.

```
(a) GENERATE        15,3,100,3,1
(b) GENERATE        100,3,200,400,3
(c) GENERATE        20,4,500,7,8
(d) GENERATE        30,0,0
```

In (a), a transaction is created every 15 ± 3 time units. The first does not enter the system until time $t = 100$. Only three of these transactions will be created. Each will have a priority level of 1. In (b), a transaction is created every 100 ± 3 time units. The first enters the system at $t = 200$ and only 400 of these transactions will be created, each having a priority of 3.

In (c), transactions are created every 20 ± 4 time units. The first enters the system at $t = 500$. Only seven such transactions are created, each with priority 8.

In (d), transactions are created every 30 time units. The first will enter the system at time 0. Contrast this with the following Block:

```
GENERATE            30
```

Here, the first transaction will enter the system at time 30, not at time 0.

If you do not wish to use all the operands, you put commas in their place—*not* blanks; that is,

```
GENERATE            100,,,1
```

will generate a single transaction at $t = 100$.

```
GENERATE              100,,,,1
```

will generate a transaction every 100 time units each having a priority level 1.

```
GENERATE              ,,,5
```

will generate five transactions immediately. This may seem a strange thing to do, but this will be very important in many of our simulations. The above is the same as:

```
GENERATE              0,0,0,5
```

If you want to study a program that will have six ships sailing from one port to another, you might start the simulation using the following Block:

```
GENERATE              ,,,6
```

This puts six ships into the system at time $t = 0$.

In the case of a GENERATE Block with no operands, an infinite amount of transactions are created at time 0. This is rarely desired, but there are examples where one might want this. In these cases, the Block after the GENERATE Block must be one that denies entrance until a specific condition is met.

8.8 The PRIORITY Block

In nearly every case, the priority of a transaction will be given by the E operand of the GENERATE Block. There will be times during the execution of a program that is desired to change the transaction's priority. This is done by the PRIORITY Block. It has a very simple form:

```
PRIORITY n
```

where n is an integer.

Thus, the Blocks

```
PRIORITY 1
PRIORITY 0
PRIORITY 25
```

will change a transaction's priority level to 1, 0, and 25, respectively. This Block will not be used until later.

Example 8.3: A Tyre Inspection Area in a Mine

A miner works in a tyre inspection area in a mine. When he arrives at work, there are three trucks waiting to be inspected. He can only inspect one truck at a time. Trucks arrive at the rate of one every 15 ± 6 minutes. It takes 12 ± 4 minutes to inspect the tyres on

```
          SIMULATE
***************************************
*  PROGRAM  CHAP8G.GPS                *
*  EXAMPLE OF TIRE INSTECTOR IN A MINE *
***************************************
          GENERATE    ,,,3
          TRANSFER    ,NEXT
          GENERATE    15,6        TRUCKS ARRIVE
 NEXT     QUEUE       WAIT        IS INSPECTOR FREE?
          SEIZE       INSPECT     YES, ENGAGE INSPECTOR
          DEPART      WAIT        LEAVE THE WAITING QUEUE
          ADVANCE     12,4        TIRE INSPECTION
          RELEASE     INSPECT     FREE THE INSPECTOR
          TERMINATE   1           TRUCK LEAVES
          START       200
          END
```

FIGURE 8.10
Listing of program CHAP8G.GPS.

a truck. Simulate for 200 trucks being inspected. Assume the miner works continuously with no breaks for shift change.

The program to simulate this is CHAP8G.GPS (Figure 8.10). When the program is run, the output will show that it takes 2987.8 minutes to inspect 200 trucks. This result is found in the file CHAP8G.LIS.

Example 8.4: Change to Example 8.3

Suppose that, in Example 8.3, the trucks arrive as before, but now it is desired to have the first truck from the mine arrive at 1 minute after the miner arrives to begin work. There are still three trucks waiting for tyre inspection. The only change in the previous program is that the first `GENERATE` Block is now

```
GENERATE 15,6,1 TRUCKS ARRIVE
```

Incorporate this change and rerun the program.

It is possible to have more than one `GENERATE` Block in a program and, in fact, this is nearly always the case. The only caution is that a transaction can never *enter* a `GENERATE` Block. Thus, one could have a program such as

```
GENERATE  1
- - - -
- - - -
- - - -
- - - -
TERMINATE
GENERATE  5000
TERMINATE 1
```

Although it is a bit early in our introduction to the GPSS/H language, the observant student will notice from the listing of the program that extra lines of code have been added. In particular, lines of code have been added to allow the mining trucks to be viewed both coming to be served and then leaving.

The next section will explain what the `TERMINATE` Block does.

8.9 The TERMINATE Block

Transactions enter the system by means of the GENERATE Block. Eventually, in most simulations, the transactions will have to leave the system. This is done by means of a Block known as the TERMINATE Block. It is quite simple in form:

```
TERMINATE     n
```

where n is a positive integer, including 0. If n is omitted, as it often is, it is taken to be 0.

Every time a transaction enters this Block, it is immediately removed from the system. Some examples of it are as follows:

```
TERMINATE     1
TERMINATE     20
TERMINATE     5
TERMINATE     0
```

The operand n *has nothing to do with the transaction being removed from the system.* Only one transaction at a time enters the TERMINATE Block. The TERMINATE Block *always* removes this one transaction. As we shall see, this operand is used to control the execution of the program in connection with another statement.

8.10 The START Statement

Every GPSS/H program must have a START statement. The simplest form of it is as follows:

```
START         n
```

where n must be a nonzero positive integer. Some examples are as follows:

```
START 1
START 10
START 200
START 66
```

The number, n, is a counter for controlling the running of the program. It has to be an integer. If it is typed as a decimal, a warning message appears on the screen, and the number is truncated to an integer and then the simulation takes place. While the program is being executed, the counter is being decremented. When it becomes zero or negative, the program stops execution. The GPSS/H processor then creates a file, (*name*). LIS, where *name* is the name of the original GPSS/H file. If the original GPSS/H program was named APRIL12.GPS, the file created will be APRIL12.LIS. This will be found in the same subdirectory where GPSS/H is found. This file contains the results of the simulation and can be viewed using the same text editor used to create the original GPSS/H file.

The way the processor knows that the program is finished is as follows:

1. The counter, n, is set aside.
2. Whenever a transaction goes through a `TERMINATE` Block that has an operand, this operand value is subtracted from n.
3. When n becomes 0 or negative, the simulation is finished and the report produced.

For example, assume that the GPSS/H programs contain only the following `TERMINATE` and `START` lines of code:

```
(a)    TERMINATE 2
       .........
       START  10

(b)    TERMINATE
       .........
       TERMINATE 3
       ........
       START  13

(c)    TERMINATE 4
       ........
       START  1
```

In (a), the program will execute until five transactions have passed through the `TERMINATE 2` Block.

In (b), the program will run until five transactions have passed through the `TERMINATE 3` Block. Any transaction that passes through the first `TERMINATE` Block will have no effect on the execution time of the program.

In (c), the program will run until one transaction has passed through the `TERMINATE 4` Block.

Most of the programs run so far have had Blocks such as the following:

```
GENERATE      480
TERMINATE     1
START         1
```

The effect of the above is to put a single transaction into the system at time 480. This transaction is immediately removed via the `TERMINATE 1` Block. The 1 in its operand causes the counter of 1, which was given by the `START 1` statement to be decremented to 0. Thus, the program stops execution at time 480.0000. Transactions that have no other effect on the program other than to stop the execution are called *timer transactions.*

A GPSS/H program starts execution when the *first* `START` statement is encountered. *If there are more statements after this first* `START` *statement, they are initially ignored.* When the program is through with execution as specified by the first `START` statement, whatever commands are given after it are then executed.

During the compiling stage, transactions are primed to move through the system at times given by the `GENERATE` Block. The transactions are placed on a time axis called the *current events chain.* When the program begins execution, the processor takes over and moves the transactions one at a time as far as each transaction can move. After it moves a transaction, it goes back to the current events chain and moves the next transaction. This continues as long as the program is running. The first transaction to be moved is the one positioned on the current events chain at the earliest time, the next is the one at the next earliest time, and so on. The transaction is moved from Block to Block in a sequential manner, unless the program specifies otherwise. A transaction is moved until one of the following things happens:

1. The transaction is removed from the system. This is done via the `TERMINATE` Block.
2. The transaction is put on another chain. This will be covered later.
3. The transaction is blocked and cannot enter a sequential Block.

8.11 The CLEAR Statement

There are ways to run a program several times with different values. These will be covered in Chapter 23. One old method will be presented here. This is using the `CLEAR` statement. This is placed after the `START` statement. When this is encountered, the program has already been run and a .LIS file is created. The `CLEAR` statement clears all statistics except for user-supplied standard numerical attributes (SNAs), which will be covered in Chapter 13. It does not reset the random number generator. If there is another `START` statement, the program will be rerun and new results obtained and sent to the .LIS file. If one wants to run the program with different data, one can list the Block with different parameters before the next `START` statement. GPSS/H will then rerun the simulation with these Blocks replacing the old ones. In order for GPSS/H to know which Blocks are to be replaced, the Blocks must have labels. The following example will illustrate this.

Example 8.5: Truck-Shovel Operation for a Small Mine

A small mine has a simple truck-shovel operation. The single shovel can load a truck in 2 ± 1 minutes. The trucks then haul to the crusher in 6 ± 2 minutes, dump in $1 \pm .5$ minute, and return to the shovel in 5 ± 1.6 minutes. Only one truck can be loaded at a time, but there is no such restriction in dumping at the crusher. The mine currently has five trucks working. They work continuously for 480 minutes. The miner wants to determine how busy the shovel is and what changes there would be in production if he purchased a sixth truck.

NOTE: This is an important example of how quickly one can simulate actual mining operations. This problem, and variations of it, has been the source of research for many years. As basic and simple as it may appear, the general case of it has no mathematical solution. Yet, an ability to solve this example is fundamental to most mining operations.

Solution

The computer program to model this is given by CHAP8H.GPS. Even though some of the code will not be presented until later, the output should be easy to understand. The program is as follows (Figure 8.11):

The following lines of code illustrate the `CLEAR` statement:

```
       START      1
       CLEAR
TRUCKS GENERATE   ,,,6
       START      1
       END
```

```
          SIMULATE
*************************
*  PROGRAM CHAP8H.GPS    *
*  ILLUSTRATION OF CLEAR *
*  STATEMENT             *
*************************
 TRUCKS GENERATE      ,,,5      TRUCKS IN THE MINE
 BACK   SEIZE         SHOVEL    USE THE SHOVEL
        ADVANCE       2,1       LOAD
        RELEASE       SHOVEL    FREE THE SHOVEL FOR THE NEXT TRUCK
        ADVANCE       6,2       HAUL
        ADVANCE       1,.5      DUMP
        ADVANCE       5,1.6     RETURN TO THE SHOVEL
        TRANSFER      ,BACK
        GENERATE      480       TIMER TRANSACTION
        TERMINATE     1
        START         1
        CLEAR                   READY FOR ANOTHER RUN
 TRUCKS GENERATE      ,,,6      RUN WITH 6 TRUCKS
        START         1
        END
```

FIGURE 8.11
Listing of program CHAP8H.GPS.

The program starts to execute when the first `START` statement is executed. This runs the simulation with five trucks in the mine (`TRUCKS GENERATE ,,,5`). The results are placed in the file CHAP8F.LIS. Then the various statistics are cleared due to the `CLEAR` statement. The original `TRUCKS GENERATE ,,,5` Block is replaced by the Block.

```
TRUCKS GENERATE ,,,6
```

The program is then rerun and the new results are added to the original .LIS file. Selected portions of this are given next.

```
RELATIVE CLOCK: 480.0000 ABSOLUTE CLOCK: 480.0000

BLOCK CURRENT     TOTAL
TRUCKS                5
BACK                166
3           1       166
4                   165
5           2       165
6                   163
7           2       163
8                   161
9                     1
10                    1

     FACILITY
     SHOVEL 0.684

     RANDOM  INITIAL   CURRENT
     STREAM  POSITION  POSITION
          1  100000    100657
```

```
RELATIVE CLOCK: 480.0000 ABSOLUTE CLOCK: 480.0000

BLOCK CURRENT     TOTAL
TRUCKS                6
BACK                193
3          1        193
4                   192
5          2        192
6                   190
7          3        190
8                   187
9                     1
10                    1

FACILITY
SHOVEL 0.815

   RANDOM  INITIAL   CURRENT
   STREAM  POSITION  POSITION
        1  100657    101422
```

The output from the program can be interpreted as follows: With five trucks, there were 163 loads dumped in the 480 minutes of work. The shovel was busy 68.4% of the time. There were 657 random numbers used. When the program was rerun with six trucks in the mine, the production increased to 190 loads. The shovel was now busy 81.5% of the time. The random number sampling started with sampling from position 100657 and was now at position 101422.

8.12 The RMULT Statement

When one examines the .LIS file for most programs, there will be a line of output such as shown in Figure 8.12.

```
RANDOM  ANTITHETIC    INITIAL    CURRENT         SAMPLE   CHI-SQUARE
STREAM    VARIATES   POSITION   POSITION          COUNT   UNIFORMITY
     1         OFF     100000     100398            398      0.93
```

FIGURE 8.12
Selected output from program CHAP8H.GPS.

We will not be concerned with this except to note the following. There are many random number streams that GPSS/H can use to sample from. By default, random number stream 1 is used. This can be changed but rarely is for mine simulations. Antithetic variates are also not used in mine simulation but are a tool of the statistician to assist in improving the rate of convergence when doing Monte Carlo simulations. While the mining engineer does Monte Carlo simulations, due to the speed and capacity of computers, it is no longer necessary to study ways to improve the convergence. The INITIAL and CURRENT POSITIONS are of interest. If one can imagine a random number stream of over two billion numbers to sample from, GPSS/H selects the one in position 100000 as the first. The example above where the .LIS file was indicates that there were 398 references to random numbers, so the current position in the random number stream is at 100398. GPSS/H also does a chi-square test to determine if the random numbers are truly random. This, too, is not needed for mine system simulations.

The random numbers used by GPSS/H are not true random numbers in that they can be repeated. This is needed for some statistics tests, although this is rarely the case for mining systems. Even so, if a simulation is required using different random numbers, there is a way to do this. This is by means of the `RMULT` statement. This statement is simply:

```
RMULT (any integer)*
```

The integer needs to be less than $2^{31}-1$ so that this should not be a problem. Some examples are as follows:

```
RMULT  1234
RMULT  333
RMULT  98765
RMULT  -123
```

In each case, the random number stream is sampled at the position starting with the operand as specified by the `RMULT` operand. In the case of a negative operand (`-123`), the random number stream is sampled starting at the position given by its absolute value. In this case, it would be position `123`.

8.13 Exercises

8.1. What will happen when the following `GENERATE` Blocks are used in a program:

```
(a) GENERATE  100,30,,5
(b) GENERATE  ,,1000,4
(c) GENERATE  1000,200,20,100
(d) GENERATE  500,400,1
(e) GENERATE  ,,,4
```

8.2. You are observing trucks at point A. Every 4 minutes, one truck passes this point. The trucks travel along a road for 5 minutes and then leave the system. In 20 minutes, how many trucks have you observed? Assume that the first truck does not pass you until 4 minutes have passed.

The GPSS/H program to simulate this is as follows:

```
SIMULATE
GENERATE   4
ADVANCE    5
TERMINATE  0
GENERATE   20
TERMINATE  1
START      1
END
```

Write this program and run it.

* If a real number is used, it will be truncated and a warning message will be printed on the screen.

8.3. In Exercise 8.2, suppose the trucks pass point A every 4 minutes but starting at $t = 0$. Change the appropriate line in your program.

8.4. In Exercise 8.2, a truck will pass point A every 4 ± 2.5 minutes. Change the appropriate line in your program.

8.5. In Exercise 8.2, you may have noticed that, after 20 minutes, the computer program said that you observed only four trucks. What happened to the truck at time 20? Suppose you change the program to give the trucks you are observing a priority of 1. What line needs to be changed? How does this affect the program results?

8.6. In Exercise 8.2, simulate for 40 minutes.

8.7. Write the GPSS/H `GENERATE` Block to have:

(a) Transactions enter the system every 5 time units.
(b) Transactions enter the system every 100.6 time units.
(c) Transactions enter the system every 10 ± 6.5 time units.
(d) Transactions enter the system every 7 time units starting at time 100.
(e) Transactions enter the system every 120 ± 35.6 time units beginning at time 200.
(f) Five transactions enter the system at time 0.
(g) Transactions enter the system every 100 time units beginning at time 80 and only 10 enter the system from this Block.
(h) Only three transactions enter the system at time 500. These transactions have priority 5.
(i) Transactions with priority 5 enter the system every 6 ± 3.4 time units, starting at time 400. Only six of these are to enter the system.

8.8. What would happen if you had the following Block?

```
GENERATE       120,130
```

8.9. What would happen if you had the following Block?

```
GENERATE       4,1,,,-1
```

8.10. What do the following lines of code do?

```
GENERATE       ,,480,1
TERMINATE      1
START          1
```

8.11. The following code is used to time the running of a GPSS/H program. What time will the simulated clock show at the completion of the program?

```
(a)   GENERATE     480
      TERMINATE    2
      START        2

(b)   GENERATE     4800
      TERMINATE    5
      START        10
```

```
(c)  GENERATE    1000
     TERMINATE   3
     START       7

(d)  GENERATE    200
     TERMINATE   10
     START       1
```

8.12. What will the simulated clock read when the following program is done running?

```
SIMULATE
GENERATE     100
TERMINATE    1
GENERATE     150
TERMINATE    2
START        10
END
```

8.13. Consider the following code:

```
GENERATE     100
TERMINATE    1
GENERATE     150
TERMINATE    3
START        20
```

Assuming no other `TERMINATE` Blocks, for how long will the program run?

9

ADVANCE and TRANSFER Blocks

Coal distribution system, Northern Queensland, Australia.

9.1 The ADVANCE Block

The ADVANCE Block is used to *hold up* a transaction while service is being performed. The processor will take a transaction that enters this Block and place it on what is called the *future events chain* (FEC). When a transaction is placed on this chain, it is no longer a candidate for being moved during a re-scan of the current events chain. For example, the time a truck is being loaded will be represented in GPSS/H by an ADVANCE Block.

There are several forms of the ADVANCE Block. They are as follows:

```
(a) ADVANCE   A
(b) ADVANCE   A,B
```

where A and B can be positive integers or variables.

In (a), the transaction will be placed on the FEC (held up) for a time equal to Operand A. Thus,

```
ADVANCE 5
ADVANCE 100
```

will hold up the transaction until 5 time units have passed in the first case and until 100 time units have passed in the second. During these times, the transaction resides on the FEC.

In (b), the transaction will be delayed by a time value from between the interval A − B and A + B. *The right end point is not included.* However, the time returned will have a six place decimal. Each time will have equal probability of occurrence.

Thus,

```
ADVANCE 12,3
```

will hold a transaction until a time between the intervals 9.00000 and 14.999999 has elapsed. Each of the possible times will happen with equal probability. Later, we shall learn how to use *any* statistical distribution in the ADVANCE Block.

ADVANCE Blocks have been used in many of our previous examples. They are one of the easiest GPSS/H Blocks to understand and use.

Example 9.1: Trucks Enter a Mine

In a very busy mine, trucks enter a haulage route at point A every 10 ± 5 seconds starting at time 0. They travel to point B in a straight line where they leave the route. The time to travel to point B is 9 ± 4 seconds. You are to determine the number of cars that pass point B each hour.

This is the first problem for which we can finally write the complete GPSS/H program. It is given by CHAP9A.GPS as follows (Figure 9.1).

The relevant portion of the output (Figure 9.2) from the .LIS file is as follows:

```
RELATIVE CLOCK: 3600.0000 ABSOLUTE CLOCK: 3600.0000
```

The number of trucks to pass point B is 363. Considering that trucks arrive with a mean interarrival time of 12 seconds, this gives us five cars per minute or an expected number of 360 for an hour. Thus, the results of the simulation are within the expected range.

The animation for this program is given by CHAP9B. The GPSS/H program is given by CHAP9B.GPS (Figure 9.3).

```
***********************************
*  PROGRAM CHAP9A.GPS             *
*  TRUCKS ARRIVE AT A AND         *
*  TRAVEL TO B                    *
***********************************
          SIMULATE
          GENERATE    10,5,0     TRUCKS ARRIVE
          ADVANCE     9,4        TRAVEL FROM A TO B
          TERMINATE
          GENERATE    3600       TIMER TRANSACTION
          TERMINATE   1
          START       1
          END
```

FIGURE 9.1
Program listing for program CHAP9A.GPS.

BLOCK	CURRENT	TOTAL
1		364
2	1	364
3		363
4		1
5		1

FIGURE 9.2
Portion of output from program CHAP9A.GPS.

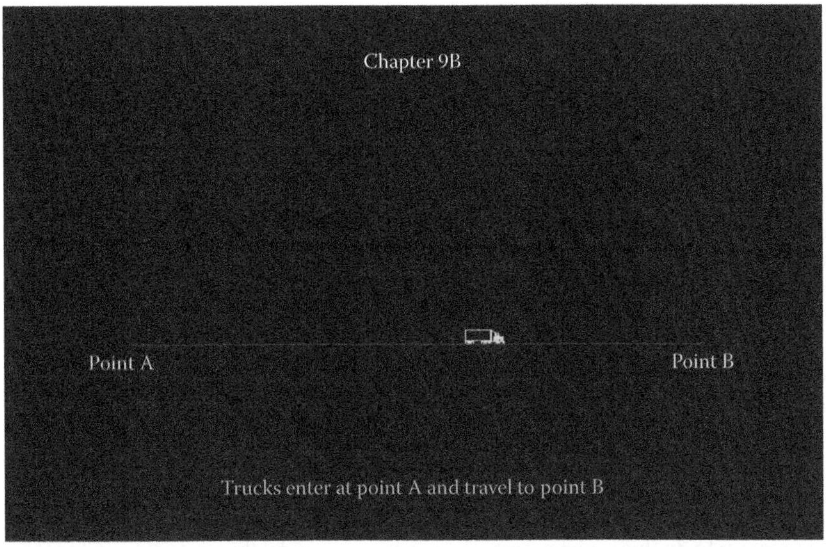

FIGURE 9.3
Screenshot from program CHAP9B.

9.2 A Caution in Writing Programs

There is a caution to keep in mind with this and other GPSS/H programs. In this program, the transactions were all created at time $t = 0$. They all left at time 0 and were put on the future events chain via the ADVANCE Block. The ADVANCE Block *always* admits transactions. Suppose we had a different problem and were going to generate transactions at various times as given by

```
GENERATE 12,3
```

Suppose, further, that the times for the first four of these transactions to enter the system are 11, 23, 35, and 44. If the Block after the GENERATE Block will not allow a transaction to enter it (some Blocks will only allow one transaction at a time to enter them, others may not allow *any* transactions to enter depending on a particular condition), what happens is that the transaction remains in the GENERATE Block until the next Block will allow it to enter. This may seem to be all right but, because the transaction cannot leave the GENERATE Block when it was originally scheduled to leave, the subsequent transactions are also delayed from leaving. Suppose that the third transaction cannot enter the next Block but must remain in the GENERATE Block for 5 time units. This means that it cannot leave until $t = 40$. The effect of this on the fourth (and subsequent transactions) is to shift them all 5 time units forward before they leave. This is normally incorrect. When a transaction is scheduled to leave the GENERATE Block, a Block that will *always* accept it should be provided. One way around this is to have the following:

```
GENERATE 12,3
ADVANCE 0
```

The `ADVANCE` Block used here is a *dummy* Block. It holds the transactions from the `GENERATE` Block for 0 time units. The only effect of it is to allow transactions to leave the `GENERATE` Block at the times they were scheduled to leave. Keep this in mind for future programs.

Example 9.2: Trucks Come for Service

Trucks come for routine service every 12 ± 3 minutes. The following are the steps that the drivers must follow:

Check in with service manager, which takes 2 ± 1 minutes.
Travel to station A, which takes 30 seconds.
Receive routine inspection here, which takes 8 ± 1 minutes.
Travel to station B, which takes 15 seconds.
Have minor service, which takes 5 ± 1.8 minutes.

After this, the truck is ready.

Simulate this system for 3 hours.

The program to do this is given next. The program to model this is stored as CHAP9C.GPS (Figure 9.4).

A portion of the output file is as shown in Figure 9.5.

```
RELATIVE CLOCK: 180.0000 ABSOLUTE CLOCK: 180.0000
```

```
          SIMULATE
******************************
*    PROGRAM CHAP9C.GPS        *
*    TRUCKS ARRIVE FOR SERVICE *
******************************
          GENERATE     12,3     TRUCKS ARRIVE
          ADVANCE      2,1      CHECK IN
          ADVANCE      .5       TRAVEL TO STATION A
          ADVANCE      8,1      INSPECTON
          ADVANCE      .25      TRAVEL TO STATION B
          ADVANCE      5,1      MINOR SERVICE
          TERMINATE
          GENERATE     180
          TERMINATE    1
          START        1
          END
```

FIGURE 9.4
Listing of program CHAP9C.GPS.

```
BLOCK CURRENT     TOTAL
1                    15
2           1        15
3                    14
4                    14
5                    14
6           1        14
7                    13
8                     1
9                     1
```

FIGURE 9.5
Portion of output from program CHAP9C.GPS.

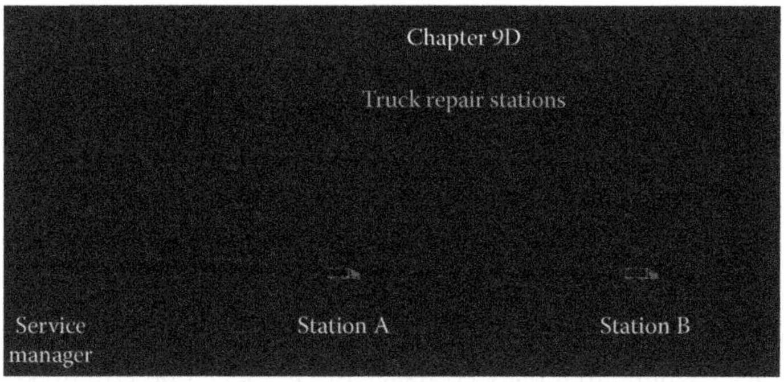

FIGURE 9.6
Screenshot of animation of CHAP9D.

The simulation ran for 3 hours (180 minutes). There were 15 trucks that entered the system. When the program ended, 13 had been finished and two were still in the system.

The animation of the problem is given by CHAP9D. A screenshot is shown in Figure 9.6.

9.3 The TRANSFER Block

By now, we know pretty much how a GPSS/H program works and how the time of the simulation is controlled. We also have learned about PROOF animation. We need just one more Block before we can start to create our own complete programs.

In general, a GPSS/H program works by selecting the transaction (event) that is the first to occur on the current events chain and takes the transaction that is to move and moves it sequentially in the program as far as it can move. Eventually, one of three things will happen to the transaction:

1. It will be destroyed or removed from the system via a TERMINATE Block.
2. It will be placed on some other chain, such as the FEC. (This is the only other chain we know of beyond the current events chain. Later, other chains will be introduced.)
3. It will be blocked and will be unable to enter the next Block until some condition changes. We have not yet been introduced to any Blocks that refuse to allow a transaction to enter it. These, too, will come later.

It is possible to route the transaction to a non-sequential Block by means of the TRANSFER Block. There are six forms of this Block but, for now, only two are going to be considered. This Block is similar to the GOTO Block in other computer languages. If you were introduced to computer programming by learning another language, you were probably taught to avoid this Block as it may lead to what is known as *spaghetti* code—that is, very difficult or almost impossible to follow. However, in GPSS/H, such Blocks are essential and in common use.

9.4 Unconditional TRANSFER Mode

The first form of the TRANSFER Block is the unconditional TRANSFER Block. It has the following form:

```
TRANSFER       ,(label)
```

In place of the (label) one could have a Block number. Blocks are numbered consecutively as you have seen in viewing the .LIS files. However, we will only use labels in our work. There are some cases where, in advanced usage of GPSS/H, one will make use of the fact that one could transfer control to the Block number.

The label can be any alphanumeric legal name up to eight spaces. Be careful not to use labels that start with GPSS/H reserved words as the compiler on some versions may take these as other GPSS/H statements. For example, do not use

STARTIT or ENDIT

for a label. Also, do not start a label with the letter X. This is because, as we will learn in a later chapter, the X is reserved for the first letter of the name of an SNA.

When a transaction enters the unconditional TRANSFER Block, it will try to enter the Block given by the label. If it can (and so far, all of the Blocks we have had, except for the GENERATE Block, allow transactions to enter them), it will do so.

The following are all valid examples of the TRANSFER Block:

```
TRANSFER       ,BACK1
TRANSFER       ,UPTOP
TRANSFER       ,DDDD
TRANSFER       ,12
TRANSFER       ,NEXT
```

To see how a TRANSFER Block might be used in a program, consider the following lines of code:

```
BACKUP         ADVANCE        12
               ADVANCE        5.5
               ADVANCE        8
               ADVANCE        1.5
               TRANSFER       ,BACKUP
```

Here, we have created an infinite loop with the transaction or transactions moving from the Block with the label BACKUP to the TRANSFER Block and back up again as long as the program runs. Obviously, there needs to be some other part of the program that stops the program from running.

Example 9.3: Excavation of a Pit

A mining contractor is excavating a pit. The machines he uses are self-loading and there never is a delay in loading them. Loading takes 1 ± .2 minutes. Once they are loaded, they travel to a waste dump in 8 ± .5 minutes, where they unload in .5 ± .1 minute. They return to the pit in 6 ± .25 minutes. He has five trucks to do this. The drivers work

450 minutes in a shift. When the shift is finished, all the trucks immediately stop. How many loads will be dumped in a shift?

The solution to this can be (almost) worked out by hand. Assume that the average times are taken. Thus, each truck will make one cycle from the pit and back in an average of 15.5 minutes. In 450 minutes we can expect each truck to make around 28 trips and there are five trucks, so the production should be around 145 loads per day.

Let us write the GPSS/H program to simulate this. The program is CHAP9E.GPS (Figure 9.7).

A portion of the .LIS file is as shown in Figure 9.8.

Running the program gives the result that there will be 143 loads dumped assuming that the trucks stop in their tracks at time 450. In an actual mine, the number of loads dumped depend on what the trucks do when the shift is over. The expected number of loads was found to be 145. If the trucks that were loaded on their way to the dump actually complete their travelling and dumping, the loads dumped would be 146.

The animation is given by CHAP9F. The program to create this animation is given by CHAP9F.GPS. Its listing is given in Figure 9.9.

A screenshot of the animation is shown in Figure 9.10.

In this example, there is no restriction on the number of trucks that can be loaded or dumped at the same time. A real mining situation will not allow this, so some code is needed to reflect this. This will be covered later. This animation is quite crude and can be greatly improved such as having the shovel's boom rotate when loading a truck, the trucks spotting, the classes better drawn, more simulation data added to the screen, and so on.

```
          SIMULATE
*******************************
*  CHAP9E.GPS  PROGRAM TO     *
*  ILLUSTRATE THE TRANSFER    *
*  BLOCK IN UNCONDITIONAL     *
*  MODE                       *
*******************************
          GENERATE     ,,,5     PROVIDE 5 TRUCKS
  UPTOP   ADVANCE      1,.2     TRUCK LOADS
          ADVANCE      8,.5     TRUCK TRAVELS TO DUMP
          ADVANCE      .5,.1    TRUCK DUMPS
          ADVANCE      6,.25    RETURN TO PIT
          TRANSFER     ,UPTOP   LOAD AGAIN
          GENERATE     450      TIMER TRANSACTION
          TERMINATE    1
          START        1
          END
```

FIGURE 9.7
Listing of program CHAP9E.GPS.

```
RELATIVE CLOCK: 450.0000   ABSOLUTE CLOCK: 450.0000

BLOCK CURRENT      TOTAL
1                      5
UPTOP         2      148
3             1      146
4                    145
5             2      145
6                    143
7                      1
8                      1
```

FIGURE 9.8
Portion of output from CHAP9E.LIS.

```
          SIMULATE
*********************************
*  CHAP9F.GPS  PROGRAM TO       *
*  ILLUSTRATE THE TRANSFER      *
*  BLOCK IN UNCONDITIONAL       *
*  TRANSFER  MODE.  THIS        *
*  MAKES THE FILE CHAP9F.ATF    *
*  FOR THE ANIMATION            *
*********************************
  ATF     FILEDEF     'CHAP9F.ATF'
          REAL        &X,&Y,&Z,&W
          INTEGER     &COUNT
          GENERATE    3,,0,5   PROVIDE 5 TRUCKS
          BPUTPIC     FILE=ATF,LINES=5,AC1,XID1,XID1,XID1,AC1
TIME *.****
CREATE TRUCK2 T*
PLACE T* ON P2
SET T* TRAVEL 6
WRITE M2 ***.**
          ADVANCE     6
  UPTOP   SEIZE       DUMMY
          BLET        &X=.4*FRN1+.8     TIME TO LOAD A TRUCK
          BPUTPIC     FILE=ATF,LINES=2,AC1,XID1,&X
TIME *.****
PLACE T* AT 26 25.1
          ADVANCE     &X
          BPUTPIC     FILE=ATF,LINES=2,AC1,AC1
TIME *.****
WRITE M2 ***.**
          BLET        &Y=1.00*FRN1+7.5   TIME TO HAUL TO DUMP
          BPUTPIC     FILE=ATF,LINES=4,AC1,XID1,XID1,XID1,&Y
TIME *.****
PLACE T* ON P1
SET T* CLASS TRUCK1
SET T* TRAVEL **.**
          ADVANCE     .5
          BPUTPIC     FILE=AT,LINES=2,AC1,AC1
TIME *.****
WRITE M2 ***.**
          RELEASE     DUMMY
          ADVANCE     &Y-.5   TRAVEL TO DUMP
          BPUTPIC     FILE=ATF,LINES=2,AC1,AC1
TIME *.****
WRITE M2 ***.**
          SEIZE       DUMMY2
          BLET        &Z=.2*FRN1+.4   TIME TO DUMP
          BPUTPIC     FILE=ATF,LINES=2,AC1,XID1,&Z
TIME *.****
PLACE T* AT 65 25
          ADVANCE     &Z      DUMP A LOAD
          BPUTPIC     FILE=AT,LINES=2,AC1,AC1
TIME *.****
WRITE M2 ***.**
```

FIGURE 9.9
Listing of program CHAP9F.GPS. *(Continued)*

```
          BLET        &COUNT=&COUNT+1
          BPUTPIC     FILE=ATF,LINES=2,AC1,&COUNT
TIME *.****
WRITE M1 **
          BLET        &W=.5*FRN1+5.75
          BPUTPIC     FILE=ATF,LINES=4,AC1,XID1,XID1,XID1,&W
TIME *.****
PLACE T* ON P2
SET T* CLASS TRUCK2
SET T* TRAVEL **.**
          ADVANCE     .5
          RELEASE     DUMMY2
          ADVANCE     &W-.5
          BPUTPIC     FILE=ATF,LINES=2,AC1,AC1
TIME *.****
WRITE M2 ***.**
          TRANSFER    ,UPTOP LOAD AGAIN
          GENERATE    ,,,1    DUMMY TRANSACTION
 BACK1    BPUTPIC     FILE=ATF,LINES=2,AC1,AC1
TIME *.****
WRITE M2 ***.**
          ADVANCE     .333
          TRANSFER    ,BACK1
          GENERATE    80  TIMER TRANSACTION
          TERMINATE   1
          START       1
          BPUTPIC     FILE=ATF,LINES=2,AC1
TIME *.****
END
          END
```

FIGURE 9.9 (Continued)
Listing of program CHAP9F.GPS.

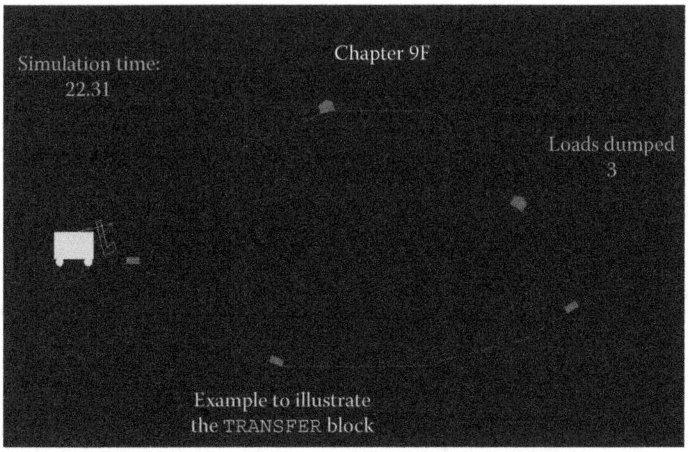

FIGURE 9.10
Screenshot from program CHAP9F.GPS.

9.5 The Conditional TRANSFER Block

This is a very important Block that will find a great many uses in GPSS/H. The general form is as follows:

```
TRANSFER            .xxx,(label1),(label2)
```

where .xxx is a fraction (up to three digits)

(label1) is an allowable label

(label2) is an allowable label

For nearly every application of this form of the `TRANSFER` Block, the (label1) is omitted. Examples of this form of the `TRANSFER` Block are as follows:

```
TRANSFER            .25,,NEXT1
TRANSFER            .5,,AWAY
TRANSFER            .3,,DDDD
TRANSFER            .055,,DUDS
```

When a transaction enters such a `TRANSFER` Block, it will generate an internal random number and route the transaction to the Block with the second label xxx% of the time; otherwise, it will go to the next sequential Block.

In the first case above, the transaction will be routed to the Block with the label, `NEXT1` 25% of the time. 75% of the time it will go to the next sequential Block.

In the second case, the transaction will be routed to the Block labelled `AWAY` 50% of the time. In the third case, it will go the Block labelled `DDDD`, 30% of the time and, in the fourth example, 5.5% of the transactions will be routed to the Block with the label `DUDS`.

Example 9.4: Two Different Mining Trucks Come to Be Loaded

Two different mining trucks come to be loaded. They arrive every 10 ± 3 minutes. 75% of the trucks will require only one load, and it takes 4.5 ± 1 minutes to load. The other trucks require two loads, and it takes 7 ± 2 minutes to load. There is no restriction on the trucks being loaded; that is, every truck can be loaded as soon as it appears. How many trucks will be loaded in an 8-hour shift?

The program to simulate this situation is given as CHAP9G.GPS. It is assumed in writing the program that each truck can be loaded when it arrives. Later, when we learn more GPSS/H, we can take queuing into account (Figure 9.11).

Selected output from the .LIS files is given in Figure 9.12.

```
RELATIVE CLOCK: 480.0000 ABSOLUTE CLOCK: 480.0000
```

The results of the simulation show that there were 44 trucks that came to be loaded. Of these, 36 needed one load and 8 two loads. These figures are slightly different from the expected ones. The expected number of trucks to arrive is 48 assuming that the slight error in the first truck's arrival time can be ignored. Of these 48, one would expect 36 to require one load and 12 to require two loads. This indicates that the simulation should have been run for a longer time. For example, to determine the expected trucks to be loaded in a shift of 8 hours, one can run the program for a much longer time and then take the average number of trucks loaded each shift. Alternately, one can run the

```
              SIMULATE
***************************
*  PROGRAM CHAP9G.GPS        *
*  EXAMPLE OF CONDITIONAL    *
*  TRANSFER BLOCK            *
***************************
          GENERATE    10,3   TRUCKS ARRIVE
          TRANSFER    .25,,BIGTCK
          ADVANCE     4.5,1 ONE LOAD
          TERMINATE
 BIGTCK   ADVANCE     7,2   TWO LOADS
          TERMINATE
          GENERATE    480
          TERMINATE   1
          START       1
          END
```

FIGURE 9.11
Listing of program CHAP9G.GPS.

```
BLOCK CURRENT      TOTAL
1                     44
2                     44
3                     36
4                     36
BIGTCK                 8
6                      8
7                      1
8                      1
```

FIGURE 9.12
Portion of output from CHAP9G.GPS.

program over and over again using different random numbers. How to do this will be covered in Chapter 23.

An animation for this example is given by CHAP9H. A listing of the program is given in Figure 9.13.

The animation is shown in Figure 9.14. The animation shows that the trucks that come to be loaded branch off in two directions. This is done to enhance the viewer's understanding of how the model works. Such things are often done with animations to better illustrate the logic of the system being modelled.

9.6 Exercises

9.1. Mining trucks come to a Y intersection in a road. Some vehicles turn to the left and some to the right. Trucks approach this intersection every 10 ± 5 seconds. 35% go to the right, whereas the rest go to the left. Simulate for 2 hours.

9.2. A car comes to let people out to attend a theatre every 15 ± 4 seconds. 50% of the cars have one passenger to let out, 30% have two, 15% have three, and the rest four people. This all takes place in an hour. Write the .GPS program to simulate this. How many people have been left off to attend the theatre in the hour?

9.3. People enter a store that has two floors: first floor and second floor. Arrivals are every 30 ± 10 seconds. 30% only go to the first floor. They take 2 ± 1 minutes to shop. The rest shop on the first floor in 25 ± 10 seconds and then go to the second

```
SIMULATE
****************************
*  PROGRAM CHAP9H.GPS       *
*  ANIMATION OF CHAP9G.GPS *
*  EXAMPLE OF CONDITIONAL   *
*  TRANSFER BLOCK           *
****************************
 ATF       FILEDEF      'CHAP9H.ATF'
           GENERATE     10,3   TRUCKS ARRIVE
           BPUTPIC      FILE=ATF,LINES=4,AC1,XID1,XID1,XID1
TIME *.****
CREATE TRUCK T*
PLACE T* ON P1
SET T* TRAVEL 5
           ADVANCE      5
           TRANSFER     .25,,BIGTCK
           BPUTPIC      FILE=ATF,LINES=3,AC1,XID1,XID1
TIME *.****
PLACE T* ON P2
SET T* TRAVEL 4
           ADVANCE      4
           BPUTPIC      FILE=ATF,LINES=2,AC1,XID1
TIME *.****
PLACE T* AT 45 40
           ADVANCE      4.5,1  SMALL LOAD
           BPUTPIC      FILE=ATF,LINES=3,AC1,XID1,XID1
TIME *.****
PLACE T* ON P4
SET T* TRAVEL 3
           ADVANCE      3
           BPUTPIC      FILE=ATF,LINES=2,AC1,XID1
TIME *.****
DESTROY T*
           TERMINATE
 BIGTCK    BPUTPIC       FILE=ATF,LINES=3,AC1,XID1,XID1
TIME *.****
PLACE T* ON P3
SET T* TRAVEL 4
           ADVANCE      4
           BPUTPIC      FILE=ATF,LINES=2,AC1,XID1
TIME *.****
PLACE T* AT 45 30
           ADVANCE      7,2 BIG LOAD
           BPUTPIC      FILE=ATF,LINES=3,AC1,XID1,XID1
TIME *.****
PLACE T* ON P5
SET T* TRAVEL 3
           ADVANCE      3
           BPUTPIC      FILE=ATF,LINES=2,AC1,XID1
TIME *.****
DESTROY T*
           TERMINATE
           GENERATE     480
           TERMINATE    1
           START        1
           BPUTPIC      FILE=ATF,LINES=2,AC1
TIME *.****
END
           END
```

FIGURE 9.13
Listing of program CHAP9H.GPS.

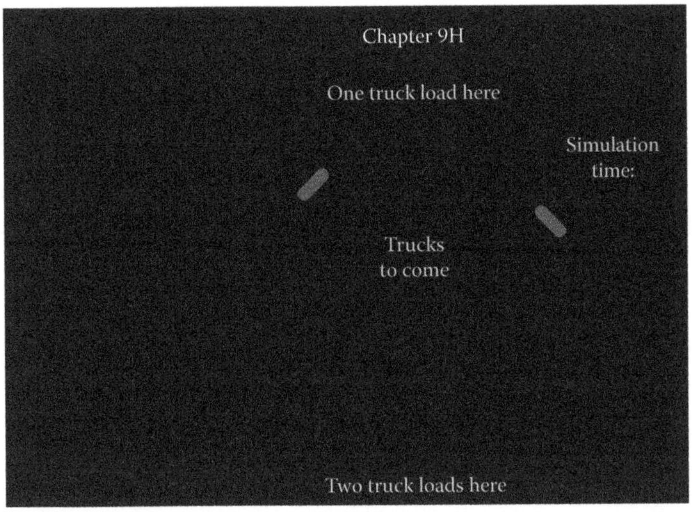

FIGURE 9.14
Screenshot from program CHAP9H.

floor. This takes them 12 ± 4 seconds to walk up (no escalator in this store). They take 4 ± 1 minutes to shop and then leave in 4 ± 1.5 seconds. In 2 hours, how many people have left the store?

9.4. Consider the program CHAP9C. Modify it to determine how long it will take until 25 trucks have left the system.

9.5. For the same example as Exercise 9.1, determine how long it would take for 50 trucks to have left the system. Is this result half the time for Exercise 9.1? Would you expect it to be?

9.6. Mining trucks come to an intersection. 25% of the trucks will go to the leach pad area, 50% to the crusher, and the rest to the waste. Write the GPSS/H to model this. Run for 1000 trucks to travel to each of the three areas to verify your results.

9.7. Mining trucks travel from the loader to the crusher in 10 ± 2 time units where they dump into a crusher in 1 time unit. After they leave the crusher, 10% of the trucks need to refuel. This takes 20 ± 8.5 time units total to return to the loader. The rest of the trucks return to the loader in 12 ± 3.4 time units. Suppose loading takes 2 time units exactly and there is no restriction on loading a truck; that is, as soon as a truck is at the loader, it loads. Also, there is no restriction on dumping trucks. Assume that the mine has five trucks. Write the program to simulate this for 480 time units.

9.8. Modify Exercise 9.7 to have six trucks in the mine, then seven, and, finally, eight.

9.9. Modify the program CHAP9E.GPS to run for six trucks, then seven trucks, and, finally, for eight trucks. Why do you think that the results are unrealistic?

10

Making .ATF Files

A bucket-wheel excavator is the largest equipment in the world.

10.1 Making .ATF Files, PUTSTRING, BPUTSTRING, BPUTPIC, PUTPIC, and Standard Numerical Attributes

This chapter will introduce the way GPSS/H creates files. This is essential to making any large .ATF file that is going to be used with the .LAY file in PROOF. After this chapter, the student will (finally) start to make his or her own complete animation programs. Other topics that are used with animation files also will be covered. These topics are essential for creating any animation .ATF file using GPSS/H.

Understanding the material presented in this chapter is essential for learning how to make animation files.

10.1.1 The PUTSTRING Statement

First, let us learn how to put text on the screen. The easiest way to put a single line of text on the screen is by means of the `PUTSTRING` statement. This name is derived from `PUT` a `STRING` (of text). This is generally placed either before any Blocks or after the `START` statement. Thus, these statements are executed either during compiling (before the program is actually run) or near the end of the program. If they are placed after the `START` statement, they will be executed after the main program is run. The general form is as follows:

```
PUTSTRING (' any text')
```

The *any text* will appear on the screen as written. For example,

```
PUTSTRING ('HELLO, MY GOOD MAN')
```

will result in the line of text being placed on the screen starting in position 1. It will be:

```
HELLO, MY GOOD MAN
```

If you want a blank line, you put:

```
PUTSTRING ('  ')
```

It may be hard to see from this example, but there are two spaces between the apostrophes. There is a way to make a blank screen and make *n* blank lines that we will learn later. There is no single statement in GPSS/H that makes a blank screen.

You may use any ASCII code in a PUTSTRING. Some other examples of a PUTSTRING might be as follows:

```
PUTSTRING ('This is a program to simulate a barber shop')
```

Notice that lower case characters are allowed in the PUTSTRING.

```
PUTSTRING (' This is goodbye, my dear Mr. Jones')
```

The space after the first quote mark will have the text printed with a space before it. Since the quote marks are used in the PUTSTRING to indicate what text is to be printed, you cannot have a quote mark printed out directly by the PUTSTRING.

The following GPSS/H program, CHAP10A.GPS, uses several PUTSTRINGs (Figure 10.1). When the program is run, a portion of the screen would be as follows shown by Figure 10.2.

```
          SIMULATE
*******************************
*  PROGRAM CHAP10A.GPS         *
*  EXAMPLES OF PUTSTRING       *
*******************************
          PUTSTRING    ('   ')
          PUTSTRING    ('   ')
          PUTSTRING    ('   ')
          PUTSTRING    ('   Intro. to simulation')
          PUTSTRING    ('   My first example')
          PUTSTRING    ('   NOTICE THAT SMALL AND CAPITAL')
          PUTSTRING    ('   LETTERS CAN BE USED')
          PUTSTRING    ('   ')
          GENERATE     10,4
          ADVANCE      9,5
          TERMINATE
          GENERATE     100
          TERMINATE    1
          START        1
          PUTSTRING    ('   ')
          PUTSTRING    ('   RESULTS IN FILE CHAPTER10A.LIS')
          PUTSTRING    ('   All done for now....')
          END
```

FIGURE 10.1
Listing of program CHAP10A.GPS.

```
Intro. to simulation
My first example
NOTICE THAT SMALL AND CAPITAL
LETTERS CAN BE USED

RESULTS IN FILE CHAP10A.LIS
All done for now....
```

FIGURE 10.2
Output from program CHAP10A.GPS.

This program is stored as CHAP10A.GPS. It can be run to verify that the output is as given here. The student is encouraged to write the program first before resorting to examining the stored file.

There is also the Block form of the `PUTSTRING`, which is simply,

```
BPUTSTRING (' text')
```

When this Block is used, whenever a transaction passes through it, the line of text, *text*, is placed on the screen. The next section gives a Block that is used more frequently than the `BPUTSTRING` Block. In fact, this Block is the one that will be used when interfacing the GPSS/H program with the animation layout files. Thus, it is a very important one.

10.1.2 The PUTPIC Statement and the BPUTPIC Block

GPSS/H allows for another way to customize the output from a program. One can direct the output as follows:

to the screen

to an output file

The `PUTSTRING` statement is used only to send output to the screen. For many programs that are only for simulation and that do not involve any animation, it is simple enough to have the output sent to the screen. However, for large simulation examples, the output may be too large, and the output may need to be sent to a file and examined when the program is done executing. In order to create a file to be used with the layout file in PROOF, it is essential that the output be directed towards an output file. In addition, one may want the results of the program also included in the output. This cannot be done using a `PUTSTRING` statement.

First, let us learn how to create output in GPSS/H using the `BPUTPIC` Block and `PUTPIC` statement. The general form of the `BPUTPIC` Block is as follows:

```
BPUTPIC FILE = (label),LINES = (integer),list
```

The `FILE = (label)` will be covered shortly.

The *list* contains the items to be a part of the output file separated by commas. The `LINES = (integer)` refers to the number of lines in the output. The statement `PUTPIC` stands for `PUT` a `PIC`ture, which is the way GPSS/H sends output to a specified place. If the `FILE = (label)` is omitted, the output will be sent to the PC screen. We will shortly learn both how to write to an output file and what some typical items might be a part of this *list* specification. Next, let us examine a few samples of the `BPUTPIC` Block:

```
        BPUTPIC        LINES = 1
Hello, this is a line that will be on the screen
```

When a transaction passes through this Block, the line of code,

```
Hello, this is a line that will be on the screen
```

is shown on the screen.

Notice that the text is in both uppercase and lowercase characters. This is acceptable for output. *The specification* `LINES = 1` *can be omitted if there is only one line of output.* It would be an error to have the Block

```
        BPUTPIC        LINE = 1
First line of output
```

as it *must* be `LINES = 1` (plural)

The following would be a correct usage of the `BPUTPIC` Block:

```
        BPUTPIC        LINES = 2
This is the first line of output
This is the second
```

To illustrate these examples, consider the following program (Figure 10.3).

This program is stored as CHAP10B.GPS. However, it should be easy to re-create. When you run it, you will see the three lines of text output on the screen. These will begin in position 3, since this is where they are located on the lines after the `BPUTPIC` Block. They can be located anywhere on the screen using spaces (keeping in mind that only about 80 characters show on a PC screen line). The output will *not* have the two blank lines. Figure 10.4 shows what the output will be. In order to have blank lines, one needs to have multiple `BPUTPIC` Blocks with a `BPUTSTRING` Block in between.

If one had the Block

```
        BPUTPIC        LINES = 1
Example of output from a BPUTPIC Block
```

```
          SIMULATE
*****************************
*   PROGRAM CHAP10B.GPS     *
*   EXAMPLE OF BPUTPIC      *
*   BLOCK                   *
*****************************
          GENERATE     1
          BPUTPIC      LINES=5
This is the first line of output


This is the second
And, this is the third line
          TERMINATE    1
          START        1
          END
```

FIGURE 10.3
Listing of program CHAP10B.GPS.

```
This is the first line of output
This is the second
And, this is the third line
```

FIGURE 10.4
Output from program CHAP10B.GPS.

and the Block

```
      BPUTPIC
Example of output from a BPUTPIC Block
```

the output from each would be identical.

10.1.3 The PUTPIC Statement

The PUTPIC statement is executed only once (unless it is in a DO loop, which will be covered later). It will be executed during compiling or after the program is over depending on where it is located. For example, let us modify program CHAP10B.GPS to become CHAP10C.GPS (Figure 10.5) as follows:

```
SIMULATE
```

When this program is run, the line of output

```
This line is on the screen before the program is executed
```

will be on the screen before any Blocks are executed.

During execution of the Blocks, there will be the lines

```
This is the first line of output
This is the second line
AND THIS IS THE THIRD
```

```
*****************************
*  PROGRAM CHAP10C.GPS      *
*  EXAMPLE OF PUTPIC        *
*  STATEMENT                *
*****************************
           PUTPIC       LINES=1
   This line is on the screen before the program is executed
           GENERATE     1
           BPUTPIC      LINES=5


   This is the first line of output
   This is the second
   And, this is the third line
           TERMINATE    1
           START        2
           PUTPIC       LINES=2
   This is placed on the screen after execution
   THE END
           END
```

FIGURE 10.5
Listing of program CHAP10C.GPS.

```
This line is on the screen before the program is executed
This is the first line of output
This is the second
And, this is the third line
This is the first line of output
This is the second
And, this is the third line
This is placed on the screen after execution
THE END
```

FIGURE 10.6
Output from CHAP10C.GPS program.

on the screen twice (because of the `START 2` statement). After the program Blocks are executed, there will be the two lines of output:

```
This is placed on the screen after execution.
THE END
```

Thus, the output will look as shown in Figure 10.6.

If one places the `PUTPIC` statement in the middle of the program, it will be executed during compiling and *not* after program execution.

The student is encouraged to experiment with different examples of both the `BPUTPIC` Block and `PUTPIC` statement to determine the results.

10.1.4 Making .ATF Files: The `FILEDEF` Statement

If one wants to have a permanent (saved) record of the output or if the output takes more than one page, it is desirable to write the output to a separate file. Making an output file is also essential for creating .ATF files that are to be used with the layout in PROOF to run the animations. Hence, the following is a very important topic. It is essential that this statement be understood and used correctly.

The way to make these output files in GPSS/H is a two-step process. The first involves defining the name of the output file. This can be any legal name with any extension desired. The statement to do this is as follows:

```
(label) FILEDEF 'output file name and destination'
```

The `FILEDEF` stands for `FILE DEF`inition.

The two quote marks are essential. The (label) is a dummy that is used when creating the file. The reason for this label will become apparent shortly.

For example, the following are all valid forms of a `FILEDEF`:

```
MYOUT FILEDEF      'DEBUG.OUT'
MYFILE FILEDEF     'JAN31B.OUT'
ABCDE FILEDEF      'C:\mysub\myresults\RESULTS.LIS'
ATFF FILEDEF       'C:\SP5\JAN28.OUT'
```

If no path is specified but only the file name, the file will be written in the subdirectory where GPSS/H is located. Thus, in the above, the files created by both `MYOUT` and `MYFILE` will be found in the subdirectory C:\GPSSH> (assuming this is where GPSS/H has been loaded). The file `RESULTS` created by reference to the label `ABCDE` will be found in

```
C:\mysub\myresults\
```

The files created by reference to the label `ATFF` will be located in the following subdirectory:

C:\SP5\

If PROOF is located in the subdirectory C:\PROOF>, the most common form of the `FILEDEF` will be as follows:

```
(label) 'C:\PROOF\(name).ATF
```

This is a statement, so it has to come anywhere before it is first referenced. In this book, this and other such specifications will generally come at the beginning of the program just after the `SIMULATE` statement.

Let us modify the program CHAP10B.GPS to become CHAP10D.GPS (Figure 10.7).

Now, when the program is run, there will be no output on the screen. Instead, the output will be located in the subdirectory C:\. The output will be in the file:

```
CHAP10D.OUT
```

The output file is shown in Figure 10.8.

In this case, there *will* be two blank lines at the beginning of the output, unlike the situation where the output is sent to the screen.

The next section will introduce the student to some of the various output data that one might want to place in the output file.

When one uses either the `BPUTPIC` or the `PUTPIC` Block, it is immaterial if one has the `LINES = (number)` or the `FILE = (label)` first or second. The Blocks

```
                SIMULATE
***********************************
*  PROGRAM CHAP10D.GPS            *
*  MODIFICATION OF CHAP10B.GPS    *
*  THIS ILLUSTRATES THE           *
*  CREATION  OF AN OUTPUT FILE    *
***********************************
   MYOUT     FILEDEF      'CHAP10D.OUT'
             GENERATE     1
             BPUTPIC      LINES=5,FILE=MYOUT


  This is the first line of output
  This is the second
  And, this is the third line
             TERMINATE    1
             START        1
             END
```

FIGURE 10.7
Listing of program CHAP10D.GPS.

```
(blank line)
(blank line)
  This is the first line of output
  This is the second
  And, this is the third line
```

FIGURE 10.8
Output from program CHAP10D.

```
       BPUTPIC       FILE = MYOUT,LINES = 2
THIS IS THE FIRST LINE OF OUTPUT
MY NAME IS MARY
```

and

```
       BPUTPIC       LINES = 2,FILE = MYOUT
THIS IS THE FIRST LINE OF OUTPUT
MY NAME IS MARY
```

would produce identical results.

10.1.5 Some Common Standard Numerical Attributes

Take a look at any list file. You will always see the clock value (actually, two values, one for the Absolute Clock and one for the Relative Clock). We will use the Absolute Clock value here. The difference between the clocks will be covered in Chapter 23. Notice also that there is always a Block count of the times a transaction has either entered or left the Block. If there are any transactions left in the system when the program is over, these are given under the current count. All of these (and many more items) are called *standard numerical attributes (SNAs)* and are given names. These are as listed below:

`AC1`	Absolute Clock
`N(label)`	Block count—total
`W(label)`	Block count—current

When a transaction is created, it is given a number. This is called `XID1` and is very important in making animations. We rarely care what this number is, only that each is unique. Recall that transactions are created either during compiling (when a transaction is scheduled to enter the system) or during execution (when a transaction leaves the `GENERATE` Block). The number of each transaction is generally not important to the program only the fact that each is unique. *Nearly every .ATF file that is created using a GPSS/H program will make use of this numbering of the transaction.*

The SNAs given above can be used in the program or written out either on the screen during the running of the program (which is also nice for debugging) or at the end of the program. How to do this will be covered next.

When the value of an SNA is to be written out either to the screen or to a file, it is necessary to give a field specification where the value is to be written. This is done using asterisks (*). These can be given with decimal points. For example, suppose that the value of the simulation time, AC1, is 56.78 time units. If one had the line of code

```
       BPUTPIC       AC1
The time is ***.**
```

when a transaction executed this Block, the output would be:

```
The time is 56.78
```

In the event that the field as specified by the asterisks is too small for the number, the value still is placed in the field, but this may distort the output. For example, suppose that the values of two variables are 2 and 234. If these are to be output as follows:

```
$$$$$$$$$$$$$$$$$$$$$$$$$$
$ The value of one is * $
$ The value of one is * $
$$$$$$$$$$$$$$$$$$$$$$$$$$
```

The output would look as follows:

```
$$$$$$$$$$$$$$$$$$$$$$$$$$
$ The value of one is 2 $
$ The value of one is 234 $
$$$$$$$$$$$$$$$$$$$$$$$$$$
```

Notice that the output box is a bit distorted. It is generally best for the appearance of the output to allow enough field length for the values of the variables to fit without having any distortion.

Let us consider an example of using these SNAs in GPSS/H programs.

Example 10.1: Study of the SNA AC1 (Clock Time)

Examine the file, CHAP10E.GPS. This is a program to output the times when transactions are placed into the simulation. These times are given by `AC1`. The program listing is given in Figure 10.9. A portion of the output is shown in Figure 10.10.

Example 10.2: The SNAs `AC1, W(label),` and `N(label)`

This example will illustrate the SNAs given by `AC1, W(`*`label`*`),` and `N(`*`label`*`)`. The GPSS/H program is given by CHAP10F.GPS.

The listing of it is as shown in Figure 10.11.

A portion of the output from the program is given in Figure 10.12.

```
                SIMULATE
******************************
*   CHAP10E.GPS    PROGRAM TO  *
*   ILLUSTRATE THE SNA AC1     *
*   WHICH IS THE CLOCK TIME    *
******************************
                GENERATE    10,,0
                BPUTPIC     AC1
  THE TIME IS   ***.**
                TERMINATE   1
                START       11
                END
```

FIGURE 10.9
Listing of program CHAP10.E.GPS.

```
THE TIME IS    0.00
THE TIME IS   10.00
THE TIME IS   20.00
THE TIME IS   30.00
THE TIME IS   40.00
THE TIME IS   50.00
THE TIME IS   60.00
THE TIME IS   70.00
THE TIME IS   80.00
THE TIME IS   90.00
THE TIME IS  100.00
```

FIGURE 10.10
Sample output from program CHAP10E.GPS.

```
           SIMULATE
*********************************
*  PROGRAM CHAP10F.GPS  THIS    *
*  SHOWS HOW THE THREE SNAs     *
*  AC1, N(LABEL) AND W(LABEL)   *
*  CAN BE USED FOR OUTPUT       *
*********************************
 PEOPLE   GENERATE     10,2,0
 INSYS    ADVANCE      11,1
 LEFT     TERMINATE    1
          START        10
          PUTPIC       LINES=4,AC1,N(PEOPLE),W(INSYS),N(LEFT)
END OF SIMULATION AT TIME ***.**
PEOPLE TO ENTER SYSTEM      **
PEOPLE LEFT IN SYSTEM        *
PEOPLE WHO LEFT THE SYSTEM **
           END
```

FIGURE 10.11
Listing of program CHAP10F.GPS.

```
END OF SIMULATION AT TIME 101.13
PEOPLE TO ENTER SYSTEM      11
PEOPLE LEFT IN SYSTEM        1
PEOPLE WHO LEFT THE SYSTEM 10
```

FIGURE 10.12
Output from program CHAP10F.GPS.

It is *essential* that the number of lines correspond exactly with the actual lines as specified in the `PUTPIC` statement. Not having the lines of output corresponding to the `LINES = (number)` specification is one of the most common errors.

If there is too much output so that the list goes beyond one line, one can use the underscore (_) to continue the `PUTPIC` to the next line. For example,

```
PUTPIC LINES = 1,N(BLOCK1),N(BLOCK2),W(BLOCK3),_
       AC1
```

Notice that, even though there is only one line in the output, it still must be the plural `LINES = 1`. If this is omitted, it is assumed that there will be only one line of output. The above could have been written as follows:

```
PUTPIC N(BLOCK1),N(BLOCK2),W(BLOCK3),_
       AC1
```

The output can often be made to look a bit fanciful. For example,

```
          PUTPIC LINES = 7,AC1,N(BLOCK2),N(BLOCK3),W(BLOCK3)
$$$$$$$$$$$$$$$$$$$$$$$$$$$$$$$$$$$$$$$$$$
^^    RESULTS OF SIMULATION                 ^^
^^    SIMULATE WAS FOR ***.** MINUTES       ^^
^^    CARS TO ENTER SYSTEM ***              ^^
^^    CARS TO NEED SERVICE ***              ^^
^^    CARS BEING SERVICED AT END ***        ^^
$$$$$$$$$$$$$$$$$$$$$$$$$$$$$$$$$$$$$$$$$$
```

Depending on the final values, the output for the above might be like the following:

```
$$$$$$$$$$$$$$$$$$$$$$$$$$$$$$$$$$$$$$$$$$
^^    RESULTS OF SIMULATION               ^^
^^    SIMULATE WAS FOR 480.00 MINUTES     ^^
^^    CARS TO ENTER SYSTEM 31             ^^
^^    CARS TO NEED SERVICE 12             ^^
^^    CARS BEING SERVICED AT END 3        ^^
$$$$$$$$$$$$$$$$$$$$$$$$$$$$$$$$$$$$$$$$$$
```

Keying in a BPUTPIC statement incorrectly can lead to one of the commonest errors in writing GPSS/H code. It is always a good idea to go back and check the LINES = n to make sure that the lines in the PUTPIC correspond to the actual lines you wrote. The same holds for making sure that the output fields correspond to the variable to be in the output.

Example 10.3: Creating Output Files

Consider the following GPSS/H program that is stored as CHAP10G.GPS (Figure 10.13).

There will be no output on the screen. The output will be in the file MYOUT.OUT in the same subdirectory the program was run in. It will look as shown in Figure 10.14.

Each time a transaction went through the BPUTPIC Block, one line of output would be written to the file MYOUT.OUT.

```
**********************************
*  PROGRAM 10G.GPS               *
*  ILLUSTRATING THE SNA XID1     *
**********************************
          SIMULATE
 MYFIRST  FILEDEF      'MYOUT.OUT'
          GENERATE     10
          BPUTPIC      FILE=MYFIRST,LINES=2,AC1,XID1
 TIME *.****
 The transaction's ID is **
          TERMINATE    1
          START        10
          END
```

FIGURE 10.13
Listing of program CHAP10G.GPS.

```
TIME 10.0000
 The transaction's ID is  1
 TIME 20.0000
 The transaction's ID is  2
 TIME 30.0000
 The transaction's ID is  3
 TIME 40.0000
 The transaction's ID is  4
 TIME 50.0000
 The transaction's ID is  5
 TIME 60.0000
 The transaction's ID is  6
 TIME 70.0000
 The transaction's ID is  7
 TIME 80.0000
 The transaction's ID is  8
 TIME 90.0000
 The transaction's ID is  9
 TIME 100.0000
 The transaction's ID is 10
```

FIGURE 10.14
Output from program CHAP10G.GPS.

When the GPSS/H program is going to be used to make an .ATF file, it is important that one always uses the clock value as the first line in the `BPUTPIC` Block. This is because PROOF needs to have the clock value at the first line in a sequence of commands. Example 10.4 shows how to create an animation file.

Example 10.4: Creating an Animation File

Write the simulation program and create the animation file that places the value of the time of the simulation and the transaction ID numbers on the screen for the following example. A person comes along every 10±1.5 time units. The person remains on the screen for 7 time units and then disappears. Simulate for 100 time units (Figure 10.15).

The file CHAP10H.LAY needs to be created separately using PROOF. Once the program CHAP10H.GPS is run, the file CHAP10H.ATF is created and can be copied to where PROOF* is stored. A screenshot is shown in Figure 10.16.

Example 10.5: A Modification of Example 10.4

Modify the last example so that the time each transaction is in the system is now 11. (If the program runs for a very long time, there eventually will be an error when too many transactions are in the system.) After the program is run, the message on the screen is to give the number of people who have been created, the number currently in the system, and the number who have left.

```
          SIMULATE
***********************************
*  PROGRAM CHAP10H.GPS            *
*  THIS MAKES A SIMPLE ANIMATION  *
***********************************
 ATF      FILEDEF     'CHAP10H.ATF'
          GENERATE    10,1.5
          BPUTPIC     FILE=ATF,LINES=3,AC1,AC1,XID1
TIME *.****
WRITE M1  ***.**
WRITE M2   **
          ADVANCE     7
          TERMINATE
          GENERATE    100
          TERMINATE   1
          START       1
          PUTPIC      FILE=ATF,LINES=2,AC1
TIME *.****
END
          END
```

FIGURE 10.15
Listing of program CHAP10H.GPS—this creates the file CHAP10H.ATF.

FIGURE 10.16
Screenshot of animation from CHAP10H.

---

* Alternatively, the statement `FILEDEF` could have had the path to where PROOF is stored. For example, one could have something such as `ATF...FILEDEF...'C:\DATA\SP5\CHAP10H.ATF'`. This would save the step of copying the file each time a new animation is to be run.

This is stored as program CHAP10I.GPS. Notice that, in order to give the desired results, it was necessary to add labels to the three Blocks:

```
GENERATE  10,,0
ADVANCE
```

and

```
TERMINATE
```

The listing of the program is shown in Figure 10.17.

The animation is easily changed. A screenshot is given in Figure 10.18.

```
          SIMULATE
****************************************
*   PROGRAM CHAP10I.GPS                 *
*   MODIFICATION OF PROGRAM CHAP10H.GPS *
****************************************
 ATF      FILEDEF     'CHAP10I.ATF'
 PEOPLE   GENERATE    10,1.5
          BPUTPIC     FILE=ATF,LINES=3,AC1,AC1,XID1
TIME *.****
WRITE M1  ***.**
WRITE M2   **
 STILLIN  ADVANCE     11
 LEFT     TERMINATE
          GENERATE    100
          TERMINATE   1
          START       1
          PUTPIC      FILE=ATF,LINES=6,AC1,N(PEOPLE),W(STILLIN),_
                      N(LEFT)
TIME *.****
WRITE M3  AT THE END OF THE PROGRAM:
WRITE M4 ***
WRITE M5 ***
WRITE M6 ***
END
          END
```

FIGURE 10.17
Listing of program CHAP10I.GPS—this is a modification of program CHAP10H.GPS.

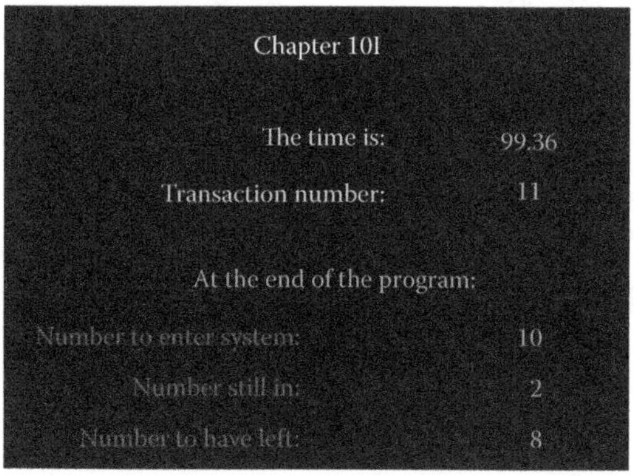

FIGURE 10.18
Screenshot of animation from program CHAP10I.

Example 10.6: A Mining Truck Travelling from A to B

Consider the case of a mining truck that is travelling on a path from A to B. Trucks appear at A every 10 time units. They travel to point B in 8 time units. The trucks then leave the system. Write the GPSS/H code to simulate and then animate this system. Simulate for 100 time units.

Solution

The program to do the simulation is quite short. It is CHAP10J.GPS and is given in Figure 10.19.

The animation program is given by CHAP10K.GPS program. This program should be studied carefully as the lines of code to create the trucks, to place them on a path, and to have them travel along this path will be repeated numerous times in the mine design examples. The listing of the program is given in Figure 10.20. The animation is shown in Figure 10.21.

```
          SIMULATE
*********************************
*  PROGRAM CHAP10J.GPS          *
*  SIMULATION FOR EXAMPLE 10.6  *
*********************************
          GENERATE    10,,0     TRUCK COMES ALONG AT A
          ADVANCE    8          TRAVEL TO B
          TERMINATE             TRUCK LEAVES
          GENERATE     100
          TERMINATE    1
          START        1
          END
```

FIGURE 10.19
Listing of program CHAP10J.GPS.

```
          SIMULATE
**************************
*  PROGRAM CHAP10K.GPS   *
*  ANIMATION OF          *
*  EXAMPLE 10.6          *
**************************
  ATF     FILEDEF     'CHAP10K.ATF'    DEFINE FILE
          GENERATE    10,,0     TRUCK COMES ALONG
          BPUTPIC     FILE=ATF,LINES=4,AC1,XID1,XID1,XID1
TIME *.****
CREATE TRUCK T*
PLACE T* ON P1
SET T* TRAVEL 8
          ADVANCE    8          TRAVEL TO B
          BPUTPIC    FILE=ATF,LINES=2,AC1,XID1
TIME *.****
DESTROY T*
          TERMINATE             TRUCK LEAVES
          GENERATE     100
          TERMINATE    1
          START        1
          PUTPIC       FILE=ATF,LINES=2,AC1
TIME *.****
END
          END
```

FIGURE 10.20
Listing of program CHAP10K.GPS.

FIGURE 10.21
Screenshot of animation from CHAP10K.GPS.

Example 10.7: Extension of Example 10.6

Once the animation of Example 10.6 is understood, it should not be difficult to add the following: Suppose point B in the example is a junction where 40% of the trucks travel to C and the rest to D. It takes 5 time units to travel from B to C and 6 time units to travel from B to D. Write the program to do the simulation and then the animation.

Solution
The program to do the simulation is easily modified from CHAP10J.GPS. This is CHAP10L.GPS program. Its listing is given in Figure 10.22.

The animation for the example is CHAP10M.GPS program. Its listing is given in Figure 10.23.

The first GENERATE Block is modified slightly to have the first truck appear at time 0. This is only so that the animation will show a truck arriving at this time, as shown in Figure 10.24.

```
          SIMULATE
*********************************
*   PROGRAM CHAP10L.GPS          *
*   SIMULATION FOR EXAMPLE 10.7  *
*********************************
          GENERATE    10        TRUCK COMES ALONG AT A
          ADVANCE     8         TRAVEL TO B
          TRANSFER    .6,,AWAYD
          ADVANCE     5
          TERMINATE             TRUCK LEAVES
 AWAYD    ADVANCE     6
          TERMINATE             TRUCK LEAVES
          GENERATE    100
          TERMINATE   1
          START       1
          END
```

FIGURE 10.22
Listing of program CHAP10L.GPS.

```
          SIMULATE
*********************************
*   PROGRAM CHAP10M.GPS         *
*  ANIMATION FOR CHAP10L.GPS    *
*                               *
*********************************
  ATF     FILEDEF     'CHAP10M.ATF'    DEFINE FILE
          GENERATE    10,,0    TRUCK COMES ALONG AT A
          BPUTPIC     FILE=ATF,LINES=4,AC1,XID1,XID1,XID1
TIME *.****
CREATE TRUCK T*
PLACE T* ON P1
SET T* TRAVEL 8
          ADVANCE     8          TRAVEL TO B
          TRANSFER    .6,,AWAYD
          BPUTPIC     FILE=ATF,LINES=3,AC1,XID1,XID1
TIME *.****
PLACE T* ON P2
SET T* TRAVEL 5
          ADVANCE     5
          BPUTPIC     FILE=ATF,LINES=2,AC1,XID1
TIME *.****
DESTROY T*
          TERMINATE
 AWAYD    BPUTPIC     FILE=ATF,LINES=3,AC1,XID1,XID1
TIME *.****
PLACE T* ON P3
SET T* TRAVEL 6
          ADVANCE     6
          BPUTPIC     FILE=ATF,LINES=2,AC1,XID1
TIME *.****
DESTROY T*
          TERMINATE
          GENERATE     100
          TERMINATE    1
          START        1
          PUTPIC       FILE=ATF,LINES=2,AC1
TIME *.****
END
          END
```

FIGURE 10.23
Listing of program CHAP10M.GPS.

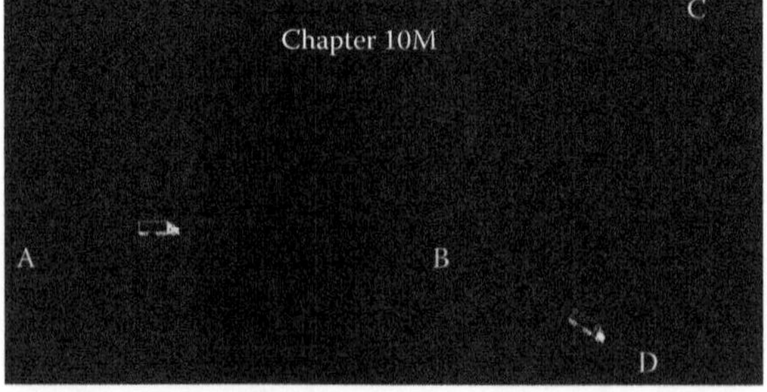

FIGURE 10.24
Screenshot of animation from CHAP10M.GPS.

10.2 Exercises

10.1. What will the output for the following `PUTPIC` statement look like? Assume that `AC1` = 450, `W(FIRST)` is 5, `W(SECOND)` is 0, and `N(BLOCK8)` is 234.

```
PUTPIC LINES = 3,AC1,W(FIRST),W(SECOND),N(BLOCK8)
THE SIMULATION RAN FOR ***.** TIME UNITS
NUMBER OF TRUCKS AT ONE ** NUMBER OF TRUCKS AT TWO **
TOTAL TRUCKS TO PASS THROUGH SYSTEM WAS ***
```

10.2. What will the output be for the following `PUTPIC` statement, assuming that the value of `AC1` is 500, `N(BLOCKA)` is 34, `N(BLOCKB)` is 56, `W(FIRST)` is 2, and `W(SECOND)` is 0?

```
PUTPIC LINES = 6,AC1,N(BLOCKA),N(BLOCKB),W(FIRST),_
W(SECOND)
THE RESULTS OF THE SIMULATION ARE AS FOLLOWS:
THE SIMULATION LASTED FOR ****.** MINUTES
THE NUMBER OF PEOPLE TO COME TO THE SHOP WAS ***
THE NUMBER OF PEOPLE TO COME TO THE HOUSE WAS ***
THE NUMBER OF PEOPLE STILL IN THE SHOP WAS **
THE NUMBER OF PEOPLE STILL IN THE HOUSE WAS **
```

10.3. Write the following program. What is the output like?

```
                 SIMULATE
            ATFF FILEDEF        'MYFIRST.LIS'
        TRUCKS   GENERATE       ,,,4
        BACKUP   ADVANCE        1
                 ADVANCE        2
                 ADVANCE        3
                 TRANSFER       ,BACKUP
                 GENERATE       6
                 TERMINATE      1
                 START          1
                 PUTPIC         FILE = ATFF,LINES = 3,AC1,N(TRUCKS)
            Here is the output
            The simulation ran for ***.** time units
            The number of trucks used was **
                   END
```

10.4. Consider the program CHAP10N.GPS whose listing is given in Figure 10.25.

Before running it, decide what the output will be like. After you run it, change the two `BPUTSTRING` statements to `PUTSTRING` and the `BPUTPIC` Block to `PUTPIC`. Now, what is the output? Why is it different?

10.5. A transaction enters the system every 10 ± 4 time units. It is put on the future event chain (FEC) for 9 ± 6 time units and then removed from the system. The simulation program is CHAP10O.GPS. The animation program is CHAP10P.GPS. This gives the transaction number and the time it entered the system. The animation is shown in Figure 10.26.

```
          SIMULATE
****************************
*  PROGRAM CHAP10N.GPS TO   *
*  GO WITH EXERCISE 10.4    *
****************************
          BPUTSTRING    ('  HELLO')
          BPUTSTRING    ('  PROGRAM CHAP10N.GPS')
          GENERATE     ,,10,1
          BPUTPIC     LINES=1,AC10
  THE TIME IS **.**
          TERMINATE    1
          START        1
          END
```

FIGURE 10.25
Listing of program CHAP10N.GPS.

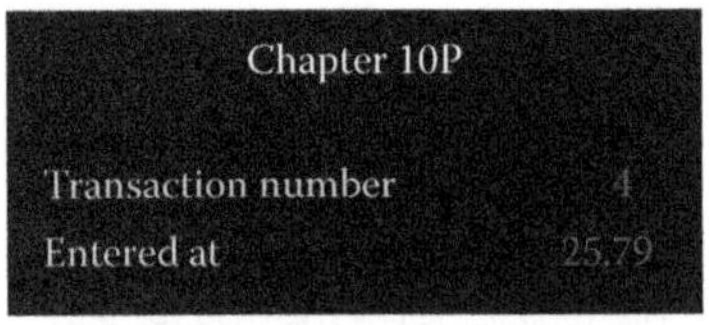

FIGURE 10.26
Screenshot of animation from program CHAP10P.GPS.

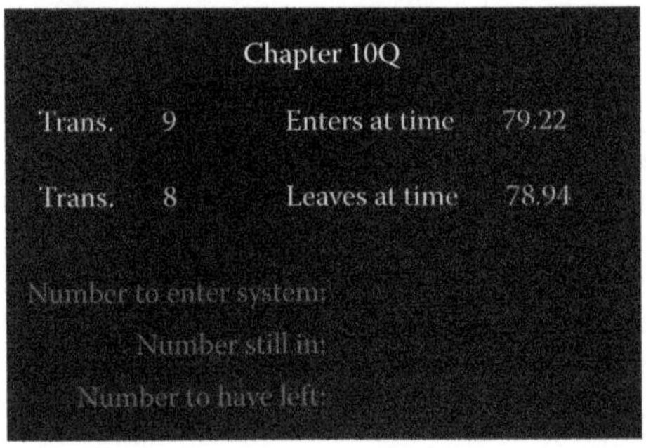

FIGURE 10.27
Screenshot of animation from CHAP10Q.GPS.

Make a new animation and program called CHAP10Q.GPS that will show the transaction number not only when one enters the system but also when a transaction leaves the system. Assume now that transactions still enter the system every 10 ± 4 time units but that they stay for 12 ± 6 time units. Your animation will look as shown in Figure 10.27.

10.6. The animation from program CHAP10Q.GPS as shown in Figure 10.27 is rather bland. It is possible to add a few icons to it. The file ICON.LAY, which is found in the subdirectory ANIMATIONS, gives multiple icons that can be used in animations. This was furnished by Wolverine Software. Add a few icons to the animation given by CHAP10Q.GPS.

10.7. Make the GPSS/H program and animation that blinks your first name on the screen for 100 time units. *Blink* means that it appears for 6 time units then disappears for 2 time units, appears for 6 time units, disappears for 2 time units, and

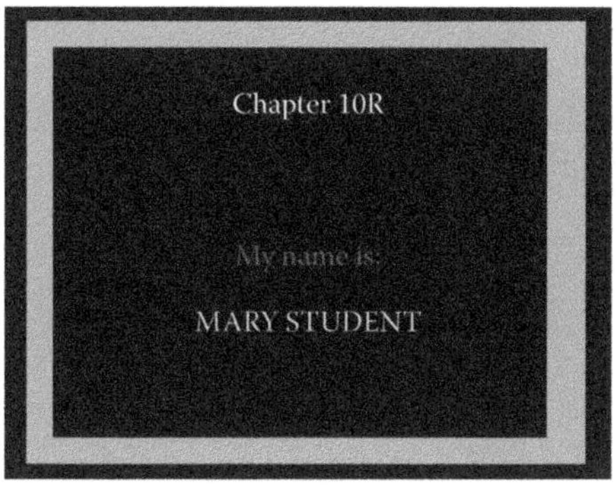

FIGURE 10.28
Screenshot of animation from CHAP10R.GPS.

so on. A sample program is found in the file CHAP10R.GPS. Try to do this before looking at the stored program CHAP10R.GPS. Your animation will look something like as shown in Figure 10.28.

10.8. Debugging GPSS/H programs is necessary from time to time when a typing error is made. Many errors are made when using the `BPUTPIC` Block. The program CHAP10S.GPS has been written with deliberate errors in them. Run them and correct the errors. The programs will be considered as being debugged when they can compile and execute with no errors.

11

QUEUE/DEPART Blocks

Trucks working in a coal mine in Borneo.

11.1 QUEUE/DEPART Blocks

11.1.1 The QUEUE Block

GPSS/H is used often for simulation of systems where queues are formed at many places. Thus, it is natural to learn how the language handles queues. A discussion of queues and the mathematical theory associated with them can be found in any textbook on operations research. In fact, there are complete books devoted to this important topic. One thing becomes quite clear when one studies queuing theory: the number of queuing problems that have exact mathematical solutions is surprisingly small. This is especially so when one is dealing with a finite number of transactions, such as the case of transactions cycling through a system. People coming to a grocery store can be considered as coming from an infinite population where the interarrival rate of one person is not dependent on how many people are in the shop. However, trucks hauling waste from a construction site cannot be considered as being from an infinite population, as there might be only a total of six trucks at the site.

There are many cases when a transaction will be denied access to a Block during a simulation. When a transaction is to use a facility that is already in use, it is denied entry and has to remain in the Block where it presently resides. In the system being simulated, this gives rise to a queue formation. Such queues are commonly found in real-life situations. These might be found in a mine with a single shovel and many trucks, a barbershop with only one barber, a checkout counter in a grocery store, a bank with many tellers, an airport with only a few runways, and so on. Often the purpose of the simulation study is to see where these queues form and how they might be eliminated or, perhaps, kept to a reasonable level. These queuing situations are handled in GPSS/H by the QUEUE Block.

The normal form of the QUEUE Block is quite simple. It is

```
QUEUE A
```

where the operand A is either a name (not more than five characters in fixed format and eight characters in free format) or a number. It could also be a variable, as we shall learn. Since GPSS/H has many reserved words, it is recommended that the operand be more than three letters. This is not a strict requirement but will always avoid the accidental use of a reserved word by mistake. Thus,

```
QUEUE 1
QUEUE FIRST
QUEUE 7
QUEUE ONE
QUEUE DUMP1
QUEUE STOPHERE
```

are examples of valid QUEUE Blocks, but

```
QUEUE -1
QUEUE LASTONEIN
```

are not.

Sometimes you will decide to use a number rather than a name for the QUEUE Block's operand. If you do choose this, the number cannot be arbitrary but will depend on the actual number of QUEUE Blocks allowed in your system. Normally, at least 50 QUEUE Blocks are allowed in most GPSS/H processors. Thus, if this is the maximum number allowed in your system, it would be all right to have

```
QUEUE 27
QUEUE 50
```

but

```
QUEUE 123
```

may not work.

Should you decide to use numbers in the operands, simply remember to start numbering the QUEUE Blocks with small numbers and you should not have any problem. One may wonder why there would ever be a need for using numbers for the operands in a QUEUE

Block, but there are some examples that make use of this. For the examples done in this chapter, only names will be considered for the operands.

Whenever a `QUEUE` Block is used, there will automatically be certain statistics printed out. This will be demonstrated by means of an example. This example contains code that will be covered later. However, in order to demonstrate the data associated with the `QUEUE` Block, it is necessary to use GPSS/H code that might delay a transaction from moving to the next Block.

11.1.2 The DEPART Block

If a transaction is in a `QUEUE` Block, it must eventually leave this Block. This is done by the `DEPART` Block. It is used as the twin to the `QUEUE` Block, and it has exactly the same operand. Thus, referring to the examples of the `QUEUE` Block, the following would be the corresponding `DEPART` Blocks.

```
DEPART 1
DEPART FIRST
DEPART 7
DEPART ONE
DEPART DUMP1
DEPART STOPHERE
```

The `DEPART` Block will not be immediately after the `QUEUE` Block but *must* appear in the program. (If it were immediately after the `QUEUE` Block, the `QUEUE` Block would give meaningless statistics since the transactions would immediately enter and leave both Blocks.) It usually appears after one or two other Blocks. These other Blocks are the ones that, for one reason or other, cause a queue to form. Just as with the `QUEUE` Block, it is possible to have a second operand.

```
DEPART NAME,2
```

In this case, the current content of the `QUEUE NAME` is decreased by 2. If the transaction was a *zero entry to the* `QUEUE`, the zero entry counter in the .LIS file is incremented by 2. This means that when the transaction arrived at the `QUEUE` Block, there were no other transactions in the `QUEUE` Block, so it moved to the next sequential Block. As with the second operand for the `QUEUE` Block, use of this operand is very rare.

Example 11.1: Trucks Arrive for Refuelling

Mine trucks arrive at a refuelling station every 18 ± 5 minutes. It takes 16 ± 6 minutes to refuel a truck. Only one truck can refuel at a time. No truck ever needs other services. Simulate for 480 minutes. This is a classic problem in queuing theory. It has to do with an infinite population (the statistics of arrivals do not change), a single server (the refuelling station), and the possibility of an infinite queue (i.e., trucks do not leave no matter how long the wait might be).

The program to simulate this is CHAP11A.GPS. A listing of it is given in Figure 11.1.

Whenever a `QUEUE` is used in a programme, output is produced and is printed out in the .LIS file. The following output is taken from the file CHAP11A.LIS and will be explained next (Figure 11.2).

The Block counts show that 26 trucks arrived for refuelling. Twenty-five trucks completed their refuelling when the program ended, so one was left in the system as shown in Block 5.

```
            SIMULATE
*****************************
*  PROGRAM CHAP11A.GPS       *
*  TRUCKS ARRIVE FOR         *
*  REFUELING                 *
*****************************
            GENERATE    18,5     Trucks come for re-fueling
            QUEUE       WAIT     Wait in the queue
            SEIZE       FUEL     Try to use facility
            DEPART      WAIT     Leave the queue
            ADVANCE     16,6     Receive fuel
            RELEASE     FUEL     Free facility
            TERMINATE            Leave
            GENERATE    480
            TERMINATE   1
            START       1
            END
            SIMULATE
```

FIGURE 11.1
Listing of program CHAP11A.GPS.

```
RELATIVE CLOCK: 480.0000   ABSOLUTE CLOCK: 480.0000

BLOCK CURRENT      TOTAL
1                    26
2                    26
3                    26
4                    26
5            1       26
6                    25
7                    25
8                     1
9                     1

 QUEUE   MAXIMUM       AVERAGE        TOTAL        ZERO      PERCENT
         CONTENTS      CONTENTS       ENTRIES      ENTIRES   ZEROS
  WAIT      1           0.088          26           14        53.6

AVERAGE          &AVERAGE    QTABLE      CURRENT

TIME/UNIT        TIME/UNIT   NUMBER      CONTENTS

 1.626           3.523         -            0
```

FIGURE 11.2
Edited output from program CHAP11A.GPS.

The lines of output below the Block count are explained next:

QUEUE
WAIT — This is the name of the queue as specified by the A operand.

MAXIMUM
CONTENTS

1 — The maximum contents of the queue at any time.

`AVERAGE`
`CONTENTS`

0.088 The average contents in the queue.

`TOTAL`
`ENTRIES`

26 The number of transactions that entered the Block.

`ZERO`
`ENTRIES`

14 Of the 26 transactions that entered the `QUEUE` Block, 14 of them immediately left and entered the next Block.

`PERCENT`
`ZEROS`

53.6 The quotient 14/26.

`AVERAGE`
`TIME/UNIT`

1.626 For *all* the transactions that entered the `QUEUE` Block, this is the average time in the Block.

`$AVERAGE`
`TIME/UNIT`

3.523 This is the average time in the `QUEUE` Block for only the transactions that were actually delayed and held in it.

`QTABLE` Later, we shall see how to construct histograms of various parameters associated with the simulation. One of these is called a `QTABLE`. If one had been used in the simulation, its name would be here.

`CURRENT`
`CONTENTS`

0 The contents of the `QUEUE` Block at the end of the simulation.

The output also contains more information, which will be covered later.

You probably do not want all of the output, but GPSS/H gives it to you regardless. Most of the time, the output will be customized to appear either on the screen or sent to a data file.

The above items are all attributes associated with having a `QUEUE` Block. In fact, they are called *standard numerical attributes (SNAs).* These all have reserved names. These are as follows:

SNA	Meaning
Q(name) or Qn	Current queue content
QA(name) or QAn	Average queue contents
QC(name) or QCn	Queue entry count
QM(name) or QMn	Maximum queue content
QT(name) or QTn	Average time spent in the queue of *all* entries
QX(name) or QXn	Average time spent in the queue excluding the zero entries
QZ(name) or QZn	Zero entries

The above SNAs can be used in the program as operands. For example, one could have

```
ADVANCE QM(WAIT)
```

The transaction entering the ADVANCE Block would be put on the future event chain (FEC) for a time given by the maximum queue length at the QUEUE WAIT.

Let us now modify the program CHAP11A.GPS to add output. The statements that are added are as follows, and the new program is stored as CHAP11B.GPS (Figure 11.3).

When the program is run, the following output will be on the screen (Figure 11.4).

Not all of the output has been covered yet but will be covered in the next chapter. Even so, we can see that the number of trucks to arrive for refuelling was 26. The maximum number waiting for service was 1; the average queue length was .09; the average time in the queue for all trucks was 1.63 minutes; and the average time for only the trucks that waited was 3.52 minutes.

As indicated, it is possible to have a second operand with the QUEUE Block, such as

```
QUEUE FIRST,2
QUEUE WAIT,3
```

```
            SIMULATE
*********************************
*  PROGRAM CHAP11B.GPS          *
*  MODIFICATION OF CHAP11A.GPS  *
*  TO INCLUDE OUTPUT ON SCREEN  *
*  TRUCKS ARRIVE FOR            *
*  REFUELING                    *
*********************************
 COMEIN   GENERATE    18,5,0    Trucks come for re-fueling
          QUEUE       WAIT      Wait in the queue
          SEIZE       FUEL      Try to use facility
          DEPART      WAIT      Leave the queue
          ADVANCE     16,6      Receive fuel
          RELEASE     FUEL      Free facility
 SERVICE  TERMINATE             Leave
          GENERATE    480
          TERMINATE   1
          START       1
          PUTSTRING   ('  ')
          PUTSTRING   ('  ')
          PUTPIC      LINES=9,AC1,N(COMEIN),W(SERVICE),_
                      FR(FUEL)/10.,_
                      QA(WAIT),QM(WAIT),QC(WAIT),QT(WAIT),QX(WAIT)
              TIME OF SIMULATION                      ***
   NO. OF TRUCKS TO COME                              ***
   NO. BEING FUELED WHEN SIMULATION OVER:              *
   UTIL. OF FUELING STATION                          **.**%
   AVERAGE QUEUE LENGTH                                *.**
   MAXIMUM IN QUEUE AT ANY TIME                        *
   TOTAL ENTRIES INTO QUEUE                           **
   AVERAGE TIME IN QUEUE (ALL TRUCKS)                 **.**
   AVERAGE TIME IN QUEUE (THOSE THAT WAITED)          **.**
              END
```

FIGURE 11.3
Listing of program CHAP11B.GPS.

```
                TIME OF SIMULATION                480
    NO. OF TRUCKS TO COME                          26
    NO. BEING FUELED WHEN SIMULATION OVER:          0
    UTIL. OF FUELING STATION                     86.19%
    AVERAGE QUEUE LENGTH                           0.09
    MAXIMUM IN QUEUE AT ANY TIME                    1
    TOTAL ENTRIES INTO QUEUE                       26
    AVERAGE TIME IN QUEUE (ALL TRUCKS)              1.63
    AVERAGE TIME IN QUEUE (THOSE THAT WAITED)       3.52
```

FIGURE 11.4
Output from program CHAP11B.GPS.

This second operand must be a positive number. If it is used, it will affect the statistics of the QUEUE Block as the B operand will cause the TOTAL ENTRY count to be increased by this amount (not 1) and the *current content* is increased by this amount.

Should you ever decide to use such a QUEUE Block, you must be very careful to interpret your results accordingly. Actually, such use of the QUEUE Block is very rare and will not appear in the rest of this book.

There is another point worth mentioning concerning the QUEUE Block, and that is the fact that a transaction can be in more than one QUEUE Block at the same time. This may seem strange, but such a situation occurs in real life. Consider a major shopping centre where a person has to take a number to purchase meat. The same person can elect to also take a different number to purchase vegetables while waiting for the first number to be called. Thus, the person is in two queues at the same time. There will be occasions when, for the purpose of gathering statistics, we will use the fact that a transaction can be in more than one QUEUE Block at the same time. In fact, a transaction can be in even more than two QUEUE Blocks at the same time. The number of QUEUE Blocks a transaction can be in at the same time is dependent on the particular processor but is around five.

Figure 11.5 gives the listing for program CHAP11C.GPS. The animation is shown in Figure 11.6.

```
          SIMULATE
*********************************
*  PROGRAM CHAP11C.GPS          *
*  ANIMATION OF CHAP11B.GPS     *
*********************************
  ATF     FILEDEF      'CHAP11C.ATF'
          INTEGER      &TDONE
          GENERATE     18,5,0   TRUCKS ARRIVE
          BPUTPIC      FILE=ATF,LINES=5,AC1,AC1,XID1,XID1,XID1
TIME *.****
WRITE M2 ***.**
CREATE TRUCK T*
PLACE T* ON P1
SET T* TRAVEL 4
          ADVANCE      4
          BPUTPIC      FILE=ATF,LINES=2,AC1,AC1
TIME *.****
WRITE M2 ***.**
          QUEUE        WAIT
          SEIZE        FUEL
          DEPART       WAIT
          BPUTPIC      FILE=ATF,LINES=2,AC1,XID1
```

FIGURE 11.5
Listing of program CHAP11C.GPS. *(Continued)*

```
TIME *.****
PLACE T* AT 45 30
          ADVANCE      16,6
          BPUTPIC      FILE=ATF,LINES=4,AC1,AC1,XID1,XID1
TIME *.****
WRITE M2 ***.**
PLACE T* ON P2
SET T* TRAVEL 4
         ADVANCE       .1
         RELEASE       FUEL
         BLET          &TDONE=&TDONE+1
         BPUTPIC       FILE=ATF,LINES=2,AC1,&TDONE
TIME *.****
WRITE M1 **
         ADVANCE       3.9
         BPUTPIC       FILE=ATF,LINES=3,AC1,AC1,XID1
TIME *.****
WRITE M2 ***.**
DESTROY T*
         TERMINATE
         GENERATE      480
         TERMINATE     1
         START         1
         PUTPIC        FILE=ATF,LINES=2,AC1
TIME *.****
END
          END
```

FIGURE 11.5 (Continued)
Listing of program CHAP11C.GPS.

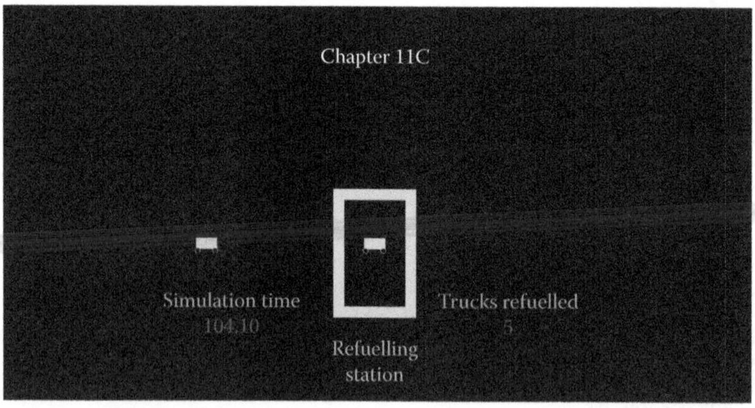

FIGURE 11.6
Screenshot of animation from program CHAP11C.

Example 11.2: Trucks Arrive at Either the Crusher or the Waste Dump

Mining trucks travel from the pit every 2 ± .6 minutes and travel to a junction in 3 minutes. At this junction, 25% travel to the waste dump in 5 minutes. There is no restriction on trucks dumping here. They dump and return to the mine. The other trucks travel to the crusher in 2.5 minutes. Only one truck can dump at a time. It takes 1.8 minutes to dump. Simulate for a shift of 450 minutes.

Solution
We do not yet know enough to fully understand all the GPSS/H Blocks that are needed to model this system as they will be given in the next chapter. Even so, the program can be run and studied. This system is modelled by program CHAP11D.GPS. Its listing is given in Figure 11.7. The output from the program is given in Figure 11.8.

The results are reasonable. A truck comes along every 2 minutes from the ore pit. This means that the expected number is 30 per hour. The simulation is for 7.5 hours, so the expected number of trucks from the pit is 225. (75% of this is 168.75 which represents the trucks hauling ore.)

The animation is given by CHAP11E.GPS. A screenshot is shown in Figure 11.9.

There are several things that can be added to the animation to enhance it, but it shows the salient features of the model as stated.

```
          SIMULATE
*******************************
*  PROGRAM CHAP11D.GPS        *
*  TRUCKS ARRIVE AT EITHER    *
*  WASTE AREA OR CRUSHER      *
*******************************
           GENERATE   2,.6         TRUCKS ARRIVE
           ADVANCE    3            TRAVEL TO JUNCTION
           TRANSFER   .25,,WASTE   25% TO WASTE
           ADVANCE    2.5          TRAVEL TO CRUSHER
           QUEUE      WAIT         JOIN QUEUE
           SEIZE      CRUSHER
           DEPART     WAIT         LEAVE THE QUEUE
           ADVANCE    1.8
 LOADSO    RELEASE    CRUSHER
           TERMINATE
 WASTE     ADVANCE    5
 LOADSW    TERMINATE
           GENERATE   450
           TERMINATE  1
           START      1
           PUTPIC     LINES=3,N(LOADSO),N(LOADSW)
                RESULTS OF SIMULATION
         NUMBER OF LOADS OF ORE DUMPED     ***
         NUMBER OF LOADS OF WASTE DUMPED ***
           END
```

FIGURE 11.7
Listing of program CHAP11D.GPS.

```
          RESULTS OF SIMULATION
   NUMBER OF LOADS OF ORE DUMPED   170
   NUMBER OF LOADS OF WASTE DUMPED  52
```

FIGURE 11.8
Results of program CHAP11D.GPS.

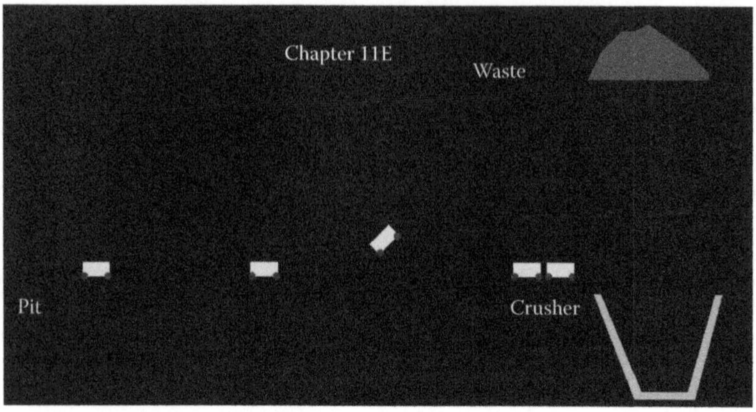

FIGURE 11.9
Screenshot of animation from CHAP11E.GPS.

11.2 Another PROOF Icon

There are several PROOF icons that were not covered in the first six chapters. One of them will be covered now. The following is the icon:

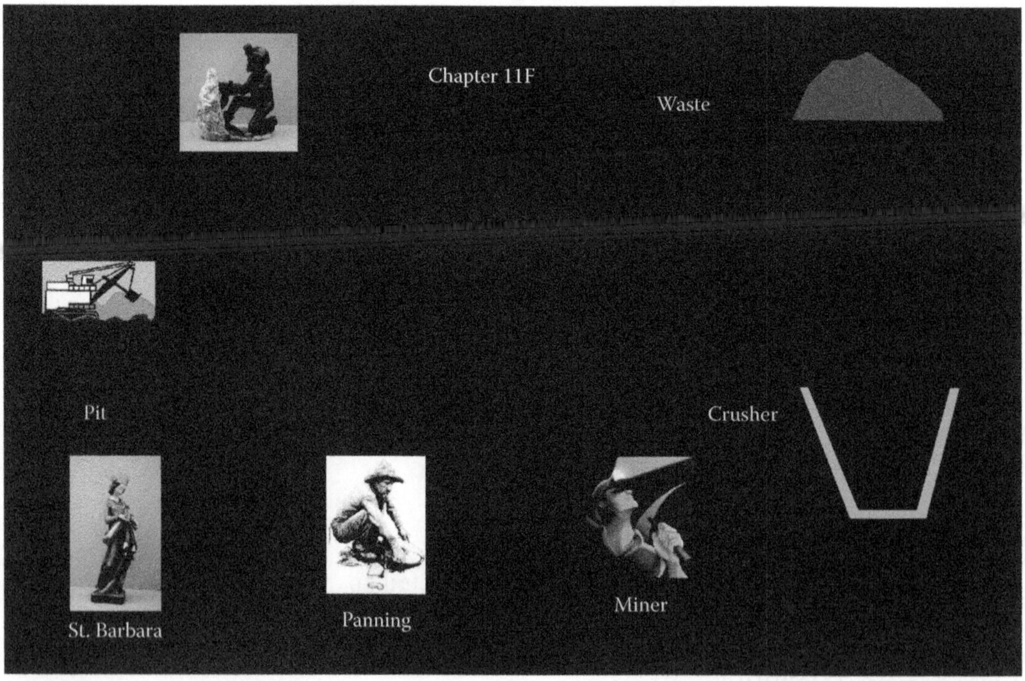

FIGURE 11.10
Animation given by CHAP11E.GPS with added mining icons (CHAP11F.LAY).

If you hold the mouse on this icon, it will shortly say *insert a bitmap image*. This allows the user to import an image into the animation, which is a very useful way to add external images to your animations. For example, the layout in Figure 11.9 is a bit bland. It should have more detail, which can be added later, but to illustrate this icon, consider the layout given by CHAP11F.LAY. This is shown in Figure 11.10. This has the original layout from CHAP11E with numerous mining icons added. These icons are found in the subdirectory Mining Icons of the directory STURGULEXAMPLES. Whether these add to the animation or not is a matter of opinion, but they are shown to illustrate how easy it is to add bitmap images to animations.

One should be prudent in the addition of these icons (and others!) to animations. They should enhance the animation and not detract from the main reason for the model. The images are easy to move around, shrink, and enlarge.

11.3 Exercises

11.1. Refer to example CHAP11A. Suppose that the first truck arrives exactly when the refuelling shop opens. Also, let the time to refuel be 17 ± 7.5 minutes and the arrival rate now be 18 ± 9 minutes. Do the statistics change much?

11.2. Consider the following program segments:

```
GENERATE    100
ADVANCE     105
QUEUE WAIT
ADVANCE     3
DEPART WAIT
```

When does the first transaction execute the `QUEUE` Block? When does it execute the `DEPART` Block?

11.3. Consider the following program segment:

```
GENERATE       100,50
QUEUE FIRST
ADVANCE        1
DEPART FIRST
```

What is the earliest the first transaction might execute the `QUEUE` Block? When is the latest time the first transaction might execute the `DEPART` Block?

11.4. Consider program CHAP11F.GPS. It is almost the same as program CHAP11D.GPS but contains an error. This is the worst type of error as the program compiles and gives *reasonable* results. Unfortunately, the results are incorrect. Run it and examine the output and correct the program. This exercise illustrates why it is important to run simulation programs with sample data where the results might be known (in this case, the program could have been run for 1 hour when 15 trucks were expected to come from the mine).

11.5. Run the program CHAP11G.GPS. This also contains errors that will be detected during compiling. Examine the file CHAP11G.LIS and determine the errors. Correct the program.

11.6. Consider the lines of code:

```
GENERATE      100
QUEUE              WAIT_1
SEIZE MACH
DEPART WAIT_1
ADVANCE       90
RELEASE MACH
TERMINATE     1
START         1
END
```

What will happen when you try to run it? Correct the errors.

12

SEIZE/RELEASE Blocks

Overview of a large gold mine known as "The Super Pit" near Kalgoorlie, Western Australia.

12.1 `SEIZE` and `RELEASE` Blocks

12.1.1 The `SEIZE` Block

This chapter will introduce the `SEIZE` Block and its twin, the `RELEASE` Block. This is the first Block that we have encountered that can deny a transaction's entrance into another Block. Knowledge of this Block will enable us to construct much more meaningful simulations. The animation can then be studied in greater detail once we have learned enough GPSS/H and PROOF to understand how to make the whole animation files.

In GPSS/H, a single server is called a *facility*. This might represent a shovel loading a truck, a crusher where only one truck can dump at a time, a barber giving a haircut, a bank clerk who waits on customers, a checkout worker in a grocery store, and so on. In order for a transaction to use a facility, a `SEIZE` Block is used. This is quite simple to use. The most common form is

```
SEIZE A
```

where the A operand is generally either a number or name but can be a variable, as we will learn later. Thus,

```
SEIZE 1
SEIZE ONE
SEIZE DUMP
SEIZE BARBER
SEIZE 33
```

are examples of the SEIZE Block.

Should a number be used for the operand of the SEIZE Block, it must be less than the number of SEIZE Blocks allowed by your processor. If you always number the Blocks starting with small numbers, this should not be a problem. In nearly all of our examples, we shall only use alphanumerics for the operand of the SEIZE Block.

When a transaction enters a SEIZE Block, no other transaction can enter it until the transaction in the SEIZE Block leaves. Transactions attempting to enter a SEIZE Block that is already being used by another transaction must (normally) remain in the Block they are in. This is analogous to the situation of a mining truck that is positioned to dump into a crusher. It is considered to have SEIZEd the crusher and no other truck can dump until the truck that has SEIZEd the crusher is done dumping and RELEASEs the crusher.

Whenever a transaction has executed the SEIZE Block, certain statistics are automatically printed out. We have seen this in some of the programs we have already written, such as the ones in the previous chapter. Since a facility can be used by only one transaction at a time, if a second transaction wishes to use it, it must wait in the previous Block until the facility is free. For example, consider the following Blocks:

```
GENERATE  10
QUEUE     WAIT
SEIZE     SHOVEL
DEPART    WAIT
ADVANCE   25
```

Here, a transaction is first generated at $t = 10$. It is moved to the QUEUE Block and immediately attempts to use the facility SHOVEL. Since SHOVEL is not being used, it does so and then enters the ADVANCE Block, where it is put in the future events chain until a time of $t = 35$ (10 + 25). A second transaction is generated at time 20. It enters the QUEUE WAIT Block and attempts to enter the SEIZE SHOVEL Block. Since it cannot, it is held up in the QUEUE Block until SHOVEL is free. Now, suppose that you did not have the QUEUE WAIT Block. Instead, suppose you had the following:

```
GENERATE  10
SEIZE     SHOVEL
ADVANCE   25
```

The second transaction would have been held up in the GENERATE Block until $t = 35$. The effect of this in GPSS/H would be to offset the time for when the third transaction leaves the GENERATE Block from $t = 30$ to $t = 60$. (Draw a time diagram to convince yourself of this.) Since this is *not* what we want, this should be avoided by either a QUEUE Block or a dummy ADVANCE Block. A dummy ADVANCE Block would look as follows:

```
       GENERATE      10
       ADVANCE
*************************************
* NOTE: YOU COULD ALSO HAVE         *
*      ADVANCE       0              *
*************************************
       SEIZE         JOE
       ADVANCE       25
```

Whenever a SEIZE Block is used, certain statistics are automatically printed out at the end of the simulation. To illustrate this, recall the program of the truck and inspector as given by program CHAP8D. The inspection takes place in 14 ± 8 minutes and trucks arrive at the rate of 14 ± 8 minutes. The program is repeated here (Figure 12.1).

Whenever a program with a SEIZE Block is used, there will be output obtained that is found in the .LIS file. A portion of the output from running program CHAP8D is shown in Figure 12.2.

The output associated with the SEIZE Block is explained as follows:

FACILITY The name of the facility was INSPECT

```
--AVG-UTIL-DURING--
TOTAL
TIME
```

The facility was busy 92.6% of the time the simulation ran.

Important note: *Even though the facility utilization is shown on the screen as 0.926, it is stored in the computer at 926 (parts per thousand). Thus, if one wants to use this value, it needs to be converted to a decimal.*

```
          SIMULATE
***************************************
*  PROGRAM  CHAP8D.GPS                *
*  EXAMPLE OF TIRE INSPECTOR IN A MINE *
***************************************
          GENERATE     15,6         TRUCKS ARRIVE
          QUEUE        WAIT         IS INSPECTOR FREE?
          SEIZE        INSPECT      YES, ENGAGE INSPECTOR
          DEPART       WAIT         LEAVE THE WAITING QUEUE
          ADVANCE      14,8         TIRE INSPECTION
          RELEASE      INSPECT      FREE THE INSPECTOR
          TERMINATE    1            TRUCK LEAVES
          START        20
          END
```

FIGURE 12.1
Listing of program CHAP8D.GPS.

```
           --AVG-UTIL-DURING--
FACILITY TOTAL AVAIL UNAVL ENTRIES AVERAGE CURRENT PERCENT SEIZING
PREEMPTING
         TIME  TIME   TIME         TIME/XACTSTATUS   AVAIL     XACT       XACT
INSPECT  0.926                20   14.942   AVAIL
```

FIGURE 12.2
Portion of output from program CHAP8D.GPS.

```
AVAIL
TIME
```

We can ignore this.

```
UNAVL
TIME
```

The time the facility was not available. This can be ignored for now.

```
ENTRIES
```

There were 20 entries. Recall that the program ran until 20 trucks finished fuelling.

```
AVERAGE
TIME/XACT
```

The average time the `SEIZE` Block was used by a transaction was 14.942. The time to give service was 14 ± 8 minutes. The mean of this distribution is 14, so this is an expected value.

```
CURRENT
STATUS
AVAIL
```

We can ignore this for now.

```
PERCENT
AVAIL
```

This can also be ignored.

The rest of the output can also be ignored until later.

Just as with the `QUEUE` Block, the above are SNAs of the system, and each will be referred to by a special reserved name.

12.1.1.1 SNAs Associated with the `SEIZE` Block

The SNAs associated with the `SEIZE` Block are given below:

SNAs	Meaning
`F(name)` or `Fn`	This will be 1 if the facility is currently in use; else its value is 0
`FC(name)` or `FCn`	Number of times the facility has been captured
`FR(name)` or `FRn`	Utilization of the facility in parts per thousand
`FT(name)` or `FTn`	Average holding time

Note that the output from the program gives the utilization as a decimal, but the SNA `FR` is in parts per thousand. Thus, if one had used the Block

```
ADVANCE FR(MACH1)
```

and the utilization of the facility `MACH1` was .432 when the transaction entered the `ADVANCE` Block, the transaction would be placed on the future events chain for 432 time units. If one wants to output the utilization of a facility named `SHOVEL` as a decimal, one would use something such as

```
FR(SHOVEL)/10
```

12.1.2 The RELEASE Block

When a facility is used via the `SEIZE` Block, it eventually must be freed for other transactions to use it. This is done by means of the *twin* of the `SEIZE` Block, the `RELEASE` Block. Some forms are as follows:

```
RELEASE   JOE
RELEASE   1
RELEASE   BARBER
RELEASE   CAR1
```

At this point, you may wish to go back to some of the previous exercises and examine the output whenever a facility is used in the program.

It is very common in our programs to have a sequence of Blocks such as

```
QUEUE     HERE
SEIZE     THERE
DEPART    HERE
ADVANCE   20,3
RELEASE   THERE
```

This sequence should be carefully examined and understood, as it is repeated a great many times in GPSS/H programs.

We are now in a position to understand nearly all of the programs we have written so far in the previous chapters. The next few examples will enable us to use what we have learned so far. Each should be carefully studied and understood.

Example 12.1: Trucks Coming to a Repair Shop

Mining trucks come to a garage for minor repairs. These repairs are of two types: routine and non-routine. There is only one mechanic who does both types of repairs. 70% of the trucks that enter need only routine repairs, whereas the rest are in for non-routine repairs. Trucks arrive every 28 ± 7 minutes. Non-routine repairs take 45 ± 15 minutes, whereas routine repairs take only 18 ± 6 minutes. The single mechanic is claiming that he is overworked. His union defines this as working more than 85% of the time. Is he justified in his claim? Simulate for 100 days of 480 minutes each. Ignore the fact that the worker leaves at the end of each shift and the effect of weekends.

Solution

The program to do this is as shown in Figure 12.3.

The results are as given in Figure 12.4.

According to the union criteria, the repairman is overworked. This program should be run and studied, as it does not contain any Blocks or statements that have not been covered. Also, since we now have enough GPSS/H and PROOF to understand

```
                SIMULATE
********************************
*  PROGRAM CHAP12A.GPS          *
*  ILLUSTRATING SEIZE/RELEASE   *
********************************
                GENERATE     28,7          TRUCKS ARRIVE
                QUEUE        WAIT          JOIN QUEUE
                TRANSFER     .70,,MINOR    70% NEED MINOR REPAIRS
                SEIZE        MECH          IS MECHANIC FREE?
                DEPART       WAIT          YES, LEAVE QUEUE
                ADVANCE      45,15         MAJOR REPAIRS
                RELEASE      MECH          FREE THE MECHANIC
                TERMINATE                  LEAVE GARAGE
     MINOR      SEIZE        MECH          IS MECHANIC FREE?
                DEPART       WAIT          YES, LEAVE QUEUE
                ADVANCE      18,6          MINOR REPAIRS
                RELEASE      MECH          FREE THE MECHANIC
                TERMINATE                  LEAVE GARAGE
                GENERATE     480*100       TIMER TRANSACTION
                TERMINATE    1
                START        1
                PUTSTRING    ('  ')
                PUTSTRING    ('  ')
                PUTPIC       LINES=3,FR(MECH)/10.,QA(WAIT)
               RESULTS OF SIMULATION
      MECHANIC WAS BUSY   ***.**% OF THE TIME
      AVERAGE NUMBER OF TRUCKS WAITING        **.**
                END
```

FIGURE 12.3
Listing of program CHAP12A.GPS.

```
            RESULTS OF SIMULATION
      MECHANIC WAS BUSY  94.11% OF THE TIME
      AVERAGE NUMBER OF TRUCKS WAITING         1.69
                1.69
```

FIGURE 12.4
Results of program CHAP12A.GPS.

all the commands in a PROOF animation trace file, it will be instructive to examine the listing of CHAP12B.GPS segment by segment to see how the animation is created (Figure 12.5).

The trucks need to be created. For the animation, the trucks will be created on the left side of the screen. Rather than have them just immediately appear in front of the repair place, a path P1 is made in the animation. The trucks will be placed on this path, and it takes them 4 time units to move to the spot in front of where they are to be repaired. In an actual simulation, the system would need to be carefully studied to obtain correct data as to arrival rates and how the trucks join the queue.

Since there are two types of repairs, minor ones and others that take longer, those trucks that require longer service are given the colour white. This is done by the lines:

```
MINOR BPUTPIC FILE = ATF,LINES = 6,AC1,XID1,XID1,XID1,XID1,AC1
TIME *.****
CREATE TRUCK T*
SET T* COLOR F3
PLACE T* ON P1
SET T* TRAVEL 4
WRITE M2 ***.**
          ADVANCE 4
```

```
           SIMULATE
***********************************
*  PROGRAM CHAP12B.GPS   THIS IS  *
*  THE ANIMATION FOR CHAP12A.GPS  *
***********************************
  ATF      FILEDEF      'CHAP12B.ATF'
           INTEGER      &TRUCKS
           GENERATE     28,7,0       TRUCKS ARRIVE FOR REPAIRS
           TRANSFER     .70,,MINOR 70% NEED MINOR REPAIRS
           BPUTPIC      FILE=ATF,LINES=5,AC1,XID1,XID1,XID1,AC1
TIME *.****
CREATE TRUCK T*
PLACE T* ON P1
SET T* TRAVEL 4
WRITE M2 ***.**
           ADVANCE      4
           BPUTPIC      FILE=ATF,LINES=2,AC1,AC1
TIME *.****
WRITE M2 ***.**
           QUEUE        WAIT       JOIN QUEUE
           SEIZE        MECH       IS MECHANIC FREE?
           DEPART       WAIT       YES, LEAVE QUEUE
           BPUTPIC      FILE=ATF,LINES=3,AC1,XID1
TIME *.****
PLACE T* AT 42 30
WRITE M3 Facility seized!!
           ADVANCE      45,15       MINOR REPAIRS TAKE PLACE
           BPUTPIC      FILE=ATF,LINES=4,AC1,XID1,XID1,AC1
TIME *.****
PLACE T* ON P2
SET T* TRAVEL           3
WRITE M2  ***.**
           ADVANCE      .5
           BPUTPIC      FILE=ATF,LINES=2,AC1,AC1
TIME *.****
WRITE M2 ***.**
           RELEASE      MECH       FREE MECHANIC
           BLET         &TRUCKS=&TRUCKS+1
           BPUTPIC      FILE=ATF,LINES=3,AC1,&TRUCKS
TIME *.****
WRITE M1 **
WRITE M3
           ADVANCE      2.5
           BPUTPIC      FILE=ATF,LINES=3,AC1,XID1,AC1
TIME *.****
DESTROY T*
WRITE M2 ***.**
           TERMINATE              LEAVE GARAGE
 MINOR     BPUTPIC      FILE=ATF,LINES=6,AC1,XID1,XID1,XID1,XID1,AC1
TIME *.****
CREATE TRUCK T*
SET T* COLOR F3
PLACE T* ON P1
SET T* TRAVEL 4
WRITE M2 ***.**
           ADVANCE      4
           QUEUE        WAIT       JOIN QUEUE
```

FIGURE 12.5
Listing of program CHAP12B.GPS. *(Continued)*

```
          SEIZE        MECH        IS MECHANIC FREE?
          DEPART       WAIT        YES, LEAVE QUEUE
          BPUTPIC      FILE=ATF,LINES=4,AC1,XID1,AC1
TIME *.****
PLACE T* AT 42 30
WRITE M2 ***.**
WRITE M3 Facility seized!!
          ADVANCE      18,6        MINOR REPAIRS TAKE PLACE
          BPUTPIC      FILE=ATF,LINES=3,AC1,XID1,XID1
TIME *.****
PLACE T* ON P2
SET T* TRAVEL          3
          ADVANCE      .5
          BPUTPIC      FILE=ATF,LINES=2,AC1,AC1
TIME *.****
WRITE M2 ***.**
          RELEASE      MECH        FREE MECHANIC
          BLET         &TRUCKS=&TRUCKS+1
          BPUTPIC      FILE=ATF,LINES=3,AC1,&TRUCKS
TIME *.****
WRITE M1 **
WRITE M3
          ADVANCE      2.5
          BPUTPIC      FILE=ATF,LINES=3,AC1,XID1,AC1
TIME *.****
DESTROY T*
WRITE M2 ***.**
          TERMINATE              LEAVE GARAGE
          GENERATE     480*100      TIMER TRANSACTION ARRIVES
          TERMINATE    1
          PUTSTRING    ('  ')
          PUTSTRING    ('  ')
          START        1
          PUTPIC       LINES=3,FR(MECH)/10.0,QA(WAIT)
        RESULTS OF SIMULATION
 THE MECHANICE WAS BUSY  **.**% OF THE TIME
 THE AVERAGE QUEUE LENGTH WAS   **.**
          PUTPIC       FILE=ATF,LINES=2,AC1
TIME *.****
END
          END
```

FIGURE 12.5 (Continued)
Listing of program CHAP12B.GPS.

In the animation, M2 is the simulation clock time. This is printed out on the screen every time there is a change in the simulation time.

```
      QUEUE    WAIT   JOIN QUEUE
      SEIZE    MECH   IS MECHANIC FREE?
      DEPART   WAIT   YES, LEAVE QUEUE
      BPUTPIC  FILE = ATF,LINES = 3,AC1,XID1
TIME *.****
PLACE T* AT 42 30
WRITE M3 Facility seized!!
```

The truck is moved from the end of path P1 to the point on the screen given by (42, 30). If desired, one could have had another path where the trucks will move from the end of path P1 and then move to some appropriate place where repairs are done.

The message *Facility seized!!* is placed on the screen. One often has many such messages that are used to enhance the animation.

```
          ADVANCE   45,15   MAJOR REPAIRS TAKE PLACE
          BPUTPIC   FILE = ATF,LINES = 4,AC1,XID1,XID1,AC1
TIME *.****
PLACE T* ON P2
SET T* TRAVEL       3
WRITE M2 ***.**
```

After a truck is repaired, it is placed on path P2 and is set to leave the shop. The travel time used for this is 3 time units. Other times may have been used depending on how the modeller wants his or her animation to appear on the screen.

```
          ADVANCE   .5
          BPUTPIC   FILE = ATF,LINES = 2,AC1,AC1
TIME *.****
WRITE M2 ***.**
          RELEASE   MECH    FREE MECHANIC
          BPUTPIC   FILE = ATF,LINES = 3,AC1
TIME *.****
WRITE M1 **
WRITE M3
```

The line of code

```
WRITE M3
```

removes the message, *Facility seized!!* from the screen.

The rest of the code is similar to that given above and there is no need to annotate it again.

```
          ADVANCE   2.5
          BPUTPIC   FILE = ATF,LINES = 3,AC1,XID1,AC1
TIME *.****
DESTROY T*
WRITE M2 ***.**
          TERMINATE           LEAVE GARAGE
MINOR    BPUTPIC    FILE = ATF,LINES = 5,AC1,XID1,XID1,XID1,AC1
TIME *.****
CREATE TRUCK T*
PLACE T* ON P1
SET T* TRAVEL 4
WRITE M2 ***.**
          ADVANCE   4
          QUEUE     WAIT    JOIN QUEUE
          SEIZE     MECH    IS MECHANIC FREE?
          DEPART    WAIT    YES, LEAVE QUEUE
          BPUTPIC   FILE = ATF,LINES = 4,AC1,XID1,AC1
TIME *.****
PLACE T* AT 43 30
WRITE M2 ***.**
WRITE M3 Facility seized!!
          ADVANCE   18,6    MINOR REPAIRS TAKE PLACE
          BPUTPIC   FILE = ATF,LINES = 3,AC1,XID1,XID1
TIME *.****
PLACE T* ON P2
```

```
SET T* TRAVEL        3
          ADVANCE   .5
          BPUTPIC   FILE = ATF,LINES = 2,AC1,AC1
TIME *.****
WRITE M2 ***.**
          RELEASE   MECH    FREE MECHANIC
          BPUTPIC   FILE = ATF,LINES = 2,AC1
TIME *.****
WRITE M1 **
          BPUTPIC   FILE = ATF,LINES = 3,AC1,XID1,AC1
TIME *.****
DESTROY T*
WRITE M2 ***.**
          TERMINATE            LEAVE GARAGE
          GENERATE 480*100         TIMER TRANSACTION ARRIVES
          TERMINATE 1
          PUTSTRING (' ')
          PUTSTRING (' ')
          START     1
          PUTPIC    LINES = 3,FR(MECH)/10.0,QA(WAIT)
          RESULTS OF SIMULATION
THE MECHANICE WAS BUSY **.**% OF THE TIME
THE AVERAGE QUEUE LENGTH WAS **.**
          PUTPIC    FILE = ATF,LINES = 2,AC1
TIME *.****
END
          END
```

The animation is as shown in Figure 12.6.

Example 12.2: Small Gold Mine Owner's Problem

The owner of a small gold mine is wondering if he has the right number of trucks to haul the ore. Figure 12.7 gives a sketch of the operation. Trucks are loaded by a single shovel and then travel to a processing plant where they dump and cycle back to the shovel. Only one truck at a time can be loaded by the shovel and dump at the processing plant. It takes 3.5 ± 1.25 minutes to load a truck, 6 minutes to haul to the dump, 2 ± .4 minutes to dump the ore, and 5 minutes to return. Financial data associated with this operation is as follows:

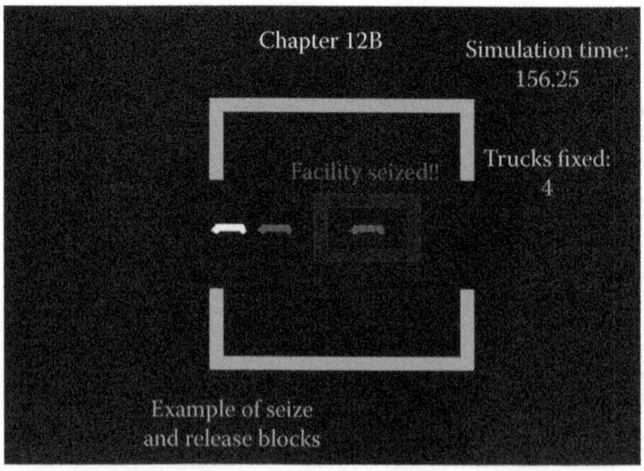

FIGURE 12.6
Screenshot of animation from program CHAP12B.GPS.

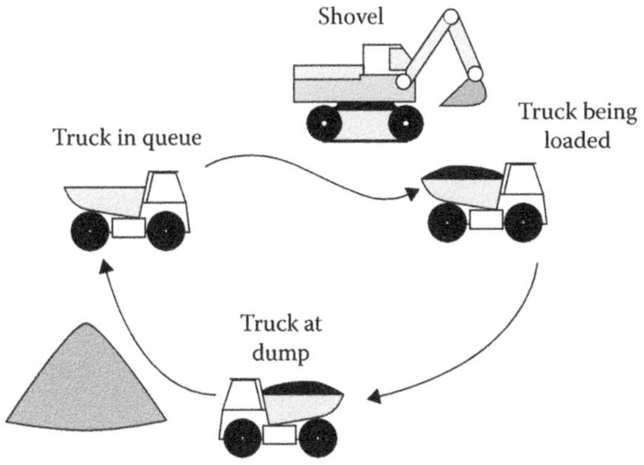

FIGURE 12.7
Sketch of the small gold mine.

Truck driver's salary	$15.75 per hour
Cost of running the shovel and dump	$275 per 8-hour day (fixed)
Profit per load (after all other expenses)	$25.50
Cost of each truck/shift	$280

Determine the correct number of trucks to have in the mine.

NOTE: This problem and its variations have long been studied by the mining engineer. It can be considered as the *fundamental problem in mine design*. While investigators have been able to solve this for particular statistical distributions such as the exponential distribution for all motion, it is not possible to obtain closed-form solutions for the general case of any distributions for motion. In addition, variables such as spotting and equipment breakdowns cannot be included in any exact mathematical solution. It is, therefore, somewhat of a surprise to the mining engineer who first encounters this problem in GPSS/H that it can be modelled in such a few lines of computer code.

The computer program that was used to obtain a solution for this problem is stored as CHAP12C.GPS (Figure 12.8).

The program was run repeatedly for two trucks, then three trucks, and so on, up to seven trucks. The results of the simulations are tabulated in Table 12.1.

All that is needed from the simulation results are the loads per day, but it is instructive to examine the other data. With only two trucks, the shovel is busy only 42.4% of the time. The average queue length is only .01. As the number of trucks in the mine is increased, the utilization of the shovel increases until it is working 100% of the time. The addition of a seventh truck has no effect on the mine other than to add another truck to the queue. In order to solve this problem, the cost data needs to be used. Table 12.2 gives the results of these calculations.

The optimum number of trucks to have is five for a maximum daily profit of $1086. Notice, however, that there would not be much of a difference in profit if four trucks were used.

The program to make the animation (Figure 12.9) is given as CHAP12D.GPS. The animation is not fancy and can be improved, especially when we learn more of GPSS/H. For example, the use of the `ROTATE` command will be used to rotate the shovel boom when a truck is being loaded. In addition, the empty trucks could be shown spotting while the full one leaves. The listing of the program is given in Figure 12.10.

```
            SIMULATE
******************************
*   CHAP12C.GPS PROGRAM TO     *
*   MODEL A MINE WITH ONE      *
*   SHOVEL AND N-TRUCKS        *
******************************
   TRUCKS  GENERATE   ,,,4       PROVIDE TRUCKS FOR THE MINE
   UPTOP   QUEUE      WAIT       QUEUE AT THE SHOVEL
           SEIZE      SHOVEL     USE THE SHOVEL
           DEPART     WAIT       LEAVE THE QUEUE
           ADVANCE    3.5,1.25   LOAD A TRUCK
           RELEASE    SHOVEL     FREE THE SHOVEL
           ADVANCE    6          TRAVEL TO DUMP
           SEIZE      DUMP       ONLY ONE TRUCK CAN DUMP
           ADVANCE    2.,.4      DUMP A LOAD
           RELEASE    DUMP       FREE THE DUMP
           ADVANCE    5         RETURN TO SHOVEL
           TRANSFER   ,UPTOP     JOIN QUEUE AGAIN
           GENERATE   480*100    SIMULATE FOR 100 SHIFTS
           TERMINATE  1
           START      1
           PUTPIC     LINES=5,N(TRUCKS),FC(DUMP)/100,FR(SHOVEL)/10.,_
                      QA(WAIT)
   RESULTS OF SIMULATION
 NUMBER OF TRUCKS USED    **
 NUMBER OF LOADS DUMPED ***  PER SHIFT
 UTIL. OF SHOVEL          **.**%
 AVERAGE QUEUE AT SHOVEL  *.**
           END
```

FIGURE 12.8
Listing of program CHAP12C.GPS.

TABLE 12.1

Results of Multiple Simulations

Loads per Day	No. of Trucks	% Util. Shovel	Avg. Queue
57	2	42.4	0.01
86	3	63.2	0.03
113	4	83.0	0.10
133	5	99.9	0.41
137	6	100.0	1.28
137	7		2.29

TABLE 12.2

Results of Calculations

No. of Trucks	Loads per Day	Salaries	Trucks	Fixed Costs	Profit
2	57	$252	$560	$275	$366.5
3	86	$378	$840	$275	$700
4	113	$504	$1120	$275	$982.5
5	133	$630	$1400	$275	$1086.5
6	137	$756	$1680	$275	$782.5
7	137	$882	$1900	$275	$367.5

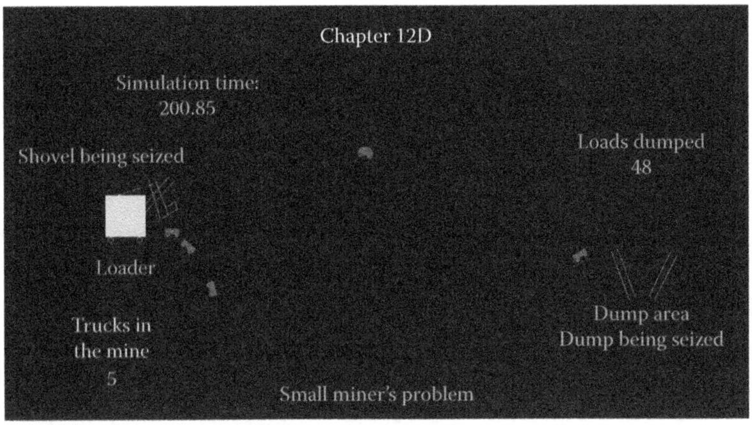

FIGURE 12.9
Screenshot of animation from program CHAP12D.GPS.

```
           SIMULATE
********************************
*  PROGRAM CHAP12D.GPS         *
*  ANIMATION OF CHAP12C.GPS    *
********************************
    ATF    FILEDEF       'CHAP12D.ATF'
           INTEGER       &LOADS
 TRUCKS    GENERATE      3,,0,5       PROVIDE TRUCKS FOR THE MINE
           BPUTPIC       FILE=ATF,LINES=6,AC1,XID1,XID1,XID1,AC1,N(TRUCKS)
TIME *.****
CREATE TRUCK T*
PLACE T* ON P1
SET T* TRAVEL 4
WRITE M1 ***.**
WRITE M5 **
           ADVANCE       4
   UPTOP   QUEUE         WAIT       QUEUE AT THE SHOVEL
           SEIZE         SHOVEL     USE THE SHOVEL
           DEPART        WAIT       LEAVE THE QUEUE
           BPUTPIC       FILE=ATF,LINES=5,AC1,XID1,XID1,AC1
TIME *.****
SET T* CLASS TRUCK1
PLACE T* AT 25 30
WRITE M1 ***.**
WRITE M3 SHOVEL BEING SEIZED
           ADVANCE       3.5,1.25   LOAD A TRUCK
           BPUTPIC       FILE=ATF,LINES=5,AC1,XID1,XID1,XID1,AC1
TIME *.****
PLACE T* ON P2
SET T* CLASS TRUCK2
SET T* TRAVEL 6
WRITE M1 ***.**
           ADVANCE       .5
           BPUTPIC       FILE=ATF,LINES=2,AC1,AC1
TIME *.****
WRITE M1 ***.**
           RELEASE       SHOVEL     FREE THE SHOVEL
           BPUTPIC       FILE=ATF,LINES=2,AC1
```

FIGURE 12.10
Listing of program CHAP12D.GPS. *(Continued)*

```
TIME *.****
WRITE M3
          ADVANCE      5.5       TRAVEL TO DUMP
          BPUTPIC      FILE=ATF,LINES=2,AC1,AC1
TIME *.****
WRITE M1 ***.**
          SEIZE        DUMP       ONLY ONE TRUCK CAN DUM
          BPUTPIC      FILE=ATF,LINES=3,AC1,XID1
TIME *.****
PLACE T* AT 59 30
WRITE M4 DUMP BEING SEIZED
          ADVANCE      2.1,.4     DUMP A LOAD
          BLET         &LOADS=&LOADS+1
          BPUTPIC      FILE=ATF,LINES=3,AC1,AC1,&LOADS
TIME *.****
WRITE M1 ***.**
WRITE M2 ***
          BPUTPIC      FILE=ATF,LINES=4,AC1,XID1,XID1,X
TIME *.****
PLACE T* ON P1
SET T* TRAVEL 5
SET T* CLASS TRUCK
          ADVANCE      .5
          BPUTPIC      FILE=ATF,LINES=2,AC1,AC1
TIME *.****
WRITE M1 ***.**
          RELEASE      DUMP       FREE THE DUMP
          ADVANCE      4.5
          BPUTPIC      FILE=ATF,LINES=3,AC1,AC1
TIME *.****
WRITE M1 ***.**
WRITE M4
          TRANSFER     ,UPTOP     JOIN QUEUE AGAIN
          GENERATE     480*10     SIMULATE FOR 10 SHIFTS
          TERMINATE    1
                      START       1
           PUTPIC      FILE=ATF,LINES=2,AC1
TIME *.****
END
            END
```

FIGURE 12.10 (Continued)
Listing of program CHAP12D.GPS.

12.2 Exercises

12.1. A factory that formerly produced only widgets is branching out into the production of squidgets. To make each squidget, a person needs to assemble various parts. It takes 30 ± 8 minutes to assemble each squidget. Then the squidgets need to be *finished*. There is only one finishing machine and only one squidget can be finished at a time by firing it in the oven. Firing takes 8 ± 3 minutes per squidget. Each squidget produced earns a tidy profit of $6.00. The finishing machine costs you $40 per day no matter what (fixed costs). You pay your workers $5 per hour. A working day is considered to consist of 450 minutes of actual production. The program should run for 10 shifts and then the production divided by 10 to obtain the profit per shift. How many workers do you hire?

12.2. Refer to Exercise 12.1. Suppose that the assemblers can now be given a special tool that decreases their production time to 25 ± 6 minutes per squidget. All other data remains the same. How does this change the model? Again, simulate for 10 shifts of 450 minutes and determine the production per shift. This is used to determine the number of workers to assign to making squidgets.

12.3. Two types of customers arrive at Joe's barbershop (from Schriber, 1974).* The first type wants only a haircut. They come every 40 ± 10 minutes. The second type wants both a haircut and a shave. They arrive every 60 ± 20 minutes. It takes Joe 18 ± 6 minutes to give a haircut and 10 ± 2 minutes for a shave. Construct a model of the shop. Run it for 20 days of 8 hours straight. Ignore the fact that, in reality, Joe would close his shop for lunch and go home at the end of each day. Determine whether Joe is working too hard. (This is defined by the union as working more than 85% of the time).

12.4. Suppose Joe decides to give preference to customers who want only a haircut. How does this change the situation? This means that, if there are multiple customers in the shop when Joe finishes a customer, Joe will ask the waiting customers what they want (just a haircut or both a haircut and a shave) and select the customer who wants only a haircut. If there is more than one customer waiting for a haircut, the one who has been waiting the longest will be served. The program should have separate queues for each type of customer, as well as a general queue for all customers (three queues in all), and the output should give the average time spent in the different queues.

12.5. Three types of mechanics arrive at a tool crib to check out tools. Only one clerk works at the crib. The arrival times and service times are as follows:

Type	Dist. of Arrival Time	Dist. of Service Time
1	30 ± 10	12 ± 5
2	20 ± 8	6 ± 3
3	15 ± 5	3 ± 1

Model the above for 20 straight shifts. The times above are minutes. Each shift lasts for 480 minutes. Determine how busy the tool crib clerk is. Determine the average number of mechanics waiting for a tool.

12.6. Refer to Exercise 12.5. Suppose that preference is given to mechanics who take the least time for service, in this case type 3 mechanics. Change the program to determine if the solution changes.

12.7. The program to do the simulation is SOL12_7.GPS. It shows that the tool crib worker was now busy 90.69% of the time and the average number of mechanics waiting was reduced to 0.79.

* Schriber, T. J., (1979), An Introduction to Simulation Using GPSS/H, John Wiley & Sons, New York.

13

Arithmetic in GPSS/H

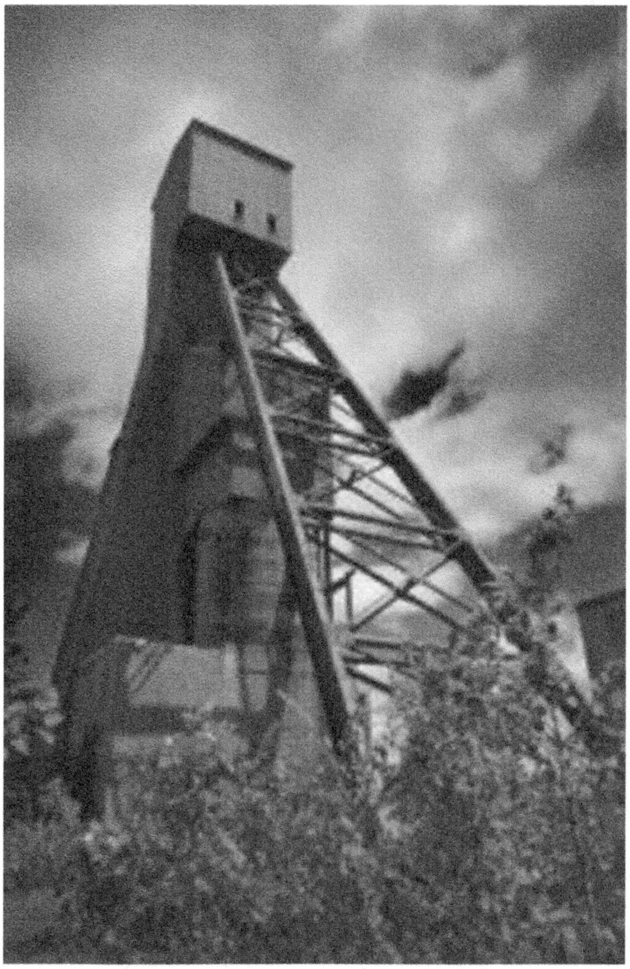

Old wooden mine head frame.

13.1 Ampervariables in GPSS/H

This chapter will further our knowledge of GPSS/H and allow us to learn how to create most of the .ATF files that we have been using in the previous chapters. It will introduce us to the concept of *ampervariables* that are used to do arithmetic in GPSS/H. It also will introduce us to some of the commonly used statements that make programming more interactive.

To use numbers or character strings in GPSS/H, you must first specify their type. As in other computer languages, there are different types of variables that can represent numbers. The common ones are real, integer, and character. These are given the name *ampervariables* since they must have an ampersand (&) as their first character. In order to use them in a program, it is necessary to first declare them before they are used. The specification is as follows:

```
REAL         &(first),&(second).........&(last)
INTEGER      &(first),&(second)........ &(last)
CHAR*n       &(first),&(second).........&(last)  (n = character length)
```

For example,

```
REAL         &X,&Y
```

states that the ampervariables &X and &Y are to be used in the program as real variables (variables with decimals).

```
INTEGER      &I,&J,&Z
```

states that the variables &I, &J, and &Z are to be used in the program as integer variables.

```
CHAR*1       &ANS
CHAR*3       &NAME
```

The first of these, CHAR*1, specifies that the string &ANS is to be a character string of length one. CHAR*3 says that the string &NAME is to be a character string whose length is three.

13.2 Arithmetic in GPSS/H

Arithmetic is done in GPSS/H using the following standard characters:

+, −, /, *, and @

These are self-explanatory except perhaps for the @, which is modular division, where only the remainder is retained. For example, with integers &X = 10 and &Y = 3, the result of the calculation of &X/&Y is 1 (10 divided by 3 gives 3 with a remainder of 1).

If only integers are used in division, the result is integer division; that is, only the dividend is retained and the remainder discarded. If one of the variables to be divided is real, the result is also. Some of the SNAs in GPSS/H are integers, so one must be careful that they do not cause integer division truncation errors when using them in arithmetic operations.

GPSS/H actually allows for two other types of these ampervariables, but in our work, only integer, real, and character ampervariables will be used. You can use subscripted variables in ampervariables such as

```
REAL         &X(50),&BINS(30)
REAL         &XX(4,4),&YYY(12,12)
```

The names of the ampervariables may be up to eight alphanumeric characters in length. Thus,

```
(a)INTEGER    &I,&JOE,&K123456,&JJJ,&XYZ
(b)REAL       &ZX,&KLMN,&TRUCKS, &SPEED
(c)CHAR*1     &ANS,&MYNAME
```

(a) would define integer ampervariables `I, JOE, K123456, JJJ,` and `XYZ;` (b) would define real ampervariables `ZX, KLMN, TRUCKS,` and `SPEED;` and (c) would define two character ampervariables `ANS` and `MYNAME`. Each would be one character long. In the event that one tried to define a character ampervariable that is more than one character in length, only the first character would be taken. How to give values to ampervariables is covered in the next section. In the main program Blocks, it is then possible to have the following Blocks:

```
QUEUE       &I
ADVANCE     &SPEED
GENERATE    &JJJ
SEIZE       &JOE
```

Integer ampervariables are commonly used in connection with the GPSS/H `DO LOOP`, which is covered shortly.

13.3 The LET Statement and the BLET Block

When an ampervariable is declared, it has the value of zero. It is also possible to initialize them by means of the `LET` statement. The form is quite simple. It is as follows:

```
(label) LET (ampervariable) = value or expression
```

For example,

```
LET &I = 12
LET &XONE = &SPEED/360.5
LET &AREA = &LENGTH*&WIDTH
```

Notice that there are no spaces in the statement. While white space is recommended in other computer languages, it is not allowed in GPSS/H.

If one wanted to have an ampervariable as a Block, the form is:

```
BLET   (ampervariable) = value or expression
```

For example,

```
     REAL          &X
     GENERATE      10
     BLET          &X = 5.25
     BPUTPIC       LINES = 1,AC1,&X,XID1
The time is **.**** X = *.** Transaction No. *** enters the system
```

Depending on the values of the ampervariables, the line of output might be something such as

```
The time is 10.0000 X = 5.25 Transaction No. 1 enters the system
```

If character ampervariables are used, the text must be separated by quote marks:

```
CHAR*1 &ANS
```

then,

```
LET  &ANS = 'Y'
```

is correct.

13.4 Another Important SNA

A very important SNA in making animations is `FRN1`. This returns a random number from 0.000000 to 0.999999 (but not 1.000000). This is useful in generating a random number. Recall that there are 32 different colours in PROOF. Often, one wants to make an object have a random colour. Several are shades of grey, so let us just assume that we want a colour from F1 to F26. This can be done as follows:

```
     INTEGER    &I
     GENERATE   10,4         A CAR COMES INTO THE SYSTEM
     BLET       &I = 26*RNF1+1 GENERATE A RANDOM NO FROM 1 to 26
     PUTPIC     FILE = MYOUT,LINES = 4,AC1,XID1,XID1,&I,XID1
TIME **.****
CREATE CAR C*
PLACE C* ON P1
SET C* COLOR F*
     . . .
     . . .
```

The expression with the ampervariable is `26*RNF1+1`. The *1* is necessary as the value returned from `26*RNF1` is a number from 0 to 25.

The .ATF file might have the following lines of code:

```
TIME  69.7894
CREATE CAR C6
PLACE C6 ON P1
SET C6 COLOR F9
```

The foreground colours are also given specifications as follow:

```
BLUE        F1
RED         F2
WHITE       F3
```

```
YELLOW      F4
PINK        F5
TAN         F6
GREEN       F7
```

The colours from F8 to F32 are not given specifications in PROOF. You can only use them by reference to their F-numbers.

The background colours are from Background (Bac) and going from L1 up to L8 and being different shades of grey. You will find it easier to use the F-number, as most of the colours are specified in large layouts using GPSS/H code.

PROOF allows you to make CLASSes multicoloured. However, if you change the colour later, the whole object takes on the single new colour.

Suppose that we wanted to have an object in the animation traverse a path for a time given in the `ADVANCE` Block obtained by sampling from the distribution 8 ± 3.5. We need to have the time `ADVANCE`d in *both* the animation and the program to be identical. This is accomplished (and has been done this way in the previous chapters as the reader may have noticed) by means of an ampervariable and the SNA `FRN1`. Consider the following line of code:

```
BLET     &X = 7.00*FRN1+4.50
```

`FRN1` will generate a random number from 0 to .999999. This is multiplied by 7 and the result added to 4.5. The minimum value given to &X is 4.5 and the maximum is 11.499999. All of these can occur with equal probability. Hence, &X returns a value from sampling from the distribution 8 ± 3.5. &X might be used in a program as follows:

```
           BLET       &X = 4.5*FRN1+7.00
           BPUTPIC    FILE = ATF,LINES = 3,AC1,XID1,XID1,&X
TIME *.****
PLACE CAR* on PATH1
SET CAR* TRAVEL **.**
           ADVANCE    &X
```

The time on the path, `PATH1`, and the time that the transaction is `ADVANCE`d are identical. Thus, if the `CAR` in the animation is placed on the path, `PATH1`, and then is placed on another path when it reaches the end of `PATH1`, it will do so in a smooth, uninterrupted manner.

Example 13.1: Generating Random Numbers

Write a GPSS/H program that gives the following as an output file that shows 20 random numbers generated using the distribution 8 ± 3.5.

The program to do this is stored at CHAP13A.GPS. Its listing is shown in Figure 13.1. The output from this program is as shown in Figure 13.2.

If the program is rerun, the same output is produced. This is because the same random numbers are used. Later, we shall learn how to rerun the program with different random numbers.

The following examples have to do with the ID numbers of the transactions. The transaction's XID1 has been introduced previously.

```
          SIMULATE
*********************************
*   PROGRAM CHAP13A.GPS         *
*   GENERATION OF RANDOM NUMBERS *
*********************************
          PUTSTRING   ('  ')
          PUTSTRING   ('  ')
          PUTSTRING    ('  ')
          REAL        &X
          GENERATE    1
          BLET        &X=7.00*FRN1+4.50
          BPUTPIC     AC1,&X
  TIME IS **    SAMPLED VALUE IS **.******
          TERMINATE   1
          START       20
          PUTSTRING  (' ')
          END
```

FIGURE 13.1
Listing of program CHAP13A.GPS.

```
TIME IS  1    SAMPLED VALUE IS  6.39207
TIME IS  2    SAMPLED VALUE IS  7.99548
TIME IS  3    SAMPLED VALUE IS  9.67840
TIME IS  4    SAMPLED VALUE IS  5.92865
TIME IS  5    SAMPLED VALUE IS  5.34467
TIME IS  6    SAMPLED VALUE IS  9.87904
TIME IS  7    SAMPLED VALUE IS  7.58276
TIME IS  8    SAMPLED VALUE IS 10.51621
TIME IS  9    SAMPLED VALUE IS 10.04656
TIME IS 10    SAMPLED VALUE IS  5.14778
TIME IS 11    SAMPLED VALUE IS  4.96015
TIME IS 12    SAMPLED VALUE IS  5.80066
TIME IS 13    SAMPLED VALUE IS  6.85358
TIME IS 14    SAMPLED VALUE IS 10.72885
TIME IS 15    SAMPLED VALUE IS  9.63710
TIME IS 16    SAMPLED VALUE IS  6.26629
TIME IS 17    SAMPLED VALUE IS 10.15337
TIME IS 18    SAMPLED VALUE IS  9.29055
TIME IS 19    SAMPLED VALUE IS  8.05013
TIME IS 20    SAMPLED VALUE IS  9.33120
```

FIGURE 13.2
Output from program CHAP13A.GPS.

Example 13.2: Study of Transaction's XID1

Suppose that a transaction is to be created every 5 time units. This transaction is to be put on the FEC for 4 time units and then destroyed. The program is to run for 25 time units, and the output file is to contain the time each transaction entered the system and the number of the transaction.

The GPSS/H program to do this is stored as CHAP13B.GPS. Its listing is as shown in Figure 13.3. The output file CHAP13B.OUT is shown in Figure 13.4.

As can be seen, the file and the transaction ID numbers are as expected. This example needs to be studied and understood, as the concepts used here are common to making most animation trace files.

```
          SIMULATE
**********************************
* CHAP13B.GPS FURTHER STUDY      *
* OF THE TRANSACTION'S XID1      *
**********************************
   ATF    FILEDEF      'CHAP13B.OUT'
          GENERATE     5,,0,,1   TRANSACTIONS CREATED
          BPUTPIC      FILE=ATF,LINES=2,AC1,XID1
TIME *.****
THE TRANSACTION'S ID IS: XID1 *
          ADVANCE      4   SAVE FOR 4 TIME UNITS
          TERMINATE
          GENERATE     25  OTHER TRANSACTIONS ARE CREATED
          BPUTPIC      FILE=ATF,LINES=2,AC1,XID1
TIME *.****
THE TRANSACTION'S ID IS: XID1 *
          TERMINATE    1
          START        1
          PUTPIC       FILE=ATF,LINES=2,AC1
TIME *.****
END
           END
```

FIGURE 13.3
Listing of program CHAP13B.GPS.

```
TIME 0.0000
THE TRANSACTION'S ID IS: XID1 1
TIME 5.0000
THE TRANSACTION'S ID IS: XID1 3
TIME 10.0000
THE TRANSACTION'S ID IS: XID1 4
TIME 15.0000
THE TRANSACTION'S ID IS: XID1 5
TIME 20.0000
THE TRANSACTION'S ID IS: XID1 6
TIME 25.0000
THE TRANSACTION'S ID IS: XID1 7
TIME 25.0000
THE TRANSACTION'S ID IS: XID1 2
TIME 25.0000
END
```

FIGURE 13.4
Listing of the output file from program CHAP13B.GPS.

Example 13.3: Creating Transactions and Destroying Them

Suppose you wanted to create three transactions at time 0. Each is placed on the FEC for 1 time unit and then removed from the system. Write the GPSS/H code that places the time and XID1 for each transaction as it is created and then destroyed.

The program to do this simulation is stored as CHAP13C.GPS. Its listing is as shown in Figure 13.5. The output is as shown in Figure 13.6.

```
SIMULATE
*******************************
*  CHAP13C.GPS  FURTHER ON     *
*  TRANSACTION'S ID            *
*******************************
  ATF      FILEDEF       'CHAP13C.OUT'
           GENERATE      3,,0,,1
           BPUTPIC       FILE=ATF,LINES=2,AC1,XID1
TIME *.****
TRANSACTION'S ID IS:     XID1 *
           ADVANCE       1    WAIT 1 TIME UNIT
           TERMINATE     1
           START         3
           PUTPIC        FILE=ATF,LINES=2,AC1
TIME *.****
END
           END
```

FIGURE 13.5
Listing of program CHAP13C.GPS.

```
TIME 0.0000
TRANSACTION'S ID IS:  XID1 1
TIME 3.0000
TRANSACTION'S ID IS:  XID1 2
TIME 6.0000
TRANSACTION'S ID IS:  XID1 3
TIME 7.0000
END
```

FIGURE 13.6
Output from program CHAP13C.GPS.

Example 13.4: Having Different Trucks in a Mine

Suppose that you have a situation where you want to create two different trucks. You are going to have three trucks of type A and two trucks of type B. The first of type A is to enter the system at time 0, the second at time 5, and the third at time 10. The first of type B is to enter at time 15 and the second at time 20. The trucks will take 50 time units to cycle through the system and continue cycling for the duration of the simulation. Simulate for 300 time units and do an animation.

The program is stored as CHAP13D.GPS. Its listing is shown in Figure 13.7 and the animation is as shown in Figure 13.8.

Example 13.5: Trucks on Different Paths

Draw a layout consisting of a single line across the screen that is 40 units long. The only path will be this line, and it will be an accumulating path. The only CLASS is one called SQUARE, and it will be a square with a side of 2 units. The hot point is at its bottom centre. The fore clearance is +1.5 and the aft clearance is +1.5 (you cannot have a negative clearance—PROOF will not allow this). Call this layout CHAP13E.

Consider trucks are being created every 10 time units and are being placed on path, P1. We are going to simulate for trucks being created and placed on this path. The trucks take 40 time units to reach the end of the path where they are destroyed. The animation is to show the number of trucks that have been created as well as the simulation time. The simulation time as shown on the animation is to change every .5 time units. The simulation ends when 10 trucks have been destroyed.

```
          SIMULATE
***********************************
*   PROGRAM CHAP13D.GPS           *
*   PUTTING TRUCKS INTO A SYSTEM  *
***********************************
  ATF     FILEDEF     'CHAP13D.ATF'
          GENERATE    5,,0,3    TYPE A TRUCKS
          BPUTPIC     FILE=ATF,LINES=3,AC1,XID1,XID1
TIME *.****
CREATE TRUCKA T*
PLACE T* ON P1
          TERMINATE
          GENERATE    5,,15,2   TYPE B TRUCKS
          BPUTPIC     FILE=ATF,LINES=3,AC1,XID1,XID1
TIME *.****
CREATE TRUCKB T*
PLACE T* ON P1
          TERMINATE
          GENERATE    300
          TERMINATE   1
          START       1
          PUTPIC      FILE=ATF,LINES=2,AC1
TIME *.****
END
          END
```

FIGURE 13.7
Listing of program CHAP13D.GPS.

FIGURE 13.8
Screenshot of animation from CHAP13D.

The following GPSS/H program will do this for us. It is stored as CHAP13E.GPS (Figure 13.9), whose animation is shown in Figure 13.10.

The animation shows the trucks appearing on the path all nicely spaced out. If the animation were to run for longer, there would be an error on the screen.

```
    SIMULATE
*******************************
* CHAP13E.GPS                 *
* PRACTICE WITH ANIMATION     *
*******************************
  ATF      FILEDEF      'CHAP13E.ATF'
           INTEGER      &I      DEFINE COUNTER
           GENERATE     10,,0
           BLET         &I=&I+1   INCREMENT
           BPUTPIC      FILE=ATF,LINES=6,AC1,XID1,XID1,XID1,AC1,&I
TIME *.****
CREATE SQUARE S*
PLACE S* ON P1
SET S* TRAVEL 40.0000
WRITE M2 ***.**
WRITE M1 *
           ADVANCE      40.0000
           BPUTPIC      FILE=ATF,LINES=2,AC1,AC1
TIME *.****
WRITE M2 ***.**
           TERMINATE   1
*****************************
*  SEGMENT TO UPDATE CLOCK  *
*  EVERY .5 TIME UNITS      *
*****************************
           GENERATE     ,,,1     DUMMY TRANSACTION
 BACK      BPUTPIC      FILE=ATF,LINES=2,AC1,AC1
TIME *.****
WRITE M2 ***.**
           ADVANCE      .5
           TRANSFER     ,BACK
           START        10
           PUTPIC       FILE=ATF,LINES=2,AC1
TIME *.****
END
         END
```

FIGURE 13.9
Listing of program CHAP13E.GPS.

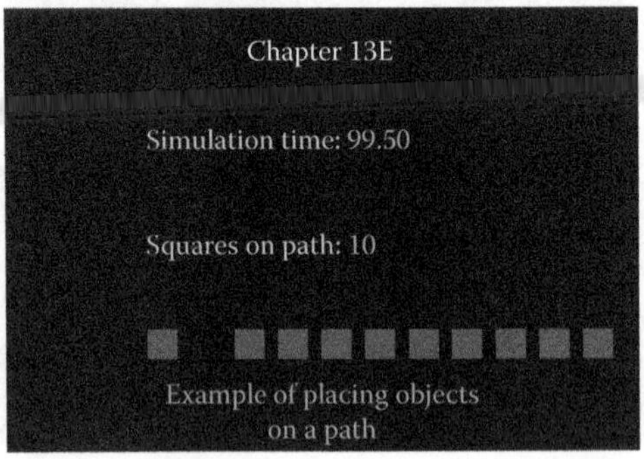

FIGURE 13.10
Screenshot of animation from CHAP13E.

Example 13.6: Parts Obtaining Different Service

Consider the following system. Parts enter at point A every 10 ± 4 minutes. They travel to the first server in 2 ± .5 minutes. The first server takes 11 ± 5 minutes to provide service. After finishing, the parts travel to the second server in 3 ± .5 minutes, which takes 10 ± 6 minutes to provide service. The parts then leave the server and travel

3 ± 1 minutes to leave the system. Simulate for 2 hours of operation. Have the parts different colours at random. The *parts* can be represented by filled circles, with a radius of .5 and clearances of .75.

Solution

Begin by drawing the layout. We have to decide how the parts will leave the paths to obtain service. We have several choices. One is to have the parts move into the centre of the servers (on a path) or to have them just *jump* to the centre of the servers. We will select the latter. Thus, when we are drawing the layout, it is necessary to make note of the coordinates of the objects where the hot point (the parts moving through the system) will be placed.

We begin by having the parts come to the system and move up to the first server. Caution must be taken here. If the first path goes right up to the edge of the server, the parts will appear to wait partially inside the server. We want the parts to be poised in front of the server. Thus, we must make the paths in several segments. Consider the first path on the left leading to the first server. This path will consist of two segments. One will go from the left of the screen almost up to the server but one unit away. If the hot point of the CLASS is at the bottom centre, then the object will come to rest at the end of the first segment, not halfway inside of the first server. The same holds for the path from the first server to the second server.

The hot point of the parts will be at the bottom and at the midpoint. There will be three paths, the first two must be accumulating paths, and the parts will have appropriate clearances. A random number will be used to give each part a separate colour.

The first section of the solution will consist of the parts being created and travelling on the first path up to the first server. Once this is correctly done, the second segment of the program can be completed. The program is stored as CHAP13F.GPS and its listing is given in Figure 13.11. The animation is given in Figure 13.12.

```
            SIMULATE
*******************************
* CHAP13F.GPS ASSEMBLY LINE    *
* ITEMS HAVE TO PASS           *
* THROUGH TWO STATIONS         *
*******************************
   ATF      FILEDEF    'CHAP13F.ATF'
            REAL       &X,&Y,&Z        DEFINE VARIABLES TO BE USED
            INTEGER    &I,&COUNT       DEFINE VARIABLES TO BE USED
            GENERATE   10,4,0          PARTS COME ALONG
*************************************************
* WANT TO MAKE EACH PART HAVE A RANDOM COLOR    *
*************************************************
            BLET       &I=24*FRN1+1
            BLET       &X=1.00*FRN1+1.5 TIME TO TRAVEL TO SERVER A
            BPUTPIC    FILE=ATF,LINES=6,AC1,XID1,XID1,XID1,&X,XID1,_
                       &I,AC1
TIME *.****
CREATE PART P*
PLACE P* ON P1
SET P* TRAVEL **.**
SET P* COLOR F*
WRITE M2 ***.**
            ADVANCE    &X      TRAVEL TO FIRST SERVER
            QUEUE      WAIT1   ENTER QUEUE
            SEIZE      FIRST   IS FIRST SERVER FREE?
            DEPART     WAIT1   LEAVE THE QUEUE
            BPUTPIC    FILE=ATF,LINES=3,AC1,XID1,AC1
```

FIGURE 13.11
Listing of program CHAP13F.GPS. *(Continued)*

```
TIME *.****
PLACE P* AT 30 29
WRITE M2 ***.**
           ADVANCE     11,5   PROVIDE SERVICE
           BLET        &Y=1.00*FRN1+2.5  TIME TO TRAVEL
           BPUTPIC     FILE=ATF,LINES=4,AC1,XID1,XID1,&Y
TIME *.****
PLACE P* ON P2
SET P* TRAVEL **.**
WRITE M2 ***.**
           ADVANCE     .5
           BPUTPIC     FILE=ATF,LINES=2,AC1,AC1
TIME *.****
WRITE M2 ***.**
           RELEASE     FIRST  FREE SERVER
           ADVANCE     &Y-.5
           QUEUE       WAIT2  ENTER QUEUE
           SEIZE       SECOND IS SECOND SERVER FREE?
           DEPART      WAIT2  LEAVE THE QUEUE
           BPUTPIC     FILE=ATF,LINES=3,AC1,XID1,AC1
TIME *.****
PLACE P* AT 47 29
WRITE M2 ***.**
           ADVANCE     10,6   PROVIDE SERVICE
           BLET        &Z=2.00*FRN1+2
           BPUTPIC     FILE=ATF,LINES=4,AC1,XID1,XID1,&Z
TIME *.****
PLACE P* ON P3
SET P* TRAVEL **.**
WRITE M2 ***.**
           ADVANCE     .5
           BLET        &COUNT=&COUNT+1   COUNT THE PARTS
           BPUTPIC     FILE=ATF,LINES=3,AC1,AC1,&COUNT
TIME *.****
WRITE M2 ***.**
WRITE M1 **
           RELEASE     SECOND
           ADVANCE     &Z-.5
           BPUTPIC     FILE=ATF,LINES=3,AC1,XID1,AC1
TIME *.****
DESTROY P*
WRITE M2 ***.**
           TERMINATE
           GENERATE    ,,,1   DUMMY TRANSACTION
  BACK     BPUTPIC     FILE=ATF,LINES=2,AC1,AC1
TIME *.****
WRITE M2 ***.**
           ADVANCE     .5
           TRANSFER    ,BACK
           GENERATE    60*2    TIMER TRANSACTION
           TERMINATE   1
           START       1
           PUTPIC      FILE=ATF,LINES=2,AC1
TIME *.****
END
           END
```

FIGURE 13.11 (Continued)
Listing of program CHAP13F.GPS.

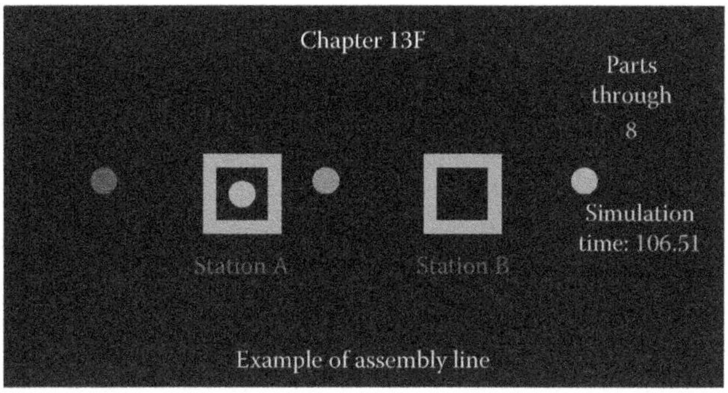

FIGURE 13.12
Screenshot of animation from CHAP13F.

Notice that, when a part is finished with having service done by either server, the program does not immediately release the facility. This is for the animation. If the server was released immediately after service was finished, the object being serviced and the one waiting (if there was one) would immediately change places. With a slight time delay of .5 time units, the animation is smoother. The way that each part is given a random colour should be carefully studied.

Example 13.7: Trucks on Path with Different Travel Times

This example is a very important one, as it will be used in many mining simulations. The trucks in a mine will travel on different paths as they move around from loaders to destinations. Each path will be broken into different segments, and each segment might have a separate travel time distribution. For example, let us consider a segment extending from A to B to C to D. Suppose mining trucks arrive at A every 15 ± 4 minutes. The time to travel from A to B is given by the distribution 8 ± 1.2; the travel time from B to C is given by the distribution 7.5 ± .8, and the travel time from C to D by the distribution 6 ± .4 (all times in minutes). The GPSS/H program to simulate this is easily written and is given by CHAP13G.GPS (Figure 13.13).

```
          SIMULATE
***************************
*  PROGRAM CHAP13G.GPS     *
*  TRUCKS TRAVEL ON        *
*  DIFFERENT SEGMENTS IN   *
*  A MINE                  *
***************************
 TRUCKS   GENERATE    15,4      TRUCK APPEARS AT A
          ADVANCE     8,1.2     TRAVEL FROM A TO B
          ADVANCE     7.5,.8    TRAVEL FROM B TO C
          ADVANCE     6.2,.4    TRAVEL FROM C TO D
 ALLGONE  TERMINATE             TRUCK LEAVES
          GENERATE    400
          START       1
          PUTPIC      LINES=2,N(TRUCKS),N(ALLGONE)
   TRUCKS TO ENTER SYSTEM **
   TRUCKS TO LEAVE SYSTEM  **
          END
```

FIGURE 13.13
Listing of program CHAP13G.GPS.

The animation needs a bit of care in developing. Consider the travel from A to B on the first segment. It is necessary to have the travel time in animation be the same as the travel time in the GPSS/H program. This can be done using the SNA FRN1. The travel time on this segment is given by the distribution 8 ± 1.2. This is a uniform distribution with any value from 6.8 to 9.2 having equal probability of occurrence. Consider the GPSS/H variable defined as follows:

```
BLET &X = 1.4*FRN1+6.8
```

If the SNA FRN1 returns 0, the value of &X is 6.8. If it returns 1 (actually 0.999999), the value of &X will be 9.2 or nearly so. This variable, &X, can then be used in both the animation and the original GPSS/H program to have the travel times the same. It should be carefully studied. The program to do the animation is given by CHAP13H.GPS. Its listing is given in Figure 13.14.

The animation is simple but important to understand. This is shown in Figure 13.15.

```
          SIMULATE
******************************
*  PROGRAM CHAP13H.GPS        *
*  ANIMATION OF CHAP13G.GPS *
******************************
 ATF      FILEDEF      'CHAP13H.ATF'
          REAL         &X,&Y,&Z
 TRUCKS   GENERATE     15,4      TRUCK APPEARS AT A
          BLET         &X=2.4*FRN1+6.8
          BPUTPIC      FILE=ATF,LINES=4,AC1,XID1,XID1,XID1,&X
TIME *.****
CREATE TRUCK T*
PLACE T* ON P1
SET T* TRAVEL **.**
          ADVANCE      &X        TRAVEL FROM A TO B
          BLET         &Y=1.6*FRN1+6.7
          BPUTPIC      FILE=ATF,LINES=3,AC1,XID1,XID1,&Y
TIME *.****
PLACE T* ON P2
SET T* TRAVEL **.**
          ADVANCE      &Y        TRAVEL FROM B TO C
          BLET         &Z=1.6*FRN1+6.7
          BPUTPIC      FILE=ATF,LINES=3,AC1,XID1,XID1,&Z
TIME *.****
PLACE T* ON P3
SET T* TRAVEL **.**
          ADVANCE      &Z
          BPUTPIC      FILE=ATF,LINES=2,AC1,XID1
TIME *.****
DESTROY T*
 ALLGONE  TERMINATE              TRUCK LEAVES
          GENERATE     400
          TERMINATE    1
          START        1
          TERMINATE    1
          PUTPIC       LINES=2,N(TRUCKS),N(ALLGONE)
    TRUCKS TO ENTER SYSTEM **
    TRUCKS TO LEAVE SYSTEM  **
          PUTPIC       FILE=ATF,LINES=2,AC1
TIME *.****
END
          END
```

FIGURE 13.14
Listing of program CHAP13H.GPS.

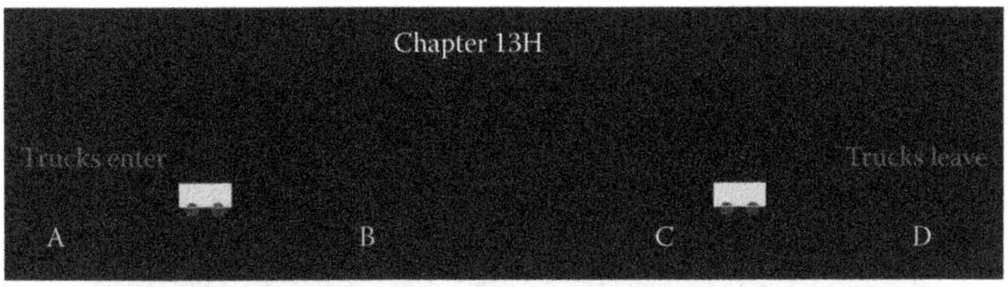

FIGURE 13.15
Screenshot of animation from program CHAP13H.GPS.

13.5 Exercises

13.1. The following ampervariables have been defined and given the values:

Real: &X = 5.5, &Y = 3.2, &Z = 4, and &W = 100

Integer: &I = 10 and &J = 5

Determine the values of the variable after the arithmetic has been done:

(a) LET &X = &Y–&J+&Z

(b) LET &I = &Y*&Z

(c) LET &W = &W+&Y–&Z

(d) LET &J = &Y/&Z

(e) LET &Z = &x*&y–&i*&J

13.2. Using the same variables as in Exercise 13.1, determine the values of the variables:

(a) LET &I = &Y*&Z@3

(b) LET &J = &Z@&W+1

(c) LET &X = &X*&J@&I

(d) LET &I = (100*&I–25)@&Z

13.3. Write a GPSS/H program to solve Exercise 13.1.

13.4. Write a GPSS/H program to solve Exercise 13.2.

13.5. Draw a layout that will have only a single CLASS, which is a star. Write the GPSS/H program to create the .ATF file that will put 100 stars at random on the screen. Each star is to have a different colour, also at random. The first star is to appear on the screen at time 1 and then another star every 1 simulated time units until all 100 are on the screen. The viewing screen grid is to have the four corners approximately at coordinates: (–10, 0), (90, 0), (–10, 60), and (90, 60) so that all the stars fit on the screen.

13.6. Write the GPSS/H program to place multiple stars on the screen. The size of the stars will be approximately 1.5 units in width. They are to appear in neat rows across the screen. The first row goes from (–10, –10) and will have 20 stars all 3 units apart. The stars will be created every 1 time unit starting at time 1. The stars will have random colours. The stars will start to appear at the bottom of the screen. Each time unit will see a new star in the row until 20 are in the row. When 20 stars are in a row, a new row is started. The first star in this row will be at (–10, –5). When completed, the animation will look as shown in Figure 13.16.

FIGURE 13.16
Animation for Exercise 13.6.

13.7. A single machine takes 8 ± 3.5 minutes to finish a part. Workers can make a raw product in 28 ± 8 minutes. As soon as a worker finishes a part, it is immediately taken to this machine to be finished. The machine can only finish one part at a time. Each part made results in a profit of $68.25. The machine costs $125 per shift (8 hours) whether it is running or not. Each worker costs the company $18.50 per hour. Determine the correct number of workers to have.

13.8. Refer to Exercise 13.7. Write the program to do the animation for the exercise. The animation should be for four workers. It should show the workers coloured green when working and coloured red when waiting for the finishing machine. They can remain stationary. When the finishing machine is working, it should jiggle. Messages on the screen should show the utilization of the finishing machine and number of parts finished. The completed animation should look as shown in Figure 13.17.

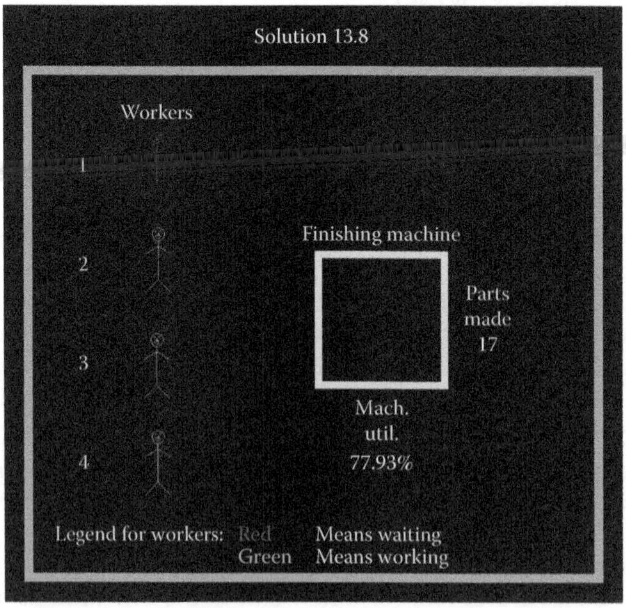

FIGURE 13.17
Possible animation for Exercise 13.8.

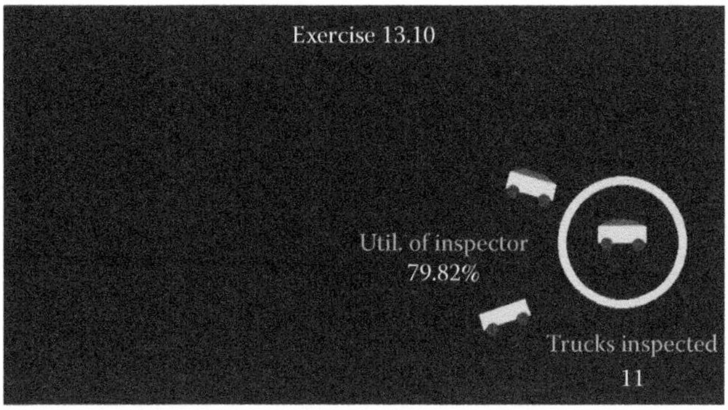

FIGURE 13.18
Screenshot of animation for Exercise 13.10.

13.9. Mining trucks arrive from the mine at point A every 8 minutes. They travel to point B in 8 ± 3 minutes. There, a single miner takes 7 ± 3.5 minutes to inspect each truck. Inspected trucks then leave via a path to point C that takes 6 ± 2.8 time units. Write the GPSS/H code to give the number of trucks inspected in a shift of 450 minutes. How busy is the miner?

13.10. Provide the animation for Exercise 13.9. The animation should match the statement of the problem. The animation will look similar to that shown in Figure 13.18.

13.11. In the animation given in Figure 13.18, when a truck is finished being inspected and another is waiting for the inspector to be free, they will instantly change places. In reality, when the truck being inspected is finished, it will move to the new path from the inspection area. Only when the inspection area is free will the waiting truck move to be inspected. This can be accomplished in several ways. A simple way is to have the truck that has just been inspected travel a slight way on the new path before the inspector is freed. For example, suppose that the new travel time from the inspection station to point B is 6.235 minutes. Here is what can be done: place the just inspected truck on the new path. Let it travel for a short time, say, .4 minutes then release the inspector. Have the inspected truck then travel to point B in a time of 5.835 minutes. Modify the program in Exercise 13.10 to do this.

13.12. Modify the program in Exercise 13.7 so that the calculations to determine the profit are done in GPSS/H.

13.13. In Exercise 13.7, once the optimum number of workers is found, suppose that the plant manager wishes to raise the price of the part to justify hiring another worker. How high should the new profit be?

14

ENTER/LEAVE Blocks

A truck for hauling coal in Venezuela.

14.1 The ENTER and LEAVE Blocks

14.1.1 Multiple Servers—The ENTER Block

It often happens that the system being studied has multiple servers that are identical. Transactions that attempt to use these will be denied access only if all the servers are busy. They do not wait at each server but are held on the current events chain (CEC) in the previous Block until one of the servers is free. Figure 14.1 illustrates this situation.

Examples of these might be the multiple berths for loading or unloading facilities in a port for ships, two shovels in the same area of a mine, and six tellers in a bank. GPSS/H handles these by means of a combination of a Block and a statement. The Block is used in the program for the transaction to actually use one of the parallel servers. The statement tells the processor how many of these servers are available. These servers will be referred

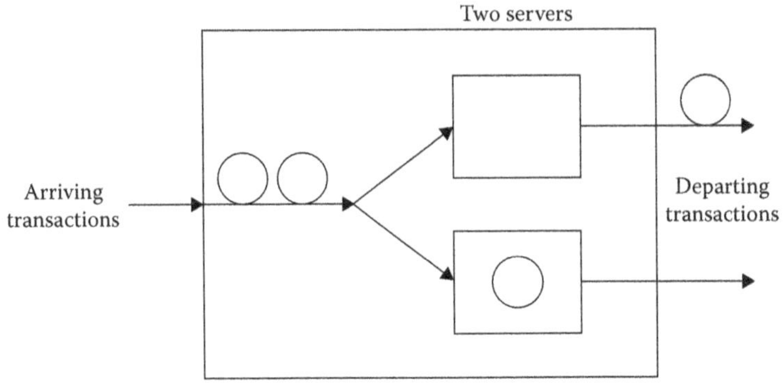

FIGURE 14.1
Sketch of multiple servers.

to as *storages*. The Block for the transaction in the programs is the ENTER Block. One form of the ENTER Block is

```
ENTER A
```

where A is the operand. This can be a name or a number. If it is a number, it cannot exceed the maximum number of such ENTER Blocks allowed in your system. It is also possible to have this operand to be a variable, as we shall see later.

Examples of the ENTER Block are as follows:

```
ENTER  DUMP
ENTER  2
ENTER  TWO
ENTER  STORE
```

The operands DUMP; 2,TWO; and STORE refer to the names of the multiple servers, or storages, as we shall now refer to them. There is no confusion in using the storages named DUMP and TWO, but care should be exercised with the one named 2. This does *not* refer to the number of storages associated with the Block ENTER 2. This number is specified by a separate statement. It is quite possible that the number of storages associated with each of the above Blocks might be 4 for DUMP, 3 for 2, and 4 for THREE. We will shortly learn how to specify the number of these. It is possible to have a second or B operand with the ENTER Block, such as

```
ENTER HARBOR,2
```

In this case, two storages are used. This might represent a large ship entering a harbour and two tug boats (in this case, the tug boats are the storages) that are needed to guide it into the harbour. Most of the time, only one of the storages is used each time a transaction enters the ENTER Block. However, there will be times when the second operand will come in very handy.

14.1.2 Defining the Number of Multiple Servers—The STORAGE Statement

Normally, when there are to be multiple servers, the number of these must be specified. This is done by the STORAGE statement. There are two forms of it. The form may look a bit strange, but if the programmer remembers that it is used in connection with the ENTER Block, it is easy to write. Suppose there are three docks in a harbour. The arriving ships will select one of them via the ENTER Block.

```
ENTER  PORTS
```

How the computer knows that there are three docks is specified by the STORAGE statement. This must be placed before the ENTER Block. Thus, it can be placed anywhere in the program as long as it is before the ENTER Block is encountered. Normally, for convenience, it is placed near the top of the program before any GENERATE Block. It can have several forms. These are as follows:

```
(a) STORAGE S(DOCKS),3
```

or

```
(b) DOCKS STORAGE 3
```

Both (a) and (b) are identical. The advantage of using (a) is that one can have multiple specifications on the same line by separating them with a /:

```
STORAGE        S(DOCKS),3/S(TUG),2/S(SHIP),10/S1,4
```

The above line specifies that the STORAGE of DOCKS is 3; the STORAGE of TUG is 2; the storage of SHIP is 10; and the STORAGE of 1 is 4.* Notice that, for STORAGES with numbers for operands, all that is needed is S1. This is identical to S(1).

Later in the program, when the Block

```
ENTER DOCKS
```

is encountered, the transaction will be able to *enter* only if there are less than three transactions in it already. If the transaction cannot move forward, it is held on the CEC until a later scan shows it can be moved. If the ENTER Block had been

```
ENTER DOCKS,2
```

and there was only one storage left, the transaction would not be able to enter but would be held in the previous Block until another storage was available.

Whenever you use the ENTER Block for parallel entities, there will be certain output created in the list file when the program is over. Consider the following example.

Example 14.1: Assembly Line with Two Identical Machines

An assembly line has two identical machines. Each can work at a rate of 13 ± 5.5 minutes when a part comes along. These parts arrive every 7 ± 2.6 minutes. Simulate for an 8-hour day.

---

* It is even possible to have a further shorthand way of specifying storages if numbers are used for storages. If one has storages S1 to S6, all of storage size 4, one could simply have the following:

```
STORAGE S1-S6,4
```

Solution

The program to do this is CHAP14A.GPS. Its listing is given in Figure 14.2 and the edited portion of the output from the program is shown in Figure 14.3.

```
RELATIVE CLOCK: 480.0000 ABSOLUTE CLOCK: 480.0000
```

The interpretation of the above is as follows:

`TOTAL TIME` The `STORAGE WORK` was busy 96.6% of the time. This is the time for both of the workers. It is not possible to tell the percentage of time each was busy. This number is stored by GPSS/H as parts per thousand (an integer!).

`ENTRIES` The number of transactions that entered the `ENTER` Block was 67.

`AVERAGE`

`TIME/UNIT` The average time a transaction was in this Block was 13.504 time units.

`CAPACITY` The storage capacity was 2. This was specified by the `STORAGE` statement.

`AVERAGE CONTENTS` The average content was 1.932. This means that, during the simulation, an average of 1.932 of the 2 storages was being used.

`CURRENT CONTENTS` When the program ended, the content of the storage was 2.

`MAXIMUM CONTENTS` The maximum number in the storage at any one time was 2.

There is more output, but it is not of concern at this time.

```
                   SIMULATE
*****************************
*   CHAP14A.GPS             *
*   EXAMPLE OF STORAGE BLOCK  *
*****************************
                   STORAGE      S(WORK),2   DEFINE STORAGES
                   GENERATE     7,2.6       PARTS ARRIVE
                   QUEUE        WAIT        WAIT IN THE QUEUE
                   ENTER        WORK        CAN WORK BE DONE?
                   DEPART       WAIT        LEAVE THE QUEUE
                   ADVANCE      13,5.5      WORK BEING DONE
                   LEAVE        WORK        FREE THE WORK AREA
                   TERMINATE                LEAVE
                   GENERATE     480         TIMER TRANSACTION
                   TERMINATE    1
```

FIGURE 14.2
Listing of program CHAP14A.GPS.

```
BLOCK CURRENT      TOTAL
1                    68
2                    68
3                    68
4                    68
5           2        68
6                    66
7                    66
8                     1
9                     1

              --AVG-UTIL-DURING--
STORAGE  TOTAL   ENTRIES   AVERAGE     CAPACITY   AVERAGE   CURRENT   MAXIMUM
         TIME              TIME/UNIT              CONTENTS  CONTENTS  CONTENTS
WORK     0.966    68        13.635         2         1.932     2         2
```

FIGURE 14.3
Edited portion of output from CHAP14A.GPS.

14.1.3 SNAs Associated with Storages

The SNAs associated with the storage are given below:

SNA	Meaning
`S(name)` or `Sn`	Current storage content
`R(name)` or `Rn`	Remaining storage content
`SA(name)` or `SAn`	Average storage content
`SC(name)` or `SCn`	Storage entry count
`SM(name)` or `SMn`	Maximum storage count
`SR(name)` or `SRn`	Utilization of storage in parts per thousand
`ST(name)` or `STn`	Average holding time

Notice that the utilization of the storage is given by `SR(name)` and is *expressed in parts per thousand.* The original storage specification is *not* an SNA. If it is desired to use this in the program, it is necessary to use `S(name)+R(name)` that always adds up to the original storage specification.

Actually, it is not necessary to specify a storage when you have an `ENTER` Block. In this case, the processor sets aside 2,147,483,677 as the storage capacity. (The number is $2^{31} - 1$.) It may appear that you would never omit giving the storage capacity, but there are examples when one actually does omit this. For example, suppose you are modelling a hardware store. Customers arrive and immediately take a shopping cart. Suppose customers arrive every 30 ± 8 seconds. If you have 20 carts available, you would model this as follows:

```
STORAGE      S(CART),20    PROVIDE 20 CARTS
GENERATE     30,4          CUSTOMERS ARRIVE
ADVANCE      0             DUMMY BLOCK
ENTER        CART          SELECT A CART
.......
.......
(rest of program)
```

Notice that an `ADVANCE` Block is used with 0 time units. This is done so that, if all the carts are taken, the transaction will not be held up in the `GENERATE` Block. This ensures that the `GENERATE` Block will produce transactions according to the specified times of 30 ± 4 seconds.

But, now suppose that you wanted to determine the maximum number of carts ever used for the simulation period. The program is then written so that arriving customers *always* were able to take a cart. One way would be to assign a very large storage for `CART`, such as

```
STORAGE S(CART),10000
```

However, we could just as easily omit the `STORAGE` statement. At the end of the program, the maximum number of entries into `CART` is listed. (It is even possible to print out a table of the statistical distribution of the carts used—this will be covered later.)

14.1.4 The `LEAVE` Block

Once a transaction uses parallel servers via the `ENTER` Block, it must indicate eventually that it is done and release or free the server. The Block to do this is the `LEAVE` Block. Just as the `QUEUE` and `DEPART` Blocks are *twins*, so are the `ENTER` and `LEAVE` Blocks. The form of the `LEAVE` Block is similar to the `ENTER` Block as it relates directly to it. This form is

```
LEAVE A
```

where A is the name (or number) of the parallel servers. As we shall see later, it can also be a variable.

Thus, you might have the following in a program:

```
STORAGE S(JOE),4
........
ENTER JOE
........
........
LEAVE JOE
........
........
```

It may appear that there is no difference between a single facility and a storage with a capacity of 1. This is almost the case, but we shall see that, when we study the concept of pre-empting, that a facility can be pre-empted, whereas a storage cannot. By pre-emption it is meant that a transaction can actually replace the one already in the `STORAGE` Block. An example of this might be a repair shop that repairs company vehicles along with important factory equipment. If a vital piece of equipment breaks down, a mechanic who is working on a vehicle immediately stops his work and starts to repair the broken equipment. Normally, but not necessarily, the vehicle waits until the mechanic is done repairing the equipment (it can be directed to another facility).

Example 14.2: Fleet of Trucks with Three Service Areas

A garage for inspecting a fleet of trucks has three identical service areas. Trucks arrive for inspection every $3 \pm .3$ minutes. These inspections take 8 ± 2 minutes. After leaving the inspection area, 70% are ready to return to service but the rest need further service, which takes 4 ± 1.5 minutes. The time to travel to the further service area is negligible. If this further service is needed, a single mechanic is assigned to do it. Simulate this system for one shift of 480 minutes.

Solution

The program to do this is CHAP14B.GPS, and its listing is as shown in Figure 14.4.

Notice that a `QUEUE` is not used, as it is not necessary to obtain statistics for this example.

A portion of the output is shown in Figure 14.5.

The output shows that 160 trucks came along each day. 106 needed the single inspection whereas 50 needed further service. There were four trucks left in the system when the simulation ended. The double inspection area was busy 88.5% of the time and the single server only 44.1% of the time.

```
          SIMULATE
**********************************
*  CHAP14B.GPS REPAIR FACILITY    *
*  MORE ON ENTER/LEAVE BLOCKS     *
**********************************
         STORAGE     S(MECH),3  DEFINE MECHANICS
         GENERATE    3,.3       TRUCKS ARRIVE
         ADVANCE     0          DUMMY ADVANCE BLOCK
         ENTER       MECH       MECHANIC FREE?
         ADVANCE     8,2        REPAIRS DONE
         LEAVE       MECH       FREE MECHANIC
         TRANSFER    .7,,AWAY   70% ARE DONE
         SEIZE       SINGLE     SINGLE REPAIR FACILITY
         ADVANCE     4,1.5      REPAIRS BEING DONE
         RELEASE     SINGLE     FREE SERVER
         TERMINATE              LEAVE SYSTEM
AWAY     TERMINATE              LEAVE SYSTEM
         GENERATE    480        TIMER TRANSACTION
         TERMINATE   1
         START       1
         END
```

FIGURE 14.4
Listing of CHAP14.B.

```
RELATIVE CLOCK: 480.0000   ABSOLUTE CLOCK: 480.0000

BLOCK CURRENT      TOTAL  BLOCK CURRENT      TOTAL
1                    160  AWAY                 106
2                    160  12                     1
3                    160  13                     1
4             3      160
5                    157
6                    157
7                     51
8             1       51
9                     50
10                    50

            --AVG-UTIL-DURING--
FACILITY    TOTAL
             TIME
  SINGLE    0.411


 STORAGE    TOTAL
            TIME
            0.885
```

FIGURE 14.5
Selected portion of the .LIS file from CHAP14B.GPS.

Example 14.3*: Shopping in a Store with Four Aisles

A hardware store consists of four aisles and a single checkout counter. Shoppers arrive with interarrival time of 85.2 ± 26.4 seconds. After arriving, each customer who plans to shop in any one or more of the aisles takes a shopping cart. However, 12% of the custom-

* This example is a modification of one from Tom Schriber's *Simulation Using GPSS,* John Wiley and Sons, New York, 1975.

ers simply go to the checkout counter where various items are for sale. These people do not take a shopping cart. The rest shop down each aisle as follows:

Aisle	Problem of Going Down	Time Required to Travel (seconds)
1	.80	125 ± 70
2	.75	144 ± 40
3	.85	150 ± 65
4	.90	175 ± 70

When a shopper is finished, he or she will join the queue in front of the counter until each is checked out. The time to check out is 45 ± 20 seconds for those who shop in the aisles and 35 ± 12 seconds for those who go directly to the checkout counter.

The store owner is concerned that he does not have enough shopping carts. Customers who arrive and find none available tend to leave and shop elsewhere. In addition, the single person who works at the checkout counter is complaining that she is working too hard and is threatening to contact her union. Union regulations forbid a person from working more than 85% of the time. Determine how many shopping carts the store should have and whether the single checkout person has a legitimate claim.

Solution

The program to solve this problem is given by CHAP14C.GPS. Its listing is given in Figure 14.6.

A portion of the .LIS file is shown in Figure 14.7.

Examining the list file gives us the results. The number of customers arriving at the store in the 10 shifts was 3272. Of these, 2769 took a cart and shopped in the aisles, and 403 went directly to the checkout counter. Not shown in Figure 14.7 but found in the list file are the following: The checkout girl was busy only 51.4% of the time, so she has no complaint of being overworked. If anything, she is underworked. The maximum number of carts ever in use was 10. At this time, it would be instructive to know how many times this happened. Later it will be shown how to make statistical distributions to provide us with this information but, for the present, we do not have this data. However, it would seem that providing something like 12 carts should be sufficient. This would take into account the maximum number obtained in the simulation, as well as provide a safety factor of two extra carts.

No animation is done in this chapter as more GPSS/H is needed to handle the concept of the ENTER Block in order to add animation. This will be covered in Chapter 15.

Example 14.4*: Spare Trucks in a Mine

A mine needs 25 trucks working at all times to maximize production. If it has fewer trucks, there is a loss in production, so the mine has spare trucks available. When a truck breaks down and a spare is available, the spare is immediately used as a replacement, so the time to make this replacement is negligible. A truck will work for 240 ± 30 hours before needing repairs. Once down for repairs, it takes 1 ± .25 hours to remove it from the mine to the repair shop. There are two repair bays and a truck takes 18 ± 8 hours to repair. Once repaired, it takes a further .5 hour to return the truck to the mine. Whenever a truck is out of production, it costs the mine $1250 per hour of downtime. Thus, if the mine has only 24 trucks working, the loss is $10,000 for an 8-hour shift. Each spare truck costs $1400 per shift to have on site. Determine the number of spare trucks to have to minimize the expected cost.

* This is a variation of an example from Schriber's textbook on GPSS. In his example, he used sewing machines and a variable number of repairmen.

```
              SIMULATE
******************************
*  CHAP14C.GPS GROCERY       *
*  STORE EXAMPLE             *
******************************
         GENERATE    85.2,26.4      CUSTOMERS ARRIVE
         TRANSFER    .12,,COUNTER   12% GO TO CHECK OUT
         ENTER       CARTS          TAKE A CART
         TRANSFER    .2,,AISLE2     SHOP AISLE 1
         ADVANCE     125,70
AISLE2   TRANSFER    .25,,AISLE3    SHOP AISLE 2
         ADVANCE     144,40
AISLE3   TRANSFER    .15,,AISLE4    SHOP AISLE 3
         ADVANCE     150,65
AISLE4   TRANSFER    .1,,CHECKOUT   SHOP AISLE 4
         ADVANCE     175,70
CHECKOUT QUEUE       LINE           AT CHECKOUT
         SEIZE       COMPLAIN
         DEPART      LINE
         ADVANCE     45,20
         RELEASE     COMPLAIN
         LEAVE       CARTS          PUT CART BACK IN STAND
         TERMINATE
COUNTER  QUEUE       LINE
         SEIZE       COMPLAIN
         DEPART      LINE
         ADVANCE     35,12
         RELEASE     COMPLAIN
         TERMINATE
         GENERATE    28000
         TERMINATE   1
         START       10           SIMULATE FOR 10 SHIFTS
         END
```

FIGURE 14.6
Listing of program CHAP14C.GPS.

RELATIVE CLOCK: 2.8000E+05   ABSOLUTE CLOCK: 2.8000E+05

BLOCK	CURRENT	TOTAL	BLOCK	CURRENT	TOTAL	BLOCK	CURRENT	TOTAL
1		3272	11	1	2585	21		403
2		3272	CHECKOUT	2	2864	22		403
3		2869	13		2862	23		403
4		2869	14		2862	24		403
5	1	2294	15	1	2862	25		10
AISLE2		2868	16		2861	26		10
7	1	2143	17		2861			
AISLE3		2867	18		2861			
9	2	2459	COUNTER		403			
AISLE4		2865	20		403			

FIGURE 14.7
Portion of the .LIS file from program CHAP14C.GPS. Only the Block count is shown.

Solution

The program to simulate this is CHAP14D.GPS. The listing is shown in Figure 14.8.

Notice that the calculations are done by GPSS/H. The variables are defined at the end of the program after the START statement. This is all right—the variables only need to be defined before they are used in the program. Recall that the SR of a STORAGE is given in parts per thousand so that it needs to be divided by 1000 in order to give the correct answer in determining the number of trucks missing from the mine due to being repaired. For example, if the utilization of the 25 trucks in the mine were to be 90%, this would mean that there would be 22.5 trucks working at all times or a total of 2.5 trucks in the repair shop. The program can be run for three spares, then four spares, and so on, up to six spares. If so, the utilization is found to be as shown in Table 14.1.

The three spare trucks cost \$4200. The lost production is \$6855. This is obtained by the following calculation:

$$[25 - 25 * (.97258)] * 1250 = \$6,855$$

Thus, the total loss associated with having three spares is \$11,055.

Similar calculation yields are tabulated in Table 14.2.

Thus, the mine should have five spare trucks.

Inventory problems are often easily solved using simulation. There will be other examples of mining inventory problems later.

```
          SIMULATE
*****************************
*   PROGRAM CHAP14D.GPS         *
*   REPAIR SHOP IN A MINE       *
*   BASED ON EXAMPLE FROM       *
*   SCHRIBER'S GPSS TEXTBOOK   *
*****************************
          STORAGE     S(MINE),25/S(REPAIR),2
 TRUCKS   GENERATE    ,,,28     PROVIDE TRUCKS (INCLUDING SPARES)
 BACK     ENTER       MINE       WORK IN MINE
          ADVANCE     240,30     TRUCKS WORKING
          LEAVE       MINE       TRUCK DOWN
          ADVANCE     1,.25      TAKE TO REPAIR
          ENTER       REPAIR     GO TO REPAIR
          ADVANCE     18,8      FIX TRUCK
          ADVANCE     .5         READY TO RETURN TO MINE
          LEAVE       REPAIR
          TRANSFER    ,BACK
          GENERATE    24*365*100    SIMULATE FOR 100 YEARS
          TERMINATE   1
          START       1
          PUTPIC      N(TRUCKS)-25,SR(MINE)/10.
  NUMBER OF SPARE TRUCKS **    UTIL. OF TRUCKS IN THE MINE   **.***
          REAL        &LESS,&LOSS
          LET         &LESS=25-25*(SR(MINE)/1000.)
          LET         &LOSS=1250*&LESS*8+1400*(N(TRUCKS)-25)
          [illegible]  (' ')
          PUTPIC      &LOSS
    THIS RESULTS IN A LOSS OF ***** PER SHIFT
          END
```

FIGURE 14.8
Listing of program CHAP14D.GPS.

TABLE 14.1
Results of Running CHAP14F Multiple Times

No. of Spares	Utilization of Trucks
3	97.258
4	98.568
5	99.259
6	99.605

TABLE 14.2

Cost Calculations for Example 14.4

No. of Spares	Expected Cost
3	$11,055
4	$9,180
5	$8,852
6	$9,387

14.2 Exercises

14.1. For Example 14.1, the program given was CHAP14A.GPS. Add the necessary code so that the relevant output is on the screen.

14.2. For Example 14.2, the program given was CHAP14B.GPS. Add the necessary code so that the relevant output is on the screen.

14.3. For Example 14.3, suppose that the store is thinking of a massive advertising campaign to attract more customers. Phase I should increase the flow of customers to an arrival rate of 75.2 ± 35 seconds, Phase II to 65 ± 25 seconds, and Phase III to 55 ± 15 seconds. At which phase, if any, will the single checkout person be considered as being overworked?

14.4. Suppose it is possible to improve the performance of the trucks in Example 14.4, so that they now work for 280 ± 40 hours before failure. How does this change the results? Run for five spare trucks.

14.5. Example 14.2 did not use `QUEUE/DEPART` Blocks. Add these and determine if there is any difference in the results obtained in the program CHAP14B.GPS. How many random numbers were referenced in the original program CHAP14B? How many were now referenced when the `QUEUE/DEPART` Blocks are added?

14.6. Refer to Exercise 14.5. Re-run both CHAP14B.GPS and the program written for this exercise but now for 100 shifts of 480 minutes each. Are the results for the utilization of the facility and storage within .001? Next, run both programs for 1000 shifts.

14.7. Consider a manufacturing system as shown in Figure 14.9.

Parts enter at A every 20 ± 6 minutes. They travel to B where a single machine takes 18 ± 6 minutes to form them. They next travel to C in 2 minutes where two identical machines take 37 ± 10 minutes to shape them. Once shaped, they travel to D in 1.5 minutes where they are moulded by a single machine in 16 ± 4 minutes. They then travel to three identical machines at E in 2 minutes. These machines paint the parts in 55 ± 10 minutes. Finally, the painted parts travel to an inspector in 1.5 minutes. The inspector takes 5 ± 1 minute for each part. The inspector rejects 10% of the parts; the rest move to the next section of the factory. Write the GPSS/H code to simulate this system to determine the amount of parts made in a shift of 480 minutes. Simulate for 300 shifts and have the average for a shift sent to the screen.

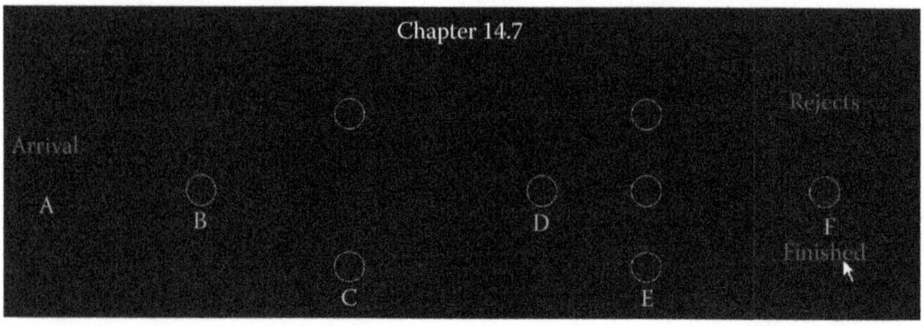

FIGURE 14.9
Manufacturing system for parts moving from machine to machine.

NOTE: Although the above may seem to be a complex system, by now the student should appreciate the tremendous power of GPSS/H to rapidly model such systems. Chapter 15 will introduce additional Blocks to assist in making the animation.

15

Other Forms of the `TRANSFER` *Block*

Herbert Hoover, former president of the United States, a famous mining engineer.

15.1 Several Other Forms of the TRANSFER Block

15.1.1 TRANSFER BOTH Block

It was noted in Chapter 14 that the ENTER Block cannot conveniently be animated. Instead, one normally uses a form of the TRANSFER Block:

```
TRANSFER BOTH,(first),(second)
```

or

```
TRANSFER ALL,(first label),(second label),(number)
```

The TRANSFER BOTH does the following.

The transaction first attempts to enter the Block with the label (first label); if it can, it does so; if it cannot, it attempts to enter the Block with the label (second label). If it can, it does so; if not, it resides in the TRANSFER Block until some later time when it can enter the next Block. If the label (first) is omitted (99% of the time it will be), it is assumed that the transaction tries to enter the next sequential Block. For example, suppose a person tries to enter a shop with a single chair to sit in. If the chair is occupied, the person leaves the shop. Consider the following:

```
              TRANSFER BOTH,CHAIR,AWAY
CHAIR         SEIZE         SITDOWN
               .........
              ............ . .
AWAY   TERMINATE
```

The above normally would be written as follows:

```
              TRANSFER          BOTH,,AWAY
              SEIZE        SITDOWN
              ..........
              ..............
AWAY          TERMINATE
```

Example 15.1: Repairing Equipment in a Mine

A mine has a machine that repairs parts in 38 ± 10 minutes. A new addition to the mine means the broken equipment will arrive at 20 ± 8 minute intervals, so the old machine cannot keep up with the repairs. A new machine is going to be purchased that will work at a rate of 32 ± 9 minutes. This, still, will not be fast enough to repair the broken parts, so the old machine will be kept but put into use only when the new machine is busy. A broken part always tries to use the new machine. Determine how busy the new and old machines will be. Here we have a situation where an ENTER Block cannot be used as the repair machines work at different rates. Simulate for 20 shifts of 480 minutes.

Solution

The program to model this system is given by CHAP15A.GPS (Figure 15.1).

```
           SIMULATE
********************************
*  CHAP15A.GPS  EXAMPLE        *
*  OF TRANSFER BOTH BLOCK      *
********************************
           GENERATE    20,8        PARTS ARRIVE
           QUEUE       HOLDIT      JOIN A QUEUE
           TRANSFER    BOTH,,OLDMACH   WHICH MACHINE IS FREE?
           SEIZE       NEWMACH     USE NEW MACHINE
           DEPART      HOLDIT      LEAVE THE QUEUE
           ADVANCE     32,9        REPAIRS MADE
           RELEASE     NEWMACH     FREE MACHINE
 FINISH1   TERMINATE               PART FINISHED
 OLDMACH   SEIZE       OLDONE      USE OLD MACHINE
           DEPART      HOLDIT      LEAVE QUEUE
           ADVANCE     38,10       REPAIRS MADE
           RELEASE     OLDONE      FREE MACHINE
 FINISH2   TERMINATE               PART FINISHED
           GENERATE    480*20      TIMER TRANSACTION
           TERMINATE   1
           START       1
           PUTSTRING   ('  ')
           PUTSTRING   ('  ')
           PUTPIC      LINES=5,FR(NEWMACH)/10.,FR(OLDONE)/10.,_
                       QA(HOLDIT),N(FINISH1),N(FINISH2)
    UTIL OF NEW MACHINE                 **.**%
    UTIL OF OLD MACHINE                 **.**%
    AVERAGE QUEUE CONTENT               **.**
    PARTS FINISHED ON NEW MACHINE     ***
    PARTS FINISHED ON OLD MACHINE     ***
            PUTSTRING  ('  ')
            END
```

FIGURE 15.1
Listing of program CHAP15A.GPS.

```
UTIL OF NEW MACHINE               85.34%
UTIL OF OLD MACHINE               88.10%
AVERAGE QUEUE CONTENT              0.17
PARTS FINISHED ON NEW MACHINE  259
PARTS FINISHED ON OLD MACHINE  218
```

FIGURE 15.2
Results of program CHAP15A.GPS.

It might be instructive to try to guess what the utilization of the machines will be before the program is run. Most of the examples done so far to learn GPSS/H have solutions that were fairly easy to estimate. Simple examples like this are not so easy. The simulation yields the following somewhat surprising results* (Figure 15.2).

The old machine, which only is put into service when the new machine is busy, actually works more than the new machine!

15.1.2 The TRANSFER ALL Block

This Block works as shown below. The general form is as follows:

TRANSFER ALL,(*first*),(*last*),*integer*

The transaction tries to enter the Block labelled (*first*). If it can, it does; if not, it counts down a number of Blocks given by the *integer* and tries to enter the next Block; if it can, it does; if not, it counts down another number of Blocks and tries to enter there, and so on.

A TRANSFER ALL Block might be used when parts come to a shop with two servers to obtain service. If both are busy, the arriving part waits until a server becomes free. One way to model this would be as shown in CHAP15B.GPS. It produces the same result as program CHAP15A.GPS. Notice that the count includes the first Block previously attempted. Thus, in program CHAP15B.GPS, the TRANSFER ALL Block is as follows:

```
TRANSFER ALL,NEWM,OLDM,5
```

The transaction first attempts to enter the Block with the label NEWM. This is the next sequential Block after the TRANSFER Block. If it can, it does; if not, the count is five Blocks below the Block with the label OLDM *including this Block* (Figure 15.3).

Since the output is identical to that from the program CHAP15A.GPS, it will not be given here.

Notice that either of the above programs can be animated with the addition of the appropriate BPUTPIC Blocks. The listing of the program CHAP15C.GPS is shown in Figure 15.4 and the animation is shown in Figure 15.5.

* The late Alan Pritsker, who was well known for his work in simulation and the development of several simulation languages, once said (in effect): A simulation study is not worth much unless it contains a surprise. He was certainly correct with examples such as this one.

```
          SIMULATE
********************************
*  CHAP15B  EXAMPLE            *
*  OF TRANSFER   ALL   BLOCK   *
********************************
          GENERATE    20,8        PARTS ARRIVE
          QUEUE       HOLDIT      JOIN A QUEUE
          TRANSFER    ALL,NEWM,OLDM,5   WHICH MACHINE IS FREE?
 NEWM     SEIZE       NEWMACH     USE NEW MACHINE
          DEPART      HOLDIT      LEAVE THE QUEUE
          ADVANCE     32,9        REPAIRS MADE
          RELEASE     NEWMACH     FREE MACHINE
 FINISH1  TERMINATE               PART FINISHED
 OLDM     SEIZE       OLDONE      USE OLD MACHINE
          DEPART      HOLDIT      LEAVE QUEUE
          ADVANCE     38,10       REPAIRS MADE
          RELEASE     OLDONE      FREE MACHINE
 FINISH2  TERMINATE               PART FINISHED
          GENERATE    480*20      TIMER TRANSACTION
          TERMINATE   1
          START       1
          PUTSTRING   ('  ')
          PUTSTRING   ('  ')
          PUTPIC      LINES=5,FR(NEWMACH)/10.,FR(OLDONE)/10.,_
                      QA(HOLDIT),N(FINISH1),N(FINISH2)
    UTIL OF NEW MACHINE                **.**%
    UTIL OF OLD MACHINE                **.**%
    AVERAGE QUEUE CONTENT              **.**
    PARTS FINISHED ON NEW MACHINE  ***
    PARTS FINISHED ON OLD MACHINE  ***
           PUTSTRING  ('  ')
           END
```

FIGURE 15.3
Listing of program CHAP15B.GPS.

```
          SIMULATE
**********************************
*  PROGRAM CHAP15C.GPS           *
*  ANIMATION FOR CHAP15B.GPS     *
**********************************
   ATF    FILEDEF     'CHAP15C.ATF'
          INTEGER     &PARTS
          GENERATE    20,8,0    PARTS ARRIVE
          BPUTPIC     FILE=ATF,LINES=4,AC1,XID1,XID1,XID1
TIME *.*****
CREATE CUST C*
PLACE C* ON P1
SET C* TRAVEL 6
          ADVANCE     6
          QUEUE       HOLDIT   JOIN A QUEUE
          TRANSFER    BOTH,,OLDMACH
          SEIZE       NEWMACH    USE NEW MACHINE
          DEPART      HOLDIT
          BPUTPIC      FILE=ATF,LINES=3,AC1,XID1,XID1
TIME *.*****
```

FIGURE 15.4
Listing of program CHAP15C.GPS.

(Continued)

```
PLACE C* ON P2
SET C* TRAVEL 3
          ADVANCE     3
          BPUTPIC     FILE=ATF,LINES=2,AC1,XID1
TIME *.*****
PLACE C* AT 43.05 33.75
          ADVANCE     32,9
          BPUTPIC     FILE=ATF,LINES=4,AC1,XID1,X
TIME *.*****
PLACE C* ON P4
SET C* TRAVEL  3
WRITE M1 **.**%
          ADVANCE     .5
          RELEASE     NEWMACH
          BLET        &PARTS=&PARTS+1
          BPUTPIC     FILE=ATF,LINES=2,AC1,&PARTS
TIME *.*****
WRITE M4 ***
          ADVANCE     2.5
          BPUTPIC     FILE=ATF,LINES=2,AC1,XID1
TIME *.*****
DESTROY C*
          TERMINATE
 OLDMACH  SEIZE       OLDONE     USE OLD MACHINE
          DEPART      HOLDIT
          BPUTPIC     FILE=ATF,LINES=3,AC1,XID1,X
TIME *.*****
PLACE C* ON P3
SET C* TRAVEL  3
          ADVANCE     3
          BPUTPIC     FILE=ATF,LINES=2,AC1,XID1
TIME *.*****
PLACE C* AT 43.05 22.88
          ADVANCE     38,10
          BPUTPIC     FILE=ATF,LINES=4,AC1,XID1,X
TIME *.*****
PLACE C* ON P4
SET C* TRAVEL  3
WRITE M2 **.**%
          ADVANCE     .5
          BLET        &PARTS=&PARTS+1
          BPUTPIC     FILE=ATF,LINES=2,AC1,&PARTS
TIME *.*****
WRITE M4 ***
          RELEASE     OLDONE
          ADVANCE     2.5
          BPUTPIC     FILE=ATF,LINES=2,AC1,XID1
TIME *.*****
DESTROY C*
          TERMINATE
          GENERATE    480*20
          TERMINATE   1
          START       1
          PUTPIC      FILE=ATF,LINES=2
TIME *.*****
END
          END
```

FIGURE 15.4 (Continued)
Listing of program CHAP15C.GPS.

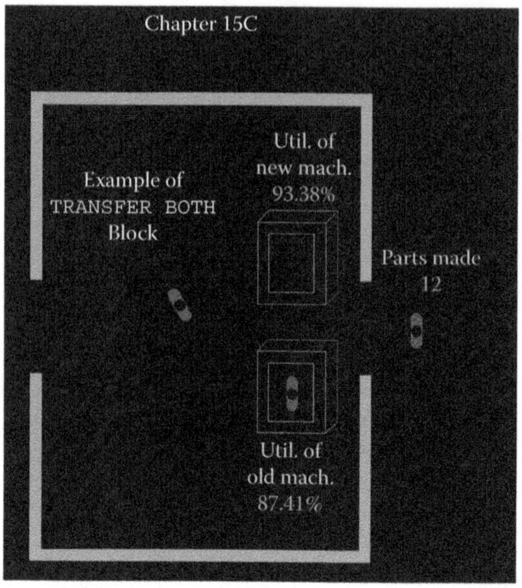

FIGURE 15.5
Animation from program CHAP15C.GPS.

Example 15.2: Shop with Five Workers

Next, we consider an example of the `TRANSFER ALL` Block. Suppose there are five workers who can give service at a rate of 40 ± 15 minutes. Customers arrive at a rate of 8 ± 4 minutes. There is only one chair for a customer to act on if all the servers are busy. If a customer arrives and finds that there is no chair to sit and wait, he or she will leave. This is undesirable, so the owner of the shop wants to purchase more waiting room chairs so that no customer ever arrives and finds the shop full. Determine what the effect of adding more waiting space to the system would be.

This is an interesting example that illustrates how difficult it can be to estimate the effects of uncertainty on a system. Before running the simulation program, it might be instructive to try to guess what percentage of people leave the shop without obtaining service. If there were no variability, the system would be *perfect* in that no waiting area is needed, and the service providers would eventually be busy 100% of the time. This is easy to see. The first customer arrives at time 8. He will finish his service at time 48. The second customer arrives at time 16, the third at time 24, and so on. The sixth customer arrives at time 48, just when the first customer is finished. From then on, each arriving customer will replace one who leaves. Unfortunately, we live in a world of uncertainty: customers do not arrive at times exactly spaced apart, ships arrive at random at a port, machines wear out at random times, trucks arrive at a shovel to be loaded at different time intervals, and so on. However, no matter how many waiting area chairs are added, the mean of the arrival rate and the mean of the service rate remain at 8 and 40, respectively.

Solution

The program to model the system is CHAP15D.GPS (Figure 15.6).

The results of running the program as given above yield the results as shown in Figure 15.7.

There was an average of 59.84 customers who arrived during each shift of 480 minutes. 9.54% of these found that there were six customers in the shop—five obtaining service and one waiting in the chair, so they left rather than wait.

```
          SIMULATE
******************************
*   PROGRAM CHAP15D.GPS        *
*   SHOP EXAMPLE               *
******************************
           REAL         &X,&Y,&Z
           STORAGE      S(SEAT),1    PROVIDE SEAT(S)
 COMEIN    GENERATE     8,4         CUSTOMERS ARRIVE
           TRANSFER     BOTH,,AWAY   IS A SEAT FREE?
           ENTER        SEAT         TAKE A SEAT
           TRANSFER     ALL,CCCC,DDDD,5
 CCCC      SEIZE        WORK1        FIRST WORKER
           LEAVE        SEAT
           ADVANCE      40,15
           RELEASE      WORK1
           TRANSFER     ,EEEE
           SEIZE        WORK2        SECOND WORKER
           LEAVE        SEAT
           ADVANCE      40,15
           RELEASE      WORK2
           TRANSFER     ,EEEE
           SEIZE        WORK3        THIRD WORKER
           LEAVE        SEAT
           ADVANCE      40,15
           RELEASE      WORK3
           TRANSFER     ,EEEE
           SEIZE        WORK4        FOURTH WORKER
           LEAVE        SEAT
           ADVANCE      40,15
           RELEASE      WORK4
           TRANSFER     ,EEEE
 DDDD      SEIZE        WORK5        FIFTH WORKER
           LEAVE        SEAT
           ADVANCE      40,15
           RELEASE      WORK5
           TRANSFER     ,EEEE
 EEEE      TERMINATE
 AWAY      TERMINATE
           GENERATE     480*500     SIMULATE FOR 500 SHIFTS
           TERMINATE    1
           START        1
           PUTSTRING    ('  ')
           PUTSTRING    ('  ')
           LET          &X=N(AWAY)
           LET          &Y=N(COMEIN)
           LET          &Z=(&X/&Y)*100
           PUTPIC       LINES=8,N(COMEIN)/500.,N(AWAY)/500.,&Z,_
                        FR(WORK1)/10.,FR(WORK2)/10.,FR(WORK3)/10.,_
                        FR(WORK4)/10.,FR(WORK5)/10.
     CUSTOMERS TO COME TO SHOP EACH SHIFT       **.**
     CUSTOMERS TO LEAVE WITHOUT SERVICE         **.**
     PERCENT TO LEAVE WITHOUT SERVICE           **.**%
     UTIL. OF FIRST WORKER   **.**%
     UTIL. OF SECOND WORKER  **.**%
     UTIL. OF THIRD WORKER   **.**%
     UTIL. OF FOURTH WORKER  **.**%
     UTIL. OF FIFTH WORKER   **.**%
           PUTSTRING  ('  ')
           END
```

FIGURE 15.6
Listing of program CHAP15D.GPS.

```
CUSTOMERS TO COME TO SHOP EACH SHIFT      59.84
CUSTOMERS TO LEAVE WITHOUT SERVICE         5.54
PERCENT TO LEAVE WITHOUT SERVICE           9.43%
UTIL. OF FIRST WORKER   94.30%
UTIL. OF SECOND WORKER  92.97%
UTIL. OF THIRD WORKER   91.49%
UTIL. OF FOURTH WORKER  88.96%
UTIL. OF FIFTH WORKER   85.38%
```

FIGURE 15.7
Output from program CHAP15D.GPS.

TABLE 15.1
Results of Simulations

Seats	1	2	3	4	5
Arrivals	29,959	29,989	29,948	29,911	29,974
Customers to leave	2824	1439	816	528	444
Util. of servers	0.943 – 0.854	0.970 – 0.936	0.981 – 0.961	0.986 – 0.972	0.991 – 0.980
Percentage to leave	9.43%	4.78%	2.72%	1.77%	1.48%

The program was run another four times for two chairs for waiting through five chairs. Table 15.1 gives the results.

As can be seen, even with five chairs for waiting, *there still will be customers who cannot be serviced.* In any design system, if the goal is to have no one ever leave without obtaining service, one should normally not have the mean arrival time the same as the mean service time, unless there is plenty of room for arrivals to wait. In fact, it is a bit of a surprise to learn that in a single server model with an infinite population, if the arrival rate were Poisson, (written as λ) and service exponential (written as μ) with the same values, the expected queue would be λ divided by $\lambda/(\lambda - \mu)$. Since if $\lambda = \mu$, the result would be division by zero, the queue would grow to an infinite number!

An exercise will be to run the program with no seats for waiting. If this is done, the number of customers who leave without obtaining service is 10.41%.

The program to provide the animation of the model is given next. It is stored as CHAP15E.GPS. Its listing is shown in Figure 15.8.

15.1.3 The PROOF Command `PLACE` *(object)* `IN` *(layout object)*

This command places an object, in this case the PERSON, in the layout onto the layout object's hot point, that is, hot point of the first object at the hot point on the layout object. The effect of this is to have the PERSON be placed in the seat.

In a later chapter, we will learn how to compact the code given above.

The animation is shown in Figure 15.9.

```
          SIMULATE
*************************************
*  PROGRAM CHAP15E.GPS              *
*  ANIMATION FOR PROGRAM CHAP15D.GPS *
*************************************
 ATF      FILEDEF     'CHAP15E.ATF'
          INTEGER     &SEATS,&WORKERS
          REAL        &PERCENT,&X
          LET         &WORKERS=5
          LET         &SEATS=1
          STORAGE     S(SEAT),&SEATS
 COMEIN   GENERATE    8,4,3
          BPUTPIC     FILE=ATF,LINES=5,AC1,XID1,XID1,AC1,XID1
TIME *.****
CREATE PERSON P*
PLACE P* ON P1
WRITE FIRST ****.**
SET P* TRAVEL 5
          ADVANCE     5
          BLET        &X=N(AWAY)
          BLET        &PERCENT=(&X/N(COMEIN))*100
          BPUTPIC     FILE=ATF,LINES=5,AC1,AC1,N(COMEIN),N(AWAY),_
                      &PERCENT
TIME *.****
WRITE FIRST ****.**
WRITE SECOND ****
WRITE THIRD ****
WRITE FOURTH ****.**%
          TRANSER     BOTH,,AWAY
          ENTER       SEAT
          BPUTPIC     FILE=ATF,LINES=2,AC1,XID1
TIME *.****
PLACE P* IN DOTT
          TRANSFER    ALL,CCCC,DDDD,7
 CCCC     SEIZE       MACH1
          LEAVE       SEAT
          BPUTPIC     FILE=ATF,LINES=4,AC1,XID1,XID1,SR(SEAT)/10.
TIME *.****
ROTATE P* 90
PLACE P* IN CHAIR1
WRITE M10 **.**%
          ADVANCE     40,15
          RELEASE     MACH1
          BPUTPIC     FILE=ATF,LINES=2,AC1,FR(MACH1)/10.
TIME *.****
WRITE M20 **.**%
          TRANSFER    ,EEEE
          SEIZE       MACH2
          LEAVE       SEAT
          BPUTPIC     FILE=ATF,LINES=3,AC1,XID1,XID1
TIME *.****
ROTATE P* 90
PLACE P* IN CHAIR2
          ADVANCE     40,15
          RELEASE     MACH2
```

FIGURE 15.8
Listing of program CHAP15E.GPS for animation of shop example. *(Continued)*

```
          BPUTPIC     FILE=ATF,LINES=2,AC1,FR(MACH2)/10.
TIME *.****
WRITE M21 **.**%
          TRANSFER    ,EEEE
          SEIZE       MACH3
          LEAVE       SEAT
          BPUTPIC     FILE=ATF,LINES=3,AC1,XID1,XID1
TIME *.****
ROTATE P* 90
PLACE P* IN CHAIR3
          ADVANCE     40,15
          RELEASE     MACH3
          BPUTPIC     FILE=ATF,LINES=2,AC1,FR(MACH3)/10.
TIME *.****
WRITE M22 **.**%
          TRANSFER    ,EEEE
          SEIZE       MACH4
          LEAVE       SEAT
          BPUTPIC     FILE=ATF,LINES=3,AC1,XID1,XID1
TIME *.****
ROTATE P* 90
PLACE P* IN CHAIR4
          ADVANCE     40,15
          RELEASE     MACH4
          BPUTPIC     FILE=ATF,LINES=2,AC1,FR(MACH4)/10.
TIME *.****
WRITE M23 **.**%
          TRANSFER    ,EEEE
 DDDD     SEIZE       MACH5
          LEAVE       SEAT
          BPUTPIC     FILE=ATF,LINES=3,AC1,XID1,XID1
TIME *.****
ROTATE P* 90
PLACE P* IN CHAIR5
          ADVANCE     40,15
          RELEASE     MACH5
          BPUTPIC     FILE=ATF,LINES=2,AC1,FR(MACH5)/10.
TIME *.****
WRITE M24 **.**%
          TRANSFER    ,EEEE
 EEEE     BPUTPIC     FILE=ATF,LINES=2,AC1,XID1
TIME *.****
PLACE P* ON P4
          ADVANCE     2
          TRANSFER    .5,,FFFF
          BPUTPIC     FILE=ATF,LINES=3,AC1,XID1,XID1
TIME *.****
PLACE P* ON P5
SET P* TRAVEL 4
          ADVANCE     4
          BPUTPIC     FILE=ATF,LINES=3,AC1,XID1,AC1
TIME *.****
DESTROY P*
WRITE FIRST ****.**
          TERMINATE
 FFFF     BPUTPIC     FILE=ATF,LINES=3,AC1,XID1,XID1
TIME *.****
PLACE P* ON P6
SET P* TRAVEL 4
          ADVANCE     4
          BPUTPIC     FILE=ATF,LINES=3,AC1,XID1,AC1
```

FIGURE 15.8 (Continued)
Listing of program CHAP15E.GPS for animation of shop example. *(Continued)*

```
TIME *.****
DESTROY  P*
WRITE FIRST ****.**
          TERMINATE
 AWAY     BPUTPIC      FILE=ATF,LINES=3,AC1,XID1,XID1
TIME *.****
PLACE P* ON P2
SET P* TRAVEL 6
          ADVANCE      6
          BPUTPIC      FILE=ATF,LINES=3,AC1,XID1,AC1
TIME *.****
DESTROY P*
WRITE FIRST ****.**
          TERMINATE
          GENERATE     480*10   TIMER TRANSACTION
          TERMINATE    1
          START        1
          PUTPIC       FILE=ATF,LINES=2,AC1
TIME *.****
END
          END
```

FIGURE 15.8 (Continued)
Listing of program CHAP15E.GPS for animation of shop example.

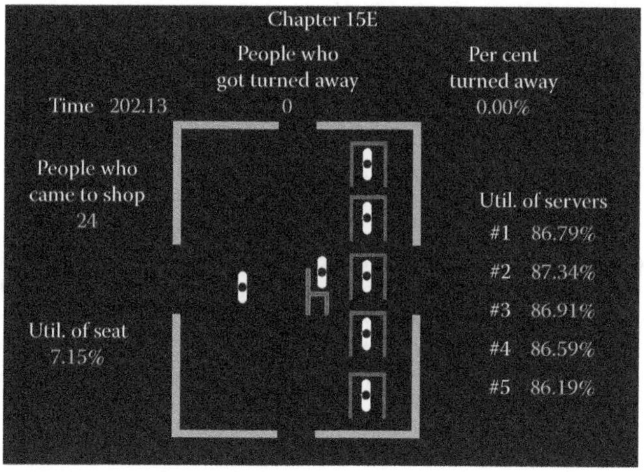

FIGURE 15.9
Screenshot of animation given by program CHAP15E.GPS.

Example 15.3: Assembly Line Problem

The following example should now be easy to model (Figure 15.10).

Valuable parts arrive at the factory every 12 ± 4.5 minutes. They travel on a conveyor belt to station A in one minute where a single machine takes 10 ± 3 minutes to form it. The parts then travel to station B in 1 minute. At B there are two machines that stamp the parts in 20 ± 5 minutes. Parts take 30 seconds to move to station C where they are painted in 31 ± 8 minutes. There are three identical painting machines at C. Finally, they move to a drying and inspection station in 30 seconds where, as they are drying, an inspector sends 10% back for complete reworking. This takes 7.5 ± 3 minutes. The plant manager would like to know where any bottlenecks are and have recommendations on

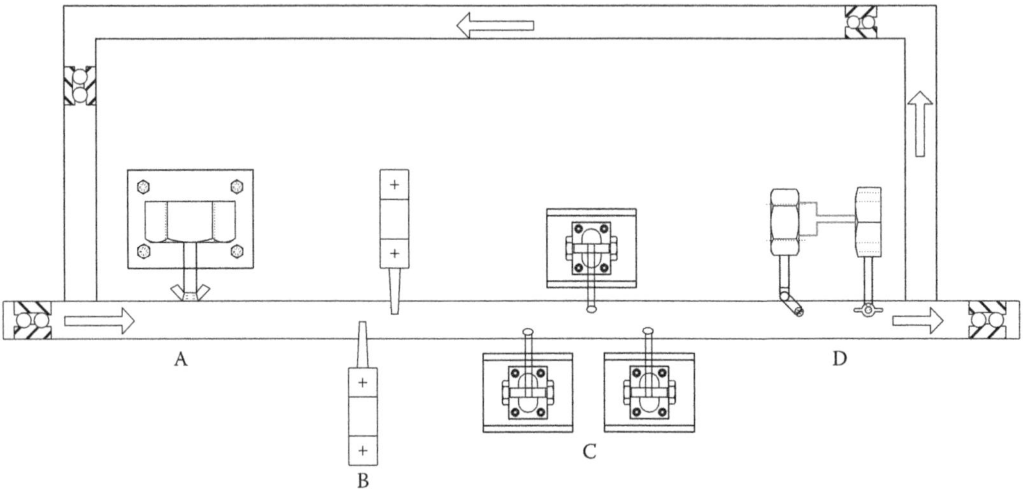

FIGURE 15.10
Schematic of parts moving through factory.

TABLE 15.2
Costs and Change of Speeds for Each Station

Change to	A	B	C	D
Costs	\$900	\$950	\$1050	\$800
New speed	9 ± 3	19 ± 5	29.5 ± 8	7.5 ± 3

Note: 4 queues times 1 (average queue length) times 480 minutes per shift times \$2.75 or \$5280 per 8-hour shift.

improving the system. He can change only one of the stations. Table 15.2 gives the costs for changing each station as well as the increased speed at each station. The parts are very valuable, and it is essential that they be moved from the stations as quickly as possible. When a part is in a queue, the cost to the plant is \$2.75 per minute; that is, if there is always one part in the queue at each of the four stations, it would cost the company.

Solution

The program to model the system is given as CHAP15F.GPS. By this time, it should be fairly easy to write the code (Figure 15.11) simply by referring to the sketch of the system.

The results of the simulation give average queue lengths of .974 at A, .577 at B, 1.213 at C, and .221 at D.

This is costing the company \$3904 per shift. The program then needs to be run four times with the changes at each station. Figure 15.12 gives the results of these runs.

This gives us a somewhat surprising result that the changes need to be made at station D, which is the drying and inspection station. As can be seen, the greatest queue before any changes is at station C, where there are three machines to paint. Making a change here does reduce the queue length from an average of 1.213 to only .337. But the change at D, results in all the queues being significantly reduced so that the average queue length is less than when the change is made at station C.

The animation will take a bit of time to add. It is strongly suggested that the best way to write the code is in parts. One can imagine the animation as being composed of the four stations separately. The code is written only for parts arriving at station A and the animation checked. Once the code is debugged and the animation working,

```
          SIMULATE
*******************************
*  PROGRAM CHAP15F.GPS         *
*  MANUFACTURING EXAMPLE       *
*******************************
          STORAGE     S(SECOND),2/S(THIRD),3
          GENERATE    12,4.5    PARTS COME ALONG
          ADVANCE     1         TRAVEL ON BELT
 BACK     QUEUE       FIRST     AT FIRST STATION
          SEIZE       MACH1     USE FIRST MACHINE
          DEPART      FIRST     LEAVE QUEUE
          ADVANCE     10,3      FORMING
          RELEASE     MACH1     FREE MACHINE
          ADVANCE     1         TRAVEL TO SECOND STATION
          QUEUE       SECMACH   QUEUE AT SECOND STATION
          ENTER       SECOND    USE ONE OF TWO STAMP MACHINES
          DEPART      SECMACH   LEAVE QUEUE
          ADVANCE     20,5      STAMP
          LEAVE       SECOND    LEAVE QUEUE
          ADVANCE     .5        TRAVEL TO THIRD STATION
          QUEUE       THIRDM    QUEUE AT THIRD STATION
          ENTER       THIRD     PAINT & INSPECTION
          DEPART      THIRDM    LEAVE QUEUE
          ADVANCE     31,8
          LEAVE       THIRD     ALL DONE
          ADVANCE     .5        TRAVEL TO LAST STATION
          QUEUE       DRYINSP   LAST STATION
          SEIZE       FOURTH    USE DRYING
          DEPART      DRYINSP   LEAVE QUEUE
          ADVANCE     7.5,3     DRY AND INSPECT
          RELEASE     FOURTH
          TRANSFER    .10,,REDO
          TERMINATE
 REDO     ADVANCE     2
          TRANSFER    ,BACK
          GENERATE    480*100
          TERMINATE   1
          START       1
          PUTSTRING   ('  ')
          PUTSTRING   ('  ')
          PUTPIC      LINES=8,QA(FIRST),QA(SECMACH),QA(THIRDM),_
                      QA(DRYINSP),FR(MACH1)/10.,SR(SECOND)/10.,_
                      SR(THIRD)/10.,FR(FOURTH)/10.
      AVERAGE QUEUE AT FIRST STATION    *.**
      AVERAGE QUEUE AT SECOND STATION   *.**
      AVERAGE QUEUE AT THIRD STATION    *.**
      AVERAGE QUEUE AT FOURTH STATION   *.**
      UTIL. OF FIRST STATION           **.**%
      UTIL. OF SECOND STATION          **.**%
      UTIL. OF THIRD STATION           **.**%
      UTIL. OF FOURTH STATION          **.**%
          PUTSTRING   ('  ')
          PUTSTRING   ('  ')
          END
```

FIGURE 15.11
Listing of program CHAP15F.GPS.

station B is added, and so on until the complete animation is done. This step-by-step approach can save the modeller a great deal of time and effort in developing the code for simulation.

The program for the animation is CHAP15G.GPS. A listing of it is as shown in Figure 15.13.

The animation for this assembly line is shown in Figure 15.14.

Factory assembly line example

	as is	Change A	Change B	Change C	Change D
Queue at A	0.947	0.794	0.265	0.873	0.726
Queue at B	0.577	1.180	0.703	0.665	0.365
Queue at C	1.213	0.223	1.282	0.337	0.792
Queue at D	0.221	0.220	0.223	0.212	0.180
Total	$3904.56	$4090.44	$4214.36	$3804.84	$3523.16

FIGURE 15.12
Results of running program CHAP15F.GPS for the changes to the stations.

```
          SIMULATE
*****************************************
* PROGRAM CHAP15G.GPS                   *
* ANIMATION OF PROGRAM CHAP15F.GPS      *
*****************************************
  ATF     FILEDEF     'CHAP15G.ATF'
          INTEGER     &COUNT
          GENERATE    12,4.5,3     PARTS COME ALONG
          SEIZE       DUMMY
          BPUTPIC     FILE=ATF,LINES=4,AC1,XID1,XID1,XID1
TIME *.****
CREATE PART P*
PLACE P* ON P1
SET P* TRAVEL 1
          ADVANCE     .1
          RELEASE     DUMMY
          BPUTPIC     FILE=ATF,LINES=3,AC1,XID1,XID1
TIME *.****
PLACE P* ON P2
SET P* TRAVEL .9
          ADVANCE     .9          TRAVEL ON BELT
 BACK     QUEUE       FIRST      AT FIRST STATION
          SEIZE       MACH1      USE FIRST MACHINE
          DEPART      FIRST      LEAVE QUEUE
          BPUTPIC     FILE=ATF,LINES=2,AC1,XID1
TIME *.****
PLACE P* AT 4 29
          ADVANCE     10,3        FORMING
          BPUTPIC     FILE=ATF,LINES=3,AC1,XID1,XID1
TIME *.****
PLACE P* ON P3
SET P* TRAVEL .25
          ADVANCE     .25          TRAVEL TO SECOND STATION
          RELEASE     MACH1
          BPUTPIC     FILE=ATF,LINES=2,AC1,FR(MACH1)/10.
TIME *.****
```

FIGURE 15.13
Listing of program CHAP15G.GPS.

(Continued)

```
WRITE M1 **.**%
          QUEUE        SECMACH   QUEUE AT SECOND STATION
          TRANSFER     BOTH,,NEXT1
          SEIZE        MACH2A
          DEPART       SECMACH
          BPUTPIC      FILE=ATF,LINES=3,AC1,XID1,XID1
TIME *.****
PLACE P* ON P4
SET P* TRAVEL .5
          ADVANCE      .5
          BPUTPIC      FILE=ATF,LINES=2,AC1,XID1
TIME *.***
PLACE P* AT 26 35
          ADVANCE      20,5      STAMP
          RELEASE      MACH2A
          BPUTPIC      FILE=ATF,LINES=4,AC1,XID1,XID1,FR(MACH2A)/10.
TIME *.****
PLACE P* ON P6
SET P* TRAVEL .25
WRITE M2 **.**%
          ADVANCE      .25        TRAVEL TO THIRD STATION
          TRANSFER     ,DDDD
 NEXT1    SEIZE        MACH2B
          DEPART       SECMACH
          BPUTPIC      FILE=ATF,LINES=3,AC1,XID1,XID1
TIME *.****
PLACE P* ON P5
SET P* TRAVEL .5
          ADVANCE      .5
          BPUTPIC      FILE=ATF,LINES=2,AC1,XID1
TIME *.****
PLACE P* AT 26 24
          ADVANCE      20,5      STAMP
          RELEASE      MACH2B
          BPUTPIC      FILE=ATF,LINES=4,AC1,XID1,XID1,FR(MACH2B)/10.
TIME *.****
PLACE P* ON P7
SET P* TRAVEL .25
WRITE M2B **.**%
          ADVANCE      .25
 DDDD     BPUTPIC      FILE=ATF,LINES=3,AC1,XID1,XID1
TIME *.****
PLACE P* ON P8
SET P* TRAVEL .25
          ADVANCE     .25
          QUEUE       THIRDM
          TRANSFER    ALL,AAAA,BBBB,10
 AAAA     SEIZE       MACH3A
          DEPART      THIRDM
          BPUTPIC     FILE=ATF,LINES=3,AC1,XID1,XID1
TIME *.****
PLACE P* ON P9
SET P* TRAVEL 1
          ADVANCE     1
          BPUTPIC     FILE=ATF,LINES=2,AC1,XID1
TIME *.*****
```

FIGURE 15.13 (Continued)
Listing of program CHAP15G.GPS.

(Continued)

```
PLACE P* AT 45 39
          ADVANCE     31,8
          RELEASE     MACH3A
          BPUTPIC     FILE=ATF,LINES=5,AC1,XID1,XID1,XID1,FR(MACH3A)/10.
TIME *.****
PLACE P*    ON P12
SET P* TRAVEL .25
SET P* CLASS PART2
WRITE M3 **.**%
          ADVANCE     .25
          TRANSFER    ,EEEE
          SEIZE       MACH3B
          DEPART      THIRDM
          BPUTPIC     FILE=ATF,LINES=3,AC1,XID1,XID1
TIME *.****
PLACE P* ON P10
SET P* TRAVEL 1
          ADVANCE     1
          BPUTPIC     FILE=ATF,LINES=2,AC1,XID1
TIME *.*****
PLACE P* AT 45 30
          ADVANCE     31,8
          RELEASE     MACH3B
          BPUTPIC     FILE=ATF,LINES=5,AC1,XID1,XID1,XID1,FR(MACH3B)/10.
TIME *.****
PLACE P* ON P13
SET P* TRAVEL .25
SET P* CLASS PART2
WRITE M3B **.**%
          ADVANCE      .25
          TRANSFER     ,EEEE
 BBBB     SEIZE        MACH3C
          DEPART       THIRDM
          BPUTPIC      FILE=ATF,LINES=3,AC1,XID1,XID1
TIME *.****
PLACE P* ON P11
SET P* TRAVEL 1
          ADVANCE      1
          BPUTPIC      FILE=ATF,LINES=2,AC1,XID1
TIME *.*****
pLACE P* AT 45 21
          ADVANCE      31,8
          RELEASE      MACH3C
          BPUTPIC      FILE=ATF,LINES=5,AC1,XID1,XID1,XID1,FR(MACH3C)/10.
TIME *.****
PLACE P*    ON P12
SET P* TRAVEL .25
SET P* CLASS PART2
WRITE M3C **.**%
          ADVANCE      .25
          TRANSFER     ,EEEE
 EEEE     QUEUE        LASTM
          BPUTPIC      FILE=ATF,LINES=3,AC1,XID1,XID1
TIME *.****
```

FIGURE 15.13 (Continued)
Listing of program CHAP15G.GPS.

(Continued)

```
PLACE P* ON P15
SET P* TRAVEL .25
          SEIZE        FOURTH
          DEPART       LASTM
          BPUTPIC      FILE=ATF,LINES=2,AC1,XID1
TIME *.****
PLACE P* AT 67 29
          ADVANCE      8,3
          TRANSFER     .1,,REDO
          BPUTPIC      FILE=ATF,LINES=3,AC1,XID1,XID1
TIME *.****
PLACE P* ON P16
SET P* TRAVEL 2
          ADVANCE      2
          RELEASE      FOURTH
          BPUTPIC      FILE=ATF,LINES=2,AC1,FR(FOURTH)/10.
TIME *.****
WRITE M4 **.**%
          BPUTPIC      FILE=ATF,LINES=3,AC1,XID1,XID1
TIME *.****
PLACE P* ON P17
SET P* TRAVEL 4
          ADVANCE      4
          BLET         &COUNT=&COUNT+1
          BPUTPIC      FILE=ATF,LINES=3,AC1,XID1,&COUNT
TIME *.****
DESTROY P*
WRITE M5 ***
          TERMINATE
  REDO    BPUTPIC     FILE=ATF,LINES=4,AC1,XID1,XID1,XID1
TIME *.****
PLACE P* ON P16
SET P* TRAVEL 2
SET P* CLASS PART
          ADVANCE      2
          RELEASE     FOURTH
          BPUTPIC      FILE=ATF,LINES=3,AC1,XID1,XID1
TIME *.****
PLACE P* ON P18
SET P* TRAVEL 5
          ADVANCE      5
          SEIZE        DUMMY
          BPUTPIC      FILE=ATF,LINES=3,AC1,XID1,XID1
TIME *.****
PLACE P* ON P19
SET P* TRAVEL 1
          ADVANCE     1
          RELEASE     DUMMY
          TRANSFER     ,BACK
          GENERATE     480
          TERMINATE    1
          START        1
          PUTPIC       FILE=ATF,LINES=2,AC1
TIME *.****
END
          END
```

FIGURE 15.13 (Continued)
Listing of program CHAP15G.GPS.

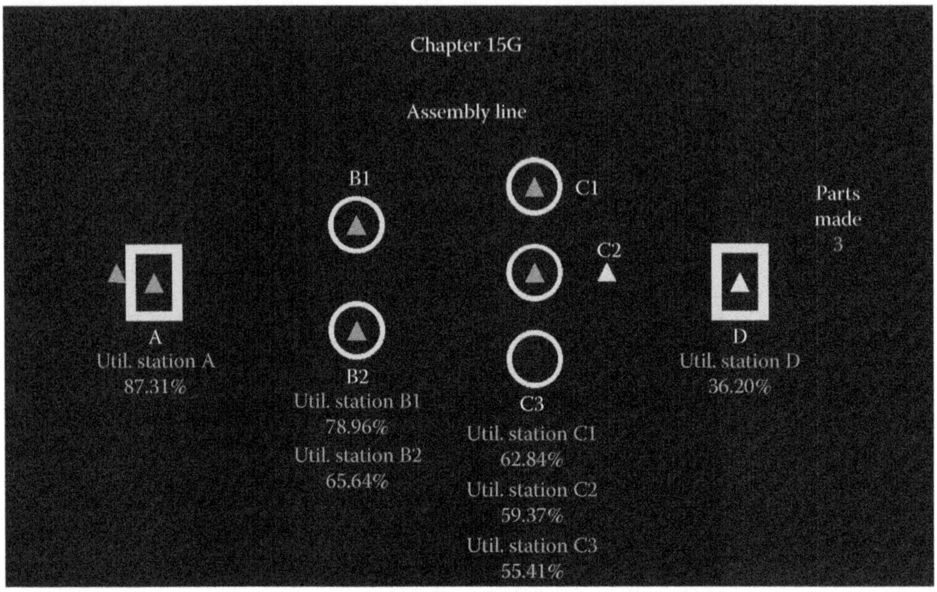

FIGURE 15.14
Animation for assembly line example, program CHAP15G.GPS.

15.2 Exercises

15.1. Vehicles come to a Y intersection in a road. Some vehicles turn to the left and some to the right. Cars approach this intersection every 10 ± 5 seconds. 35% go to the right, whereas the rest go to the left. Simulate for 2 hours.

15.2. A car comes to let people out for a theatre every 15 ± 4 seconds. 50% of the cars have one passenger to let out, 30% have two, 15% have three, and the rest four people. This all takes place within 1 hour. How many people have been left off to attend the theatre?

15.3. People enter a store that has two floors. They arrive every 30 ± 10 seconds. 30% only shop on the first or ground floor. They take 2 ± 1 minutes to shop. They then leave in 15 ± 6 seconds. The rest of the people shop on the ground floor in 1.5 ± .6 minutes and then go to the second floor. It takes them 12 ± 4 second to walk up (no escalator in this store). They take 4 ± 1 minutes to shop and then leave in 20 ± 8.5 seconds. In 2 hours, how many people have left the store?

15.4. In Example 15.1, the new machine worked at a rate of 32 ± 9 minutes. Suppose that the engineer in charge of ordering equipment feels that the old machine cannot possibly work more than 50% of the time. How fast would the new machine have to work?

15.5. GPSS/H is used a great deal in modelling traffic flow. There are many situations where it can be used. Consider the following example. Figure 15.15 shows the streets where traffic can only flow one way. Your house is where the X is. Traffic enters at A every 14 ± 5 seconds. It takes 12 ± 4 seconds to travel to B. At B, 70% of the traffic turns to C. It takes 10 ± 3.5 seconds to travel to C. 60% of the traffic

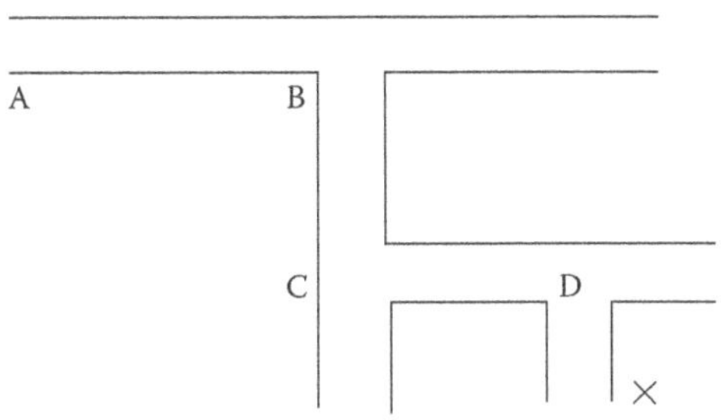

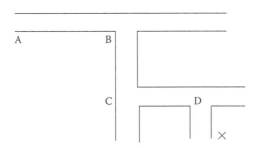

FIGURE 15.15
Transportation example.

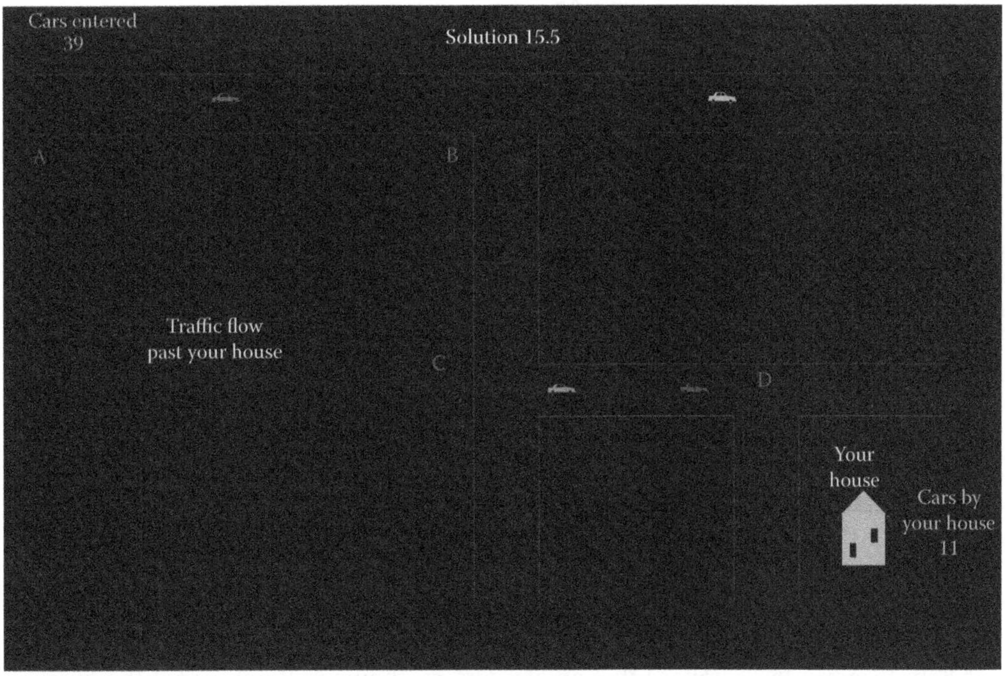

FIGURE 15.16
Cars past your house.

travels to D. It takes 8 ± 2.4 seconds to travel from C to D. At D, 75% turn to drive past your house. It takes 6 ± 1.5 seconds to drive from D past your house. Simulate for 2 hours to determine how many cars drive past your house.

Your animation should look something like that shown in Figure 15.16.

15.6. The following lines of code have errors. Determine where the errors are. The errors are only with the `TRANSFER`, `GENERATE`, `ADVANCE`, or `QUEUE/DEPART` Blocks.

(a)

```
GENERATE    12,3,4
QUEUE       FIRST
ADVANCE     12,13
DEPART      FIRST
```

(b)

```
        GENERATE   12,3,4,0
        QUEUE      WAITUP
        ADVANCE    12,-3
        DEPART     WAITUP
```

(c)

```
        QUEUE      LAST
        TRANSFER   AWAY
        ADVANCE    5
        TERMINATE
AWAY    ADVANCE 3
        TERMINATE
```

(d)

```
        GENERATE   300,,,1
        TRANSFER   .4,AWAY
        ADVANCE    50,49
AWAY    ADVANCE    30
```

(e)

```
        GENERATE   19,12
        QUEUE      JOHNSON
        SEIZE      MACH1
        DEPART     JOHNSEN
        ADVANCE    17,.4
        RELEASE    MACH1
```

15.7. Three types of mechanics arrive at a tool crib to check out tools. Only one clerk works at the crib. The arrival times and service times are as follows:

Type	Arrival Time Dist.	Service Time Dist.
1	30 ± 10	12 ± 5
2	20 ± 8	6 ± 3
3	15 ± 5	3 ± 1

Model this for 100 shifts of 100 minutes each.

15.8. In Exercise 15.7, suppose that an arriving mechanic will not wait if the queue waiting for tools is 1 but will leave. How many mechanics leave each shift without obtaining the needed tool?

15.9. The animation program for Example 15.3 is given by CHAP15G.GPS. Add the necessary code to this as well as modify the animation so that each part will be slightly different as it leaves each station.

15.10. Write the program to do the animation for Exercise 15.1. The animation should look something like as shown in Figure 15.17.

15.11. Write the program to do the animation for Exercise 15.2. The animation should look something like as shown in Figure 15.18.

15.12. Write the program to do the animation for Exercise 15.3. In order to do this at this stage, some of the data need to be changed: The customers still arrive

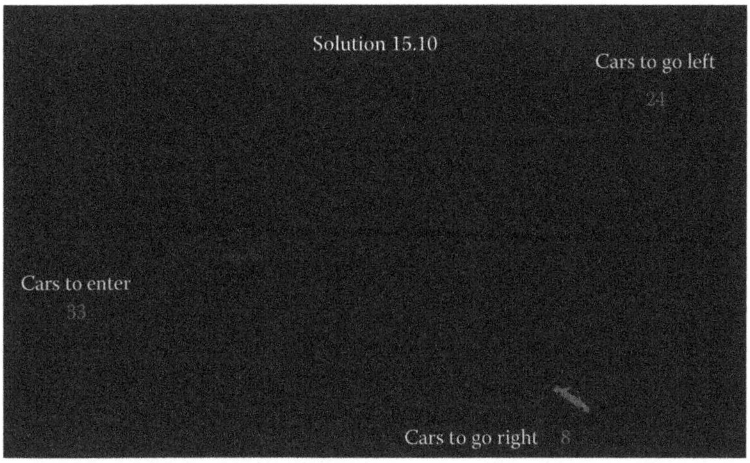

FIGURE 15.17
Screenshot of animation for Exercise 15.10.

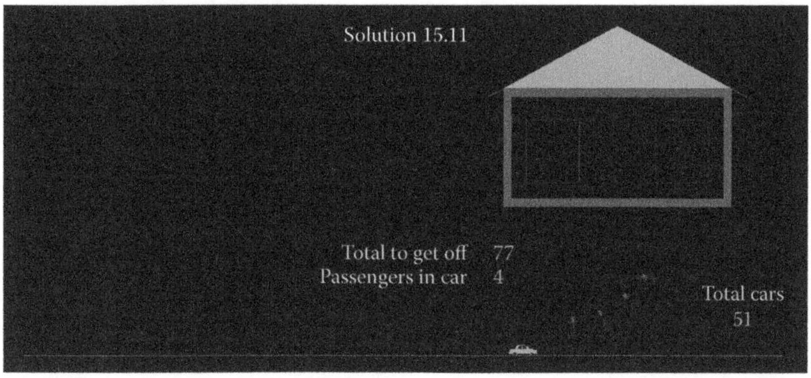

FIGURE 15.18
Screenshot of animation for Exercise 15.11.

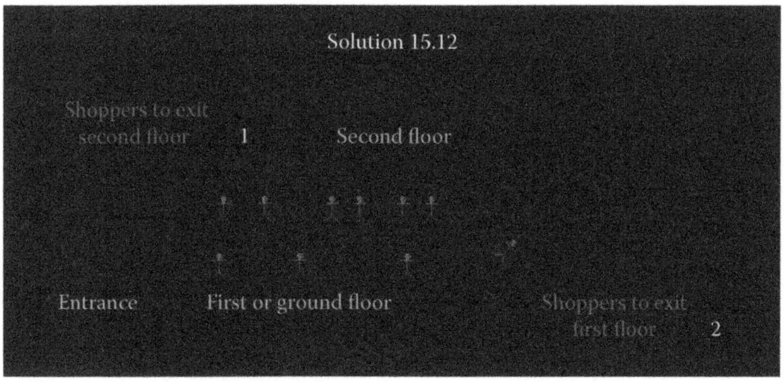

FIGURE 15.19
Screenshot of possible animation for Exercise 15.12.

at every 30 ± 10 seconds and 30% of the shoppers still shop on the first floor. These customers now take 2 minutes to shop. They then leave in 15 seconds. The rest of the people shop on the first floor in 1.5 minutes and then go to the second floor. It takes them 12 seconds to walk up. They take 4 minutes to shop and then leave in 20 seconds.

The animation should look as shown in Figure 15.19.

16

The `TEST` *Block*

Coal distribution system (Queensland, Australia).

16.1 The `TEST` Block

Up to this point, transactions moved through the various systems sequentially from Block to Block. The only method we had to route them to different Blocks was via the `TRANSFER` Block. There are many times in a model when the programmer will want to route a transaction to one Block or another depending on some aspect of the system. There will also be times when the programmer will want to keep transactions from moving forward until a specific condition is met. Both of these things are done using the `TEST` Block.

It is possible to do a test on two SNAs and then route the transaction to one or another of two Blocks depending on the result of the test. Examples of where a `TEST` Block might be used arise frequently during a mine simulation. Some possible examples where a `TEST` Block might be used are as follows:

1. If the queue at a shovel is greater than 5, trucks move to another area of the mine.
2. After 12 hours of working, a machine is shut down for half an hour for maintenance.
3. At 5 o'clock, the barber locks the door, but customers already in are still served.

GPSS/H can do the above and can also perform a test on two SNAs and hold the transaction at the Block doing the test until the test is true. Examples of these might be as follows:

1. Ships cannot enter the harbour until a tug boat is free to guide it into the berth.
2. A part will not be moved from one machine to the next unless the next machine has been used less than 75% of the time.
3. If it is between noon and 12:30 p.m., no truck can enter a repair facility.
4. Once a machine has finished making 500 parts, it is taken out of service for repairs and maintenance. This downtime lasts for 2 hours. Parts arriving have to wait until the repairs and maintenance are finished.
5. In the animation of a mine, the boom of a shovel will swing back and forth when a truck is being loaded. It stops moving when the truck is loaded.
6. A mine will continue to haul from an area until a set amount of ore is reached, then the trucks will be directed to another area.

We shall see that the use of `TEST` Blocks will greatly expand our programming ability.

There are two basic forms of the `TEST` Block that we will learn in this chapter.

16.2 The TEST Block in Refusal Mode

This form of the `TEST` Block is as follows:

```
TEST R  A,B
```

where *R* is called a *conditional operator* that is one of the following:

Symbol	Meaning
L	Less than
LE	Less than or equal
E	Equal
NE	Not equal
G	Greater than
GE	Greater than or equal

The conditional operator must be placed only one space after the word `TEST`. `A` and `B` are any SNAs to be tested via the conditional operator. Some examples of the `TEST` Block are as follows:

```
(a) TEST E   Q(TOM),Q(BILL)
(b) TEST NE  R(DOCK),4
(c) TEST L   FR(MACH),400
(d) TEST G   W(BACK1),1
(e) TEST E   N(BLOCKA),N(BLOCKB)
```

The explanation for the way the `TEST` Block works when a transaction enters is that the first SNA is compared with the second using the conditional operator. If the test is true, the transaction moves to the next sequential Block. If the test is false, the transaction *must wait in the* TEST *Block until some future time when the test becomes true.* In addition, the transaction remains on the current events chain.

Thus, in (a), the test *is the length of the queue named* TOM *equal to the length of the queue named* BILL? If the answer is yes, the transaction will move to the next sequential Block; but if the answer is no, the transaction will remain in the Block until such time that the test is true. In (b), the remaining storage of DOCK must be not equal to 4 before the transaction can move to the next Block. Similarly, for (c), the fractional utilization of the facility MACH must be less than .400 or the transaction will reside in the TEST Block until it is. In example (d), the transaction will test to see if the current count of the Block BACK1 is greater than 1. Unless it is greater than 1, the transaction will not leave the TEST Block. In (e), the transaction will be held until the total Block count of the Block labelled BLOCKA is equal to the count of the Block labelled BLOCKB.

The next example should be studied to understand how a TEST Block in refusal mode can be used.

Example 16.1: A Refuelling Depot in a Mine

A refuelling depot in a mine takes 12 ± 8 minutes per truck. Trucks arrive every 13 ± 6 minutes. At 5 o'clock, the superintendant locks the door and stops any refuelling no matter if a truck is being refuelled or if any trucks are waiting. The shop operates for 8 hours straight with no breaks for lunch. Simulate for a typical day.

The program to do the simulation is given below. First, the program will be given for the refuelling shop working for 8 hours straight and then immediately closing. This is program CHAP16A.GPS. Its listing is shown in Figure 16.1.

The results of the program are shown in Figure 16.2.

There were 37 trucks that came to the shop for refuelling in the 8 hours. After 8 hours, the shop closed and there were two trucks left at the shop. One truck was waiting in the queue and one was being refuelled. Next, the program was modified using two TEST

```
          SIMULATE
********************************
*  PROGRAM CHAP16A.GPS          *
*  TRUCK REFUELING EXAMPLE      *
********************************
 COMEIN   GENERATE    13,6    TRUCKS ARRIVE
          QUEUE       WAIT
          SEIZE       REFUEL  IS REFUELING AVAILABLE?
          DEPART      WAIT
          ADVANCE     12,8    REFUEL
          RELEASE     REFUEL
 ALLD     TERMINATE
          GENERATE    480
          TERMINATE   1
          START       1
          PUTSTRING   ('  ')
          PUTSTRING   ('  ')
          PUTPIC      LINES=5,N(COMEIN),Q(WAIT),F(REFUEL),N(ALLD),_
                      FR(REFUEL)/10.
   TRUCKS TO COME TO SHOP                **
   NUMBER LEFT WAITING AT 5 O'CLOCK      **
   NUMBER REFUELING AT 5                 **
   NUMBER DONE REFUELING                 **
   FACILITY WAS BUSY                     **.**% OF THE TIME
          END
```

FIGURE 16.1
Listing of program CHAP16A.GPS.

```
TRUCKS TO COME TO SHOP              37
NUMBER LEFT WAITING AT 5 O'CLOCK     1
NUMBER REFUELING AT 5                1
NUMBER DONE REFUELING               35
FACILITY WAS BUSY                   95.38% OF THE TIME
```

FIGURE 16.2
Results from program CHAP16A.GPS.

Blocks to make sure that no truck was left at the refuelling depot until all were done. This is program CHAP16B.GPS. A listing is given in Figure 16.3.

The results of the program are shown in Figure 16.4.

Two lines of code were added. Each was placed after a GENERATE Block. The first is

```
TEST NE   N(OVER),1
```

This is placed just below the GENERATE Block where trucks arrive at the shop. OVER is the label of the timer transaction Block. Until time 480, this is 0 and the test is true, so transactions pass through. At time 480, this becomes false and the program ends. The

```
           SIMULATE
********************************
*   PROGRAM CHAP16B.GPS        *
*   TRUCK REFUELING EXAMPLE    *
********************************
 COMEIN    GENERATE     13,6    TRUCKS ARRIVE
           TEST NE      N(OVER),1    AFTER 8 HOURS, NO MORE TRUCKS
           QUEUE        WAIT
           SEIZE        REFUEL  IS REFUELING FREE?
           DEPART       WAIT
           ADVANCE      12,8    REFUEL
           RELEASE      REFUEL
 ALLD      TERMINATE
 OVER      GENERATE     480
           TEST E       QC(WAIT),N(ALLD) WAIT UNTIL SHOP IS CLEAR
           TERMINATE    1
           START        1
           PUTSTRING    ('  ')
           PUTSTRING    ('  ')
           PUTPIC       LINES=6,N(COMEIN),Q(WAIT),F(REFUEL),N(ALLD),
                        FR(REFUEL)/10.,AC1
    TRUCKS TO COME TO SHOP              **
    NUMBER LEFT WAITING AT 5 O'CLOCK    **
    NUMBER REFUELING AT 5               **
    NUMBER DONE REFUELING               **
    FACILITY WAS BUSY                   **.**% OF THE TIME
    SHOP CLOSED AFTER                  ***.** MINUTES
           END
```

FIGURE 16.3
Listing of program CHAP16B.GPS.

```
TRUCKS TO COME TO SHOP              38
 NUMBER LEFT WAITING AT 5 O'CLOCK    0
 NUMBER REFUELING AT 5               0
 NUMBER DONE REFUELING              37
 FACILITY WAS BUSY                  95.49% OF THE TIME
 SHOP CLOSED AFTER                 491.91 MINUTES
```

FIGURE 16.4
Results of program CHAP16B.GPS.

relational operator could have been *L* and the same result obtained. The second TEST Block is placed after the time transaction GENERATE Block. This is

```
TEST E     QC(WAIT),N(ALLD)
```

Every truck that enters the shop contributes to the queue count. In program CHAP16A.GPS, there were 37 trucks that entered the shop. The number of trucks that had been refuelled after 8 hours was 35. In the program CHAP16B, the TEST Block now holds the timer transaction until these numbers are equal. The program ends at time 491.91. Thus, the refuelling depot had to work an extra 11.91 minutes after closing the shop. Notice that, now, 38 trucks arrived for refuelling. This is because the depot was open for those extra 11.91 minutes. One arrived after the facility was closed and was held in the first TEST Block.

Example 16.2: Expansion of Example 16.1

This is an expansion of the previous example. In an inspection shop, a single inspector can inspect a truck in 16 ± 6 minutes. Trucks arrive every 12 ± 5 minutes so, in the long run, the inspector cannot handle the traffic. He does have a single bay for trucks to wait if he is busy. However, if a truck arrives and finds the bay taken, it will leave. 70% of those who leave will not return, but the remaining 30% will wait for 30 ± 12 minutes and, then, once again attempt to gain entrance to the inspection area. When the inspector has worked for 8 hours, the doors are closed, but the inspector must remain working until all the trucks have been served.

The program to simulate this is CHAP16C.GPS. A listing is shown in Figure 16.5.

```
                SIMULATE
***********************************
*  PROGRAM CHAP16C.GPS    FURTHER  *
*  ON THE TEST BLOCK               *
***********************************
          STORAGE    S(WAIT),1     PROVIDE ONE WAITING BAY
          GENERATE   12,5          TRUCKS ARRIVE
          TEST L     N(TIME),1     IS IT CLOSING TIME?
BACK      TRANSFER   BOTH,,AWAY    IS THERE ROOM IN THE SHOP?
INSHOP    ENTER      WAIT          YES, ENTER
          SEIZE      INSPECT       IS THE INSPECTOR FREE?
          LEAVE      WAIT          LEAVE THE WAITING BAY
          ADVANCE    16,6          RECEIVE INSPECTION
NDONE     RELEASE    INSPECT       FREE
          TERMINATE                LEAVE THE SHOP
AWAY      TRANSFER   .7,,GONE      70% LEAVE
          ADVANCE    30,12         30% WAIT
          TRANSFER   ,BACK         RETURN TO THE SHOP
GONE      TERMINATE                LEAVE THE SYSTEM
TIME      GENERATE   480           5 O'CLOCK COMES
          TEST E     N(NDONE),N(INSHOP)  IS SHOP EMPTY?
          TERMINATE  1
          START      1
          PUTSTRING  ('  ')
          PUTSTRING  ('  ')
          PUTPIC     LINES=4,N(INSHOP),FR(INSPECT)/10.,AC1
   RESULTS OF STUDY OF INSPECTION AREA
   TRUCKS TO RECEIVE INSPECTION              **
   INSPECTOR WAS BUSY                        **.**% OF THE TIME
   INSPECTION AREA OPEN FOR                 ***   MINUTES
             END
```

FIGURE 16.5
Listing of program CHAP16C.GPS.

```
RESULTS OF STUDY OF INSPECTION AREA
TRUCKS TO RECEIVE INSPECTION           33
INSPECTOR WAS BUSY                     97.71% OF THE TIME
INSPECTION AREA OPEN FOR              500  MINUTES
```

FIGURE 16.6
Result of program CHAP16C.GPS.

The output from the program is shown in Figure 16.6.

The inspector was busy for 97.71% of the time. A total of 33 trucks entered the shop and received inspection. The inspector worked 20 minutes past five to finish the inspection of the trucks in the shop.*

Whenever a TEST Block in refusal mode is used in a program, great care must be exercised that the transaction does not remain in the Block forever if this is not the programmer's desire. There is another caution in using this Block that we have not been too concerned with up to this time. Whenever a transaction is in a blocked condition at a TEST Block, it remains on the current events chain. Whenever the processor does a re-scan, this Block must be tested. This can be quite costly in terms of execution time. In some cases there will be ways to avoid using such inefficient Blocks. The TEST Block is both convenient and easy to understand. However, if it is possible to avoid using it, alternate programming should be used. Some other Blocks that might be used in its place will be introduced later. In some cases, there is no other method available.

Example 16.3: Animation of Inspection Shop

It will be instructive to view the animation of the program written for Example 16.2. This was program CHAP16C.GPS. The animation program is CHAP16D.GPS. Its listing is given in Figure 16.7.

```
           SIMULATE
************************************
*   PROGRAM CHAP16D.GPS            *
*   FURTHER ON THE TEST BLOCK      *
*   ANIMATION OF PROGRAM CHAP16C.GPS *
************************************
 ATF       FILEDEF    'CHAP16D.ATF'
           INTEGER    &COUNT1,&COUNT2,&COUNT3
           STORAGE    S(WAIT),1     PROVIDE ONE WAITING BAY
           GENERATE   12,5          TRUCKS ARRIVE
           TEST L     N(TIME),1     IS IT CLOSING TIME?
           BLET       &COUNT1=&COUNT1+1
           BPUTPIC    FILE=ATF,LINES=5,AC1,XID1,XID1,XID1,&COUNT1
TIME *.****
CREATE TRUCK T*
PLACE T* ON P1
SET T* TRAVEL 4
WRITE M1 **
           ADVANCE    4
 BACK      TRANSFER   BOTH,,AWAY    IS THERE ROOM IN THE SHOP?
 INSHOP    ENTER      WAIT        YES, ENTER
           BPUTPIC    FILE=ATF,LINES=2,AC1,XID1
```

FIGURE 16.7
Listing of program CHAP16D.GPS. *(Continued)*

* The results of this simulation will vary depending on the random numbers used. Later we will learn how to run the simulations with different random number streams.

```
TIME *.****
PLACE T* AT 35.5 29
           SEIZE      INSPECT      IS THE INSPECTOR FREE?
           LEAVE      WAIT         LEAVE THE WAITING BAY
           BPUTPIC     FILE=ATF,LINES=2,AC1,XID1
TIME *.****
PLACE T* AT 41 38
           ADVANCE    16,6         RECEIVE INSPECTION
           BPUTPIC    FILE=ATF,LINES=3,AC1,XID1,XID1
TIME *.****
PLACE T* ON P5
SET T* TRAVEL  4
           ADVANCE    .5
 NDONE     RELEASE    INSPECT      FREE INSPECTOR
           BPUTPIC    FILE=ATF,LINES=2,AC1,FR(INSPECT)/10.
TIME *.****
WRITE M4 **.**%
           ADVANCE    3.5
           BLET       &COUNT2=&COUNT2+1
           BPUTPIC    FILE=ATF,LINES=3,AC1,XID1,&COUNT2
TIME *.****
DESTROY T*
WRITE M2 **
           TERMINATE               LEAVE THE SHOP
 AWAY      BPUTPIC     FILE=ATF,LINES=3,AC1,XID1,XID1
TIME *.****
PLACE T* ON P2
SET T* TRAVEL 5
           ADVANCE     5
           TRANSFER   .7,,GONE     70% LEAVE
           BPUTPIC    FILE=ATF,LINES=3,AC1,XID1,XID1
TIME *.****
PLACE T* ON P3
SET T* TRAVEL 7
           ADVANCE    7
           ADVANCE    23,9         30% WAIT
           BPUTPIC    FILE=ATF,LINES=3,AC1,XID1,XID1
TIME *.****
PLACE T* ON P1
SET T* TRAVEL 4
           ADVANCE    4
           TRANSFER   ,BACK        RETURN TO THE SHOP
 GONE      BPUTPIC    FILE=ATF,LINES=3,AC1,XID1,XID1
TIME *.****
PLACE T* ON P4
SET T* TRAVEL 3
          ADVANCE     3
          BLET       &COUNT3=&COUNT3+1
          BPUTPIC    FILE=ATF,LINES=3,AC1,XID1,&COUNT3
TIME *.****
DESTROY T*
WRITE M3 **
           TERMINATE               LEAVE THE SYSTEM
 TIME      GENERATE   480          5 O'CLOCK COMES
           TEST E     N(NDONE),N(INSHOP)   IS SHOP EMPTY
           TERMINATE  1
           START      1
           PUTPIC    FILE=ATF,LINES=2,AC1
TIME *.****
```

FIGURE 16.7 (Continued)
Listing of program CHAP16D.GPS.

(Continued)

```
        END
                 PUTSTRING  ('  ')
                 PUTSTRING  ('  ')
                 PUTPIC     LINES=4,N(INSHOP),FR(INSPECT)/10.,AC1
            RESULTS OF STUDY OF INSPECTION AREA
            TRUCKS TO RECEIVE INSPECTION                **
            INSPECTOR WAS BUSY                          **.**% OF THE TIME
            INSPECTION AREA OPEN FOR                   ***   MINUTES
                  END
```

FIGURE 16.7 (Continued)
Listing of program CHAP16D.GPS.

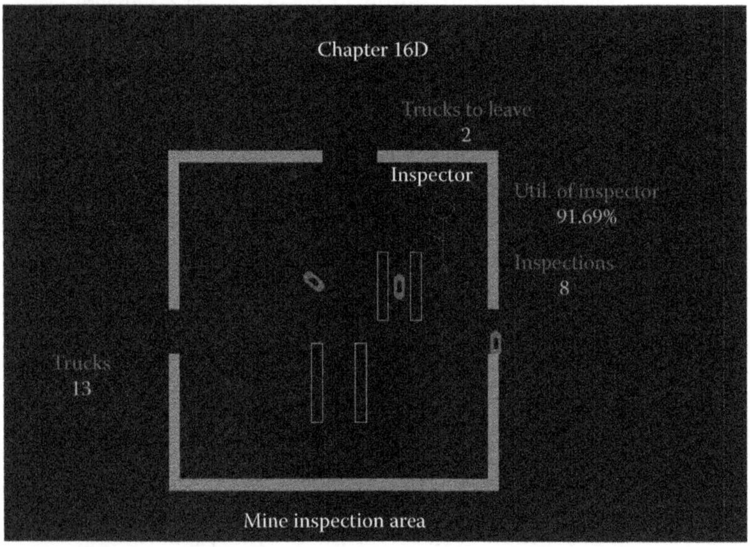

FIGURE 16.8
Screenshot of animation given by program CHAP16D.GPD.

The animation will look as shown in Figure 16.8. The animation program will yield slightly different results than the original program. That is because additional time has been provided for the trucks to move through the system.

16.3 `TEST` Block in Normal Mode

The other form of the `TEST` Block has a C operand. The form of it is simply

```
TEST R   A,B,C
```

where `C` is the name of a Block the transaction is routed to if the test is false.

If the test is true, the transactions continue to the next sequential Block. Thus,

```
TEST E   Q(TOMMY),Q(SALLY),DDDD
```

will test the queue length of the queue TOMMY and the queue length at SALLY. If they are equal, the transaction will go to the next sequential Block. If they are unequal, the transaction will go to the Block named DDDD. For some programmers, who are used to the logic of Fortran, the way GPSS/H works for the TEST Block is going to seem to be quite the opposite to what one would expect. Thus, *great care is required when using this Block.*

Example 16.4: A Manufacturing Process with Two Machines

In a manufacturing process, parts come to a machine for forming. The interarrival rate is 14 ± 7.5 minutes. There are two machines available for forming. The first can form in 16 ± 5.4 minutes and the other takes considerably longer as it takes 24 ± 8 minutes to form. In fact, this second machine is in such poor condition that it is not used until the first machine is utilized to its fullest. This faster machine cannot be used more than 85% of the time or it may overheat. Parts enter the room where both machines are located and use the faster machine until it reaches 85% utilization, at which time the slower machine is used until the utilization is again below 85%. Build a GPSS/H model to represent the system. Simulate for 20 straight shifts of 8 hours (480 minutes). Determine the utilization of both machines and the average number of parts waiting for service.

Solution

The GPSS/H program to do the simulation is as shown in Figure 16.9. The results of the simulation are as given by Figure 16.10.

Notice that machine 1 was busy slightly more than the maximum allowable time, namely, 85.04%. Machine 2 was only busy 43.98% of the time. It can be shown that the maximum queue was 4 but the average queue length was only 0.62.

The animation for this example is done by CHAP16F.GPS. Its listing is given in Figure 16.11.

The animation is shown in Figure 16.12.

```
          SIMULATE
************************************
*   PROGRAM CHAP16E.GPS            *
*   STUDY OF MANUFACTURING SYSTEM  *
************************************
          GENERATE    14,7.5
          QUEUE       WAIT
          TEST LE     FR(MACH1),850,DOWN1
          SEIZE       MACH1
          DEPART      WAIT
          ADVANCE     16,5.4
          RELEASE     MACH1
          TERMINATE
 DOWN1    SEIZE       MACH2
          DEPART      WAIT
          ADVANCE     24,8
          RELEASE     MACH2
          TERMINATE
          GENERATE    480*20
          TERMINATE   1
          START       1
          PUTPIC      LINES=4,FR(MACH1)/10.,FR(MACH2)/10.,QA(WAIT)
        RESULTS OF SIMULATION
    MACHINE 1 WAS BUSY    **.**% OF THE TIME
    MACHINE 2 WAS BUSY    **.**% OF THE TIME
    THE AVG. QUEUE WAS     *.**
          END
```

FIGURE 16.9
Listing of program CHAP16E.GPS.

```
          RESULTS OF SIMULATION
      MACHINE 1 WAS BUSY   85.04% OF THE TIME
      MACHINE 2 WAS BUSY   43.98% OF THE TIME
      THE AVG. QUEUE WAS    0.62
```

FIGURE 16.10
Results of program CHAP16E.GPS.

```
          SIMULATE
***********************************
*  PROGRAM CHAP16F.GPS             *
*  ANIMATION OF CHAP16E.GPS        *
***********************************
   ATF    FILEDEF      'CHAP16F.ATF'
          GENERATE     14,7.5
          BPUTPIC      FILE=ATF,LINES=4,AC1,XID1,XID1,XID1
TIME *.****
CREATE TRUCK T*
PLACE T* ON P1
SET T* TRAVEL 3
          ADVANCE      3
          QUEUE        WAIT
          TEST LE      FR(MACH1),850,DOWN1
          SEIZE        MACH1
          BPUTPIC      FILE=ATF,LINES=2,AC1,XID1
TIME *.****
PLACE T* AT 1.5 10
          DEPART       WAIT
          ADVANCE      16,5.4
          BPUTPIC      FILE=ATF,LINES=3,AC1,XID1,XID1
TIME *.****
PLACE T* ON P2
SET T* TRAVEL 3
          ADVANCE      .5
          RELEASE      MACH1
          BPUTPIC      FILE=ATF,LINES=2,AC1,FR(MACH1)/10.
TIME *.****
WRITE M1 **.**%
          ADVANCE      2.5
          BPUTPIC      FILE=ATF,LINES=2,AC1,XID1
TIME *.****
DESTROY T*
          TERMINATE
 DOWN1    SEIZE        MACH2
          BPUTPIC      FILE=ATF,LINES=2,AC1,XID1
TIME *.****
PLACE T* AT 1.5 -12
          DEPART       WAIT
          ADVANCE      24,8
          BPUTPIC      FILE=ATF,LINES=3,AC1,XID1,XID1
TIME *.****
PLACE T* ON P2
SET T* TRAVEL 3
          ADVANCE      .5
          RELEASE      MACH2
          BPUTPIC      FILE=ATF,LINES=2,AC1,FR(MACH2)/10.
TIME *.****
WRITE M2 **.**%
          ADVANCE      2.5
          BPUTPIC      FILE=ATF,LINES=2,AC1,XID1
TIME *.****
DESTROY T*
          TERMINATE
```

FIGURE 16.11
Listing of program CHAP16F.GPS.

(Continued)

```
              GENERATE     480*20
              TERMINATE    1
              START        1
              PUTPIC       LINES=4,FR(MACH1)/10.,FR(MACH2)/10.,QA(WAIT)
            RESULTS OF SIMULATION
        MACHINE 1 WAS BUSY    **.**% OF THE TIME
        MACHINE 2 WAS BUSY    **.**% OF THE TIME
        THE AVG. QUEUE WAS     *.**
              PUTPIC       FILE=ATF,LINES=2,AC1
    TIME *.****
    END
               END
```

FIGURE 16.11 (Continued)
Listing of program CHAP16F.GPS.

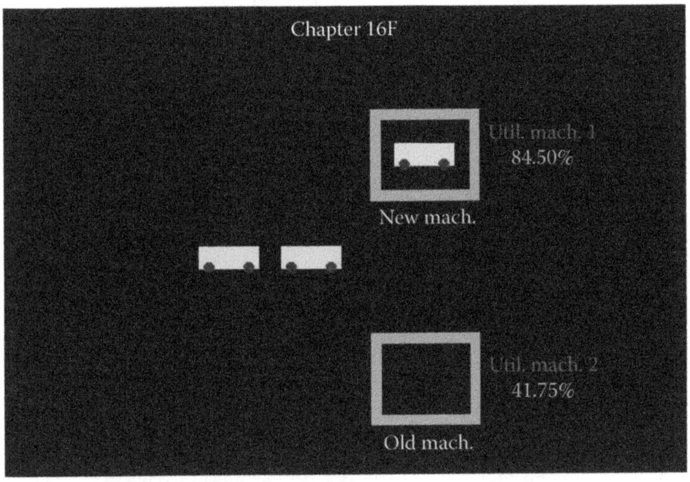

FIGURE 16.12
Screenshot of animation given by CHAP16F.GPS.

Example 16.5: A Mine Doing Stripping and Then Ore Hauling

A mine has five trucks that haul waste before they start to haul ore. Each truck holds 90 tons. A single loader takes 1.5 ± .5 minutes to load a truck with waste material. Loaded trucks travel to the waste dump in 15 ± 4 minutes. These trucks dump in 1 ± .3 minutes and return to the mine in 12.5 ± 3 minutes. There is no restriction on dumping waste, so more than one truck can dump at a time. This mining process continues until 7000 tons of waste has been removed. Once this amount of waste has been removed, the trucks start to haul ore. If a truck has been loaded with waste and more than 7000 tons of waste has been dumped, the loaded truck will still travel to the waste dump, dump, and then travel to the ore shovel. Travel from the waste dump to the ore shovel takes 13 ± 3 minutes. The single ore shovel loads a truck in 1.2 ± .2 minutes. It takes 12 ± 4 minutes to travel to the crusher where only one truck can dump at a time. It takes 1 ± .2 minutes to dump into the crusher. The return to the ore shovel for a truck takes 10.5 ± 2.4 minutes. Determine how many loads of waste and of ore are hauled each shift. A shift is for 10 hours of 60 minutes.

Solution

The program to simulate this is CHAP16G.GPS. Its listing is given in Figure 16.13.

The output from this program is given in Figure 16.14.

The changeover from hauling waste to hauling ore does not occur until time 499.25, which is after 8 hours. Thus, the mine spends most of the time hauling waste. The mine owner may well consider adding another truck. This is left as an exercise.

```
                    SIMULATE
*****************************
*   PROGRAM CHAP16G.GPS      *
*****************************
                INTEGER     &TOTAL,&TORE
                REAL        &T
                GENERATE    ,,,5        TRUCKS IN THE MINE
        UPTOP   QUEUE       SHOVEL
                SEIZE       SHOVEL
                DEPART      SHOVEL
                ADVANCE     1.5,.5
                RELEASE     SHOVEL
                ADVANCE     15,4        TRAVEL TO WASTE DUMP
                ADVANCE     1.,.3       DUMP
                BLET        &TOTAL=&TOTAL+90
                TEST L      &TOTAL,7000,BACK
                ADVANCE     12.5,3      RETURN TO WASTE
                TRANSFER    ,UPTOP
        BACK    BLET        &T=AC1
                ADVANCE     13,3        TRAVEL TO ORE SHOVEL
        BACK2   QUEUE       WAITO
                SEIZE       ORESH
                DEPART      WAITO
                ADVANCE     1.2,.2      LOAD A TRUCK WITH ORE
                RELEASE     ORESH
                ADVANCE     12,4        TRAVEL TO CRUSHER
                SEIZE       CRUSH
                ADVANCE     1.,.2       DUMP INTO CRUSHER
                RELEASE     CRUSH
                BLET        &TORE=&TORE+90
                ADVANCE     10.5,2.4 RETURN TO SHOVEL
                TRANSFER    ,BACK2
                GENERATE    600
                TERMINATE   1
                START       1
                PUTPIC     LINES=4,&T,&TOTAL,&TORE
                RESULTS OF SIMULATION
          TIME CHANGE OVER OCCURRED ***.**
          TOTAL WASTE HAULED         ****
          TOTAL ORE HAULED           ****
                 END
```

FIGURE 16.13
Listing of program CHAP16G.GPS.

```
                    RESULTS OF SIMULATION
          TIME CHANGE OVER OCCURRED 499.25
          TOTAL WASTE HAULED         7380
          TOTAL ORE HAULED           1800
```

FIGURE 16.14
Output from program CHAP16G.GPS.

The animation is done with the assumption that all hauling times are constant. Thus, the time for a truck to haul from the waste shovel to the waste dump is 15 minutes, and the time to return to the waste shovel is 12 minutes. In Chapter 17, we shall learn how to handle distributions in the animations for these various times.

The program to do the animation is CHAP16H.GPS. Its listing is given in Figure 16.15. The animation is given in Figure 16.16.

```
            SIMULATE
*******************************
*  PROGRAM CHAP16H.GPS         *
*  ANIMATION FOR CHAP16G.GPS   *
*******************************
            INTEGER     &TOTAL,&TORE
            REAL        &T
  ATF       FILEDEF     'CHAP16H.ATF'
            GENERATE    3,,0,5   TRUCKS IN THE MINE
            BPUTPIC     FILE=ATF,LINES=2,AC1,XID1
TIME *.****
CREATE TRUCKE T*
   UPTOP    QUEUE       SHOVEL
            SEIZE       SHOVEL
            BPUTPIC     FILE=ATF,LINES=3,AC1,XID1,XID1
TIME *.****
PLACE T* AT -48 -10
SET T* CLASS TRUCKS
            DEPART      SHOVEL
            ADVANCE     1.5,.5
            BPUTPIC     FILE=ATF,LINES=4,AC1,XID1,XID1,XID1
TIME *.****
PLACE T* ON P1
SET T* TRAVEL 15
SET T* CLASS TRUCKF
            ADVANCE     .5       TRAVEL TO WASTE DUMP
            RELEASE     SHOVEL
            ADVANCE     14.5
            SEIZE       DUMMY
            BPUTPIC     FILE=ATF,LINES=2,AC1,XID1
TIME *.****
PLACE T* AT 11 -11
            ADVANCE     1.,.3    DUMP
            RELEASE     DUMMY
            BLET        &TOTAL=&TOTAL+90
            TEST L      &TOTAL,7000,BACK
            BPUTPIC     FILE=ATF,LINES=5,AC1,XID1,XID1,XID1,&TOTAL
TIME *.****
PLACE T* ON P2
SET T* TRAVEL 12
SET T* CLASS TRUCKE
WRITE M1 ****
            ADVANCE     12       RETURN TO WASTE SHOVEL
            TRANSFER    ,UPTOP
  BACK      BLET        &T=AC1
            BPUTPIC     FILE=ATF,LINES=5,AC1,XID1,XID1,XID1,&T
TIME *.****
PLACE T* ON P3
SET T* TRAVEL 13
SET T* CLASS TRUCKE
WRITE M2 ***.**
            ADVANCE     13
  BACK2     QUEUE       WAITO
            SEIZE       ORESH
            BPUTPIC     FILE=ATF,LINES=3,AC1,XID1,XID1
TIME *.****
PLACE T* AT -48 8
SET T* CLASS TRUCKS
            DEPART      WAITO
            ADVANCE     1.2,.2   LOAD A TRUCK WITH ORE
            BPUTPIC     FILE=ATF,LINES=4,AC1,XID1,XID1,XID1
```

FIGURE 16.15
Listing of program CHAP16H.GPS. *(Continued)*

```
TIME *.****
PLACE T* ON P4
SET T* TRAVEL 12
SET T* CLASS TRUCKF
          ADVANCE      .5
          RELEASE      ORESH
          ADVANCE      11.5      TRAVEL TO CRUSHER
          SEIZE        CRUSH
          BPUTPIC      FILE=ATF,LINES=2,AC1,XID1
TIME *.****
PLACE T* AT 15 13
          ADVANCE      1.,.2     DUMP INTO CRUSHER
          BLET         &TORE=&TORE+90
          BPUTPIC      FILE=ATF,LINES=5,AC1,XID1,XID1,XID1,&TORE
TIME *.****
PLACE T* ON P5
SET T* TRAVEL 10
SET T* CLASS TRUCKE
WRITE M3 ****
          ADVANCE      .5
          RELEASE      CRUSH
          ADVANCE      9.5
          TRANSFER     ,BACK2
          GENERATE     600
          TERMINATE    1
          START        1
          PUTPIC      LINES=4,&T,&TOTAL,&TORE
          RESULTS OF SIMULATION
    TIME CHANGE OVER OCCURRED ***.**
    TOTAL WASTE HAULED          ****
    TOTAL ORE HAULED            ****
          PUTPIC      FILE=ATF,LINES=2,AC1
TIME *.****
END
           END
```

FIGURE 16.15 (Continued)
Listing of program CHAP16H.GPS.

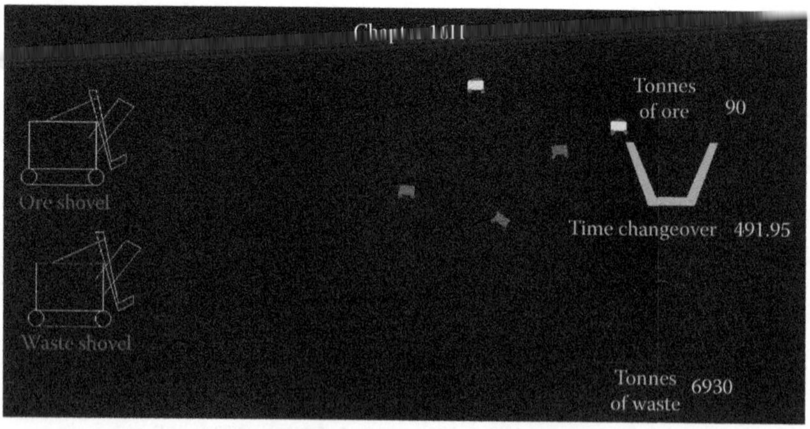

FIGURE 16.16
Screenshot of animation from CHAP16H.GPS.

Example 16.6: Making the Shovel Booms Rotate

In the animations done for mining operations thus far, the shovels have remained stationary. Figure 16.15 is an example of this. In an actual mine, the booms of the shovels will move while the trucks are being loaded. Animation showing the movement of the boom can be done using a layout object for the boom of the shovel. The actual GPSS/H program can be modified by adding a segment with a dummy transaction that controls the motion of the boom. When a truck transaction is being loaded, the shovel is `SEIZE`d and the dummy transaction circulates to cause the boom to rotate. This is best illustrated by a simple, short program that has trucks coming to a shovel to be loaded. Trucks arrive at the shovel every 15 ± 5 time units and take 1.5 ± 3 time units to load. While loading, the boom rotates in a 40° arc, first clockwise and then counterclockwise.

The program to do the simulation is CHAP16I.GPS. Its listing is given in Figure 16.17. The animation is given in Figure 16.18.

```
             SIMULATE
*******************************
*  PROGRAM CHAP16I.GPS        *
*  ANIMATION OF A SHOVEL BOOM *
*******************************
  ATF        FILEDEF      'CHAP16I.ATF'
             GENERATE     15,5    TRUCKS COME
             BPUTPIC      FILE=ATF,LINES=4,AC1,XID1,XID1,XID1
TIME *.****
CREATE TRUCKE T*
PLACE T* ON P1
SET T* TRAVEL 10
             ADVANCE      10
             QUEUE        SHOVEL
             SEIZE        SHOVEL
             BPUTPIC      FILE=ATF,LINES=3,AC1,XID1,XID1
TIME *.****
PLACE T* AT 19.5 7.5
SET T* CLASS TRUCKS
             DEPART       SHOVEL
             ADVANCE      1.5,.3 LOAD A TRUCK
             BPUTPIC      FILE=ATF,LINES=4,AC1,XID1,XID1,XID1
TIME *.****
PLACE T* ON P2
SET T* TRAVEL 15
SET T* CLASS TRUCKL
             ADVANCE      .5      TRAVEL TO WASTE DUMP
             RELEASE      SHOVEL
             ADVANCE      12
             BPUTPIC     FILE=ATF,LINES=2,AC1,XID1
TIME *.****
DESTROY T*
            TERMINATE
            GENERATE      ,,,1    DUMMY TRANSACTION
 BACK1      TEST E        F(SHOVEL),1
 BACK2      BPUTPIC       FILE=ATF,LINES=2,AC1
TIME *.****
ROTATE BB1 40 STEP 5 TIME .3
            ADVANCE       .3
            BPUTPIC       FILE=ATF,LINES=2,AC1
```

FIGURE 16.17
Listing of program CHAP16I.GPS. *(Continued)*

```
TIME *.****
ROTATE BB1 -40 STEP 5 TIME .3
          ADVANCE      .3
          TEST E       F(SHOVEL),1,BACK1
          TRANSFER     ,BACK2
          GENERATE     200
          TERMINATE    1
          START        1
          PUTPIC       FILE=ATF,LINES=2,AC1
TIME *.****
END
            END
```

FIGURE 16.17 (Continued)
Listing of program CHAP16I.GPS.

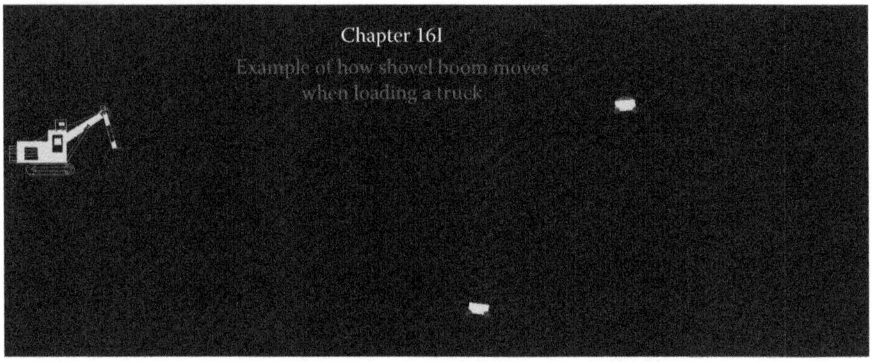

FIGURE 16.18
Screenshot of animation given by program CHAP16I.GPS.

16.4 Exercises

16.1. State what the following `TEST` Blocks do:

```
(a) TEST E   Q(FIRST),Q(SECOND)
(b) TEST G   FR(MACH1),500
(c) TEST NE          &X,&Y
(d) TEST LE          AC1,480,AWAY
(e) TEST L   S(TUGS),Q(DOCK),BYEBYE
(f) TEST LE          R(SILO),10000,PART2
(g) TEST G   SR(WORKERS),700
(h) TEST NE          Q(WAIT1),2
```

16.2. A factory makes large equipment. As a piece of equipment comes along an assembly line at one stage, it must wait on a machine that can work on only one piece of equipment at a time. There is room for only one piece of equipment to wait. If the machine at this stage is busy and one piece is waiting, the arriving equipment must be moved to another part of the factory. The cost to the company each time this happens is $28.50. Equipment arrives every 14 ± 5 minutes.

The server takes 15 ± 6 minutes to work on a piece. The engineer in charge has determined that it would cost an average of $75 per shift to enlarge the waiting area to accommodate two pieces of equipment. Should she do this? The solution should have an animation to present to management. It is only necessary to make an animation for the system as it is presently working.

NOTE: Often an animation is used mainly to illustrate that the logic of the system is correct. Assuming the above data is correct, the animation will have paths leading from the waiting area to the machine. If time for this travel is added to the animation, the solution will differ slightly from the original GPSS/H program. For this example, the animation need only show the logic.

16.3. If the mine owner in Example 16.5 decides to purchase a sixth truck, when do the trucks start to carry ore?

16.4. A mine has a single crusher. Trucks approach this crusher every 6 ± 1.5 minute. They cannot dump unless there is less than 200 tons in the crusher. Each truck holds exactly 130 tons and it takes 2 ± 1 minutes to dump into the crusher. The crusher can crush 24 tons per minute so it works whenever there are more than 24 tons to be crushed. Determine how many loads will be dumped in a shift of 480 minutes. How busy is the crusher?

16.5. Animate Exercise 16.4.

16.6. The mine owner for Exercise 16.4 wants to add another truck to the mine. This will result in trucks approaching the crusher at a rate of one every 5.5 ± 1 minute. If the crusher has to work more than 90% of the time, its motor will not work properly. Should the mine owner purchase this new truck?

16.7. Suppose the mine owner in Exercise 16.4 decides to purchase a new crusher that can crush ore at a rate of 50 tons per minute. Should the owner then purchase a sixth truck?

16.8. Refer to Example 16.2. Two `TEST` Blocks were used to control the running time of the program. Change the two `TEST` Blocks used to ones that make use of the simulation clock, `AC1`. The first will replace the Block `TEST NE N(OVER),1`. Your results should be identical to those obtained previously.

17

Standard Numerical Attributes

From a mining museum, Northern Spain.

17.1 More on Standard Numerical Attributes

In previous chapters, the concept of standard numerical attributes (SNAs) was introduced. They were used mainly by the `PUTPIC` statement and `BPUTPIC` Block for either creating output or creating animation files. It is possible to use SNAs in a program as an operand. Reference to them is made by enclosing them with parentheses. Some examples are given here. Assume that the SNAs have been defined as follows: `&TRUCKS = 4; &CARS = 6; &MEAN = 10; &MODE = 4.5; &TIME = 123.45; &PATH = 4; Q(ONE) = 15; Q(TWO) = 8;` and `XID1 = 35`

```
(a) GENERATE &TRUCKS
(b) GENERATE ,,,&CARS
(c) ADVANCE  &MEAN,&MODE
(d) ADVANCE  &TIME
(e) BPUTPIC  FILE = ATF,LINES = 2,AC1,XID1,&PATH
    TIME *.****
    PLACE CAR* ON PATH P*
(f) ADVANCE  Q(ONE),Q(TWO)
```

The above Blocks will be the same as follows:

```
(a) GENERATE 4
(b) GENERATE ,,,6
(c) ADVANCE  10,4.5
(d) ADVANCE  123.45
(e) BPUTPIC  FILE = ATF,LINES = 2, AC1,XID1,&PATH
    TME *.****
    PLACE CAR* ON P*
(f) ADVANCE  15,8
```

NOTE: (e) has not been changed. The line of code sent to the .ATF file would be

```
PLACE CAR(XID1) ON P4
```

where `CAR(XID1)` would have the class `CAR` given the value of `XID1`. For example, it might be `CAR54`, `CAR123`, and `CAR12`.

17.2 Other SNAs

There are many other SNAs. Some have been encountered already without specifically referring to them as such. These are known as system SNAs. A listing of some of these, which we had before, as well as a few new ones, is as follows:

`W(FIRST)` is the number of transactions currently at the Block with the label `FIRST`. If there are four transactions in the Block at the time one used `W(FIRST)`,

```
FIRST     ADVANCE   12,4
```

this SNA is 4.

If the Block was

```
FIRST     QUEUE  LINE
```

and there were four transactions in the `QUEUE` with the operand `LINE`, then `W(FIRST)` is 4. This is exactly the same as the SNA `Q(LINE)`. However, most Blocks do not have such an SNA and the `W(name)` must be used.

`N(SECND)` is the total number of transactions that have entered the Block with the label `SECND`. If 54 transactions have entered the BLOCK with this label, the value of the SNA is 54. This SNA has been introduced before and used in `PUTPIC` statements.

`C1` is the Relative Clock (given before).

`AC1` is the Absolute Clock (given before).

`TG1` is the current value of the termination counter.

`RNj` Random number from 0 to 1. This will be used in defining functions. This also can be written as `RN(j)`. If this is used in connection with a function, the value returned is from the interval [0, 1), that is, from 0.000000 to 0.999999. If used in any other context, the value returned is from 000 to 999.

`FRN1` Random number from 0.000000 to 0.999999 (given before). Actually, the *1* in `FRN1` is the number of the random number stream to be sampled from. Any number up to $2^{31} - 1$ could be used. The student version limits the number to 1000. It is rare that more than one or two random number streams are used.

Although reference to SNAs is by parenthesis, it also is possible to reference them by means of the single dollar sign ($). Thus, one could write `Q$FIRST`, which is the same as `Q(FIRST)`. This will not work when the entity is given a number and not a name. For example, if the queue Block is `QUEUE 5`, then reference to the queue length is given by either `Q(5)` or `Q5` but not `Q$5`. This is an old method of referencing and will not be used here.

Constants are SNAs and have their own *family name*. This is the letter K in front of them, that is, 3 is `K3`, 501, is `K501`, and so on. This option is used rarely. Actually, about the only time students ever use this is to see if it really does work. It is not natural to write a number with a K in front of it, so this is considered obsolete.

`M1` whenever a transaction enters the system, it is tagged with the time of entry. Whenever `M1` is then referenced, this value is subtracted from the current clock value. `M1` is the difference between these two times. Suppose the time of entry was 5040 and, when `M1` is referenced, the clock is now at 5880. `M1` will be 840 or 5880 – 5040. `M1` is a *floating point number*.

The total number of SNAs may seem a bit staggering, especially since we have not yet learned all that we can do with the SNAs. However, as we learn more GPSS/H statements and Blocks, the use of the SNAs will become apparent. Other SNAs will be introduced as more Blocks are presented.

Several examples of the possible use of SNAs are given next.

Any SNA can be used in a program as an operand. The use of SNAs will greatly expand one's ability to build meaningful simulation models. As additional Blocks are introduced, it will become even more apparent as to how useful they are in writing simulation models. Some of the following examples might appear to be quite fanciful, but they illustrate the extreme power and flexibility of the language.

```
(a) TRUCK GENERATE           ,,,4
          ADVANCE            60*N(TRUCK)-60
```

Four transactions will be scheduled to leave the `GENERATE` Block at time 0. The first is put on the future events chain (FEC) for a time of `60*N(TRUCK)-60`. The Block count when the first transaction has left is 1. Therefore, the time on the FEC is 0. The second transaction will be put on the FEC for a time of 60, the third for a time of 120, and the fourth for 180 time units. The effect of this is to delay the entry of the ensuing transactions by a factor of 60 time units. This is the same as having the Block.

```
              GENERATE    60,,0,4
```

```
(b) BLOCKA    SEIZE    TOMMY
       ...    ......
       ...    ......
              ADVANCE N(BLOCKA)*2
```

Transactions entering the ADVANCE Block will be put on the FEC for a time equal to two times the number of transactions that have entered the SEIZE Block with the label BLOCKA. The above ADVANCE Block could have been written as

```
ADVANCE                    FC(TOMMY)*2
```

because FC(TOMMY) gives the number of times the facility TOMMY has been captured.

```
(c) TERMINATE              W(BLOCKC)
```

The counter given by the START n statement is decremented by the amount equal to the current Block count at the Block with the label BLOCKC.

```
(d) ADVANCE                2.5*AC1
```

The transaction will be put on the FEC for a time equal to 2.5 times the Absolute Clock value.

```
(e) ADVANCE                QA(LINE)
```

The transaction will be placed on the FEC for a time equal to the integer portion of the average queue content of the queue named LINE.

```
(f) ADVANCE                Q(STORE)
```

A transaction entering the ADVANCE Block will be placed on the FEC for a time equal to the length of the queue STORE. Most of the programs from now on will make use of SNAs.

Example 17.1: Another Truck Inspection Station

Trucks arrive at an inspection station every 15 ± 6.5 minutes. This arrival distribution is constant throughout the day. If no trucks are waiting, the inspector will tend to take his time doing the inspection. As trucks arrive and fill up the shop, the inspector will speed up his inspection. The time it takes for an inspection is given by the following:

People in Queue	Time for Inspection
0	16
1	14
2	12
3	10
4	8
5	6
6	4

Simulate the operation of this inspection shop for 100 straight shifts of 480 minutes each.

The program to do the simulation is CHAP17A.GPS. A listing of it is shown in Figure 17.1.

The output from the program is given in Figure 17.2.

The output shows that the inspector was very busy, as his utilization was .997. The maximum queue was 3 with an average queue length of .96.

The animation program is CHAP17B.GPS. Its listing is given in Figure 17.3.

```
           SIMULATE
***************************
*  PROGRAM CHAP17A.GPS     *
***************************
  COMEIN  GENERATE     15,6.5  TRUCKS ARRIVE
          QUEUE        WAIT    SIT AND WAIT
          SEIZE        INSPECT    USE INSPECTOR
  DLEAVE  DEPART       WAIT
          ADVANCE      16-2*Q(WAIT) INSPECTION TIME
          RELEASE      INSPECT
          TERMINATE
          GENERATE     480*100
          TERMINATE    1
          START        1
          PUTSTRING    ('  ')
          PUTSTRING    ('  ')
          PUTPIC       LINES=3,FR(INSPECT)/10.,N(COMEIN),N(DLEAVE),QM(WAIT),_
                       QA(WAIT)
           RESULTS OF SIMULATION
     UTIL  **.**%    COMEIN ****  LEAVE ******
     MAXIMUM QUEUE   **        AVERAGE QUEUE  *.**
          PUTSTRING    ('  ')
          PUTSTRING    ('  ')
          PUTPIC
   ALL DONE
          END
```

FIGURE 17.1
Listing of program CHAP17A.GPS.

```
           RESULTS OF SIMULATION
     UTIL  99.68    COMEIN 3192  LEAVE   3191
     MAXIMUM QUEUE    3       AVERAGE QUEUE  0.96
```

FIGURE 17.2
Output from program CHAP17A.GPS.

```
          SIMULATE
******************************
*  PROGRAM CHAP17B.GPS        *
*  ANIMATION OF CHAP17A.GPS   *
******************************
 ATF      FILEDEF      'CHAP17B.ATF'
          INTEGER      &GOOUT
  COMEIN  GENERATE     15,6.5  TRUCKS ARRIVE
          BPUTPIC      FILE=ATF,LINES=4,AC1,XID1,XID1,XID1
TIME *.****
CREATE TRUCK T*
PLACE T* ON P1
SET T* TRAVEL 3
          ADVANCE      3
          QUEUE        WAIT
          SEIZE        INSPECT    USE INSPECTOR
          BPUTPIC      FILE=ATF,LINES=2,AC1,XID1
TIME *.****
PLACE T* AT 44 31
  DLEAVE  DEPART       WAIT
          ADVANCE      16-2*Q(WAIT) INSPECT
          BPUTPIC      FILE=ATF,LINES=3,AC1,XID1,XID1
```

FIGURE 17.3
Listing of program CHAP17B.GPS.

(Continued)

```
TIME *.****
PLACE T* ON P2
SET T* TRAVEL 3
          ADVANCE     .25
          RELEASE     INSPECT
          BLET        &GOOUT=&GOOUT+1
          BPUTPIC     FILE=ATF,LINES=3,AC1,FR(INSPECT)/10.,&GOOUT
TIME *.****
WRITE M4 **.**%
WRITE M2 ***
          ADVANCE     2.75
          TRANSFER    .5,,AWAY
          BPUTPIC     FILE=ATF,LINES=3,AC1,XID1,XID1
TIME *.****
PLACE T* ON P6
SET T* TRAVEL 3
           ADVANCE    3
          BPUTPIC     FILE=ATF,LINES=2,AC1,XID1
TIME *.****
DESTROY T*
          TERMINATE
 AWAY     BPUTPIC     FILE=ATF,LINES=3,AC1,XID1,XID1
TIME *.****
PLACE T* ON P7
SET T* TRAVEL 3
           ADVANCE    3
          BPUTPIC     FILE=ATF,LINES=2,AC1,XID1
TIME *.****
DESTROY T*
          TERMINATE
          GENERATE    480*10
          TERMINATE   1
          START       1
          PUTPIC      FILE=ATF,LINES=2,AC1
TIME *.****
END
          PUTSTRING   ('  ')
          PUTSTRING   ('  ')
          PUTPIC      LINES=3,FR(INSPECT)/10.,N(COMEIN),N(DLEAVE),QM(WAIT),_
                      QA(WAIT)
           RESULTS OF SIMULATION
     UTIL  **.**    COMEIN ****   LEAVE ******
     MAXIMUM QUEUE    **        AVERAGE QUEUE   *.**
          PUTSTRING   ('  ')
          PUTSTRING   ('  ')
          PUTPIC
   ALL DONE
          END
```

FIGURE 17.3 (Continued)
Listing of program CHAP17B.GPS.

As is often the case with animations, a slight modification has been done to enhance the viewing. In this case, when a truck is served, it must choose one of two paths each having equal chance of occurring. This is shown in Figure 17.4.

When a truck travels on a path or a path segment, the GPSS/H program and the animation travel times need to be synchronized. Thus, if a truck is to travel from A to B in the simulation program in time 123.456 seconds, it needs to also travel in the animation in the exact same time of 123.456 seconds. We have avoided doing this

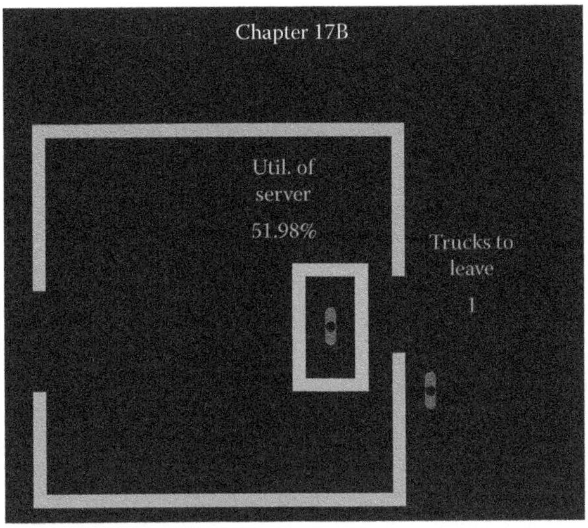

FIGURE 17.4
Screenshot of animation from program CHAP17B.GPS.

until now. In the animation, the travel time is usually given by a line of animation code such as

```
SET T* TRAVEL **.**
```

The corresponding GPSS/H Block will be something like

```
ADVANCE &X
```

It is necessary to have the times given in the `SET TRAVEL` and `ADVANCE` equal. The next example will illustrate how this can be done.

Example 17.2: Travel Times Sampled from Distributions

A mining truck enters a road at point A every 12 time units. It travels to point B in 10 ± 2 time units. There is a bridge at point B that allows only one-way traffic. It takes $1.1 \pm .3$ minutes for a truck to cross this bridge. Once across the bridge trucks travel to point C in 8 ± 3.5 time units. There is a second bridge at point C that also allows only one-way traffic. This bridge takes 2 ± 1 minute to cross. Once across, the trucks travel for $4 \pm .5$ minutes and leave at point D. Mining trucks are also coming from the opposite direction on this road. They arrive at point D every 15 minutes. It takes them $3.5 \pm .8$ minutes to arrive at the bridge at point C. Their travel times across the bridges are the same as for trucks coming from the other direction. These trucks travel to the bridge at point B in 7.6 ± 2.3 minutes. From the bridge at point B, the trucks take 9 ± 1.5 minutes to exit at point A.

The GPSS/H program is written and is called CHAP17C.GPS. The listing is given in Figure 17.5.

The output from the program is given in Figure 17.6.

There were 118 trucks that entered from the left and 94 from the right. The bridge on the left was used 16.26% of the time and the bridge on the right was used 29.72%. These results are as expected. Even so, it is instructive to do the animation that gives conclusive proof that the program is correct. This is provided by program CHAP17D.GPS. A screenshot of the animation is given in Figure 17.7.

```
          SIMULATE
*************************
*  PROGRAM CHAP17C.GPS   *
*************************
          GENERATE      12       TRUCKS ARRIVE AT A
          ADVANCE       10,2     TRAVEL TO B
          SEIZE         BRIDGE1
          ADVANCE       1.1,.3      CROSS THE BRIDGE
          RELEASE       BRIDGE1
          ADVANCE       8,3.5    TRAVEL TO C
          SEIZE         BRIDGE2
          ADVANCE       2,1      CROSS THE BRIDGE
          RELEASE       BRIDGE2
          ADVANCE       4,.5
 BYEBYE1  TERMINATE
          GENERATE      15        TRUCKS ARRIVE FROM THE RIGHT
          ADVANCE       3.5,.8   TRAVEL TO C
          SEIZE         BRIDGE2
          ADVANCE       2,1       CROSS THE BRIDGE
          RELEASE       BRIDGE2
          ADVANCE       7.6,2.3 TRAVEL TO B
          SEIZE         BRIDGE1
          ADVANCE       1.1,.3   CROSS THE BRIDGE
          RELEASE       BRIDGE1
          ADVANCE       9,1.5
 BYEBYE2  TERMINATE
          GENERATE 480*3
          TERMINATE     1
          START         1
          PUTPIC        LINES=4,FR(BRIDGE1)/10.,FR(BRIDGE2)/10.,_
                      N(BYEBYE1),N(BYEBYE2)
       UITL. OF BRIDGE 1 **.**%
       UTIL. OF BRIDGE 2 **.**%
       TRUCKS FROM LEFT TO EXIT ON RIGHT ***
       TRUCKS FROM RIGHT TO EXIT ON LEFT  ***
          END
```

FIGURE 17.5
Listing of program CHAP17C.GPS.

```
UTIL. OF BRIDGE 1 16.26%
UTIL. OF BRIDGE 2 29.72%
TRUCKS FROM LEFT TO EXIT ON RIGHT 118
TRUCKS FROM RIGHT TO EXIT ON LEFT   94
```

FIGURE 17.6
Output from program CHAP17C.GPS.

FIGURE 17.7
Screenshot of animation from program CHAP17D.GPS.

17.3 Exercises

17.1. Suppose that the SNAs have the values: `&X` = 5; `&Y` = 6; `Q(WAIT)` = 3; `FR(MACH1)` - 500; `AC1` = 35.600; `XID1` = 3. What will be the result when a transaction passes through the Blocks?

```
(a) ADVANCE     &X*&Y
(b) TEST E      Q(WAIT),2
(c) TEST G      FR(MACH1),500,AWAY1
(d) ADVANCE     FR(MACH1)+&X
(e) TEST NE     XID1,3,WAIT
(f) GENERATE    ,,,&X
(g) GENERATE    &Y,,&X,,&X
(h) ADVANCE     XID1*AC1
```

17.2. The animation for Example 17.2 can be improved on. For example, the trucks can be made to stop slightly before the bridges by moving their hot points to the fronts. Also, one could put stop signs on either side of the bridges that turn from green to red when a truck is on the bridge. Add these to the animation and make the corresponding changes to the program CHAP17D.GPS.

17.3. Go back to Example 16.4 and the corresponding program CHAP16H.GPS. Modify the program to include the travel time distributions that were used in program CHAP16G.GPS.

17.4. Use the SNA `FRN1` to give a different colour to each of the trucks in Example 17.2. Assume that the colour of the trucks can be from F2 to F25.

17.5. Assume that there are three types of truck that the mine has for Example 17.2. Change the program to reflect this. Suppose that 20% are type 1, 50% type 2, and the rest type 3.

18

Functions

Coal handling system, Maracaibo, Venezuela.

18.1 Functions

18.1.1 Commonly Used Built-In Functions

So far the only built-in GPSS/H function used was the uniform distribution* (and that was not the general form, as it was restricted to having the mean less than the spread). GPSS/H has over 20 built-in functions that can be used. Two distributions that are very commonly used in simulations are the Poisson (exponential) and normal (Gaussian) distributions. These arise in the simulation studies of a great many systems. For example, the interarrival rates of telephone calls are often given by the exponential distribution; the time to travel from point A to point B by truck is normally distributed; the time for a ship to return to a port is often exponential; the time between storms might be exponential; and so on. Both of these functions are well known to engineering students.

Since these functions are so commonly referenced in simulation studies, they are built-in to GPSS/H and sampling from them is quite easy.

* The uniform distribution can also be referenced by the built-in function RVUNI (*rns, mean, spread*) where *rns* is the random number stream, *mean* is the mean value, and *spread* is the spread.

18.1.2 Poisson (Exponential) Distribution

The Poisson distribution is a one-parameter distribution being completely specified by its mean value. The built-in function to be used in sampling from it is given by

```
RVEXPO(random number stream, mean)
```

where the random number stream is the number of the random number stream to be used in obtaining the sample. Recall that GPSS/H has nearly an infinite number of these, but, normally, one only uses small numbers such as 1, 2, and 3.

Examples of this are as follows:

```
(a) ADVANCE          RVEXPO(1,12.3)
(b) GENERATE         RVEXPO(1,3.4)
(c) ADVANCE          RVEXPO(2,2)
```

In (a), the transaction is placed on the FEC for a time given by sampling from the exponential distribution with a mean of 12.3.

In (b), transactions are generated at times given by sampling from the exponential distribution with a mean of 3.4.

In (c), the transactions are placed on the FEC for a time given by sampling from the exponential distribution with a mean of 2. Random number stream 2 is used.

This is an important distribution for the mining engineer, as loading times in a mine are often given by this distribution. Equipment downtime and repair time also are represented often by this distribution.

18.1.3 Normal (Gaussian) Distribution

The normal distribution is a two-parameter distribution and is specified by the mean and standard deviation. The GPSS/H built-in function to sample from the normal distribution is given by

```
RVNORM(random number stream, mean, std. dev.)
```

where random number stream refers to the number of the random number stream to sample from (as in RVEXPO).

Examples of this might be as follows:

```
(a) ADVANCE          RVNORM(1,20,2.3)
(b) GENERATE         RVNORM(1,30,5.5)
```

In (a), the transaction is placed on the FEC for a time that is obtained by sampling from a normal distribution with a mean of 20 and a standard deviation of 2.3.

In (b), transactions are generated according to the normal distribution with a mean of 30 and a standard deviation of 5.5.

GPSS/H professional samples from a distribution bounded by 44 standard deviations above and below the mean, so, while it is theoretically possible to obtain samples that are negative, this is rare. However, with the student versions of GPSS/H, the standard deviation needs to be less than 20% of the mean. Thus,

`RVNORM(1,20,4)` is correct but
`RVNORM(1,20,5)` might not be

What to do for the case that the standard deviation is greater than 20% of the mean will be covered later in this chapter.

18.1.4 The Triangular Distribution

GPSS/H has a built-in function that represents the triangular distribution. This distribution is one that looks like a triangle with one side on the x-axis. The side extends from a minimum value to a maximum value. The most likely value is the mode. A triangular distribution with a minimum of 10, a maximum of 100, and a mode of 20 will be skewed to the left. A triangular distribution with minimum value of 10, maximum of 100, and a mode of 80 will be skewed to the right. If the mode of this distribution was 55, the distribution would be symmetrical. To sample from a triangular distribution in GPSS/H, one uses the following built-in function:

```
RVTRI(random no. stream, min., mode, max.)
```

Thus,

```
ADVANCE RVTRI(2,0,3,10)
```

will sample from the triangular distribution having minimum value of 0, mode of 3, and maximum value of 10.

The ones given here are those most commonly used by the mining engineer. It should be noted that, in addition to the three distributions given here, GPSS/H has 22 other built-in distributions. These can be found in the GPSS/H users' manual. For nearly all mining engineering applications, either the built-in functions given here or the ones that the user constructs himself/herself are sufficient.

Example 18.1: A Second Look at Example 15.2

Recall Example 15.2, which was of a system where workers performed service at a rate of 40 ± 15 units. Customers arrived at a rate of 8 ± 4 units. There were five servers and one seat for waiting customers to use. If a customer arrived and found the shop full, he or she would leave. The program was later run for 2, 3, 4, and 5 seats. The results were given in Table 15.1 that showed that, for 1 seat, the percentage of customers to leave without obtaining service was 9.43%. For five seats, the percentage dropped to 1.48%. How does the system perform with arrivals and service normally distributed, and how will it perform with the arrivals and service exponential?

Computer program CHAP15B.GPS is easily changed to study the system with arrivals and service normally distributed. Since the means must be less than 20% of the standard deviation, the arrival rate was changed to

```
GENERATE RVNORM(1,8,1.6)
```

and the service rate to

```
ADVANCE RVNORM(1,40,8)
```

The new program is given by CHAP18A.GPS. When the exponential distribution was used, the changes were for the arrivals

```
GENERATE RVEXPO(1,8)
```

and for service

```
ADVANCE RVEXPO(1,40)
```

A listing of program CHAP18A.GPS is shown in Figure 18.1.

The results for the normal distribution are given in Table 18.1. The results for using the exponential distribution are given in Table 18.2.

A comparison of the results shows the difference in using the different distributions. Even with five seats for customers to wait, the percentage of customers who leave without obtaining service is 12.12% when arrivals and service are both exponential. For the original example, when arrival and service was obtained from the uniform distribution, the percentage who left without obtaining service was only 1.48%. With the normal distribution, it was only .99%. Since many systems in real life can be modelled using the exponential distribution, this example shows that obtaining the correct data can be critical.

```
            SIMULATE
*******************************
*   PROGRAM CHAP18A.GPS         *
*   MODIFICATION OF PROGRAM     *
*   CHAP15B.GPS                 *
*******************************
          RMULT       12345        RANDOM NUMBER SEED
          REAL        &X
          STORAGE     S(SEAT),1    PROVIDE SEAT(S)
 COMEIN   GENERATE    RVNORM(1,8,1.6)
          TRANSFER    BOTH,,AWAY   IS A SEAT FREE?
          ENTER       SEAT         TAKE A SEAT
          TRANSFER    ALL,CCCC,DDDD,5
 CCCC     SEIZE       SERV1        FIRST SERVER
          LEAVE       SEAT
          ADVANCE     RVNORM(1,40,8)
          RELEASE     SERV1
          TRANSFER    ,EEEE
          SEIZE       SERV2        SECOND SERVER
          LEAVE       SEAT
          ADVANCE     RVNORM(1,40,8)
          RELEASE     SERV2
          TRANSFER    ,EEEE
          SEIZE       SERV3        THIRD SERVER
          LEAVE       SEAT
          ADVANCE     RVNORM(1,40,8)
          RELEASE     SERV3
          TRANSFER    ,EEEE
          SEIZE       SERV4        FOURTH SERVER
          LEAVE       SEAT
          ADVANCE     RVNORM(1,40,8)
          RELEASE     SERV4
          TRANSFER    ,EEEE
 DDDD     SEIZE       SERV5        FIFTH SERVER
          LEAVE       SEAT
          ADVANCE     RVNORM(1,40,8)
          RELEASE     SERV5
          TRANSFER    ,EEEE
 EEEE     TERMINATE
 AWAY     TERMINATE
          GENERATE    480*500    SIMULATE FOR 500 SHIFTS
          TERMINATE   1
          START       1
```

FIGURE 18.1
Listing of program CHAP18A.GPS.

(Continued)

```
          LET            &X=N(AWAY)
          PUTSTRING      ('  ')
          PUTSTRING      ('  ')
          PUTPIC         LINES=8,N(COMEIN),N(AWAY),S(SEAT)+R(SEAT),_
                         (&X/N(COMEIN))*100,SR(SEAT)/10.,_
                         FR(SERV1)/10.,FR(SERV2)/10.,FR(SERV3)/10.,_
                         FR(SERV4)/10.,FR(SERV5)/10.
         NUMBER TO COME TO SHOP  ******
         NUMBER TO LEAVE MAD       ****
         NUMPBER OF SEATS TO WAIT    **
         PERCENT WHO LEFT MAD        **.**%
         UTIL. OF SEAT OR SEATS      **.**%
         UTIL. OF FIRST SERVER       **.**%   UTIL. OF SECOND SERVER   **.**%
         UTIL. OF THIRD SERVER       **.**%   UTIL. OF FOURTH SERVER   **.**%
         UTIL. OF FIFTH SERVER       **.**%
          END
```

FIGURE 18.1 (Continued)
Listing of program CHAP18A.GPS.

TABLE 18.1
Results Using Normal Distribution for Example 18.2

Customers	30,002	29,969	29,989	29,958	29,982
Those who left	2376	1019	576	353	298
Percentage left	7.92	3.4	1.92	1.18	0.99
Seats	1	2	3	4	5
Percentage util. of seat(s)	21.57	27.90	33.19	34.92	39.17
Percentage util. serv1	94.90	97.41	98.52	98.89	99.19
Percentage util. serv2	93.79	96.96	98.20	98.82	99.11
Percentage util. serv3	92.38	96.49	98.10	98.58	98.93
Percentage util. serv4	91.05	96.01	97.74	98.32	98.77
Percentage util. serv5	88.35	95.21	97.44	98.04	98.57

TABLE 18.2
Results Using Exponential Distribution for Example 18.2

Customers	29,941	30,002	29,895	29,926	29,882
Those who left	6731	5525	4668	4074	3623
Percentage left	22.61	18.44	15.61	13.61	12.12
Seats	1	2	3	4	5
Percentage util. of seat(s)	22.80	27.98	31.35	34.31	35.81
Percentage util. mach1	86.49	89.65	90.99	92.42	93.22
Percentage util. mach2	83.49	86.88	88.43	90.22	91.25
Percentage util. mach3	79.50	83.17	85.75	87.77	88.71
Percentage util. mach4	73.64	79.00	81.71	86.44	85.82
Percentage util. mach5	67.30	72.79	77.50	80.56	82.34

18.1.5 User-Supplied Functions

It is possible to build functions from data. There are two such functions that will be described next. The first has the function returning only a finite number of values and is called a *discrete function*. It is normally built up from a statistical distribution of possible values that needs to be a cumulative distribution that has been normalized. In most mining simulations, one has different cycle times that are obtained from other computerized systems. However, it is an easy matter to obtain the statistical distributions desired for various times such as loading, hauling on segments, dumping, and returning. Hence, this is an important topic.

The function needs to be defined as follows:

```
(label or number)    FUNCTION      SNA,Dn
```

The (`label or number`) is any allowable label or number in GPSS/H. Labels have been used before for statements but not numbers. The word `FUNCTION` must appear as the operation. `RN1` is the random number stream to be used. In many cases, the SNA is given as `RN1`, `RN2`, `RN3`, and so on. When a function is referenced with RNn, RNn returns a number from 0.000000 to 0.999999. *If RNn is used in any other context, it returns a number from 000 to 999.* The *D* stands for *discrete* and must be this. The *n* in Dn is the number of pairs that make up the function. In the case here, it is 6. Several representations of the function given here might be as follows:

```
FIRST    FUNCTION    RN1,D6
MYFN1    FUNCTION    RN1,D6
    1    FUNCTION    RN2,D6
```

The pairs of values that make up the D6 specification come from the cumulative statistical distribution. These are of the following form:

```
cum. dist.1,value1/cum. dist2,value2/cum. dist3,value3/.../cum.
dist6,value6/
```

The last slash (/) is optional. The values can be continued on the following line or lines if there are many values. If so, the first character cannot be a slash. One could even have only one value per line if desired. The first number can appear in position 1 on the line below the `FUNCTION` specification. The `FUNCTION` specification *must* come before it is referenced. Usually, these specifications are placed before the first `GENERATE` Block but do not need to be.

The following example illustrates how one would build a user-supplied function.

Example 18.2: User-Supplied Functions

The following values have been found to occur from the study of a system.

Percentage of Times	Value
10	2
15	3.5
25	6
30	7
15	8
5	12

No other values are possible. The above data need to be turned into a cumulative distribution curve. This is done by adding the percentages up to 100. Thus, the above becomes

Percentage of Times	Value	Cumulative Probability (%)
10	2	10
15	3.5	25
25	6	50
30	7	80
15	8	85
5	12	100

For the example given here, one could have

```
   FIRST FUNCTION RN1,D6
.1,2/.25,3.5/.5,6/.8,7/.88,8/1,12
```

It could have been written as

```
   FIRST FUNCTION RN1,D6
.1,2/.25,3.5
.5,6
.8,7
.88,8
1,12
```

Notice that no slash was used for several values. This is all right. There is no particular advantage in writing the values this way.

A `FUNCTION` is referenced in a Block by the following*:

`FN(`*`label or number`*`)`

If a number is used for the `FUNCTION`, the parentheses can be omitted. Here is how a `FUNCTION` works when it is referenced. A random number from 0.000000 to 0.999999 is generated. If the random number is less than or equal to .100000, the value returned is 2. If the random number is between .1000001 and .250000, the value returned is 3.5, and so on, up to 0.999999 where the value 12 is returned. Since the value of RN1 is never 1.000000, there is a slight error of one part in a million.

The program CHAP18B.GPS illustrates a function made up for Example 18.2. A total of 20 values are printed to the screen. The function gives the values for returning both `REAL` and `INTEGER` values when the function is referenced. It also shows what is returned when `RN1` is used outside of the `FUNCTION`. Its listing is given in Figure 18.2.

The output that will be shown on the screen is shown in Figure 18.3.

The output shows that the values for `&X`, `&I`, and `&Y` are different for each `BPUTPIC`. This is because, every time the function `MYFIRST` is referenced, a different random number is obtained.

As noted, the first operand for a function can be any SNA. This can be very useful and will be used many times in these chapters. For example, suppose Joe can cut hair in 10 minutes if there are no customers waiting for him to finish. However, if there is one person waiting, he cuts hair in 9 minutes; if there are two or more people waiting, he cuts hair in 8 minutes. Consider the following `FUNCTION` and corresponding `ADVANCE` Block:

* An old way of referencing functions is by use of the *dollar* ($) sign. One could write FN$*label*. Thus, FN(FIRST) and F$FIRST are the same. This still works but is rarely used.

```
           SIMULATE
****************************
*    PROGRAM CHAP18B.GPS     *
*    ILLUSTRATING FUNCTIONS *
****************************
           REAL          &X,&Y
           INTEGER       &I
*******************************
*  FUNCTON SPECIFICATION NEXT *
*******************************
  MYFIRST FUNCTION      RN1,D6
.1,2/.25,3.5/.5,6/.8,7/.95,8/1,12/
           PUTSTRING     ('  ')
           PUTSTRING     ('  ')
           GENERATE      1
           BLET          &X=FN(MYFIRST)
           BLET          &I=FN(MYFIRST)
           BLET          &Y=RN1
           BPUTPIC       &X,&I,&Y
   X IS **.**            I IS **       Y is ******.**
           TERMINATE     1
           START         20

    END
```

FIGURE 18.2
Listing of program CHAP18B.GPS.

```
X IS  6.00          I IS  6      Y is     739.00
X IS  3.50          I IS  3      Y is     768.00
X IS  6.00          I IS  8      Y is     792.00
X IS  2.00          I IS  2      Y is     185.00
X IS  6.00          I IS  8      Y is     733.00
X IS  6.00          I IS  8      Y is     684.00
X IS  7.00          I IS  7      Y is     691.00
X IS  6.00          I IS  8      Y is      97.00
X IS  7.00          I IS  7      Y is     315.00
X IS  6.00          I IS  8      Y is     626.00
X IS  7.00          I IS 12      Y is     911.00
X IS  2.00          I IS 12      Y is     665.00
X IS  3.50          I IS  7      Y is     280.00
X IS  6.00          I IS  7      Y is     842.00
X IS  3.50          I IS  7      Y is     413.00
X IS  7.00          I IS  3      Y is     676.00
X IS  2.00          I IS  3      Y is     363.00
X IS  8.00          I IS  7      Y is     566.00
X IS  8.00          I IS  7      Y is      87.00
X IS 12.00          I IS  7      Y is     241.00
```

FIGURE 18.3
Output from program CHAP18B.GPS.

```
CUTHAIR FUNCTION      Q(WAIT),D3
0,10/1,9/2,8
      ..............
      ..............
      QUEUE            WAIT
      ADVANCE          FN(CUTHAIR)
```

In the event that there are three people waiting, so that `Q(WAIT)` is 3, the function returns a value of 8. Similarly, if you used an SNA that could return a value less than the minimum value, the function would return the first value. For example,

```
   MYSEC   FUNCTION &X,D3
1,3/2,4/3,7
```

If &X was 0 or any value less than 1, the value returned would always be 3.

18.1.6 A Caution

At one time, before GPSS/H, GPSS did not have built-in functions and one had to use piecewise functions as shown here. In order to have functions such as the normal or exponential, it was necessary to use a method that made use of the fact that, when two functions are used (one for the mean and the other for the spread), the result was to multiply the first by the second. This still exists, so if one had the Block

```
ADVANCE              FN(ONE),FN(TWO)
```

and FN(ONE) returned a 10 while FN(TWO) returned a 4, the result would be an ADVANCE of 40, not as sampled from the distribution 10 ± 4! Returning to Joe the barber, if Joe could cut hair in 10 ± 4 minutes if there were no customers waiting, 9 ± 3 minutes if there was one person waiting and 8 ± 2.5 if two or more people were waiting, the code could be something like this:

```
      FIRST FUNCTION               Q(WAIT),D3
0,10/1,9/2,8
      SECOND FUNCTION              Q(WAIT),D3
0,4/1,3/2,2.5
              REAL                 &MEAN,&SPREAD
              . . .
              . . .
              QUEUE                WAIT
              BLET                 &MEAN = FN(FIRST)
              BLET                 &SPREAD = FN(SECOND)
              ADVANCE              &MEAN,&SPREAD
```

18.1.7 Continuous Functions

There are many times when a continuous function is needed. This is a function that can return an infinite number of values. This, too, is made from a cumulative statistical distribution. The general form is similar to a discrete function with the only difference being a Cn as the second operand, instead of Dn, that is,

```
(label or number)FUNCTION RN1,Cn
```

A continuous function is built up as shown in the following example.

Suppose the following times have been found for a truck to travel from point A to point B:

Percentage of Times	Time
0	<4.00
18	4.00–5.00
30	5.00–5.50
20	5.50–5.75
22	5.75–5.90
10	5.90–6.10

The above is converted to a cumulative distribution:

Percentage of Times	Time	Cumulative Percentage (%)
0	<4.00	0
18	4.00–5.00	18
30	5.00–5.50	48
20	5.50–5.75	68
22	5.75–5.90	90
10	5.90–6.10	100

Imagine a series of line segments drawn connecting the points:

(0,4), (.18,5), (.38,5.5), (.68,5.75), (.9,5.9), and (1,6.1)

The function is referenced in the same manner as was the discrete function by first giving it a name and then writing it as follows:

```
      FAST FUNCTION RN1,C6
0,4/.18,5/.38,5.5/.68,5.75/.9,5.9/1,6.1
```

Suppose one had for travel time

```
ADVANCE FN(FAST)
```

the function `FAST` is referenced as follows.

A random number is generated internally by the GPSS/H processor from the random number stream number 1. This number will be from 0.000000 to 0.999999. Suppose it is .3000000. The corresponding *y* value is then obtained by linearly interpolating between 5 and 5.5. This is the value that is returned. Because the random number is never equal to 1, the value of 6.1 cannot be returned but 6.099999 can. This slight error is normally of no consequence.

To be more nearly accurate when using functions, more points are taken to make up the function. For example, the early use of the normal distribution is sampled from a function consisting of 24 line segments and the exponential distribution from one with 25 line segments.

The uniform distribution can be written as a continuous function. Consider the Block

```
ADVANCE             8,4
```

the corresponding continuous function can be*:

```
MY3RD FUNCTION      RN1,C2
0,4/1,12
```

18.1.8 Other Forms of Functions

It is possible to have functions that return an SNA. These are called *attribute-valued* functions. These are distinguished from discrete and continuous functions by having their second operand written as the letter E.

* If the returned value is to be an integer from 4 to 12, one must be careful to write the function as going from 4 to 13, and not 12. This is because GPSS/H *truncates* and does not round off.

Some examples are as follows:

```
      TEST1 FUNCTION Q(FIRST),E3
1,&X/2,&Y/3,&Z
```

The variables `&X, &Y,` and `&Z` need to be specified first and given values. Depending on the queue length at the Block `QUEUE FIRST`, the function will return either the value of `&X, &Y,` or `&Z`.

```
      OTHER FUNCTION S(TUGS),E2
1,S(BOATS1)/2,S(BOATS2)
```

Depending on the `STORAGE` of `TUGS`, the function `OTHER` will return either the `STORAGE` of `BOATS1` or `BOATS2`.

```
      STILL FUNCTION F(MYWORK),E2
0,R(SHIP)/1,R(BOAT)
```

The value of `F(WORK)` is either a 0 or 1 if the facility `MYWORK` is currently captured or not. If it is free, the value returned by the function is the remaining storage of the `STORAGE SHIP` else it returns the remaining storage of `BOAT`.

```
      MYLAST RN1,E3
.1,Q(JOHN)/.5,Q(MARY)/1,Q(TOM)
```

The function `MYLAST` will return the value of the queues at `JOHN, MARY,` or `TOM` depending on the value returned by `RN1.`

Functions also can reference other functions. For example,

```
ANOTHER FUNCTION Q(WAITUP),E3
0,FN(ONE)/1,FN(TWO)/2,FN)THREE)
```

Depending on the value of the queue `WAITUP`, the value returned will be obtained by sampling from the function `ONE, TWO,` or `THREE.`

Example 18.3: Machine with Variable Work Rates

Parts arrive at a machine that has an adjustable work rate. Parts arrive every 11 ± 7 seconds. The machine can sense how many parts are waiting for service. If no parts are waiting, it works according to the normal distribution with a mean of 11 and a standard deviation of 2; if one part is in the queue, it works according to the normal distribution with a mean of 10 and a standard deviation of 1.5; if two or more parts are in the queue, it works according to the normal distribution with a mean of 9 and a standard deviation of 1.3. The plant manager has noticed that the machine has been breaking down often, as it is not supposed to work more than 90% of the time. Determine whether he should consider purchasing a faster machine.

The program to simulate this is given next. The use of the attribute-valued function eliminates the need for the use of the `TEST` Block. This is program CHAP18C.GPS. Its listing (Figure 18.4) is as follows.

The results of the simulation are shown in Figure 18.5.

As can be seen, the machine is working beyond its capacity and should be replaced.

```
                SIMULATE
*******************************
*  PROGRAM CHAP18C.GPS        *
*  EXAMPLE SHOWING ENTITY     *
*  FUNCTION                   *
*******************************
 WORK      FUNCTION     Q(WAITUP),E3
0,RVNORM(1,11,2)/1,RVNORM(1,10,1.5)/2,RVNORM(1,9,1.3)
          GENERATE     11,7       PARTS ARRIVE
          QUEUE        WAITUP     WAIT FOR MACHINE
          SEIZE        MACH1      SEIZE MACHINE
          DEPART       WAITUP
          ADVANCE      FN(WORK)
          RELEASE      MACH1
          TERMINATE
          GENERATE     3600*24*100
          TERMINATE    1
          START        1
          PUTSTRING    ('  ')
          PUTSTRING    ('  ')
          PUTPIC       LINES=2,FR(MACH1)/10.
             RESULTS OF SIMULATION
        THE MACHINE WAS BUSY    **.**% OF THE TIME
          END
```

FIGURE 18.4
Listing of program CHAP18C.GPS.

```
RESULTS OF SIMULATION
THE MACHINE WAS BUSY 95.73% OF THE TIME
```

FIGURE 18.5
Results of program CHAP18C.GPS.

The animation of the program is a bit more complex than the program as it will show the following:

1. When the machine is working at a different rate, this is shown on the screen.
2. The utility of the machine.
3. When the machine is working on a part, it jiggles. This is done by making the machine to be a layout object and a small circular path to place it on when a part is being worked on.

A listing of the program, CHAP18D.GPS, is given in Figure 18.6.
The animation is shown in Figure 18.7.

18.1.9 A Special Case of Functions

It is common to have the first term of the pairs of values start with the integer 1 and increase by 1. This can only be for discrete and attribute-valued functions. In these cases, the second operand is changed to `L` for discrete functions and `M` for attribute-type functions. For example, the discrete function,

```
      MYFUNC FUNCTION &I,D4
1,2/2,4/3,6/4,7
```

```
          SIMULATE
********************************
*  ANIMATION OF CHAP18C.GPS    *
********************************
 WORK     FUNCTION     Q(WAITUP),E3
0,RVNORM(1,11,2)/1,RVNORM(1,10,1.5)/2,RVNORM(1,9,1.3)
 ATF      FILEDEF      'CHAP18D.ATF'
          GENERATE     11,7      PARTS ARRIVE
          BPUTPIC      FILE=ATF,LINES=4,AC1,XID1,XID1,XID1
TIME *.****
CREATE PART P*
PLACE P* ON P1
SET P* TRAVEL 3
          ADVANCE      3
          QUEUE        WAITUP    WAIT FOR MACHINE
          SEIZE        MACH1     SEIZE MACHINE
          DEPART       WAITUP
          TEST E       Q(WAITUP),0,NEXT
          BPUTPIC     FILE=ATF,LINES=4,AC1,XID1
TIME *.****
PLACE MACH1 ON P2
WRITE M1 SLOW RATE
PLACE P* AT -11 0
          ADVANCE      FN(WORK)
          RELEASE      MACH1
          BPUTPIC      FILE=ATF,LINES=6,AC1,XID1,XID1,FR(MACH1)/10.
TIME *.****
PLACE MACH1 AT -11 0
PLACE P* ON P3
SET P* TRAVEL 3
WRITE M1
WRITE M4 **.**%
          ADVANCE      3
          BPUTPIC      FILE=ATF,LINES=2,AC1,XID1
TIME *.***
DESTROY P*
          TERMINATE
 NEXT     TEST E       Q(WAITUP),1,NEXT1
          BPUTPIC      FILE=ATF,LINES=4,AC1,XID1
TIME *.****
PLACE MACH1 ON P2
WRITE M1 MEDIUM RATE
PLACE P* AT -11 0
          ADVANCE      FN(WORK)
          RELEASE      MACH1
          BPUTPIC      FILE=ATF,LINES=6,AC1,XID1,XID1,FR(MACH1)/10.
TIME *.****
PLACE MACH1 AT -11 0
PLACE P* ON P3
SET P* TRAVEL 3
WRITE M1
WRITE M4 **.**%
          ADVANCE      3
          BPUTPIC      FILE=ATF,LINES=2,AC1,XID1
TIME *.***
DESTROY P*
          TERMINATE
```

FIGURE 18.6
Listing of program CHAP18D.GPS. *(Continued)*

```
 NEXT1    BPUTPIC      FILE=ATF,LINES=4,AC1,XID1
TIME *.****
PLACE MACH1 ON P2
WRITE M1 FAST RATE
PLACE P* AT -11 0
          ADVANCE      FN(WORK)
          RELEASE      MACH1
          BPUTPIC      FILE=ATF,LINES=6,AC1,XID1,XID1,FR(MACH1)/10.
TIME *.****
PLACE MACH1 AT -11 0
PLACE P* ON P3
SET P* TRAVEL 3
WRITE M1
WRITE M4 **.**
          ADVANCE      3
          BPUTPIC      FILE=ATF,LINES=2,AC1,XID1
TIME *.***
DESTROY P*
          TERMINATE
          GENERATE     3600
          TERMINATE    1
          START        1
          PUTSTRING    ('  ')
          PUTSTRING    ('  ')
          PUTPIC       LINES=2,FR(MACH1)/10.
             RESULTS OF SIMULATION
        THE MACHINE WAS BUSY    **.**% OF THE TIME
          PUTPIC      FILE=ATF,LINES=2,AC1
TIME *.****
END
           END
```

FIGURE 18.6 (Continued)
Listing of program CHAP18D.GPS.

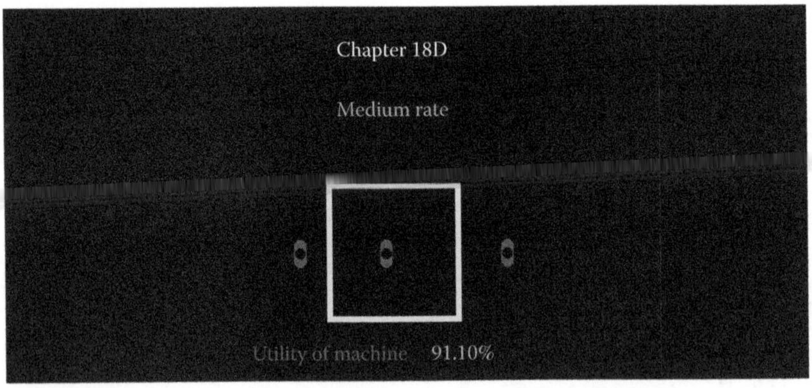

FIGURE 18.7
Screenshot of animation from CHAP18D.GPS.

could have been written:

```
        MYFUNC FUNCTION &I,L4
1,2/2,4/3,6/4,7
```

The advantage of this is that, if the SNA is out of range, there will be an error message. Thus, if the SNA, `&I`, has the value of 5, there would be an error; whereas for the discrete function, it would return the value 7 for values of &I greater than 4. This feature is especially useful when using parameters as operands, which will be covered in Chapter 19.

In the case of entity functions, the corresponding SNA is M*n*.

Example 18.4: Importance of Using Correct Distributions

Let us again consider a small mine operator's problem. The mine is simple as it has only a single shovel that can load one truck at a time. Loaded trucks travel to a crusher where one truck at a time can dump and the trucks return to the shovel. The following are average times for each operation:

Operation	Time (minutes)
Load	2.5
Haul	10
Dump	1.5
Return	8

Each shift is 450 minutes in length. The program to model the mine with these figures is quickly written. The assumption is made that there are eight trucks in the mine. The program is CHAP18E.GPS. The simulation is run for 100 shifts. Its listing is given in Figure 18.8.

The results of running the program for 8, 9, and then 10 trucks are given in Table 18.3. The changes to these times are given next.

As can be seen, maximum production of 39.99 loaded per shift is achieved for nine trucks. With nine trucks, the shovel is working 100% of the time.

```
          SIMULATE
****************************
* PROGRAM CHAP18E.GPS      *
* SMALL MINE EXAMPLE       *
****************************
 TRUCKS   GENERATE     ,,,8     PUT TRUCKS IN THE MINE
 BACK     QUEUE        WAIT     QUEUE AT SHOVEL
          SEIZE        SHOVEL   USE SHOVEL
          DEPART       WAIT     LEAVE QUEUE
          ADVANCE      2.5      LOAD
          RELEASE      SHOVEL   FREE SHOVEL
          ADVANCE      10       HAUL
          SEIZE        CRUSH    AT CRUSHER
          ADVANCE      1.5      CRUSH
          RELEASE      CRUSH    FREE CRUSHER
          ADVANCE      8        RETURN
          TRANSFER     ,BACK
          GENERATE     450*100   SIMULATE FOR 450 SHIFTS
          TERMINATE    1
          START        1
          PUTPIC       LINES=4,N(TRUCKS),FR(SHOVEL)/10.,FC(CRUSH)/450.
          RESULTS OF SIMULATION
      TRUCKS IN THE MINE    *
      UTIL. OF SHOVEL     **.**%
      PRODUCTION PER SHIFT ***.**
          END
```

FIGURE 18.8
Listing of program CHAP18E.GPS.

TABLE 18.3
Results from Running Program CHAP18E.GPS

Trucks	Shovel Utility	Loads per Shift
8	90.91	34.31
9	100.00	39.99
10	100.00	39.99

The next change to the program is to add randomness to the various times. The first assumption is that the various times are given by uniform distributions:

Operation	Time (minutes)
Load	2.5 ± 1
Haul	10 ± 2
Dump	$1.5 \pm .5$
Return	8 ± 1.5

The mean times remain the same. The program to simulate this is CHAP18F.GPS. Its listing is given in Figure 18.9.

The output from this program is shown in Table 18.4.

The results of the program show that an additional truck would be needed for maximum production due to the variability of the times.

```
            SIMULATE
****************************
* PROGRAM CHAP18F.GPS      *
* SMALL MINE EXAMPLE       *
* NOW USING UNIFORM DIS.   *
* FOR MINING OPERATIONS    *
****************************
 TRUCKS     GENERATE    ,,,8     PUT TRUCKS IN THE MINE
 BACK       QUEUE       WAIT     QUEUE AT SHOVEL
            SEIZE       SHOVEL   USE SHOVEL
            DEPART      WAIT     LEAVE QUEUE
            ADVANCE     2.5,1    LOAD
            RELEASE     SHOVEL   FREE SHOVEL
            ADVANCE     10,2     HAUL
            SEIZE       CRUSH    AT CRUSHER
            ADVANCE     1.5,.5   CRUSH
            RELEASE     CRUSH    FREE CRUSHER
            ADVANCE     8,1.5    RETURN
            TRANSFER    ,BACK
            GENERATE    450*100   SIMULATE FOR 450 SHIFTS
            TERMINATE   1
            START       1
            PUTPIC      LINES=4,N(TRUCKS),FR(SHOVEL)/10.,FC(CRUSH)/450.
            RESULTS OF SIMULATION
        TRUCKS IN THE MINE    *
        UTIL. OF SHOVEL     **.**%
        PRODUCTION PER SHIFT ***.**
            END
```

FIGURE 18.9
Listing of program CHAP18F.GPS.

TABLE 18.4

Results from Running Program CHAP18F.GPS

Trucks	Shovel Utility	Loads per Shift
8	85.58	34.14
9	93.77	37.37
10	98.5	39.26

TABLE 18.5

Results from Running Program CHAP18G.GPS

Trucks	Shovel Utility	Loads per Shift
8	77.12	30.64
9	83.22	33.05
10	88.00	35.08
11	92.71	36.20
12	94.93	37.55
13	96.97	38.34

Suppose, however, that the loading time was changed to something more realistic for mining operations and that the loading time was found to be given by the Poisson distribution with a mean of 12. No other changes in the times are made. What, then, is the production? The program to do the simulation is CHAP19G.GPS and its only change is one line, namely,

```
      ADVANCE 2.5,1 LOAD
to    ADVANCE RVEXPO(1,2.5) LOAD
```

The results of running this program may seem a bit surprising but they are instructive. Table 18.5 gives the results for running the program for the number of trucks from 8 to 13.

It can been seen that, even with 13 trucks, production is still not as high as when the program ran with nine trucks and no variation in the different times.

This is a simple example but illustrates the importance of using exact data distributions for correct mine design and production estimation.

18.2 Exercises

18.1. Redo Example 18.1 using the triangular distribution. Do the results differ much?

18.2. State what the following Blocks will do.

```
(a) GENERATE    10,2,,RVNORM(1,10,2)
(b) GENERATE    RVNORM(1,20,2),2
(c) GENERATE    RVNORM(1,20,2),,,3
(d) GENERATE    RVEXPO(11),,8
(e) GENERATE    ,,,RVNORM(1,10,2)
```

18.3. State what the following `FUNCTION`s will do.

```
(a) FIRST FUNCTION    RN1,D3
    .2,8/6,9/1,20
(b) SECOND FUNCTION   RN1,C2
    0,5/1,10
(c) THIRD FUNCTION    &X,D4
    0,4/1,5/2,6/10,20
(d) FOURTH FUNCTION &Y,C3
    1,10/2,13/3,16
```

18.4. Construct a continuous function to replace the operands in the Blocks.

```
(a) ADVANCE 10,6
(b) ADVANCE 20,20
(c) ADVANCE 100,65
(d) GENERATE 30,3
```

18.5. Add a clock to the program CHAP18D.GPS, which is the animation for Example 18.3. Also, add a message to the screen that gives the time when the utilization of the machine exceeds 85%.

18.6. Add the animation for Example 18.4. This can be quite simple showing only the trucks as they travel from shovel to crusher and return. No messages or other motion need be on the screen.

18.7. Suppose that the data distributions for Example 18.4 were all from the normal distributions with means as given and standard deviations of 10% of these. How does the production change?

18.8. Suppose that the data distributions for Example 18.4 were all from the exponential distribution with means as given. How does the production change? How many trucks are needed to achieve a production level of 38 loads per shift?

18.9. For the animation done for Exercise 18.6, add the following messages:

Utility of the shovel

Loads dumped

19

Parameters

Alfred Nobel, the inventor of dynamite.

19.1 Introduction

As each transaction travels from Block to Block, it carries with it several things. For example, we already know that transactions can have different levels of priority. In addition, there are other ways to make each transaction different. One important way to do this is given next. In fact, to become an efficient programmer with GPSS/H, *it is essential that one knows how to use the subject matter in this chapter.*

Each transaction possesses a set of abstract entities known as *parameters*. These are carried with the transaction as it moves through the simulation and can be modified during the program. The values of these are not usually a part of the normal output report (the .LIS file) but can be used during the program by the programmer. They can be printed out using the `BPUTPIC` or `PUTPIC` commands. Just as transactions can be viewed conceptually as *stick people*, it is convenient to think of parameters as pockets on the pants of the stick people. You can put numbers inside each of the pockets to differentiate between the transactions. Each pocket is numbered. You can give pocket number 12 the value 4, pocket number 7 the value −234, and so on. How to do this will be explained below.

Each transaction can have four different kinds of parameters. There can be up to 100 of each of these, although it is rare that one would use more than a few in a typical program. Parameters can be thought of as a collection of SNAs that the transaction owns. These parameters are normally numbers, although one can give them names also. For our purposes, they will be used mainly as numbers, but one example will be given to show them used as names. The different types of parameters in GPSS/H are as follows:

1. *Half-word parameters.* These can be a number that ranges from –32,768 to +32,767. These must be integers. These are referenced by PH.
2. *Full-word parameters.* These can range from -2^{31} to $+2^{31} - 1$. These values range from –2,147,483,648 and +2,147,483,647 and they are integers. These are referenced by PF.
3. *Bit-word parameters.* These can range from only –128 to +127 (-2^7 to $+2^7 - 1$). These, too, are integers. These are referenced by PB.
4. *Floating point or real (decimal) parameters.* The size of these are machine dependent but can be as large (or small) as $\pm 10^{35}$. These are referenced by PL.

Initially, every transaction is assigned 12 half-word parameters by default. Thus, although we didn't know it, all of our transactions so far had 12 of these half-word parameters. The number and types of parameters are given in the GENERATE Block in position F through I. Half-word parameters are indicated by nPH, full-word parameters by mPF, bit-word parameters by iPB, and floating point parameters by jPL. It makes no difference where in positions F through I you indicate the number of each type of parameter.

Some examples of these are as follows:

```
(a) GENERATE      12,2,,,,4PH
(b) GENERATE      ,,,5,,12PF,20PH
(c) GENERATE      ,,,12,,5PF
(d) GENERATE      12,4,,,,0PH,1PF
(e) GENERATE      100,3,,,,3PH,4PF,5PB,6PL
(f) GENERATE      ,,,10,1,12PL
(g) GENERATE      100,,,,,20PH,50PB
```

These will GENERATE transactions with:

(a) four half-word parameters

(b) 12 full-word parameters and 20 half-word parameters

(c) only five full-word parameters (no half-word parameters)

(d) no half-word parameters and one full-word parameter (this is the same as `GENERATE 12,4,,,,1PF`)

(e) three half-word parameters, four full-word parameters, five bit-word parameters, and six floating point parameters

(f) 12 floating point parameters

(g) 20 half-word transactions and 50 bit-word parameters

It is important to remember that, *once you specify parameter types and numbers via the F–I operands of the* GENERATE *Block, you no longer have the 12 half-word parameters by default.*

Thus,

```
GENERATE   ,,,1,,1PL
```

creates a single transaction with only one floating point parameter and no other parameters.

Due to storage constraints, it is best to use half-word parameters unless the numbers used as parameter values are beyond the ranges. In addition, although it may not seem obvious, it is preferable to have all the transactions in a program have the same number and type of parameters. For example, you may have the main GENERATE Block as `GENERATE ,,,5,,20PH`. Later in the program, the timer transaction enters via `GENERATE 480*100`. It is preferable to have this as `GENERATE 480*100,,,,,20PH`, even though it is going to be terminated immediately.

How we assign values to the parameters is given next.

19.2 The ASSIGN Block

There are several ways to give values to parameters. The way shown next is the original way, and is still used by many programmers. Initially, the values of all parameters are zero. The value of a transaction's parameter can be modified by the ASSIGN Block.

```
ASSIGN     (parameter number or variable or name),(SNA),parameter type
```

where parameter number is the number of the parameter such as 1, 6, and 8. This can be a variable. If a variable is used here, it must have a value that will be evaluated and correspond to a parameter. It could also be a name.

SNA is the value the parameter is to be given. This can be a constant, a variable, or even a function as we shall see.

Parameter type—either PH, PF, PB, or PL for half-word parameter, full-word parameter, bit-word parameter, or real parameter. This can be omitted for certain cases, but it is not considered good programming to do so. For example, if the transaction is used with only the 12 half-word default parameters, it would be acceptable to omit the parameter type. This is not done in this book.

Thus, when a transaction leaves the Block

```
ASSIGN        1,5,PH
```

half-word parameter 1 will have the value of 5. If parameter 1 had a previous value, this value is deleted.

```
GENERATE      ,,,3
ASSIGN        2,100,PH
```

Three transactions are generated, and the value of their second parameter is set to 100.

```
VALUE FUNCTION      RN1,D3
.2,4/.5,7/1,8
      GENERATE        ,,,1
      ASSIGN          2,FN(VALUE),PH
```

The transaction's second parameter will have the value of 4 20% of the time. For 30% of the time, the value will be 7 and it will have the value of 8 for the rest of the time.

An example of this might be to distinguish trucks approaching a refuelling depot. Suppose that there are three types of trucks that are in the mine: 20% are type 1, 50% are type 2, and 30% are type 3. They arrive at the depot every 12 ± 4 hours. The code might be as follows:

```
TYPE FUNCTION       RN1,D3
.2,1//7,2/1,3
      GENERATE        12,4 TRUCKS ARRIVE
      ASSIGN          1,FN(TYPE),PH
```

Since the operands of the `ASSIGN` Block are SNAs, it is possible to have the following:

```
ASSIGN PH1,3,PH
```

Now, what happens depends on the value already in the transaction's parameter 1. If it is 4, then parameter 4 will have the value 3; if it is 6, then parameter 6 will have the value 3.

```
TIMES  GENERATE      ,,,5
       ASSIGN        1,N(TIMES),PH
```

Five transactions are created at time 0. The first will have a 1 in parameter 1, the second a 2, the third a 3, and so on. This is one way that the programmer can generate transactions each having sequential numbers in parameter 1.

Later, it would be possible to have an `ADVANCE` Block with a `FUNCTION` such as

```
     TRAVEL     FUNCTION     PH1,M5
1,RVNORM(1,20,3)/2,RVNORM(1,18,1.5)/3,RVNORM(1,15,2.5)/
4,RVNORM(1,22,4)/5,RVNORM(1,15,2)
                ...
                ...
                ADVANCE      FN(TRAVEL)
```

where M5 is the specification for the entity list function.

Each of the five transactions would have a travel time obtained by sampling from a different normal distribution.

```
ASSIGN           Q(WAIT)+1,1.23,PL
```

will assign the value of 1.23 to the floating point parameter given by `Q(WAIT)+1`.

It is also possible to have the following:

```
ASSIGN           TOM,10,PH
```

Later, when reference to the parameter named TOM is made, it is done as follows:

```
ADVANCE           PH(TOM)
```

The transaction will be put on the FEC for a time of 10 since the value of the parameter TOM is 10.

Since parameters are SNAs, they can be used as operands. For example, consider the following lines of code.

```
(a) ADVANCE      PF4
(b) TEST E       PH1,PH4,DDOWN1
(c) QUEUE        PH1
(d) ENTER        TUGS,PB1
```

In (a), the transaction will be placed on the FEC for a time given by the transaction's full-word parameter number 4.

In (b), a test is made to see if the first and fourth half-word parameters are equal. If so, the transaction moves sequentially to the next Block. It they are not equal, the transaction is routed to the Block with the label DDOWN1.

In (c), the transaction joins the queue given by its first half-word parameter. This will be a number. Recall that, when the QUEUE Block was introduced, it was mentioned that the operand could be either a name or a number.

In (d), the transaction will enter the storage TUGS and use a storage amount as specified by its first bit-word parameter. If PB1 was 3, the storage would be decremented by 3.

In addition, consider the following:

```
  TIMES FUNCTION PH1,D4
1,100/2,125/3,150/4,175
```

Now, when a transaction enters the Block

```
ADVANCE    FN(TIMES)
```

it will be placed on the FEC for a time of either 100, 125, 150, or 175 time units depending on the value of its first half-word parameter.

19.3 A Caution in Using the ADVANCE Block

Suppose we wanted to have an ADVANCE Block* that would put a transaction on the FEC for a time given by 10 ± 2 but sampling from the function, and we had written it as

```
ADVANCE    FN(FIRST),FN(SECOND)
```

* This is basically a repeat of the caution for Chapter 18 where functions were used in the ADVANCE Blocks. However, it is important enough to bear repeating with parameters.

where the function FIRST gave the mean and the function SECOND gave the spread. This might be written as:

```
     FIRST FUNCTION    RN1,C2
0,8/1,12
     SECOND FUNCTION   RN1,C2
0,1/1,4
```

It would appear that the Block

```
ADVANCE         FN(FIRST),FN(SECOND)
```

would result in sampling from the correct distribution. However, GPSS/H will return the result of sampling from the function FIRST and then sampling from the function SECOND. However, the results are then *multiplied*. There are several ways to avoid this. One of the ways to avoid this is to use parameters. Consider the following code:

```
ASSIGN          1,FN(FIRST),PL
ASSIGN          2,FN(SECOND),PL
ADVANCE         PL1,PL2
```

This would put the transaction on the FEC for the required time.

19.4 The ASSIGN Block in Increment or Decrement Mode

If a parameter already has a value, the ASSIGN Block deletes this value and replaces it with the new one. You can add to (or subtract from) the value of a parameter by putting a plus (or minus) before the first comma in the operands:

```
ASSIGN     4+,5,PH
```

This will take the value in parameter 4 and add 5 to it.

```
ASSIGN     3-,6,PH
```

This will subtract 6 from the value in parameter 3.

```
ASSIGN     1+,Q(WAIT),PH
```

This will add the length of the queue WAIT to the transaction's first parameter. If the queue length was 4 and the value of PH1 was 12, its new value is now 16.

```
ASSIGN     7+,QX(WAIT1),PL
```

This will add the average waiting time of the non-zero entry transactions for the queue WAIT1 to the transaction's seventh floating point parameter.

19.5 The `TRANSFER ,FN`(*label*) Block

The third form of the TRANSFER Block is called the TRANSFER Function Block. Its form is easy to understand. It is

```
TRANSFER    ,FN(label)
```

where *(label)* refers to a label where the function is defined. This label can be either a legal alphanumeric or a number (SNA). For example,

```
        FIRST FUNCTION PH1,D3
1,BLOCK1/5,BLOCK2/6,BLOCK3
```

If half-word parameter 1 has the value 1, the transaction is routed to the Block with the label BLOCK1; if it has the value 5, the transaction is routed to the Block with the label BLOCK2; and if it has the value 6, to the Block with the label BLOCK3. In this example, care must be taken to ensure that the value of the half-word parameter is only one of the three values given in the function specification else an error might be obtained. The interested student may experiment with different values assigned to the half-word parameters to see what happens.

If one uses numbers in the function specification, these refer to Block numbers with the numbering being as given in the .LIS file. In the example given above, if one has written:

```
        FIRST FUNCTION PH1,D3
1,3/5,7/6,18
```

The transaction would be routed to the 3rd, 7th, and 18th Block in the program depending on the value of the half-word parameter. Again, care must be taken if this form of the TRANSFER Block is ever used.

The function specification is normally given at the beginning of the program where other statements and specifications are found, but GPSS/H allows them to be any place in the program—even after where the TRANSFER Block is located.

Example 19.1: Use of Parameters to Route Trucks in a Mine

This example illustrates how one might use parameters. Suppose a mine has trucks that come to a junction with three possible paths. A truck is to be routed to the Block with the label FIRST 25% of the time, to the Block with the label SECOND 45% of the time, and to the Block with the label THIRD 30% of the time. Each takes 5 time units on their respective paths. All transactions then meet up at the Block with the label NEXT. Assume trucks arrive at the junction every 6 time units.

The example uses the third form of the TRANSFER Block.

The program to do this is given by CHAP19A.GPS (Figure 19.1).

Output for the program is shown in Figure 19.2.

In the 300 time units the program ran, the Block FIRST was executed 11 times, the Block SECOND 23 times, and the Block THIRD 15 times. The expected numbers are 12, 22, and 15, respectively. As each transaction is generated, the value given to parameter

```
            SIMULATE
*****************************
*   PROGRAM CHAP19A.GPS       *
*   EXAMPLES OF PARAMETERS    *
*****************************
 WHAT       FUNCTION    RN1,D3
.25,1/.7,2/1,3
 WHERE      FUNCTION    PH1,L3
1,FIRST/2,SECOND/3,THIRD
            GENERATE    6,,,,,1PH    TRANSACTIONS CREATED
            ASSIGN      1,FN(WHAT),PH   PARAMETER 1 IS TRUCK TYPE
            TRANSFER    ,FN(WHERE)
 FIRST      ADVANCE     5
            TRANSFER    ,NEXT
 SECOND     ADVANCE     5
            TRANSER     ,NEXT
            TERMINATE
 THIRD      ADVANCE     5
 NEXT       TERMINATE
            GENERATE    300
            TERMINATE   1
            START       1
            PUTPIC      LINES=3,N(FIRST),N(SECOND),N(THIRD)
  TIMES TO FIRST    **
  TIMES TO SECOND   **
  TIMES TO THIRD    **
            END
```

FIGURE 19.1
Listing of program CHAP19A.GPS.

```
TIMES TO FIRST    11
TIMES TO SECOND   23
TIMES TO THIRD    15
```

FIGURE 19.2
Output from program CHAP19A.GPS.

1 is either a 1, a 2, or a 3. The function `WHERE` routes the transaction to the correct Block using the value of PH1 as the operand.

The animation is easily added. It is given by program CHAP19B.GPS (Figure 19.3). The animation is shown in Figure 19.4.

Example 19.2: A Small Harbour Problem

A small harbour has a single dock to unload and load ships. This dock can handle only one ship at a time. Ships arrive every 10.1 ± 5 hours, 24 hours a day. There are three types of ships: 30% are type 1, 50% type 2, and the rest type 3. Type 1 ships take 8 ± 4 hours to unload and load; type 2 ships take 10 ± 5 hours to unload and load; and type 3 ships take 12 ± 5.5 hours to unload and load. The owner of the dock has noticed that there always seem to be ships waiting in a queue for the dock to become free. The cost in lost time is $125 per hour for ships to wait for the dock to become free. He has obtained an estimate to modify the dock so that it can service two ships at a time. The cost of the modification would amount to the owner having to pay $6000 per day. Should he go ahead with these modifications?

Solution

Program CHAP19C.GPS illustrates how parameters can be used to greatly shorten the code in this example. Its listing is given in Figure 19.5.

```
           SIMULATE
*****************************
*  PROGRAM CHAP19B.GPS      *
*  ANIMATION OF CHAP19A.GPS *
*****************************
 ATF       FILEDEF      'CHAP19B.ATF'
 WHAT      FUNCTION     RN1,D3
.25,1/.7,2/1,3
 WHERE     FUNCTION     PH1,L3
1,FIRST/2,SECOND/3,THIRD
           GENERATE     6,,,,,1PH   TRANSACTIONS CREATED
           BPUTPIC      FILE=ATF,LINES=4,LINES=4,AC1,XID1,XID1,XID1
TIME *.****
CREATE TRUCK T*
PLACE T* ON P1
SET T* TRAVEL 5
           ADVANCE      5
           ASSIGN       1,FN(WHAT),PH   PARAMETER 1 IS TRUCK TYPE
           TRANSFER     ,FN(WHERE)
 FIRST     BPUTPIC      FILE=ATF,LINES=3,AC1,XID1,XID1
TIME *.****
PLACE T* ON P2
SET T* TRAVEL 5
           ADVANCE      5
           BPUTPIC      FILE=ATF,LINES=2,AC1,XID1
TIME *.****
DESTROY T*
           TERMINATE
 SECOND    BPUTPIC      FILE=ATF,LINES=3,AC1,XID1,XID1
TIME *.****
PLACE T* ON P3
SET T* TRAVEL 5
           ADVANCE      5
           BPUTPIC      FILE=ATF,LINES=2,AC1,XID1
TIME *.****
DESTROY T*
           TERMINATE
 THIRD     BPUTPIC     FILE=ATF,LINES=3,AC1,XID1,XID1
TIME *.****
PLACE T* ON P4
SET T* TRAVEL 6
           ADVANCE      6
           BPUTPIC     FILE=ATF,LINES=2,AC1,XID1
TIME *.****
DESTROY T*
           TERMINATE
           GENERATE     300
           TERMINATE    1
           START        1
           PUTPIC       FILE=ATF,LINES=2,AC1
TIME *.****
END
      PUTSTRING    ('  ')
      PUTSTRING    ('  ')
      PUTPIC       LINES=3,N(FIRST),N(SECOND),N(THIRD)
TIMES TO FIRST    **
TIMES TO SECOND   **
TIMES TO THIRD    **
      END
```

FIGURE 19.3
Listing of program CHAP19B.GPS.

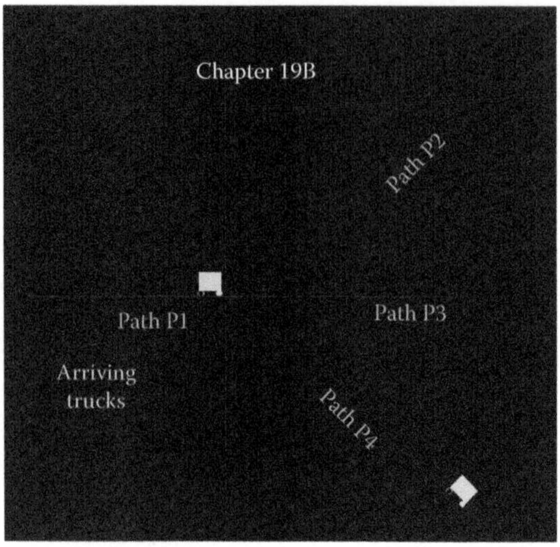

FIGURE 19.4
Animation from program CHAP19B.GPS.

```
              SIMULATE
**********************************
*  PROGRAM CHAP19C.GPS          *
*  EXAMPLE USING PARAMETERS     *
**********************************
 TYPE    FUNCTION    RN1,D3
.3,1/.8,2/1,3
 MEAN     FUNCTION    PH1,L3
1,8/2,10/3,12
 SPREAD   FUNCTION    PH1,L3
1,4/2,5/3,5.5
          GENERATE    10.1,5,,,,1PH,2PL    SHIPS ARRIVE
          ASSIGN      1,FN(TYPE),PH   WHAT TYPE SHIP?
          QUEUE       HARBOR
          SEIZE       DOCK
          DEPART      HARBOR
          ASSIGN      1,FN(MEAN),PL
          ASSIGN      2,FN(SPREAD),PL
************************************
*  NOTICE HOW THE ADVANCE BLOCK     *
*  IS WRITTEN                       *
************************************
          ADVANCE     PL1,PL2   UNLOAD AT THE DOCK
          RELEASE     DOCK
          TERMINATE
          GENERATE    480*3*100
          TERMINATE   1
          START       1
          PUTSTRING   ('  ')
          PUTSTRING   ('  ')
          PUTPIC      LINES=4,QA(HARBOR),QM(HARBOR),FR(DOCK)/10.
            RESULTS OF SIMULATION
        THE AVERAGE QUEUE IN THE HARBOR WAS **.**
        THE MAXIMUM QUEUE IN THE HARBOR WAS **
        THE DOCK WAS BUSY                   **.**% OF THE TIME
          END
```

FIGURE 19.5
Listing of program CHAPTER19C.GPS.

```
        RESULTS OF SIMULATION
    THE AVERAGE QUEUE IN THE HARBOR WAS  2.29
    THE MAXIMUM QUEUE IN THE HARBOR WAS 12
    THE DOCK WAS BUSY                   97.02% OF THE TIME
```

FIGURE 19.6
Results of simulation program CHAPTER19C.GPS.

As each ship arrives at the harbour, it is assigned a 1, 2, or 3 in half-word parameter 1 to indicate which type it is. This is done by the function TYPE. When a ship is ready to unload and load, two other ASSIGN Blocks are used to place the mean and spread of its unloading and loading times into real parameters 1 and 2, respectively. The Block ADVANCE PL1,PL2 represents its time in the dock. This form of the ADVANCE Block is necessary, because, as mentioned previously in Chapter 18, if the ASSIGN Block has two functions as operands, it multiples their results. This would lead to an erroneous result.

The output from the program is shown in Figure 19.6.

The results of the program indicate that there is an average of 2.29 ships waiting for the dock to become free. This represents a loss of 2.29 times $125 times 24 or $6870 each day. Before accepting this as justification for making the changes to the dock, it is necessary to modify the program to determine if there will be ships waiting for the dock *after* the modifications. This modification is rapidly done by changing the dock from a facility to storage. CHAP19D.GPS gives these changes. Its listing (Figure 19.7) is as follows:

```
          SIMULATE
*******************************
*  PROGRAM CHAP19D.GPS        *
*  MODIFICATION OF PROGRAM    *
*  CHAP19C.GPS                *
*******************************
 TYPE     FUNCTION    RN1,D3
.3,1/.8,2/1,3
 MEAN     FUNCTION    PH1,L3
1,8/2,10/3,12
 SPREAD   FUNCTION    PH1,L3
1,4/2,5/3,5.5
          STORAGE     S(DOCK),2         DOCK CAN HANDLE TWO SHIPS
          GENERATE    10.1,5,,,,1PH,2PL     SHIPS ARRIVE
          ASSIGN      1,FN(TYPE),PH   WHAT TYPE SHIP?
          QUEUE       HARBOR
          ENTER       DOCK
          DEPART      HARBOR
          ASSIGN      1,FN(MEAN),PL
          ASSIGN      2,FN(SPREAD),PL
          ADVANCE     PL1,PL2      AT THE DOCK
          LEAVE       DOCK
          TERMINATE
          GENERATE    480*3*100
          TERMINATE   1
          START       1
          PUTSTRING   ('  ')
          PUTSTRING   ('  ')
          PUTPIC      LINES=4,QA(HARBOR),QM(HARBOR),SR(DOCK)/10.
            RESULTS OF SIMULATION
        THE AVERAGE QUEUE IN THE HARBOR WAS **.**
        THE MAXIMUM QUEUE IN THE HARBOR WAS **
        THE DOCK WAS BUSY                   **.**% OF THE TIME
          END
```

FIGURE 19.7
Listing of program CHAP19D.GPS.

```
        RESULTS OF SIMULATION
    THE AVERAGE QUEUE IN THE HARBOR WAS  0.00
    THE MAXIMUM QUEUE IN THE HARBOR WAS  1
    THE DOCK WAS BUSY                   48.89% OF THE TIME
```

FIGURE 19.8
Results of program CHAP19D.GPS.

Figure 19.8 gives the results of running this program.

It can be seen that there will not be a queue if the dock can handle two ships. Thus, the owner of the dock should make the modifications.

The program to do the animation of program CHAP19C is stored as CHAP19E.GPS. Its listing is given in Figure 19.9.

The animation of this example is given in Figure 19.10.

Notice that there are six Classes in the animation for the three ships in the program to make the .ATF file. That is because, when the ships are done at the dock and ready to leave, they are put on path P2. They are to go from left to right on the screen so that they

```
          SIMULATE
************************************
*  PROGRAM CHAP19E.GPS             *
*  ANIMATION FOR SHIP EXAMPLE      *
************************************
  ATF      FILEDEF     'CHAP19E.ATF'
  TYPE     FUNCTION    RN1,D3
.3,1/.8,2/1,3
 MEAN      FUNCTION    PH1,L3
1,8/2,10/3,12
 SPREAD    FUNCTION    PH1,L3
1,4/2,5/3,5.5
          GENERATE     10.1,5,,,,1PH,2PL     SHIPS ARRIVE
          ASSIGN       1,FN(TYPE),PH   WHAT TYPE SHIP?
          BPUTPIC      FILE=ATF,LINES=4,AC1,PH1,XID1,XID1,XID1
TIME *.****
CREATE SHIP* S*
PLACE S* ON P1
SET S* TRAVEL 2
          ADVANCE      2
          QUEUE        HARBOR
          SEIZE        DOCK
          DEPART       HARBOR
          BPUTPIC      FILE=ATF,LINES=3,AC1,XID1,QA(HARBOR)
TIME *.****
PLACE S* AT 36 25
WRITE M2 *.**
          ASSIGN       1,FN(MEAN),PL
          ASSIGN       2,FN(SPREAD),PL
          ADVANCE      PL1,PL2       AT THE DOCK
          BPUTPIC      FILE=ATF,LINES=4,AC1,XID1,XID1,PH1,XID1
TIME *.****
PLACE S* ON P2
SET S* CLASS SHIP*A
SET S* TRAVEL 3
          ADVANCE      .2
          RELEASE      DOCK
          BPUTPIC      FILE=ATF,LINES=2,AC1,FR(DOCK)/10.
```

FIGURE 19.9
Listing of program CHAP19E.GPS.

(*Continued*)

```
TIME *.****
WRITE M1 **.**%
          ADVANCE      2.8
          BPUTPIC      FILE=ATF,LINES=2,AC1,XID1
TIME *.****
DESTROY S*
          TERMINATE
          GENERATE     480*3*100
          TERMINATE    1
          START        1
          PUTSTRING    ('  ')
          PUTSTRING    ('  ')
          PUTPIC       LINES=4,QA(HARBOR),QM(HARBOR),FR(DOCK)/10.
            RESULTS OF SIMULATION
        THE AVERAGE QUEUE IN THE HARBOR WAS **.**
        THE MAXIMUM QUEUE IN THE HARBOR WAS **
        THE DOCK WAS BUSY                   **.**% OF THE TIME
          PUTPIC       FILE=ATF,LINES=2,AC1
TIME *.****
END
          END
```

FIGURE 19.9 (Continued)
Listing of program CHAP19E.GPS.

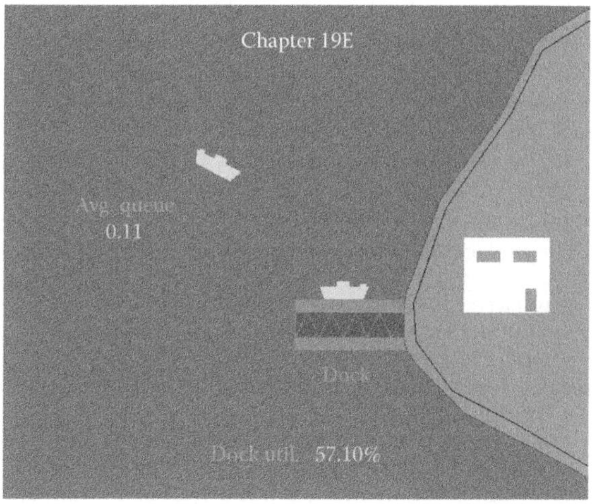

FIGURE 19.10
Animation of Example 19.2. This is given by program CHAP19E.GPS.

would be positioned upside down in the animation. The first ship has the Class `SHIP1`. The Class `SHIP1A` is the same ship but upside down so that it will be right side up on the path. The code to do this is worth examining. It is:

```
BPUTPIC   FILE = ATF,LINES = 4,AC1,XID1,XID1,PH1,XID1
TIME *.****
PLACE S* ON P2
SET S* CLASS SHIP*A
SET S* TRAVEL 3
```

The line: `SET S* CLASS SHIP*A`. This sets the ship, which is now the object, S* equal to the new Class, namely, `SHIP(PH1)A`, where `PH1` is either 1, 2, or 3.

Example 19.3: Shopping in a Warehouse

A warehouse has four long tables in a series where customers walk along with a shopping trolley and select different merchandise. There is one checkout person at the end of the fourth table who checks out each customer one at a time. Further, 15% of the people entering the warehouse go immediately to the checkout counter where they purchase from 2 to 5 items. This takes 50 ± 10 seconds. Everyone who shops in the warehouse, except those who go immediately to the checkout counter, takes a shopping trolley from the trolley stand. The checkout takes 4.25 seconds per item. Customers arrive according to the normal distribution with a mean of 90 seconds and a standard distribution of 15.

The number of items selected by each person who goes down each aisle is given by the following:

Bench	No. Items Selected	Time to Shop
1	3 ± 2	120 ± 10
2	4 ± 3	Normal distribution mean of 80 and standard deviation of 8
3	3 ± 1	Normal distribution mean of 115 and standard deviation of 12
4	5 ± 4	115 ± 12

The 15% of the customers who bypass the tables go to a separate table and select 5 ± 2 items. They then join the queue at the checkout counter. Checkout time is 4.25 seconds per item no matter what this item is.

Determine how busy the checkout person is. Also, the manager of the warehouse would like to know how many shopping carts to have so that no customer enters the store and finds the cart stand empty. He is worried that customers who arrive to find no shopping carts will either leave and go elsewhere to shop or not purchase as many goods as they would if they had a shopping cart.

The program to do the simulation is given next as CHAP19F.GPS (Figure 19.11).

Notes on the program are as follows:

The customers who do not shop along the four tables are transferred to the Block with the label AROUND. They select the items from the function FIFTH. The others take a cart.

```
         SIMULATE
****************************************
*  PROGRAM CHAP19F.GPS    SHOPPING      *
*  IN A STORE - BASED ON SIMILAR        *
*  MODIFIED BY SCHRIBER, SIMULATION     *
*  USING GPSS, JOHN WILEY & SONS, NY    *
****************************************
         STORAGE     S(CARTS),10000  PROVIDE PLENTY OF CARTS
  FIRST  FUNCTION    RN1,C2
0,1/1,6
  SECOND FUNCTION    RN1,C2
0,1/1,8
 THIRD   FUNCTION    RN1,C2
0,2/1,5
 FOURTH  FUNCTON     RN1,C2
0,1/1,10
 FIFTH   FUNCTION    RN1,C2
0,3/1,8
         REAL        &X
         GENERATE    RVNORM(1,90,15),,,,,3PH
         TRANSFER    .15,,AROUND
         ENTER       CARTS
```

FIGURE 19.11
Listing of program CHAP19F.GPS. *(Continued)*

```
          ADVANCE     120,10              FIRST BENCH
          ASSIGN      1,FN(FIRST),PH      ADD TO CARTS
          ADVANCE     RVNORM(1,80,8)      SECOND BENCH
          ASSIGN      1+,FN(SECOND),PH
          ADVANCE     RVNORM(1,115,12)    THIRD BENCH
          ASSIGN      1+,FN(THIRD),PH
          ADVANCE     115,10              FOURTH BENCH
          ASSIGN      1+,FN(FOURTH),PH
BACKUP    QUEUE       CHECKOUT
          SEIZE       CHECK
          DEPART      CHECKOUT
          BLET        &X=PH1
          ADVANCE     4.25*&X
          RELEASE     CHECK
          TEST E      PH2,0,AWAY
          LEAVE       CARTS                PUT CART BACK
          TERMINATE
AWAY      TERMINATE                        PEOPLE WITH NO CARTS
AROUND    ADVANCE     50,10
          ASSIGN      1,FN(FIFTH),PH
          ASSIGN      2,1,PH
          TRANSFER    ,BACKUP
          GENERATE    3600*1000*8
          TERMINATE   1
          START       1
          PUTSTRING   ('  ')
          PUTSTRING   ('  ')
          PUTPIC      LINES=3,FR(CHECK)/10.,SM(CARTS)
              RESULTS OF SIMULATION OF WAREHOUSE
       CHECKOUT PERSON WAS BUSY     **.**% OF THE TIME
       MAXIMUM CARTS IN USE WAS ***
          END
```

FIGURE 19.11 (Continued)
Listing of program CHAP19F.GPS.

Since the maximum number of carts is to be determined, an arbitrarily large number is provided. In this case, 10,000. If no storage had been specified, the following message would appear on the screen:

```
***Warning: The following Storage may be undefined: (Capacity of 2**31-1
assumed)
            Storages: CARTS
```

With an assigned storage of 10,000, this message is avoided. The results would have been the same had no storage been used.

The total number of items selected from each table is given by the four functions and the `ASSIGN` Block to store the amounts in each transaction's half-word parameter number 1. If real (floating point) parameters had been used, the number of items taken from each table would be some fraction and would have taken slightly more time to check out than the actual case. This would lead to an incorrect answer. Such errors are the hardest to find. At the checkout, the number of items selected is converted to a real number and the time taken to check out is a real number as it takes 4.25 seconds per item. The results of the simulation are shown in Figure 19.12.

```
                RESULTS OF SIMULATION OF WAREHOUSE
        CHECKOUT PERSON WAS BUSY     63.77% OF THE TIME
        MAXIMUM TROLLEYS IN USE WAS   8
```

FIGURE 19.12
Results of program CHAP19F.GPS.

The checkout person was busy 63.77% of the time, and the maximum number of carts in use at any one time was 8. The simulation was run for 1000 8-hour shifts, so it appears to be safe to say that 10 shopping carts are sufficient to have available. Later, a method will be given to show the statistical distribution of the number of carts in use during the simulation.

The animation is obtained from the program CHAP19G.GPS. The program is as follows (Figure 19.13):

```
           SIMULATE
**********************************
*   ANIMATION OF CHAP19G.GPS.    *
**********************************
  ATF      FILEDEF      'CHAP19G.ATF'
           STORAGE      S(CARTS),100000  PROVIDE PLENTY OF CARTS
  FIRST    FUNCTION     RN1,C2
0,1/1,6
  SECOND   FUNCTION     RN1,C2
0,1/1,8
 THIRD     FUNCTION     RN1,C2
0,2/1,5
 FOURTH    FUNCTON      RN1,C2
0,1/1,10
 FIFTH     FUNCTION     RN1,C2
0,3/1,8
 TIMEOUT   FUNCTION     PH2,D2
0,NEXT2/1,NEXT3
           INTEGER      &I,&TOTAL,&J,&K,&L,&SERVED
           REAL         &X,&Y,&Z,&U,&V,&XX
           GENERATE     RVNORM(1,90,15),,20,,,3PH
           BLET         &I=24*FRN1+1
           BPUTPIC      FILE=ATF,LINES=5,AC1,XID1,XID1,XID1,&I,XID1
TIME *.****
CREATE CUST C*
PLACE C* ON P1
SET C* COLOR F*
SET C* TRAVEL  20
           ADVANCE      20
           TRANSFER     .15,,AROUND
           ENTER        CARTS
           BPUTPIC      FILE=ATF,LINES=2,AC1,S(CARTS)
TIME *.****
WRITE M10
* FIRST BENCH
           BLET         &X=20*FRN1+110
           BPUTPIC      FILE=ATF,LINES=3,AC1,XID1,XID1,&X
TIME *.****
PLACE C* ON P2
SET C* TRAVEL **.**
           ADVANCE      &X     FIRST BENCH  BENCH 1
           ASSIGN       1,FN(FIRST),PH    WHAT IS IN BASKET?
           BLET         &TOTAL=&TOTAL+PH1
           BPUTPIC      FILE=ATF,LINES=3,AC1,PH1,&TOTAL
TIME *.****
WRITE M1 *
WRITE M10 ***
**************************** END OF FIRST BENCH
* SECOND BENCH
           BLET         &Y=RVNORM(1,80,8)
           BPUTPIC      FILE=ATF,LINES=3,AC1,XID1,XID1,&Y
```

FIGURE 19.13
Listing of program CHAP19G.GPS. *(Continued)*

```
TIME *.****
PLACE C* ON P3
SET C* TRAVEL **.**
          ADVANCE      &Y
          BLET         &J=FN(SECOND)  HOW MANY FROM SECOND
          BLET         &TOTAL=&TOTAL+&J     ADD TO TOTAL
          ASSIGN       1+,&J,PH     ADD TO CART
          BPUTPIC      FILE=ATF,LINES=3,AC1,&J,&TOTAL
TIME *.****
WRITE M2 *
WRITE M10 ***
***************************** END OF SECOND BENCH
* THIRD BENCH
          BLET         &Z=RVNORM(1,115,12)    THIRD BENCH
          BPUTPIC      FILE=ATF,LINES=3,AC1,XID1,XID1,&Z
TIME *.****
PLACE C* ON P4
SET C* TRAVEL **.**
          ADVANCE      &Z     THIRD BENCH
          BLET         &K=FN(THIRD)
          ASSIGN       1+,&K,PH
          BLET         &TOTAL=&TOTAL+&K
          BPUTPIC      FILE=ATF,LINES=3,AC1,&K,&TOTAL
TIME *.****
WRITE M3 *
WRITE M10 ***
***************************** END OF THIRD BENCH
* FOURTH BENCH
          BLET         &U=24*FRN1+103
          BPUTPIC      FILE=ATF,LINES=3,AC1,XID1,XID1,&U
TIME *.****
PLACE C* ON P5
SET C* TRAVEL **.**
          ADVANCE      &U     FOURTH BENCH
          BLET         &L=FN(FOURTH)
          ASSIGN       1+,&L,PH
          BLET         &TOTAL=&TOTAL+&L
          BPUTPIC      FILE=ATF,LINES=3,AC1,&L,&TOTAL
TIME *.****
WRITE M4 *
WRITE M10 ***
          SEIZE        DUMMY
          BPUTPIC      FILE=ATF,LINES=3,AC1,XID1,XID1
TIME *.****
PLACE C* ON P6
SET C* TRAVEL 10
          ADVANCE      10
          RELEASE      DUMMY
 BACKUP   QUEUE        CHECKOUT
          BPUTPIC      FILE=ATF,LINES=3,AC1,XID1,XID1
TIME *.****
PLACE C* ON P7
SET C* TRAVEL 20
          ADVANCE      20
          SEIZE        CHECK
          DEPART       CHECKOUT
          TRANSFER     ,FN(TIMEOUT)
```

FIGURE 19.13 (Continued)
Listing of program CHAP19G.GPS.

(Continued)

```
  NEXT2   BLET        &XX=PH1
          ADVANCE     4.25*&XX
          TRANSFER    ,NEXT5
  NEXT3   BLET         &XX=PH3
          ADVANCE     4.25*&XX
  NEXT5   RELEASE      CHECK
          BPUTPIC     FILE=ATF,LINES=4,AC1,XID1,XID1,FR(CHECK)/10.
TIME *.****
PLACE C* ON P8
SET C* TRAVEL 15
WRITE M11 **.**%
          ADVANCE     15
          TEST E      PH2,0,AWAY
          LEAVE       CARTS           PEOPLE WITH CARTS
          BPUTPIC     FILE=ATF,LINES=2,AC1,SM(CARTS)
TIME *.****
WRITE M21 **
 AWAY     TRANSFER    .5,,AWAY2
          BPUTPIC     FILE=ATF,LINES=3,AC1,XID1,XID1
TIME *.****
PLACE C* ON P10
SET C* TRAVEL 30
          ADVANCE     30
          BLET        &SERVED=&SERVED+1
          BPUTPIC     FILE=ATF,LINES=3,AC1,XID1,&SERVED
TIME *.****
DESTROY C*
WRITE M12 ***
          TERMINATE
 AWAY2    BPUTPIC     FILE=ATF,LINES=3,AC1,XID1,XID1
TIME *.****
PLACE C* ON P11
SET C* TRAVEL 30
          ADVANCE     30
          BLET        &SERVED=&SERVED+1
          BPUTPIC     FILE=ATF,LINES=3,AC1,XID1,&SERVED
TIME *.****
DESTROY C*
WRITE M12 ***
          TERMINATE                PEOPLE WITH NO CARTS
 AROUND   BLET        &V=30*FRN1+10
          BPUTPIC     FILE=ATF,LINES=3,AC1,XID1,XID1,&V
TIME *.****
PLACE C* ON P9
SET C* TRAVEL **.**
          ADVANCE     &V
          BPUTPIC     FILE=ATF,LINES=3,AC1,XID1,XID1
TIME *.****
PLACE C* ON P12
SET C* TRAVEL 14
          ADVANCE     14
          ASSIGN      3,FN(FIFTH),PH
          BPUTPIC     FILE=ATF,LINES=2,AC1,PH3
TIME *.****
```

FIGURE 19.13 (Continued)
Listing of program CHAP19G.GPS.

(Continued)

```
WRITE M13 *
          ASSIGN       2,1,PH
          SEIZE        DUMMY
          BPUTPIC      FILE=ATF,LINES=3,AC1,XID1,XID1
TIME *.****
PLACE C* ON P13
SET C* TRAVEL 10
          ADVANCE      10
          RELEASE      DUMMY
          TRANSFER     ,BACKUP
          GENERATE     3600
          TERMINATE    1
          START        1
          PUTPIC       FILE=ATF,LINES=2,AC1
TIME *.****
END
         PUTPIC       LINES=3,FR(CHECK)/10.,SM(CARTS)
              RESULTS OF SIMULATION OF WAREHOUSE
      CHECKOUT PERSON WAS BUSY      **.**% OF THE TIME
      MAXIMUM CARTS IN USE WAS ***
          END
```

FIGURE 19.13 (Continued)
Listing of program CHAP19G.GPS.

This is quite a complex program and should be studied carefully. The customers who bypass the four tables need to be separated from the other customers. They are given the value of 1 in parameter 2. Thus, at the checkout counter, the function `TIMEOUT` is introduced to route the customers so that the checkout person takes the correct amount of time. The program was run for 3600 time units, as only the animation is desired. The output code has also been removed.

The animation is shown in Figure 19.14.

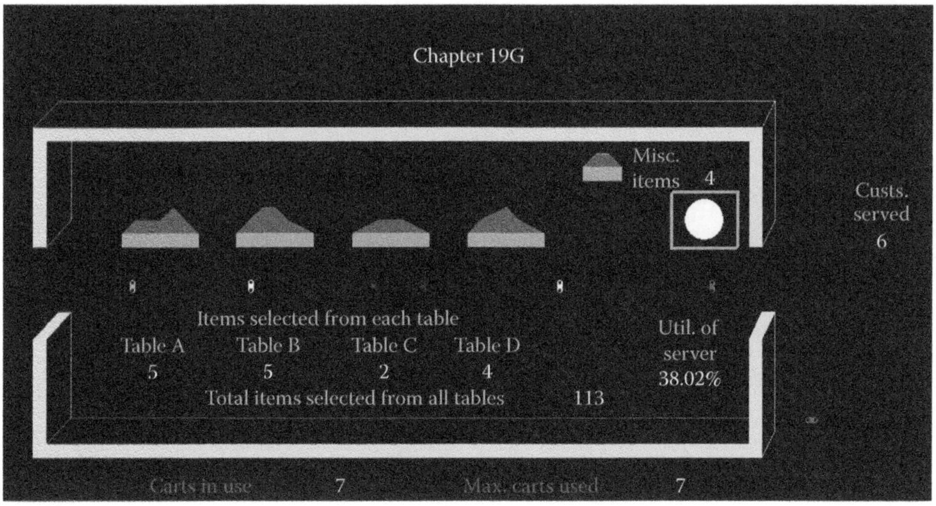

FIGURE 19.14
Animation of Example 19.4. This is given by program CHAP19G.GPS.

Example 19.4: People Spending Money at Four Stations

People enter a system where they are given an amount of money to spend as they travel through the system. The system consists of four stations where the people purchase items. To begin with, 10% of the people who pass through the system have 10 units of money to spend, 30% have 12 units, 35% have 13 units and 25%, have 14 units. All people pass through the stations in consecutive order. Once all their money is gone, they must leave the system. Table 19.1 gives the number of items a person would purchase at each of the four stations and the probabilities associated with each purchase:

For 1000 people who enter this system, how many make it all the way through with either some or zero money left?

Solution

Computer program CHAP19H.GPS simulates the system for 1000 people. This program uses names for the parameters instead of a number to illustrate how this form is used. A listing of the program is shown in Figure 19.15.

The solution is given in Figure 19.16.

The output shows that, for 1000 people to enter the system, only 25 made it all the way through without spending all their money. None left after the first station. This has to be the case, since everyone has at least 10 units of money to spend and the most that can be spent at the first station is 7 units. However, this solution, while correct, is hardly illuminating. The animation is a much better way to present the system and the solution. The program for the animation is CHAPTER19I.GPS. Its listing is as follows (Figure 19.17).

The animation is shown in Figure 19.18. This animation shows the power of presenting simulation results in animation form.

TABLE 19.1

Number of Items and Associated Probabilities for Example 19.5

Station	No. Items to Purchase	Probability (percent)
First	3	20
	5	50
	7	30
Second	5	40
	6	60
Third	2	25
	3	15
	4	40
	6	10
Fourth	2	33
	3	33
	5	33

```
          SIMULATE
          GENERATE    10          PEOPLE COME EVERY 10 TIME UNITS
 AMOUNT   FUNCTION    RN1,D4
.1,10/.4,12/.75,13/1,14
 FIRST    FUNCTION    RN1,D3
.2,3/.7,5/1,7
 SECOND   FUNCTION    RN1,D2
.4,5/1,6
 THIRD    FUNCTION    RN1,D4
.25,2/.5,3/.9,4/1,6
 FOURTH   FUNCTION    RN1,D3
.333,2/.666,3/1,5
          ASSIGN      TOTAL,FN(AMOUNT),PH  HOW MUCH IS GIVEN
          ADVANCE     5,2   TO STATION A
          BLET        PH(TOTAL)=PH(TOTAL)-FN(FIRST) FIRST STATION
          TEST GE     PH(TOTAL),0,AWAY1
          ADVANCE     5,2   TO STATION B
          BLET        PH(TOTAL)=PH(TOTAL)-FN(SECOND) SECOND STATION
          TEST GE     PH(TOTAL),0,AWAY2
          ADVANCE     5,2
          BLET        PH(TOTAL)=PH(TOTAL)-FN(THIRD)  THIRD STATION
          TEST GE     PH(TOTAL),0,AWAY3
          ADVANCE     5,2
          BLET        PH(TOTAL)=PH(TOTAL)-FN(FOURTH)  FOURTH STATION
          TEST GE     PH(TOTAL),0,AWAY4
 MADEIT   TERMINATE   1
 AWAY1    TERMINATE   1
 AWAY2    TERMINATE   1
 AWAY3    TERMINATE   1
 AWAY4    TERMINATE   1
          START       1000
          PUTSTRING   ('  ')
          PUTSTRING   ('  ')
          PUTPIC      LINES=5,N(AWAY1),N(AWAY2),N(AWAY3),N(AWAY4),N(MADEIT)
        NUMBER TO MAKE IT ONLY UP TO STATION A  ***
        NUMBER TO MAKE IT ONLY UP TO STATION B  ***
        NUMBER TO MAKW IT ONLY UP TO STATION C  ***
        NUMBER TO MAKE IT ONLY UP TO STATION D  ***
        NUMBER TO MAKE IT ALL THE WAY THROUGH   ***
          END
```

FIGURE 19.15
Listing of program CHAP19H.GPS.

```
NUMBER TO MAKE IT ONLY UP TO STATION A    0
NUMBER TO MAKE IT ONLY UP TO STATION B  103
NUMBER TO MAKE IT ONLY UP TO STATION C  606
NUMBER TO MAKE IT ONLY UP TO STATION D  266
NUMBER TO MAKE IT ALL THE WAY THROUGH    25
```

FIGURE 19.16
Output from program CHAPTER19H.GPS.

```
          SIMULATE
***********************************
*  PROGRAM CHAP19I.GPS            *
*  ANIMATION OF EXAMPLE 19.4      *
***********************************
  ATF    FILEDEF     'CHAPTER19I.ATF'
          INTEGER     &I,&MADEIT,&COUNTA,&COUNTB,&COUNTC,&COUNTD
          REAL        &X,&Y,&Z,&PCT
  AMOUNT  FUNCTION    RN1,D4
.1,10/.4,12/.75,13/1,14
 FIRST    FUNCTION    RN1,D3
.2,3/.7,5/1,7
 SECOND   FUNCTION    RN1,D2
.4,5/1,6
 THIRD    FUNCTION    RN1,D4
.25,2/.5,3/.9,4/1,6
 FOURTH   FUNCTION    RN1,D3
.333,2/.666,3/1,5
 PEOPLE   GENERATE    10,,0
          BLET        &I=24*FRN1+1
          BPUTPIC     FILE=ATF,LINES=5,AC1,XID1,XID1,XID1,XID1,&I
TIME *.****
CREATE PERSON P*
PLACE P* ON P1
SET P* TRAVEL 3
SET P* COLOR F*
          ADVANCE     3
          ASSIGN      TOTAL,FN(AMOUNT),PH
          BLET        PH(STOTAL)=PH(TOTAL)
          BPUTPIC     FILE=ATF,LINES=3,AC1,N(PEOPLE),PH(TOTAL)
TIME *.****
WRITE M6 ***
WRITE M7  **
          BLET        PH(TOTAL)=PH(TOTAL)-FN(FIRST)
          TEST GE     PH(TOTAL),0,AWAY1
          BLET        &X=5*FRN1+3
          BPUTPIC     FILE=ATF,LINES=3,AC1,XID1,XID1,&X
TIME *.****
PLACE P* ON P2
SET P* TRAVEL *.**
          ADVANCE     &Y    TO STATION B
          BLET        PH(TOTAL)=PH(TOTAL)-FN(SECOND)
          TEST GE     PH(TOTAL),0,AWAY2
          BLET        &Y=5*FRN1+2
          BPUTPIC     FILE=ATF,LINES=3,AC1,XID1,XID1,&Y
TIME *.****
PLACE P* ON P3
SET P* TRAVEL *.**
          ADVANCE     &Y
          BLET        PH(TOTAL)=PH(TOTAL)-FN(THIRD)
          TEST GE     PH(TOTAL),0,AWAY3
          BLET        &Z=5*FRN1+2
          BPUTPIC     FILE=ATF,LINES=3,AC1,XID1,XID1,&Z
TIME *.****
```

FIGURE 19.17
Listing of program CHAP19I.GPS.

(Continued)

```
PLACE P* ON P4
SET P* TRAVEL *.**
          ADVANCE      &Z
          BLET         PH(TOTAL)=PH(TOTAL)-FN(FOURTH)
          TEST GE      PH(TOTAL),0,AWAY4
          BLET         &MADEIT=&MADEIT+1
          BLET         &PCT=&MADEIT
          BLET         &PCT=&PCT/N(PEOPLE)
          BPUTPIC      FILE=ATF,LINES=7,AC1,XID1,XID1,&MADEIT,PH(STOTAL),_
                       PH(TOTAL),&PCT*100
TIME *.****
PLACE P* ON P5
SET P* TRAVEL 3
WRITE M5 **
WRITE M8 **
WRITE M9 **
WRITE M10 **.**%
          ADVANCE      3
          BPUTPIC      FILE=ATF,LINES=2,AC1,XID1
TIME *.****
DESTROY P*
          TERMINATE    1
 AWAY1    BPUTPIC      FILE=ATF,LINES=3,AC1,XID1,XID1
TIME *.****
PLACE P* ON P6
SET P* TRAVEL 3
          ADVANCE      3
          BLET         &COUNTA=&COUNTA+1
          BPUTPIC      FILE=ATF,LINES=3,AC1,&COUNTA,XID1
TIME *.****
WRITE M1 **
DESTROY P*
          TERMINATE    1
 AWAY2    BPUTPIC      FILE=ATF,LINES=3,AC1,XID1,XID1
TIME *.****
PLACE P* ON P7
SET P* TRAVEL 3
          ADVANCE      3
          BLET         &COUNTB=&COUNTB+1
          BPUTPIC      FILE=ATF,LINES=3,AC1,&COUNTB,XID1
TIME *.****
WRITE M2 **
DESTROY P*
          TERMINATE    1
 AWAY3    BPUTPIC      FILE=ATF,LINES=3,AC1,XID1,XID1
TIME *.****

PLACE P* ON P8
SET P* TRAVEL 3
          ADVANCE      3
          BLET         &COUNTC=&COUNTC+1
          BPUTPIC      FILE=ATF,LINES=3,AC1,&COUNTC,XID1
TIME *.****
WRITE M3 **
```

FIGURE 19.17 (Continued)
Listing of program CHAP19I.GPS.

(Continued)

```
DESTROY P*
          TERMINATE   1
 AWAY4    BPUTPIC     FILE=ATF,LINES=3,AC1,XID1,XID1
TIME *.****
PLACE P* ON P9
SET P* TRAVEL 3
          ADVANCE    3
          BLET       &COUNTD=&COUNTD+1
          BPUTPIC    FILE=ATF,LINES=3,AC1,&COUNTD,XID1
TIME *.****
WRITE M4 **
DESTROY P*
          TERMINATE   1
          START       100
          PUTPIC      FILE=ATF,LINES=2,AC1
TIME *.****
END
          PUTSTRING   ('  ')
          PUTSTRING   ('  ')
          PUTPIC      LINES=5,N(AWAY1),N(AWAY2),N(AWAY3),N(AWAY4),&MADEIT
      NUMBER TO MAKE IT ONLY UP TO STATION A  ***
      NUMBER TO MAKE IT ONLY UP TO STATION B  ***
      NUMBER TO MAKW IT ONLY UP TO STATION C  ***
      NUMBER TO MAKE IT ONLY UP TO STATION D  ***
      NUMBER TO MAKE IT ALL THE WAY THROUGH   ***
          END
```

FIGURE 19.17 (Continued)
Listing of program CHAP19I.GPS.

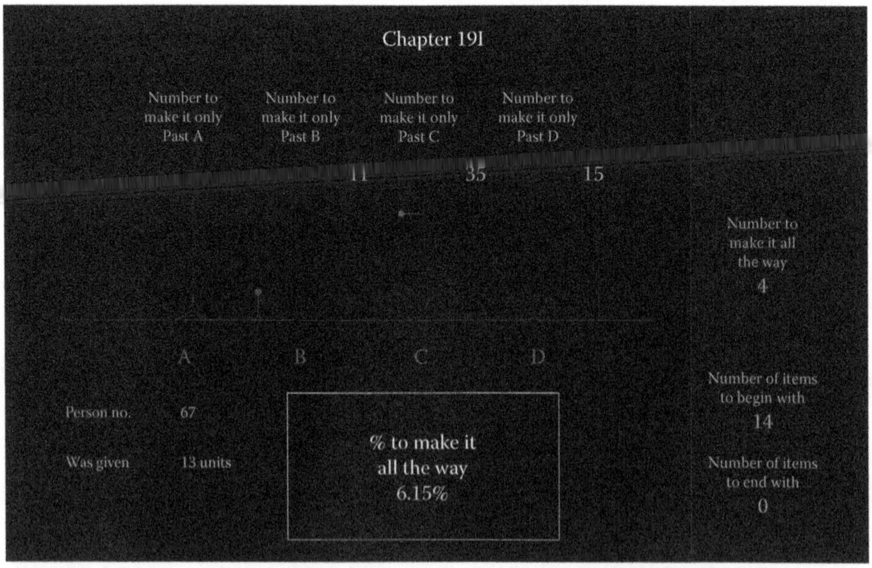

FIGURE 19.18
Animation of Example 19.4.

19.6 General Form of the ASSIGN Block

There is a more general form of the `ASSIGN` Block that is not used often. At one time, common distributions, such as the normal and the exponential, were not built in. This was then the only way to assign values to operand B from such functions. It will be presented here for the sake of being complete.

This is as follows:

```
ASSIGN   A,B,C,D
```

Operands A and B have their usual meaning. C, however, is the name or number of a function. D is of the type that parameter A is. If C is omitted, then D takes its place and we have the `ASSIGN` Block presented previously. If one used all four operands, the effect is as follows:

1. The function as specified by the C operand is evaluated. If it returns a decimal, the value is truncated.
2. This value is then multiplied by the number in the B operand.
3. The result of the multiplication in (2) is placed in the transaction's parameter as specified by the A operand.

For example,

```
ASSIGN   3,6,5,PH
```

The function defined with the label 5 is evaluated. Suppose the result is 2. This is multiplied by 6 and the result, namely, 12 is placed in the transaction's third half-word parameter.

```
ASSIGN   1,3,FIRST,PF
```

The function `FIRST` is evaluated. Suppose the number returned is 2.9543. This is truncated to 2 and multiplied by 3. The result, 6, is placed in the transaction's 1st full-word parameter.

Notice that the form is as given above. If you had put

```
ASSIGN   1,3,FN(FIRST),PF
```

a run time error would result.

19.7 Exercises

19.1. A transaction has values 1 and 2 in half-word parameters 1 and 2, respectively. State what will happen when the transaction enters each Block. Also, the functions `FIRST` and `SECOND` have been defined as follows:

```
FIRST FUNCTION   PH1,D2
1,6/2,10
```

```
SECOND FUNCTION   PH2,D2
1,2/2,3
```

(a) `ADVANCE` `PH1*60`
(b) `ADVANCE` `FN(FIRST)*FN(SECOND)`
(c) `QUEUE` `PH2`
(d) `ENTER` `TUGS,PH1`
(e) `LEAVE` `SHIP,PH1*PH2`
(f) `ADVANCE` `FN(FIRST),PH2`
(g) `ADVANCE` `10,FN(FIRST)`
(h) `ASSIGN` `3,PH1,PH`
(i) `ASSIGN` `4,PH2,PH`
(j) `ADVANCE` `PH3*PH4`
(k) `ASSIGN` `3,PH1+PH2,PH`
(l) `ASSIGN` `4,PH2,PH`
(m) `ASSIGN` `PH2,PH2+PH1*PH2,PH`

19.2. Assume that the following entities hold at a particular time in a program.

Entity	Value
F(MACH1)	1
Q(WAIT1)	2
Q(WAIT2)	3
FR(MACH1)	578
S(TUGS)	5
SC(TUGS)	123
R(TUGS)	2

A transaction has the value of its first six half-word parameters as follows. All other parameter values are 0.

Parameter Number	Value
1	3
2	–1
3	–2
4	5
5	4
6	6

State what happens if the transactions were to enter each of the following independent Blocks:

(a) `ASSIGN` `4,F(MACH1),PH`
(b) `ASSIGN` `4+,F(MACH1),PH`
(c) `ASSIGN` `PH4,F(MACH1),PH`
(d) `ASSIGN` `PH4-,F(MACH1),PH`
(e) `ASSIGN` `1,PH1*PH5,PH`
(f) `ASSIGN` `1-,PH1*PH5,PH`
(g) `ASSIGN` `Q(WAIT1),PH3,PH`
(h) `ASSIGN` `PH5,FR(MACH1)/S(TUGS),PH`
(i) `ASSIGN` `1,Q(WAIT1)+Q(WAIT2),PH`

(j) `ASSIGN     S(TUGS),SC(TUGS),PH`
(k) `ASSIGN     5-,PH1+PH6,PH`
(l) `ASSIGN     PH5+,SC(TUGS)-F(MACH1),PH`

19.3. A construction job is using a single shovel and two types of trucks, 60% are type 1 and 40% are type 2. Trucks arrive at the shovel every 2.75 ± 2 minutes. The shovel can load only one truck at a time. Each truck is loaded in 1.2 ± .5 minutes. Both types of trucks then travel to a junction. Type 1 trucks take a time given by the normal distribution with a mean of 5 minutes and a standard deviation of .5 minutes. Type 2 trucks take a time given by the normal distribution with a mean of 4.5 and a standard deviation of .45. At the junction, trucks of type 1 travel to Area A in 3 ± .4 minutes. Type 2 trucks travel to Area B in 2.5 ± .3 minutes. All of the trucks take 1 ± .2 minutes to dump. Simulate for 100 continuous shifts of 480 minutes each. You are to use the same `ADVANCE` Blocks for both truck types travelling from the loader to the junction.

19.4. Do the animation that will show the utilization of the shovel and the number of loads dumped by each type of truck.

19.5. Consider the animation given by program CHAP19B.GPS. Add the necessary code so that the trucks have random colours. Also, count the number of trucks that leave by each of the three paths.

19.6. Three types of trucks come to a repair shop in a mine for repairs. The repair shop can repair only one truck at a time. The trucks arrive at a rate of one every 4.5 ± 2.5 hours. Further, 50% are type 1 trucks that are quite valuable and, so, need to be repaired first. This means that, if any other trucks are waiting for repairs, the type 1 trucks are taken to be repaired before any other type of waiting trucks. They can be repaired in a time given by the normal distribution with a mean of 3 hours and a standard deviation of .6 hours. 30% of the trucks are less important and they take 5 ± 1 hours for repairs. The remaining trucks are the least important and their repairs take a time given by the exponential distribution with a mean of 6 hours. Write a program to simulate this system and determine the utility of the repair ship and how many trucks are repaired each day. A *day* consists of three shifts with 21 actual hours of work.

19.7. Redo Exercise 19.6 but this time with no priority for the trucks; that is, they are repaired on a first-come first-served basis. Is there a difference in the utilization of the repair shop? What about the number of trucks repaired each day?

19.8. Do the animation for Exercise 19.6. Have the repair shop jiggle when it is in use. The animation can just show the different trucks coming for repairs and then leaving.

19.9. Refer to Exercise 19.8. Add the following to the animation:

Count all the trucks coming to the repair shop

Have the utility of the repair shop shown

Give the number of each type trucks repairs as they leave

The average trucks in the queue

Have a slight delay when a truck is repaired before any waiting truck replaces it in the repair shop.

20

More on Parameters

Coal distribution system for the Curragh coal mine, Queensland, Australia.

20.1 More on Parameters: The LOOP Block EQU Statement

The use of parameters in GPSS/H is so important (and often the most difficult part of learning GPSS/H) that further discussion of them will be presented in this chapter.

Some examples of the `ASSIGN` Block are given next. Assume that the values stored in the parameters are as follows:

```
PH1 = 3                PL1 = 1.370
PH2 = 4                PL2 = -4.60
PH3 = 8                PL3 = 9.870
PH4 = -20              PL4 = 1.77
```

(a) `ASSIGN     1,PH2,PH`
(b) `ASSIGN     PH3,PH4,PH`
(c) `BLET       PH(JOE) = PL4`
(d) `ASSIGN     2-,PH3,PL`

```
(e) ASSIGN    PH1,PH4,PH
(f) ASSIGN    3,PL1+PL2+PL3+PL4,PH
(g) ASSIGN    PH2,PH1*PL2,PH
(h) ASSIGN    3-,PH3+PH4,PH
```

In (a), the value stored in half-word parameter 1 is now 4.

In (b), the value stored in half-word parameter 8 is now –20.

In (c), the value stored in half-word parameter JOE is now 1. GPSS/H truncates numbers.

In (d), the value of real parameter 2 is now –4. Its value was 4, and the value stored in PH3 was 8 and this is subtracted from 4.

In (e), the value stored in half-word parameter 3 is now –20.

In (f), the value stored in half-word parameter 3 is now 8. The algebraic sum of the four real parameters is 8.41 and GPSS/H truncates this to 8.

In (g), the value stored in half-word parameter 4 is now –14

In (h), the value stored in half-word parameter 3 is now 20. (PH3 + PH4 is –12 and this is subtracted from the current value of 8 to obtain 20.)

Remember that the B operand in an `ASSIGN` statement *can* be any SNA. Thus, Blocks such as

```
ASSIGN Q(WAIT),S(TUG)+R(DOCK),PH
```

may look strange but will work as long as the B operand does not return values that are incorrect such as Q(WAIT) that is zero or less.

Example 20.1: Picking Up Miners at End of a Shift

At the end of a shift, an underground mining train comes along to pick up miners. There are from 5 to 12 miners waiting when a train arrives and always has room for them. It takes 4 ± 3 seconds for each miner to board. Simulate for all the miners getting on board. Have the output show how long it took for each miner to board.

Solution

It will be sufficient to simulate for a single train trip. The program to do the simulation is given as CHAP20A.GPS. Its listing is shown in Figure 20.1.

The train is represented as a single transaction. When it arrives in this example at time 0, the number of miners who are waiting are placed in its first half-word parameter. Notice that the continuous function to do this, `GETON`, has the ordered pairs (0,5) and (1,13) and not (0, 5) and (1, 12). This is because `RN1` generates a random number from 0.000000 to 0.999999, but never 1. Thus, the value returned is from 5.000000 to 12.999999. When this is converted to an integer, GPSS/H truncates, not rounds, so the number placed in the (train) transaction's first half-word parameter is a number from 5 to 12.

The output from the program is given in Figure 20.2.

This example shows a common use of the `ASSIGN` Block.

The animation is given by CHAP20B.GPS. Its listing is shown in Figure 20.3. This example should be studied closely as it is not as straightforward as it may seem at first.

This is an important example to study. The train is a transaction, and, when it arrives, a function `GETON` determines the number of miners who will board. A variable, &I, is used to count the miners. In the case shown here, the number is 7. The variable &I is decremented as each miner is placed on the path to wait for the train. Notice in

```
          SIMULATE
****************************
*  PROGRAM CHAP20A.GPS     *
*  TRAIN PICKS UP MINERS   *
****************************
  GETON   FUNCTION    RN1,C2
0,5/1,13
          REAL        &X
          GENERATE    ,,,1   TRAIN ARRIVES
          ASSIGN      1,FN(GETON),PH   HOW MANY MINERS WAITING?
          BPUTSTRING  ('  ')
          BPUTSTRING  ('  ')
          BPUTPIC     LINES=2,PH1
       RESULTS OF TRAIN SIMULATION
     PEOPLE WAITING TO BOARD TRAIN **
          BPUTSTRING  ('  ')
  BACK    SEIZE       DOOR
          ADVANCE     4,3
  ONBUS   RELEASE     DOOR
          BLET        &X=AC1-&X
          BPUTPIC     N(ONBUS),&X
   PEOPLE ON TRAIN  **   TIME TO GET ON *.**
          BLET        &X=AC1
          ASSIGN      1-,1,PH     ANOTHER MINER ON
          TEST LE     PH1,0,BACK EVERYONE ON?
          TERMINATE   1
          START       1
          END
```

FIGURE 20.1
Listing of program CHAP20A.GPS.

```
             RESULTS OF TRAIN SIMULATION
             PEOPLE WAITING TO BOARD TRAIN  7

           PEOPLE ON TRAIN   1   TIME TO GET ON 4.00
           PEOPLE ON TRAIN   2   TIME TO GET ON 5.44
           PEOPLE ON TRAIN   3   TIME TO GET ON 2.22
           PEOPLE ON TRAIN   4   TIME TO GET ON 1.72
           PEOPLE ON TRAIN   5   TIME TO GET ON 5.61
           PEOPLE ON TRAIN   6   TIME TO GET ON 3.64
           PEOPLE ON TRAIN   7   TIME TO GET ON 6.16
```

FIGURE 20.2
Output from program CHAP20A.GPS.

the animation that no time is given for the miners to travel on the path. Instead, the miner was given a speed of 1 unit when each was created. The animation is shown in Figure 20.4.

Example 20.2: Machine Manufactures Three Different Parts

A manufacturing facility has a single machine to assist in the manufacturing of three different parts. These parts arrive at the machine from an assembly process every 12 ± 5 minutes. A sensor in the machine detects which part it is and works as follows: type 1 parts take 10 ± 4 minutes, whereas type 2 parts take a time given by sampling from the normal distribution with a mean of 9 and a standard deviation of 1.5; type 3 parts take a time given by sampling from the exponential distribution with a mean of 9. 30% of the parts are type 1, 45% are type 2, and the remaining are type 3. The machine can work on only one part at a time and is not supposed to work for more than 80% of

```
              SIMULATE
*******************************
*  PROGRAM CHAP20B.GPS        *
*  ANIMATION FOR EXAMPLE 20.1 *
*******************************
  ATF     FILEDEF      'CHAP20B.ATF'
  GETON   FUNCTION     RN1,C2
0,5/1,13
          REAL         &X
          INTEGER      &I
          GENERATE     ,,,1   TRAIN ARRIVES
          ASSIGN       1,FN(GETON),PH   HOW MANY MINERS WAITING?
          BPUTPIC      FILE=ATF,LINES=2,AC1,PH1
TIME *.****
WRITE M1 *
          BLET         &I=PH1
BACKA     BPUTPIC      FILE=ATF,LINES=3,AC1,&I,&I
TIME *.****
CREATE MINER M*
PLACE M* ON P2
          BLET         &I=&I-1
          ADVANCE      2
          TEST E       &I,0,BACKA
          BPUTPIC      FILE=ATF,LINES=4,AC1,XID1,XID1,XID1
TIME *.****
CREATE TRAIN T*
PLACE T* ON P1
SET T* TRAVEL 5
          ADVANCE      5
          BPUTSTRING  ('  ')
          BPUTSTRING  ('  ')
          BPUTPIC      LINES=2,PH1
       RESULTS OF TRAIN SIMULATION
     PEOPLE WAITING TO BOARD TRAIN **
          BPUTSTRING  ('  ')
          BLET         &I=PH1
  BACK    SEIZE        DOOR
          BPUTPIC      FILE=ATF,LINES=2,AC1,&I
TIME *.****
PLACE M* AT 50 30
          ADVANCE      4,3
          BPUTPIC      FILE=ATF,LINES=2,AC1,&I
TIME *.****
DESTROY M*
  ONTRAIN RELEASE      DOOR
          BLET         &I=&I-1
          BLET         &X=AC1-&X
          BPUTPIC      N(ONTRAIN),&X
    MINERS ON TRAIN  **    TIME TO GET ON *.**
          BLET         &X=AC1
          ASSIGN       1-,1,PH     ANOTHER MINER ON
          TEST LE      PH1,0,BACK EVERYONE ON?
          BPUTPIC      FILE=ATF,LINES=3,AC1,XID1,XID1
TIME *.****
PLACE T* ON P3
SET T* TRAVEL 4
          ADVANCE      4
```

FIGURE 20.3
Listing of program CHAP20B.GPS. *(Continued)*

```
              BPUTPIC      FILE=ATF,LINES=2,AC1,XID1
TIME *.****
DESTROY T*
              TERMINATE    1
              START        1
              PUTPIC       FILE=ATF,LINES=2,AC1
TIME *.****
END
              END
```

FIGURE 20.3 (Continued)
Listing of program CHAP20B.GPS.

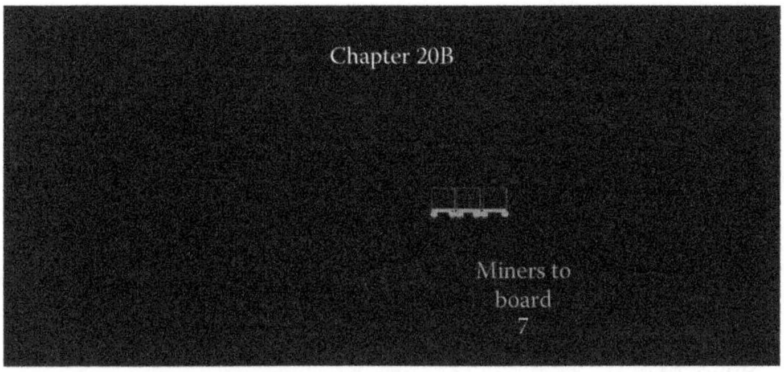

FIGURE 20.4
Screenshot of animation from program CHAP20B.GPS.

the time or it will overheat. The company makes its largest profit from type 3 parts so it is considering purchasing equipment, so that the type 3 parts will now cause the arrival of *all* parts to be every 11 ± 5 minutes. Should it purchase this equipment? The machine is available 100% of the time and a shift is 480 minutes.

Solution

The program to simulate this system as it is presently working is given by CHAP20C. GPS. The simulation was run for 100 shifts of 480 minutes. Its listing is as shown in Figure 20.5.

The output is shown in Figure 20.6.

In the 1000 shifts of the simulation, the machine made 39990 parts. These were 29.75% of type 1 parts, 44.8% of type 2 parts, and 25.4% of type 3 parts. The per cent of each type to arrive at the machine were 30%, 45%, and 25%, respectively, so this result is quite reasonable. The machine is busy 77.92% of the time, so it will not overheat.

The change to the program to consider an overall arrival rate of 11 ± 5 minutes is easily made by changing the `GENERATE 12,5` Block. Doing this produces the output as shown in Figure 20.7.

As can be seen, the machine would now be working 84.54% of the time which is too much, so the proposed new equipment should not be purchased.

The animation is given by the program CHAP20D.GPS. Its listing is as shown in Figure 20.8. The animation is given in Figure 20.9.

The animation shows the machine jiggling when a part is being worked on. This is done by having the red machine as a layout object and a small rough-shaped circular path in its centre. When a part is to be worked on, the machine is placed on this path and given a speed. When the machine is free, the speed is set equal to zero. Three

```
          SIMULATE
*******************************
*  PROGRAM CHAP20C.GPS        *
*  MANUFACTURING EXAMPLE      *
*******************************
 WHICH    FUNCTION    RN1,D3
.3,1/.75,2/1,3
 WORKON   FUNCTION    PH1,M3
1,FN(FIRST)/2,RVNORM(1,9,1.5)/3,RVEXPO(1,9)
  FIRST   FUNCTION    RN1,C2
0,6/1,14
          INTEGER     &TYPE1,&TYPE2,&TYPE3
          GENERATE    12,5  PARTS COME ALONG
          ASSIGN      1,FN(WHICH),PH  WHICH TYPE IS IT?
          QUEUE       FRONT
          SEIZE       MACH1
          DEPART      FRONT
          ADVANCE     FN(WORKON)
          RELEASE     MACH1
          TEST E      PH1,1,AWAY1
          BLET        &TYPE1=&TYPE1+1
          TRANSFER    ,AWAY3
  AWAY1   TEST E      PH1,2,AWAY2
          BLET        &TYPE2=&TYPE2+1
          TRANSFER    ,AWAY3
  AWAY2   BLET        &TYPE3=&TYPE3+1
  AWAY3   TERMINATE
          GENERATE    1000*480
          TERMINATE   1
          START       1
          PUTSTRING   ('  ')
          PUTSTRING   ('  ')
          PUTPIC      LINES=6,FR(MACH1)/10.,FC(MACH1),&TYPE1,&TYPE2,_
                      &TYPE3
       RESULTS OF SIMULATION
    MACHINE WAS BUSY  **.**% OF THE TIME
    IT MADE   *** PARTS
    NO TYPE 1 PARTS MADE    ***
    NO TYPE 2 PARTS MADE    ***
    NO TYPE 3 PARTS MADE    ***
          END
```

FIGURE 20.5
Listing of program CHAP20C.GPS.

```
   RESULTS OF SIMULATION
MACHINE WAS BUSY  77.92% OF THE TIME
IT MADE  39990 PARTS
NO TYPE 1 PARTS MADE    11897
NO TYPE 2 PARTS MADE    17928
NO TYPE 3 PARTS MADE    10165
```

FIGURE 20.6
Output from program CHAP20C.GPS.

```
   RESULTS OF SIMULATION
MACHINE WAS BUSY  84.54% OF THE TIME
IT MADE  43525 PARTS
NO TYPE 1 PARTS MADE    13003
NO TYPE 2 PARTS MADE    19513
NO TYPE 3 PARTS MADE    11008
```

FIGURE 20.7
Output from making the change to CHAP20C.GPS.

```
          SIMULATE
*********************************
*  PROGRAM CHAP20D.GPS          *
*  ANIMATION FOR EXAMPLE 20.2   *
*********************************
 ATF      FILEDEF     'CHAP20D.ATF'
 WHICH    FUNCTION    RN1,D3
.3,1/.75,2/1,3
 WORKON   FUNCTION    PH1,M3
1,FN(FIRST)/2,RVNORM(1,9,1.5)/3,RVEXPO(1,9)
  FIRST   FUNCTION    RN1,C2
0,6/1,14
          INTEGER     &TYPE1,&TYPE2,&TYPE3
          GENERATE    12,5  PARTS COME ALONG
          ASSIGN      1,FN(WHICH),PH  WHICH TYPE IS IT?
          TEST E      PH1,1,NEXT1
          BPUTPIC     FILE=ATF,LINES=4,AC1,XID1,XID1,XID1
TIME *.****
CREATE TYPE1 T*
PLACE T* ON P1
SET T* TRAVEL 5
          ADVANCE     5
          TRANSFER    ,NEXT2
 NEXT1    TEST E      PH1,2,NEXT3
          BPUTPIC     FILE=ATF,LINES=4,AC1,XID1,XID1,XID1
TIME *.****
CREATE TYPE2 T*
PLACE T* ON P1
SET T* TRAVEL 5
          ADVANCE     5
          TRANSFER    ,NEXT2
  NEXT3   BPUTPIC     FILE=ATF,LINES=4,AC1,XID1,XID1,XID1
TIME *.****
CREATE TYPE3 T*
PLACE T* ON P1
SET T* TRAVEL 5
          ADVANCE     5
  NEXT2   QUEUE       FRONT
          SEIZE       MACH1
          DEPART      FRONT
          BPUTPIC     FILE=ATF,LINES=4,AC1,XID1
TIME *.****
PLACE T* AT 44 30
PLACE M1 ON P2
SET M1 SPEED 20
          ADVANCE     FN(WORKON)
          RELEASE     MACH1
          BPUTPIC     FILE=ATF,LINES=2,AC1,FR(MACH1)/10.
TIME *.****
WRITE M4 **.**%
          TEST E      PH1,1,AWAY1
          BLET        &TYPE1=&TYPE1+1
          BPUTPIC     FILE=ATF,LINES=2,AC1,&TYPE1
TIME *.****
```

FIGURE 20.8
Listing of program CHAP20D.GPS. *(Continued)*

```
WRITE M1 ***
          TRANSFER      ,AWAY3
  AWAY1   TEST E        PH1,2,AWAY2
          BLET          &TYPE2=&TYPE2+1
          BPUTPIC       FILE=ATF,LINES=2,AC1,&TYPE2
TIME *.****
WRITE M2 ***
          TRANSFER      ,AWAY3
  AWAY2   BLET          &TYPE3=&TYPE3+1
          BPUTPIC       FILE=ATF,LINES=2,AC1,&TYPE3
TIME *.****
WRITE M3 ***
  AWAY3   BPUTPIC       FILE=ATF,LINES=5,AC1,XID1,XID1,XID1
TIME *.****
SET M1 SPEED 0
PLACE T* ON P3
SET T* TRAVEL 4
SET T* CLASS TYPE4
          ADVANCE       4
          BPUTPIC       FILE=ATF,LINES=2,AC1,XID1
TIME *.****
DESTROY T*
          TERMINATE
          GENERATE      10*480
          TERMINATE     1
          START         1
          PUTSTRING     ('  ')
          PUTSTRING     ('  ')
          PUTPIC
LINES=6,FR(MACH1)/10.,FC(MACH1)/10,&TYPE1/10,&TYPE2/10,_
                        &TYPE3/10
       RESULTS OF SIMULATION
    MACHINE WAS BUSY  **.**% OF THE TIME
    IT MADE  *** PARTS PER DAY
    NO TYPE 1 PARTS MADE PER SHIFT  ***
    NO TYPE 2 PARTS MADE PER SHIFT  ***
    NO TYPE 3 PARTS MADE PER SHIFT  ***
          PUTPIC        FILE=ATF,LINES=2,AC1
TIME *.****
END
          END
```

FIGURE 20.8 (Continued)
Listing of program CHAP20D.GPS.

separate parts enter the system and, when the machine is finished, they all leave as one orange ball.

20.1.1 The LOOP Block

GPSS/H does not have DO loops for Blocks that are common in other languages such as Fortran and Pascal.* There is, however, a Block that acts similarly to the DO loop but is a bit restricted. This is the LOOP Block, and it acts in connection with a transaction's specified parameter. The general form of it is shown as

* It is possible to have DO loops that are used in control statements. These DO loops are very useful in running programs multiple times with selected variables changed. They will be presented later.

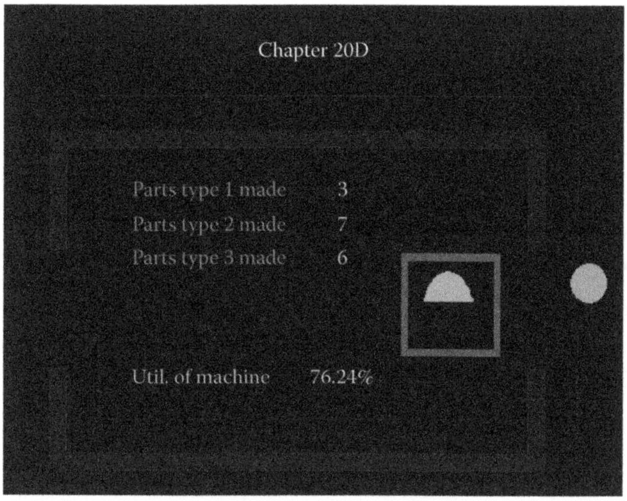

FIGURE 20.9
Animation of CHAP20D.GPS.

```
LOOP A,(label)
```

where A is an integer parameter, that is, `PH`, `PB`, or `PF` but not `PL`.

For example, `LOOP 1PH,BACK1` would be such a block.

The way the `LOOP` Block works is as follows:

The transaction's parameter specified in the A operand is evaluated. This is decremented by 1 and the result compared with 0. If the value is zero, the transaction is routed to the next sequential Block. If the value is greater than zero, the transaction is routed to the Block given by the B operand. This Block is *always* before the `LOOP` Block. Some examples of this Block are as follows*:

(a) `LOOP 3PH,BACK1`
(b) `LOOP PH1,UPTOP`
(c) `LOOP 5PF,OVER`

In (a), suppose that the value of the transaction's third half-word parameter is 6. Then the looping is done for values of 5, 4, 3, 2, and 1.

In (b), the transaction's first parameter is given. Suppose this is 4 and the value of parameter 4 is 5. Looping is done for 3, 2, and 1.

In (c), the looping is done depending on the value of the transaction's parameter number 5.

Looping is done only by a decrement of 1 and cannot be done by increments. As restrictive as this Block may seem, there are many uses for it.

---

* It is possible to write the `LOOP` Block without reference to the parameter type if only one parameter type is used. For example, the code

```
       GENERATE     ,,,1
       ASSIGN       1,4
BACK   ADVANCE      1
       LOOP         1,BACK
```

would compile and execute as expected. However, this type of programming will not be done here.

Example 20.3: Illustrations of the LOOP Block

The following two short programs are given to illustrate the LOOP Block. They are CHAP20E.GPS and CHAP20F.GPS. The listing of program CHAP20E.GPS is given in Figure 20.10. The output from the program is shown in Figure 20.11.

Here the LOOP Block looks at what is stored in half-word parameter 1, which is 3. Thus, it executes from the Block with the label BACK to the LOOP Block three times. The variable &X is incremented each time the looping is done. The half-word parameter 1 is decremented each time.

CHAP20F.GPS has the LOOP Block changed from LOOP 1PH,BACK to:

```
LOOP        PH1PH,BACK
```

The program (Figure 20.12) is as follows.

The output is shown in Figure 20.13.

```
                 SIMULATE
*****************************
*  PROGRAM CHAPTER20E.GPS   *
*  EXAMPLE OF LOOP BLOCK    *
*****************************
                 REAL          &X
                 GENERATE      ,,,1,,1PH
                 ASSIGN        1,3,PH
 BACK            BLET          &X=&X+1
                 BPUTPIC       PH1,&X
         PH1 = *   &X = *
                 LOOP          1PH,BACK
                 TERMINATE     1
                 START         1
                 END
```

FIGURE 20.10
Listing of program CHAP20E.GPS.

```
PH1 = 3   &X = 1
PH1 = 2   &X = 2
PH1 = 1   &X = 3
```

FIGURE 20.11
Output from program CHAP20E.GPS.

```
                 SIMULATE
*****************************
*  PROGRAM CHAP20F.GPS      *
*  EXAMPLE OF LOOP BLOCK    *
*****************************
                 REAL          &X
                 GENERATE      ,,,1,,3PH
                 ASSIGN        1,3,PH
                 ASSIGN        3,6,PH
 BACK            BLET          &X=&X+1
                 BPUTPIC       PH1,&X
     PH1 =       *    &X   = *
                 LOOP          PH1PH,BACK
                 TERMINATE     1
                 START         1
                 END
```

FIGURE 20.12
Listing of program CHAP20F.GPS.

```
PH1 =  3    &X  = 1
PH1 =  3    &X  = 2
PH1 =  3    &X  = 3
PH1 =  3    &X  = 4
PH1 =  3    &X  = 5
PH1 =  3    &X  = 6
```

FIGURE 20.13
Output from program CHAP20F.GPS.

Here the value stored in parameter 1 is 3. This says to look in the transaction's parameter for the value, which is 6. Figure 20.13 shows the six values.

These two examples should be studied closely, as it is easy for the person learning GPSS/H to have the two examples confused.

Example 20.4: Determining How Busy a Worker Is

Joe is supposed to work only half time on special projects. He comes to work and finds that he has one special project to do 10% of the time, two special projects 40% of the time, and three special projects 50% of the time. He must do every one of the special projects before he can do his regular work. Joe can finish working on a special project in 1.5 ± .5 hours. A shift is 8 hours. Determine how busy Joe is working on special projects by simulating for 1000 days. Is his work load above 50%?

The program to do the simulation is CHAP20G.GPS. When a new day begins, Joe is assigned the number of special projects to do. This number is placed in Joe's first parameter. After Joe finishes these projects, he can work on his regular projects until 8 hours have passed. Joe's working on the projects is simulated using a `LOOP` Block. After Joe finishes his special projects for the day, the time Joe can work on his regular projects is given by the Block:

```
ADVANCE 8*N(BACK1)-AC1
```

The listing of program CHAP20G.GPS is given in Figure 20.14.

```
               SIMULATE
*****************************
*  PROGRAM CHAPTER20G.GPS    *
*****************************
  MANY     FUNCTION    RN1,D3
.1,1/.5,2/1,3
          GENERATE     ,,,1       JOE COMES TO WORK
 BACK1    ASSIGN       1,FN(MANY)    HOW MANY SPECIAL PROJECTS
 UPTOP    SEIZE        JOEW
          ADVANCE      1.5,.5
          RELEASE      JOEW
          LOOP         1PH,UPTOP
          ADVANCE      8*N(BACK1)-AC1
          TRANSFER     ,BACK1
          GENERATE     8000
          TERMINATE    1
          START        1
          PUTPIC       LINES=2,FR(JOEW)/10.
       RESULTS OF SIMULATION
  JOE WAS BUSY  ***.**% OF THE TIME
          END
```

FIGURE 20.14
Listing of program CHAP20G.GPS.

```
RESULTS OF SIMULATION
JOE WAS BUSY   44.59% OF THE TIME
```

FIGURE 20.15
Results of program CHAP20G.GPS.

The output from the program is as shown in Figure 20.15.

This shows that Joe was busy 44.6% of the time working on special projects.

The program to do the animation is CHAP20H.GPS. Its listing is given in Figure 20.16. The animation is shown in Figure 20.17.

20.1.2 The `EQU` Compiler Directive

One can use parameters as operands in Blocks such as the `QUEUE` and `SEIZE` Blocks. In fact, this is often done. Thus, one could have Blocks such as

```
QUEUE      PH2
QUEUE      PH1+PH4
SEIZE      PH4
DEPART     PB5
RELEASE    PF5
```

The use of such code can greatly compress the number of lines of GPSS/H. For example, suppose there are three types of ships entering a harbour. Type 1 requires one tug boat to berth it, type 2 requires two tug boats, and type 3 requires four tugboats. Rather than have three nearly identical segments, one could have each ship type have a different number in one of its parameters, say parameter 5. Thus, type 1 ships might have 1 in parameter 5; type 2 ships might have 2 in parameter 5; and type 3 ships might have 3 in parameter 5. Then, you could have

```
ENTER TUGS,PH5
```

where TUGS is the storage to represent the number of tug boats available. In addition, you could have the ships enter separate queues by:

```
QUEUE PH5
```

If you wanted the ships to be in a global queue, it is tempting to write:

```
QUEUE    WAIT
QUEUE    PH5
```

Unfortunately, there is a problem. Assuming there are no other queues in the program during compiling, the queue `WAIT` is assigned to the first queue. Now, when a type 1 ship enters queue 1, this is not only `QUEUE 1` but also `QUEUE WAIT`. The statistics will be incorrect. One way around this is to have the queue `WAIT` renamed as, say,

```
QUEUE 10
```

```
                  SIMULATE
*************************************
*  PROGRAM CHAPT20H.GPS             *
*  ANIMATION FOR EXAMPLE 20.4       *
*************************************
  ATF      FILEDEF      'CHAP20H.ATF'
  MANY     FUNCTION     RN1,D3
.1,1/.5,2/1,3
           GENERATE     ,,,1       JOE COMES TO WORK
           BPUTPIC      FILE=ATF,LINES=4,AC1,XID1
TIME *.****
CREATE JOEW J*
CREATE STUFF S1
PLACE S1 AT 37 33
 BACK1     ASSIGN       1,FN(MANY)    HOW MANY SPECIAL PROJECTS
           BPUTPIC      FILE=ATF,LINES=8,AC1,XID1,XID1,PH1,N(BACK1)
TIME *.****
SET J* CLASS JOEW
WRITE M10 WORKING ON
WRITE M1 SPECIAL PROJECTS
PLACE J* AT 31 32
SET S1 CLASS STUFF
WRITE M2 **
WRITE M4 ***
 UPTOP     SEIZE        JOEW
           BPUTPIC      FILE=ATF,LINES=2,AC1,PH1
TIME *.****
WRITE M3 *
           ADVANCE      1.5,.5
           RELEASE      JOEW
           BPUTPIC      FILE=ATF,LINES=2,AC1,FR(JOEW)/10.
TIME *.****
WRITE M12 **.**%
           LOOP         1PH,UPTOP
           BPUTPIC      FILE=ATF,LINES=8,AC1,XID1,XID1
TIME *.****
SET J* CLASS JOEW2
WRITE M5 REGULAR
WRITE M6 PROJECTS
SET S1 CLASS STUFF1
WRITE M1
WRITE M10
PLACE J* AT 43 31
           ADVANCE      8*N(BACK1)-AC1
           BPUTPIC      FILE=ATF,LINES=3,AC1
TIME *.****
WRITE M5
WRITE M6
           TRANSFER     ,BACK1
           GENERATE     8000
           TERMINATE    1
           START        1
           PUTPIC       FILE=ATF,LINES=2,AC1
TIME *.****

END
           PUTPIC       LINES=2,FR(JOEW)/10.
       RESULTS OF SIMULATION
  JOE WAS BUSY  ***.**% OF THE TIME
           END
```

FIGURE 20.16
Listing of program CHAP20H.GPS.

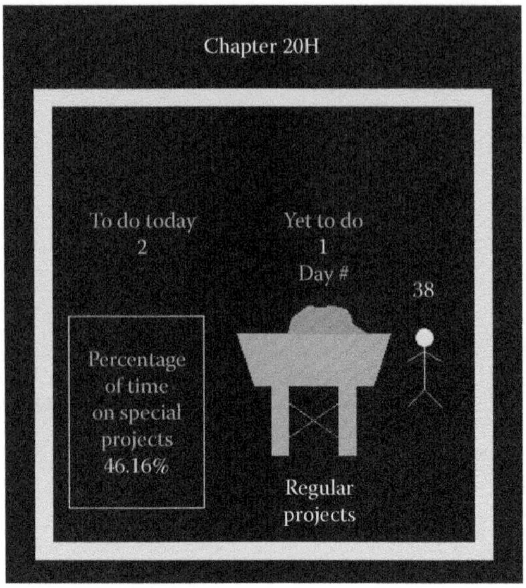

FIGURE 20.17
Screenshot of animation for Example 20.4.

This will work, but it is better to use mnemonics that represent names that are more meaningful than just numbers. The `EQU` compiler allows the analyst to use names for operands but assigns them specific numbers. The general form of this is as follows:

```
(label) EQU (number),(entity type)
```

The (entity type) is the family name of the entity. These are

Entity	Family Name
Facility	F
Function	7
Parameter	PH, PB, PF, or PL
Queue	Q
Random number	RN
Storage	S

The (label) is the mnemonic you want to use. The (number) is the integer number you want to assign to the entity.

Some examples of this are as follows:

```
(a) WAIT     EQU    7,Q
(b) MACH1    EQU    10,F
(c) HALTIT   EQU    4,Q
(d) TRAVEL   EQU    100
```

In (a), the queue `WAIT` is specified as being the seventh queue of all of the possible queues GPSS/H has. In a program, if you had the following:

```
QUEUE  WAIT
QUEUE  PH1
```

and `PH1` was only 1, 2, or 3, there would be no problems in having the queue WAIT specified as queue 7.

In (b), the facility `MACH1` is specified as being the 10th facility.

In (c), the queue `HALTIT` is specified as being the fourth queue.

In (d), if you had `ADVANCE TRAVEL`, the transaction would be on the FEC for a time of 100 units (this is rarely done).

It is possible to have the following:

```
HALT   EQU    10,Q,F
       - - - - - -
       - - - - - -
       QUEUE  HALT
       SEIZE  HALT
```

Now, both queue `HALT` and facility `HALT` are designated as the 10th queue and facility, respectively. The next example will illustrate this.

Example 20.5: Parts Come along an Assembly Line

Three different parts come along an assembly line every 10 ± 4 time units. 25% of the parts are type 1; 60% of the parts are type 2; and the remaining parts are type 3. There are three identical machines that work on them in 15 ± 7 time units. Simulate this for 100 shifts of 480 minutes each. The relevant statistics are the utilizations of the three machines and the maximum individual queues, the maximum overall queue, and the average queue for each machine.

Solution

The GPSS/H program is CHAP20I.GPS. It uses the `EQU` compiler directive to assign the overall queue as queue number 10. The parts are assigned their types in their first half-word parameter (Figure 20.18).

The results of the simulation are shown in Figure 20.19.

The results are not surprising. The machine that has to work on 60% of the parts (type 2) is busy 89.61% of the time. It once has a queue of 15 and the maximum overall queue was 15. Even so, the average queues were not large.

The animation for this example needs to be written with much more code, as each of the three machines needs to have separate code written for it. The program for the animation is CHAP20J.GPS. Its listing is given in Figure 20.20. The animation is shown in Figure 20.21.

```
          SIMULATE
******************************
*  PROGRAM CHAPT20I.GPS        *
*  EXAMPLE OF EQU STATEMENT    *
******************************
 TYPE     FUNCTION    RN1,D3
.25,1/.85,2/1,3
  WAIT     EQU         10,Q    WAIT IS THE 10TH QUEUE
           GENERATE    10,4    PARTS COME ALONG
           ASSIGN      1,FN(TYPE),PH    WHAT TYPE IS IT?
           QUEUE       WAIT    OVERALL QUEUE
           QUEUE       PH1     INDIVIDUAL QUEUE
           SEIZE       PH1     USE FACILITY
           DEPART      PH1     LEAVE INDIVIDUAL QUEUE
           DEPART      WAIT    LEAVE OVERALL QUEUE
           ADVANCE     15,7    WORK IN PROGRESS
           RELEASE     PH1     FREE MACHINE
           TERMINATE
           GENERATE    480*100
           TERMINATE   1
           START       1
           PUTSTRING   ('  ')
           PUTSTRING   ('  ')
           PUTPIC      LINES=5,FR1/10.,FR2/10.,FR3/10.,QM1,QM2,QM3,_
                        QM(WAIT),QA1,QA2,QA3,QA(WAIT)
      UTIL MACH 1  **.**%  UTIL MACH 2 **.**%  UTIL MACH 3  **.**%
      MAX  QUEUE 1      **
      MAX  QUEUE 2      **
      MAX  QUEUE 3      **
      MAX OVERALL QUEUE **
           END
```

FIGURE 20.18
Listing of program CHAP20I.GPS.

```
UTIL MACH 1  38.73%  UTIL MACH 2 89.61%  UTIL MACH 3  22.46%
   MAX  QUEUE 1       2
   MAX  QUEUE 2      15
   MAX  QUEUE 3       2
MAX OVERALL QUEUE 16
```

FIGURE 20.19
Results of program CHAP20I.GPS.

```
          SIMULATE
******************************
*  PROGRAM CHAP20J.GPS         *
*  ANIMATION OF CHAP20I.GPS    *
******************************
 TYPE     FUNCTION    RN1,D3
.25,1/.85,2/1,3
  ATF     FILEDEF     'CHAP20J.ATF'
 WHERE    FUNCTION    PH1,L3
1,FIRST/2,SECOND/3,THIRD
          GENERATE    10,4    PARTS COME ALONG
          ASSIGN      1,FN(TYPE),PH    WHAT TYPE IS IT?
          TRANSFER    ,FN(WHERE)
 FIRST    BPUTPIC     FILE=ATF,LINES=4,AC1,XID1,XID1,XID1
TIME *.****
```

FIGURE 20.20
Listing of program CHAP20J.GPS. *(Continued)*

```
CREATE PART1 P*
PLACE P* ON P1
SET P* TRAVEL  3
          ADVANCE      3
          BPUTPIC      FILE=ATF,LINES=3,AC1,XID1,XID1
TIME *.****
PLACE P* ON P2
SET P* TRAVEL 3
          ADVANCE      3
          QUEUE        WAIT1
          QUEUE        OVERALL  OVERALL QUEUE
          SEIZE        MACH1
          DEPART       WAIT1
          DEPART       OVERALL
          BPUTPIC      FILE=ATF,LINES=5,AC1,XID1,QA(WAIT1),QA(OVERALL)
TIME *.****
PLACE P* AT 0 28
SET MACH1 COLOR F2
WRITE M4 **.**
WRITE M7 **.**
          ADVANCE      15,7
          RELEASE      MACH1
          BPUTPIC      FILE=ATF,LINES=5,AC1,XID1,XID1,FR(MACH1)/10.
TIME *.****
PLACE P* ON P5
SET P* TRAVEL 3
SET MACH1 COLOR F7
WRITE M1 **.**%
          ADVANCE      3
          BPUTPIC      FILE=ATF,LINES=3,AC1,XID1,XID1
TIME *.****
PLACE P* ON P8
SET P* TRAVEL          3
          ADVANCE      3
          BPUTPIC      FILE=ATF,LINES=2,AC1,XID1
TIME *.****
DESTROY P*
          TERMINATE
 SECOND   BPUTPIC      FILE=ATF,LINES=4,AC1,XID1,XID1,XID1
TIME *.****
CREATE PART2 P*
PLACE P* ON P1
SET P* TRAVEL  3
          ADVANCE      3
          BPUTPIC      FILE=ATF,LINES=3,AC1,XID1,XID1
TIME *.****
PLACE P* ON P3
SET P* TRAVEL 1.5
          ADVANCE      1.5
          QUEUE        WAIT2
          QUEUE        OVERALL    OVERALL QUEUE
          SEIZE        MACH2
          DEPART       WAIT2
          DEPART       OVERALL
          BPUTPIC      FILE=ATF,LINES=5,AC1,XID1,QA(WAIT2),QA(OVERALL)
TIME *.****
PLACE P* AT 0 -2
SET MACH2 COLOR F2
WRITE M5 **.**
WRITE M7 **.**
```

FIGURE 20.20 (Continued)
Listing of program CHAP20J.GPS.

(Continued)

```
          ADVANCE     15,7
          RELEASE     MACH2
          BPUTPIC     FILE=ATF,LINES=5,AC1,XID1,XID1,FR(MACH2)/10.
TIME *.****
PLACE P* ON P6
SET P* TRAVEL 1
SET MACH2 COLOR F7
WRITE M2 **.**%
          ADVANCE     1
          BPUTPIC     FILE=ATF,LINES=3,AC1,XID1,XID1
TIME *.****
PLACE P* ON P8
SET P* TRAVEL 3
          ADVANCE     3
          BPUTPIC     FILE=ATF,LINES=2,AC1,XID1
TIME *.****
DESTROY P*
          TERMINATE
 THIRD    BPUTPIC     FILE=ATF,LINES=4,AC1,XID1,XID1,XID1
TIME *.****
CREATE PART3 P*
PLACE P* ON P1
SET P* TRAVEL  3
          ADVANCE     3
          BPUTPIC     FILE=ATF,LINES=3,AC1,XID1,XID1
TIME *.****
PLACE P* ON P4
SET P* TRAVEL 3
          ADVANCE     3
          QUEUE       WAIT3
          QUEUE       OVERALL   OVERALL QUEUE
          SEIZE       MACH3
          DEPART      WAIT3
          DEPART      OVERALL
          BPUTPIC     FILE=ATF,LINES=5,AC1,XID1,QA(WAIT3),QA(OVERALL)
TIME *.****
PLACE P* AT 0 -32
SET MACH3 COLOR F2
WRITE M6 **.**
WRITE M7 **.**
          ADVANCE     15,7
          RELEASE     MACH3
          BPUTPIC     FILE=ATF,LINES=5,AC1,XID1,XID1,FR(MACH3)/10.
TIME *.****
PLACE P* ON P7
SET P* TRAVEL 3
SET MACH3 COLOR F7
WRITE M3 **.**%
          ADVANCE     3
          BPUTPIC     FILE=ATF,LINES=3,AC1,XID1,XID1
TIME *.****
PLACE P* ON P8
SET P* TRAVEL         3
          ADVANCE     3
          BPUTPIC     FILE=ATF,LINES=2,AC1,XID1
TIME *.****
```

FIGURE 20.20 (Continued)
Listing of program CHAP20J.GPS.

(Continued)

```
DESTROY P*
          TERMINATE
          GENERATE     480*100
          TERMINATE    1
          START        1
          PUTSTRING    ('  ')
          PUTSTRING    ('  ')
          PUTPIC       LINES=5,FR(MACH1)/10.,FR(MACH2)/10.,FR(MACH3)/10.,_
                       QM(WAIT1),QM(WAIT2),QM(WAIT3),_
                        QM(OVERALL),QA(WAITA),QA(WAIT2),QA(WAIT3),QA(OVERALL)
     UTIL MACH  1  **.**%  UTIL MACH 2 **.**%  UTIL MACH 3  **.**%
     MAX  QUEUE 1       **
     MAX  QUEUE 2       **
     MAX  QUEUE 3       **
     MAX OVERALL QUEUE **
          PUTPIC       FILE=ATF,LINES=2,AC1
TIME *.****
END
          END
```

FIGURE 20.20 (Continued)
Listing of program CHAP20J.GPS.

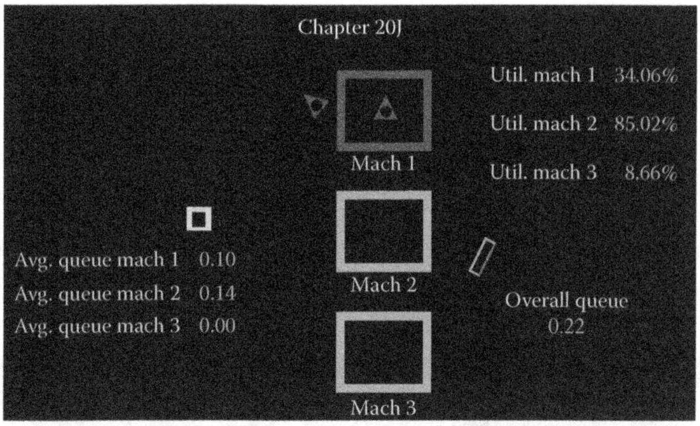

FIGURE 20.21
Screenshot of animation from program CHAP20J.GPS.

20.2 Exercises

20.1. In Example 20.1, the simulation was for only one train to come to pick up the miners. Simulate for 20 trains to come along. Have the output show the average number of miners waiting. The results will be lengthy so it is best to have them placed in an output file.

20.2. In Example 20.1, add the code to the program CHAP20B.GPS so that the animation now has each miner a different colour. Have the animation give the time for each miner to board and then the total time for all the miners.

20.3. What would the output be from the following programs?

```
    (a)     SIMULATE
            REAL           &X
            GENERATE       ,,,1
            ASSIGN         1,4,PH
BACK        BLET           &X = 2*&X-1
            BPUTPIC        LINES = 1,&X
     THE VALUE OF X IS **
            LOOP           1PH,BACK
            TERMINATE      1
            START          1
            END

    (b)     SIMULATE
            REAL           &X
            GENERATE       ,,,1
            ASSIGN         1,4,PH
            ASSIGN         PH1,PH1+6,PH
BACK        BPUTPIC        LINES = 1,&X
THE VALUE OF X IS **
            BLET           &X = &X+PH4
            LOOP           4PH,BACK
            TERMINATE      1
            START          1
            END
```

You may wish to write these programs and then run them to verify your results.

20.4. Mrs. Jones shops in a store with three aisles. It takes her 2 ± 1 minutes to shop in aisle 1, where she select 5 ± 1 items. Next, she shops in aisle 2, where she selects

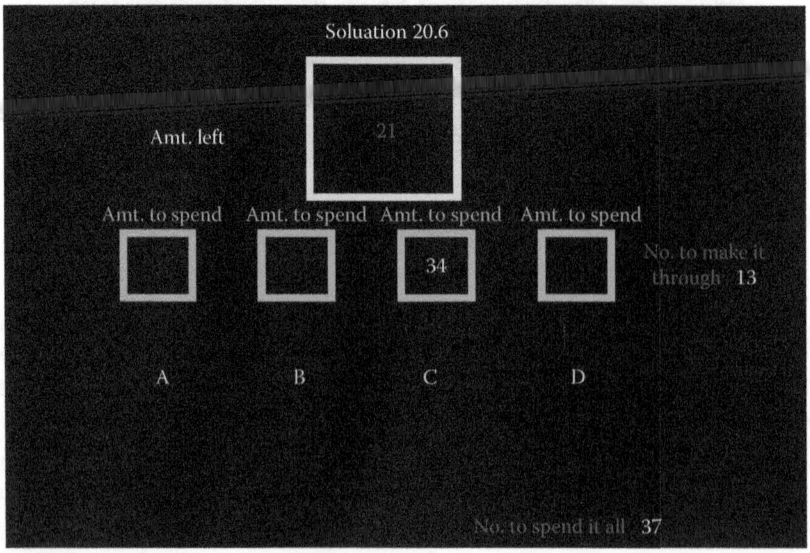

FIGURE 20.22
Screenshot of possible animation for Exercise 20.6.

items based on the normal distribution with a mean of 7 and a standard deviation of 1.2 (these items need to be integers). This takes her 1.5 ± .25 minutes. In the last aisle she shops in, she selects 8 ± 1 items, which takes her 1 ± .1 minutes. Checkout time is 8.5 seconds per item and there is never a line for checking out. The checkout clerk is always there. Determine the expected shopping time for Mrs. Jones by simulation for 50 shopping visits.

20.5. People are given coins to spend at four stations. The number of coins received is 100 ± 20. At the first station, they will spend 30 ± 10 coins; at the second 35 ± 10 coins; at the third, if possible, they will spend 25 ± 15 coins; and at the last, if possible, they will spend 20 ± 10 coins. If a person arrives at a station and does not have enough coins to purchase all the items, he or she will purchase as many items as possible and then must leave the system. Write the program to determine how many people out of 100 can pass through the four stations and still have money left over. The program is to have as output the number of people who did not make it with money to spare, the number who did not make it all the way through, and the average coins left of those who made it through.

20.6. Add the animation to Exercise 20.5. The animation should look as shown in Figure 20.22.

21

TABLES

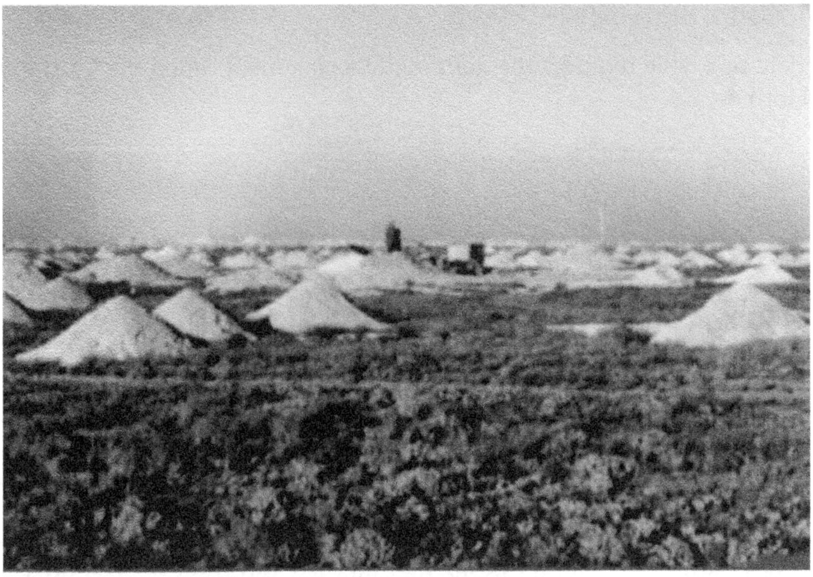

Opal mines, Australia.

21.1 Introduction

GPSS/H has the ability to make statistical tables of *any* standard numerical attribute (SNA). These tables are in the form of histograms and might be the length of a queue, the value of a parameter, the per cent of time a machine is working, the number of shopping carts in use at a grocery store, and so on. Often the time a transaction has been in the system is tabulated (how long it took the transaction to go from one point in the program to another point). For example, in most mining operations, the time for trucks to cycle from the shovels and return is of interest. In studying people entering a grocery store and using the shopping carts, the simulation might give a maximum number of carts in use as 14. The mean number of carts in use at any time might be 4.5. It would be instructive to see how often 14 carts were in use as well as the distribution of the usage of the carts during the simulation. This can be done easily in GPSS/H.

In making tables, GPSS/H also computes the sample mean of the data, the sample distribution, the number of samples that fall into each of the ranges, and the percentage of values in the sample that fall into each of the ranges in the series. This is all done automatically.

Recall that a histogram has intervals. Each of these intervals records the number of times a variable falls into each interval. Since a person doing a simulation is often concerned with tabulating data, GPSS/H provides a very simple method of doing this. In fact, to make a histogram for any SNA takes only two lines of code (!). One of these is a Block and the other a statement.

21.2 The TABLE Statement

To record data in a table requires the defining of a statement called the `TABLE` statement. Its general form is

```
(label or number) TABLE    SNA,B,C,D
```

where:

- `SNA` is any standard numerical attribute
- `B` is the starting value of the values
- `C` is the class interval
- `D` is the number of the class intervals

Figure 21.1 gives an example of a histogram.

The values go from 5 to 35. The class interval is 5 and the number of intervals as shown is 6. There were six values between 5 and 10, 9 between 10 and 15, 24 between 15 and 20, and so on. Tables in GPSS/H actually have two more intervals, as GPSS/H considers the regions from minus infinity up to the first interval and from the last interval to plus infinity as intervals. Thus, GPSS/H would have eight intervals in the histogram shown in Figure 21.1.

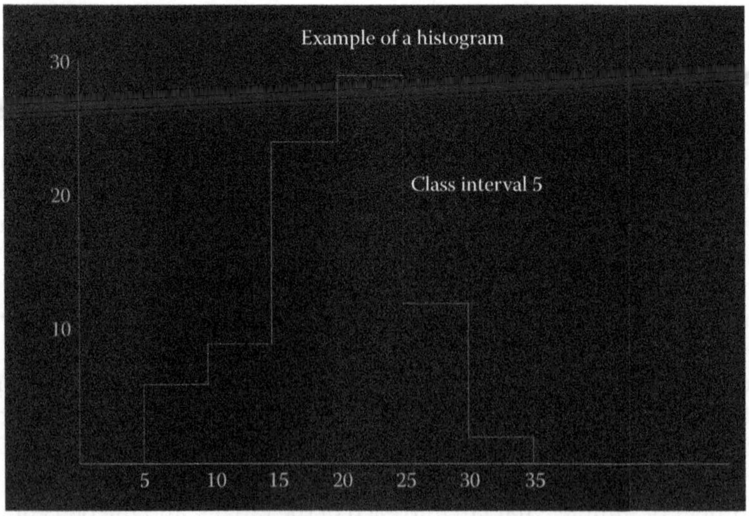

FIGURE 21.1
Example of a histogram.

Some examples of the TABLE statement might be as follows:

```
(a) FIRST     TABLE    S(CARTS),0,1,20
(b) MARK1     TABLE    Q(WAIT),0,1,15
(c) NEXT      TABLE    FR(MACH1),0,50,22
(d) 1         TABLE    &X,10,2,25
```

Example (a) will record the amount of storage used in the storage CARTS and put them in a table with intervals from 0 to 1. The table will have 20 intervals. The first will go from minus infinity to 0 (even though there will never be an entry here), the second from 0 to 1, and the last end point from 18 to plus infinity. GPSS/H will count the intervals you specify as well as the one before the data and the one after. In general, these are the regions from minus infinity to the first data point of the histogram as specified in the B operand and from the last data point of the histogram as specified by the C operand to + infinity. *This is important to keep in mind when using the* TABLE *statement.*

The second table, (b), will give the distribution of the number of people who were in the queue called WAIT. The table will have 15 intervals that include the number from minus infinity to zero to those greater than 13.

The third table, (c), gives the utilization of the facility called MACH1. Recall that this is given in parts per thousand. Thus, the table will go from 0 to 50, 50 to 100, 100 to 150, and so on (not counting the end intervals).

The fourth table, (d), will record the values of the ampervariable &X, and the class intervals will start at 10 and have a size of 2. There will be 25 of these intervals.

21.3 The TABULATE Block

To make an entry in a table, the TABULATE Block is used.

```
TABULATE name
```

Every time a transaction enters this Block, an entry is made in the TABLE with the label name. Thus, the combination of a TABLE statement and the TABULATE Block, such as

```
FIRST TABLE       Q(WAIT),0,1,10
      .....       ..............
      .....       ..............
      TABULATE    FIRST
```

would make a table (histogram) of the people in the queue named WAIT.

Example 21.1: Table of People in a Queue

To illustrate what a table might look like, consider the program to study people using a single facility. People arrive every 20 ± 8 minutes. The service time is 19 ± 10 minutes. If the server is busy, people wait in a queue until the server is free. Simulate for 100 days of 480 minutes per day. Obtain a table of the number of people who had to wait

```
              SIMULATE
************************************
*   PROGRAM CHAP21A.GPS                *
*   EXERCISE ON THE TABLE STATEMENT    *
*   AND THE TABULATE BLOCK             *
************************************
 INQUE     TABLE         Q(WAIT),0,1,15
           GENERATE      20,8      PEOPLE COME
           QUEUE         WAIT
           SEIZE         MACH
           TABULATE      INQUE     TABLE ENTRY
           DEPART        WAIT
           ADVANCE       19,10
           RELEASE       MACH
           TERMINATE
           GENERATE      480*100
           TERMINATE     1
           START         1
           PUTSTRING     ('  ')
           PUTSTRING     ('  ')
           PUTPIC        LINES=4,FR(MACH)/10.,QM(WAIT),QA(WAIT)
        RESULTS OF SIMULATION
      UTIL. OF MACHINE     **.**%
      MAXIMUM QUEUE        **
      AVERAGE QUEUE        **.**
     END
```

FIGURE 21.2
Listing of program CHAP21A.GPS.

for service. The program to do this simulation is CHAP21A.GPS. Its listing is given in Figure 21.2. Output from the program is show in Figure 21.3.

The output shows that the maximum queue was 7 and that the average queue was 1.1. However, how often during the 100 shifts of the simulation did a maximum queue of 7 occur? The histogram is found in the .LIS files. While it is possible to have a histogram sent to the screen, with today's graphical software, this is no longer done. Instead, the .LIS file gives the histogram that is shown in Figure 21.4.

The histogram shows the observed frequency of times there were one or less people in the queue to be 1362. Two people were in the queues 584 times, and there were seven people in the queue only one time out of 2397 entries into the table. Not shown in Figure 21.4 is the fact that there were 407 zero entries as the figure on 1362 includes both the zero entries and the time there was one person in the queue. The .LIS file also gives other data, including the percentage of the total entries and the cumulative percentage.

There is nothing else to do to make tables of SNAs. This ability to make tables of any SNA so easily and rapidly is often hard for people to believe the first time they are introduced to it.

```
      RESULTS OF SIMULATION
  UTIL. OF MACHINE    95.14%
  MAXIMUM QUEUE        7
  AVERAGE QUEUE        1.10
```

FIGURE 21.3
Output from program CHAP21A.GPS.

```
TABLE    INQUE

ENTRIES IN TABLE  MEAN ARGUMENT STANDARD DEVIATION  SUM OF ARGUMENTS
      2397.0000       1.7359        1.0504  4161.0000   NON-WEIGHTED

  UPPER  OBSERVED      PERCENT  CUMULATIVE  CUMULATIVE    MULTIPLE    DEVIATION
  LIMIT  FREQUENCY     OF TOTAL PERCENTAGE  REMAINDER     OF MEAN     FROM MEAN
           ...
  1.0000 1362.0000     56.8210    56.82        43.18        0.5761      -0.7006
  2.0000  584.0000     24.3638    81.18        18.82        1.1521       0.2514
  3.0000  241.0000     10.0542    91.24         8.76        1.7282       1.2034
  4.0000  155.0000      6.4664    97.71         2.29        2.3043       2.1554
  5.0000   43.0000      1.7939    99.50         0.50        2.8803       3.1073
  6.0000   11.0000      0.4589    99.96         0.04        3.4564       4.0593
  7.0000    1.0000      0.0417   100.00         0.00        4.0324       5.0113
```

FIGURE 21.4
Histogram from program CHAP21A.GPS.

21.4 Standard Numerical Attributes Associated with Tables

There are several SNAs associated with tables. These are as follows:

- `TB(name)` or `TBj` Sample mean. For the previous table, `INQUE`, this would be 1.7359.
- `TC(name)` or `TCj` Number of observations. This is 2397 for the table `INQUE`.
- `TD(name)` or `TDj` Standard Deviation. For the table, `INQUE`, this would be 1.0504.

21.4.1 The SNA `M1`

There is an SNA that is quite useful when constructing tables. This is called `M1` and gives the time a transaction has been in the system. Whenever a transaction is created, it is marked with the time it enters the system (using the Absolute Clock!). The SNA `M1` takes the time of the Absolute Clock and subtracts the entry time from it. This gives the time that the transaction has been in the system. For example, suppose a transaction entered the system at time 404.56 and now the Absolute Clock is at 625.77, the SNA `M1` is equal to 231.21 (625.77 minus 404.56). Consider the following lines of code:

```
TIMES TABLE M1,0,25,30EDIT CHAP21E.GPS

      ......
      ......
      TABULATE TIMES
```

The table `TIMES` will give the tabulation of times that the transactions were in the system from the time they entered up to the time they entered the Block `TABULATE TIMES`. The times will be in intervals of 25, and there will be 30 such intervals.

Going back to the program CHAP21A.GPS, if one desired to see a table of the times people were in the system, all that is needed to be done is to add two lines of code as shown in the program CHAP21B.GPS. This program now will make a table of the times a person is in the system starting at time 8 and going up to time 37. The listing of the program is given in Figure 21.5.

```
               SIMULATE
******************************
*   PROGRAM CHAP21B.GPS        *
*   TABLES IN GPSS/H           *
******************************
 INQUE     TABLE       Q(WAIT),0,1,15
 TIMEIN    TABLE       M1,8,1,30
           GENERATE    20,8      PEOPLE COME
           QUEUE       WAIT
           SEIZE       MACH
           TABULATE    INQUE
           DEPART      WAIT
           ADVANCE     19,10
           RELEASE     MACH
           TABULATE    TIMEIN
           TERMINATE
           GENERATE    480*100
           TERMINATE   1
           START       1
           PUTSTRING   ('  ')
           PUTSTRING   ('  ')
           PUTPIC      LINES=4,FR(MACH)/10.,QM(WAIT),QA(WAIT)
       RESULTS OF SIMULATION
     UTIL. OF MACHINE      **.**%
     MAXIMUM QUEUE          **
     AVERAGE QUEUE          **.**
           END
```

FIGURE 21.5
Listing of program CHAP21B.GPS.

The histogram of the time in the system is obtained from the .LIS file. A portion of it is shown in Figure 21.6.

There were not enough intervals, so GPSS/H takes these 1173 overflow values (out of the 2397) and calculates the mean value as 59.0664. While this is not large, the tabled values comprise only 51.06% of the values. Thus, one should either change the operands on the `TABLE` statement to have larger intervals or have more of them or both. Even so, it is clear from the output that the times in the system are rather flatly distributed.

```
TABLE    TIMEIN

ENTRIES IN TABLE MEAN ARGUMENT STANDARD DEVIATION    SUM OF ARGUMENTS
   2397.0000         41.1454           22.7188          98625.5517

  UPPER     OBSERVED       PERCENT  CUMULATIVE  CUMULATIVE     MULTIPLE
  LIMIT    FREQUENCY      OF TOTAL  PERCENTAGE   REMAINDER      OF MEAN
   ...
  10.0000    18.0000       0.7509    0.75          99.25         0.2430
  11.0000    31.0000       1.2933    2.04          97.96         0.2673
  12.0000    20.0000       0.8344    2.88          97.12         0.2916
  13.0000    26.0000       1.0847    3.96          96.04         0.3160
  14.0000    31.0000       1.2933    5.26          94.74         0.3403
  15.0000    33.0000       1.3767    6.63          93.37         0.3646
  16.0000    41.0000       1.7105    8.34          91.66         0.3889
  17.0000    31.0000       1.2933    9.64          90.36         0.4132
```

FIGURE 21.6
Portion of output from program CHAP21B.GPS.

(Continued)

18.0000	39.0000	1.6270	11.26	88.74	0.4375
19.0000	44.0000	1.8356	13.10	86.90	0.4618
20.0000	56.0000	2.3363	15.44	84.56	0.4861
21.0000	48.0000	2.0025	17.44	82.56	0.5104
22.0000	54.0000	2.2528	19.69	80.31	0.5347
23.0000	47.0000	1.9608	21.65	78.35	0.5590
24.0000	57.0000	2.3780	24.03	75.97	0.5833
25.0000	52.0000	2.1694	26.20	73.80	0.6076
26.0000	67.0000	2.7952	28.99	71.01	0.6319
27.0000	60.0000	2.5031	31.50	68.50	0.6562
28.0000	69.0000	2.8786	34.38	65.62	0.6805
29.0000	80.0000	3.3375	37.71	62.29	0.7048
30.0000	44.0000	1.8356	39.55	60.45	0.7291
31.0000	49.0000	2.0442	41.59	58.41	0.7534
32.0000	48.0000	2.0025	43.60	56.40	0.7777
33.0000	57.0000	2.3780	45.97	54.03	0.8020
34.0000	40.0000	1.6688	47.64	52.36	0.8263
35.0000	39.0000	1.6270	49.27	50.73	0.8506
36.0000	43.0000	1.7939	51.06	48.94	0.8749
OVERFLOW	1173.0000	48.94	100.00	0.00	

AVERAGE VALUE OF OVERFLOW IS 59.0664

FIGURE 21.6 (Continued)
Portion of output from program CHAP21B.GPS.

21.5 The MARK Block

Suppose you want a tabulation of times for the transaction to go from point A to point B in a system. It is first necessary to make a record of the time the transaction is at point A. This is done by making a record of the clock time by storing the ABSOLUTE CLOCK value in a transaction's parameter. Which parameter this time is to be stored in must be specified. The Block that does this is the `MARK` Block. The general form of the Block is

```
MARK   P(type)j
```

where (type) can be either PH, PF, PL or PB and `j` is the number of the parameter.

NOTE: `MARK` (number of parameter)$(parameter type) will also work. This form will not be used here.

When a transaction leaves this Block, the effect is to put the ABSOLUTE CLOCK time in the transaction's parameter as specified by the operand number. Since the ABSOLUTE CLOCK is a real number, if the parameter number does not refer to a floating point parameter, it is truncated to an integer. When this happens, the following message appears:

```
"IN STATEMENT 9 - WARNING 393 - Clock value (floating point) will be
truncated to an integer value."
```

Thus, if the clock was at 1234.567 and you had:

```
MARK   4PH
```

the half-word parameter 4 would have the value of 1234. This is normally not desired, and when using this Block, it is recommended that floating point parameters be used to avoid this round-off error. For example, you could have:

```
GENERATE    ,,,1,,5PL
.........   ........
MARK        5PL
```

If you had omitted the PL in the MARK Block, you would get an error message.

NOTE: Don't forget that, since you are now specifying that the transaction has five floating point parameters, it does not have *any* half-word parameter.

The SNA that goes with the MARK Block is

```
MP(number)(type)
```

where (number) is the parameter number. The effect of this is to subtract this value from the current ABSOLUTE CLOCK value. Referring to the same example, if the clock value was now 2344.777, the value of MPL4 would be 1000.210.

Examples of this are as follows:

```
FIRST   TABLE      MP1PL,0,50,20
SECOND  TABLE      MP2PL,0,40,22
THIRD   TABLE      MP3PL,0,30,25
        GENERATE   30,12,,,,12PH,3PL
        .......
        MARK       1PL
        .......
        MARK       2PL
        .......
        MARK       3PL
        TABULATE   FIRST
        TABULATE   SECOND
        TABULATE   THIRD
```

The above will give three tables of times it takes the transactions to travel from the three points in the program where it has passed through the Blocks.

```
MARK  1PL
MARK  2PL
MARK  3PL
```

Example 21.2: A Second Look at the Small Gold Miner's Problem

Recall Example 12.2, which was the example of a gold mine owner. His problem was to find the correct number of trucks to have for his mine. The program to model his situation was CHAP12C.GPS. Let us now make the additions to the program to determine the following times:

Cycle time from shovel to dump and return
Time from shovel to dump
Time from dump to shovel

Program CHAP12C.GPS needs to be modified. The GENERATE Block now needs to have PL parameters. The rest of the modifications are easily made. The listing of the program, CHAP21C.GPS, is given in Figure 21.7. The output is shown in Figure 21.8.

The cycle time for the trucks is 16.95 minutes. Considering that the average of all times (loading, hauling, dumping, and returning) is 16.5, this is reasonable. The difference in actual cycle time and this average is due to the slight queuing. The other two times divide the total cycle times into the time that the trucks arrive at the shovel, time they are loaded, and the time they arrive at the dump and then the time for the truck to dump and return to the shovel. Notice that they equal the complete cycle time as they should.

A large mining example will have many such times determined.

```
         SIMULATE
******************************
*  CHAP21C.GPS  PROGRAM TO    *
*  MODEL A MINE WITH ONE      *
*  SHOVEL AND N-TRUCKS        *
******************************
   FIRST   TABLE       MP1PL,10,2,20
   SECOND  TABLE       MP2PL,6,1,20
   THIRD   TABLE       MP3PL,6,1,20
   TRUCKS  GENERATE    ,,,4,,3PL        PROVIDE TRUCKS FOR THE MINE
   UPTOP   MARK        1PL         FIRST RECORD
           MARK        2PL
           QUEUE       WAIT        QUEUE AT THE SHOVEL
           SEIZE       SHOVEL      USE THE SHOVEL
           DEPART      WAIT        LEAVE THE QUEUE
           ADVANCE     3.5,1.25    LOAD A TRUCK
           RELEASE     SHOVEL      FREE THE SHOVEL
           ADVANCE     6,.5        TRAVEL TO DUMP
           MARK        3PL         SECOND RECORD
           TABULATE    SECOND      ENTRY ON SECOND TABLE
           SEIZE       DUMP        ONLY ONE TRUCK CAN DUMP
           ADVANCE     2.,.4       DUMP A LOAD
           RELEASE     DUMP        FREE THE DUMP
           ADVANCE     5,.3        RETURN TO SHOVEL
           TABULATE    FIRST       ENTRY IN SECOND TABLE
           TABULATE    THIRD       ENTRY IN THIRD TABLE
           TRANSFER    ,UPTOP      JOIN QUEUE AGAIN
           GENERATE    480*100     SIMULATE FOR 100 SHIFTS
           TERMINATE   1
           START       1
           PUTPIC      LINES=5,N(TRUCKS),FC(DUMP)/100,FR(SHOVEL)/10.,_
                       QA(WAIT)
   RESULTS OF SIMULATION
  NUMBER OF TRUCKS USED      **
  NUMBER OF LOADS DUMPED ***   PER SHIFT
  UTIL. OF SHOVEL            **.**%
  AVERAGE QUEUE AT SHOVEL   *.**
            PUTPIC     LINES=3,TB(FIRST),TB(SECOND),TB(THIRD)
     CYCLE TIME FOR TRUCKS - SHOVEL TO DUMP AND BACK          **.**
     TIME FOR TRUCKS TO LOAD AND ARRIVE AT DUMP               **.**
     TIME FOR TRUCKS TO ARRIVE AT DUMP AND RETURN TO SHOVEL **.**
      END
```

FIGURE 21.7
Listing of program CHAP21C.GPS.

```
  RESULTS OF SIMULATION
NUMBER OF TRUCKS USED    4
NUMBER OF LOADS DUMPED 113   PER SHIFT
UTIL. OF SHOVEL          82.67%
AVERAGE QUEUE AT SHOVEL  0.10
   CYCLE TIME FOR TRUCKS - SHOVEL TO DUMP AND BACK          16.95
   TIME FOR TRUCKS TO LOAD AND ARRIVE AT DUMP                9.94
   TIME FOR TRUCKS TO ARRIVE AT DUMP AND RETURN TO SHOVEL   7.01
```

FIGURE 21.8
Results from program CHAP21C.GPS.

21.6 Additional Tables

There are several tables that GPSS/H can make for you with only a minor change to the program. These tables are as follows:

21.6.1 IA Mode Table

Whenever a `GENERATE` Block is used, the distribution of interarrival times is required as data. Often, the interarrival times at points *interior* to a model are required. The logic of doing this is clear. At a particular point in a model, whenever a transaction arrives, a record is made of the arrival time, say, 456.78 time units. At a later time, a second transaction arrives at the same point at, say, 666.99 time units. The interarrival time is then determined as: 666.99 – 456.78 or 210.21.

This time can be tabulated by the following lines of code:

```
FIRST   TABLE     IA,0,20,25
        .....
        ......
        TABULATE  FIRST
```

The `IA` in the `TABLE` statement is essential for the first operand.

Example 21.3: The IA Mode Table

Trucks in a mine come for two services. They arrive for the service every 12 ± 5 minutes. This service takes 11 ± 6 minutes. They then travel to the second service (no time passes), which takes 10 ± 3.5 minutes. The mining engineer wishes to determine the interarrival times it takes for the trucks to arrive from the first service to the second. Simulate for 100 shifts of 480 minutes to provide this information.

Solution
The program to do the simulation is CHAP21D.GPS. Its listing is given in Figure 21.9. The output is shown in Figure 21.10.

The average interarrival time of the trucks from machine 1 to machine 2 is 12.07 minutes. The actual histogram of the times is found in the file CHAP21D.LIS.

```
****************************
*  PROGRAM CHAP21D.GPS      *
*  ILLUSTRATE THE IA TABLE  *
****************************
  ARRIV   TABLE       IA,5,1,20
          GENERATE    12,5      TRUCKS ARRIVE
          QUEUE       FIRST     FIRST INSPECTION
          SEIZE       MACH1
          DEPART      FIRST
          ADVANCE     11,6
          RELEASE     MACH1
          TABULATE    ARRIV     TABLE OF INTERARRIVAL TIMES
          QUEUE       SECOND    SECOND INSPECTION
          SEIZE       MACH2
          DEPART      SECOND
          ADVANCE     10,3.5
          RELEASE     MACH2
          TERMINATE
          GENERATE    480*100
          TERMINATE   1
          START       1
          PUTPIC,     LINES=3,FR(MACH1)/10.,FR(MACH2)/10.,TB(ARRIV)
    UTIL. OF MACH1  **.**%
    UTIL. OF MACH2  **.**%
    AVERAGE INTERARRIVAL TIME AT MACH 2 **.**
          END
```

FIGURE 21.9
Listing of program CHAP21D.GPS.

```
UTIL. OF MACH1  90.72%
UTIL. OF MACH2  82.84%
AVERAGE INTERARRIVAL TIME AT MACH 2 12.07
```

FIGURE 21.10
Output from program CHAP21D.GPS.

21.6.2 RT Mode Table

Closely related to the IA mode is the *rate* of the arrivals at a point. For example, the arrivals per 10 seconds, the arrivals per minute, and so on. A table having these values is constructed as follows:

```
SECOND      TABLE     RT,0,15,25,10
```

Notice that there is an *E* operand in the TABLE statement. This gives the time span to be used. In the above example, the TABLE will give arrivals per 10 time units. To make an entry in the table, the Block

```
TABULATE    SECOND
```

is used.

21.6.3 Q-Table Mode

The average residence time in a queue is often required. The average time is given in the ordinary output. The Q-Table gives the table of average times in the queue. Amazingly enough, this table requires only a single line of code:

```
NAME   QTABLE   WAIT,0,600,20
```

Whenever a transaction leaves the queue `WAIT`, an entry is made in the table of the time it spent in the queue.

Example 21.4: Assembly Line with Five Machines in Series

Another example of the `MARK` Block is presented next. Consider an assembly line that is to be built. It will consist of five machines in series. These are called A, B, C, D, and E. Parts come every 20 ± 4 seconds. They take 2 seconds to arrive at machine A and 2 seconds to travel from machine to machine. The times each machine takes is given in Table 21.1.

The plant foreman would like to determine how long it takes the parts to travel from machine to machine as well as how long they take to travel from the start of the assembly line to the end. This means that there are 15 average times desired. These are the 15 times:

Times Desired				
A to B	B to C	C to D	D to E	E to out
A to C	B to D	C to E	D to out	
A to D	B to E	C to out		
A to E	B to out			
A to out				

The assembly line has not yet been installed, and the foreman is wondering where any bottlenecks might occur. He notes that if each part moved along through the assembly line with each machine using only the average times, (i.e., machine A takes 18 seconds, machine B takes 17 seconds, etc.), it would result in an expected average time for a part to be finished of 87 seconds at all five stations. This, of course, assumes no queuing. He has the option to purchase different machines for stations B and D. These machines work at rates sampled from the distributions `RVNORM(1,17,2)` at B and `RVNORM(1,16,2)` at D. These are a bit more expensive, but he would like to know whether they would improve the efficiency of the system.

The program to simulate the system is given by CHAP21E.GPS. Its listing is as shown in Figure 21.11. The results of the simulation are as shown in Figure 21.12.

TABLE 21.1

Times for Each Machine to Work on a Part

Machine	Time to Work on Part
A	18 ± 6
B	RVEXPO(1,17)
C	RVNORM(1,17,2.5)
D	RVEXPO(1,16)
E	RVNORM(1,19,2)

```
          SIMULATE
********************************
*  PROGRAM CHAP21E.GPS          *
*  EXAMPLE OF THE MARK BLOCK    *
********************************
 1        TABLE       MP1PL,12,2,20
 2        TABLE       MP1PL,24,2,20
 3        TABLE       MP1PL,36,2,20
 4        TABLE       MP1PL,48,2,20
 5        TABLE       MP1PL,60,2,20
 6        TABLE       MP2PL,20,2,20
 7        TABLE       MP2PL,40,2,20
 8        TABLE       MP2PL,60,2,20
 9        TABLE       MP2PL,80,2,20
 10       TABLE       MP3PL,20,2,20
 11       TABLE       MP3PL,40,2,20
 12       TABLE       MP3PL,60,2,20
 13       TABLE       MP4PL,20,2,20
 14       TABLE       MP4PL,40,2,20
 15       TABLE       MP5PL,20,2,20
          GENERATE    20,4,,,,10PL   PARTS ARRIVE
          ADVANCE     2              TRAVEL TO MACHINE A
          MARK        1PL
          QUEUE       WAIT1
          SEIZE       MACHA
          DEPART      WAIT1
          ADVANCE     18,6
          RELEASE     MACHA
          ADVANCE     2             TRAVEL TO MACHINE B
          MARK        2PL
          TABULATE    1               A TO B
          QUEUE       WAIT2
          SEIZE       MACHB
          DEPART      WAIT2
          ADVANCE     RVEXPO(1,17)
          RELEASE     MACHB
          ADVANCE     2             TRAVEL TO MACHINE C
          MARK        3PL
          TABULATE    2              A TO C
          TABULATE    6              B TO C
          QUEUE       WAIT3
          SEIZE       MACHC
          DEPART      WAIT3
          ADVANCE     RVNORM(1,17,2.5)
          RELEASE     MACHC
          ADVANCE     2             TRAVEL TO MACHINE D
          MARK        4PL
          TABULATE    3             A TO D
          TABULATE    7             B TO D
          TABULATE    10            C TO D
          QUEUE       WAIT4
          SEIZE       MACHD
          DEPART      WAIT4
          ADVANCE     RVEXPO(1,16)
          RELEASE     MACHD
          ADVANCE     2             TRAVEL TO MACHINE E
          MARK        5PL
          TABULATE    4             A TO E
          TABULATE    8             B TO E
          TABULATE    11            C TO E
          TABULATE    13            D TO E
          QUEUE       WAIT5
```

FIGURE 21.11
Listing of program CHAP21E.GPS.

(Continued)

```
         SEIZE       MACHE
         DEPART      WAIT5
         ADVANCE     RVNORM(1,19,2)
         RELEASE     MACHE
         ADVANCE     2              PART FINISHED
         TABULATE    5            A TO OUT
         TABULATE    9            B TO OUT
         TABULATE    12           C TO OUT
         TABULATE    14           D TO OUT
         TABULATE    15           E TO OUT
         TERMINATE
         GENERATE    3600*100
         TERMINATE   1
         START       1
         PUTSTRING   ('  ')
         PUTSTRING   ('  ')
         PUTPIC      LINES=12,TB1,TB6,TB10,TB13,_
                     TB2,TB7,TB11,TB14,_
                     TB3,TB8,TB12,_
                     TB4,TB9,_
                     TB5,TB15,_
                     QA(WAIT1),QA(WAIT2),QA(WAIT3),QA(WAIT4),QA(WAIT5)
             RESULTS OF SIMULATION  (MEAN TIMES)
A TO B    ***.**   B TO C    ***.**   C TO D    ***.**  D TO E    ***.**
A TO C    ***.**   B TO D    ***.**   C TO E    ***.**  D TO OUT  ***.**
A TO D    ***.**   B TO E    ***.**   C TO OUT  ***.**
A TO E    ***.**   B TO OUT  ***.**
A TO OUT  ***.**                                        E TO OUT  ***.**
             AVERAGE QUEUES AT EACH MACHINE
 MACH  A    *.***
 MACH  B    *.***
 MACH  C    *.***
 MACH  D    *.***
 MACH  E    *.***
   END
```

FIGURE 21.11 (Continued)
Listing of program CHAP21E.GPS.

The results are somewhat surprising, as the mean time for a part to move from the first machine, A, until it is finished is 297.24 seconds. This is considerably higher than the average of 87 seconds using just the mean times for the machines. An inspection of the average queue lengths shows that there are long queues at the two machines that work according to the exponential distribution. At machine B, the average queue is 2.526 and at machine E, the average queue length is 4.043. Program CHAP21F.GPS shows what the results would be if the more expensive machines were purchased. These changes result in considerable savings in the times for parts to move through the system. Figure 21.13 gives the results of running program CHAP21F.

The average time for a part to go through the system is now reduced from, 297.24 to 108.26 seconds, which is a difference of over 3.15 minutes. The queues are also considerably reduced. The next step for the foreman is to do a simple economic analysis to see if the increased production justifies the more expensive equipment. He now has the data he needs to do these calculations.

The animation is given by program CHAP21F.GPS. Its listing is shown in Figure 21.14.

The results from this program will differ from the original one since the facilities are not RELEASEd immediately when a part is finished at a machine. This is done only for the animation. In a real case, the actual data needs to be used. The animation is shown in Figure 21.15.

```
         RESULTS OF SIMULATION  (MEAN TIMES)
A TO B     22.31   B TO C     69.81   C TO D     48.16  D TO E     54.90
A TO C     92.11   B TO D    117.97   C TO E    103.06  D TO OUT 156.96
A TO D    140.27   B TO E    172.87   C TO OUT 205.12
A TO E    195.18   B TO OUT 274.94
A TO OUT 297.24                                         E TO OUT 102.04
         AVERAGE QUEUES AT EACH MACHINE
 MACH  A    0.116
 MACH  B    2.526
 MACH  C    1.455
 MACH  D    1.842
 MACH  E    4.043
```

FIGURE 21.12
Results of simulation of assembly line example CHAP21E.GPS.

```
         RESULTS OF SIMULATION  (MEAN TIMES)
A TO B     22.61   B TO C     20.24   C TO D     20.20  D TO E     18.59
A TO C     42.86   B TO D     40.44   C TO E     38.79  D TO OUT   45.21
A TO D     63.05   B TO E     59.03   C TO OUT   65.41
A TO E     81.64   B TO OUT   85.65
A TO OUT 108.26                                         E TO OUT   26.62
         AVERAGE QUEUES AT EACH MACHINE
 MACH  A    0.129
 MACH  B    0.064
 MACH  C    0.059
 MACH  D    0.030
 MACH  E    0.282
```

FIGURE 21.13
Results of using different machines in the assembly line.

```
            SIMULATE
****************************************
*  ANIMATION OF CHAP21E.GPS            *
*  ORIGINAL ASSEMBLY LINE MODEL        *
*  PROGRAM CHAP21G.GPS                 *
*  ANIMATION GIVES SLIGHTLY DIFFERENT  *
*  RESULTS DUE TO DELAY IN USING       *
*  NEW MACHINES FOR EACH PART          *
*  SEE TEXT FOR EXPLANATION            *
****************************************
 ATF      FILEDEF     'CHAP21G.ATF'
 1        TABLE       MP1PL,12,2,20
 2        TABLE       MP1PL,24,2,20
 3        TABLE       MP1PL,36,2,20
 4        TABLE       MP1PL,48,2,20
 5        TABLE       MP1PL,60,2,20
 6        TABLE       MP2PL,20,2,20
 7        TABLE       MP2PL,40,2,20
 8        TABLE       MP2PL,60,2,20
 9        TABLE       MP2PL,80,2,20
 10       TABLE       MP3PL,20,2,20
 11       TABLE       MP3PL,40,2,20
 12       TABLE       MP3PL,60,2,20
 13       TABLE       MP4PL,20,2,20
 14       TABLE       MP4PL,40,2,20
 15       TABLE       MP5PL,20,2,20
          GENERATE    20,4,0,,,10PL   PARTS ARRIVE
          BPUTPIC     FILE=ATF,LINES=4,AC1,XID1,XID1,XID1
```

FIGURE 21.14
Program CHAP21F.GPS to animate CHAP21E.GPS. *(Continued)*

```
TIME *.****
CREATE PART1 P*
PLACE P* ON P1
SET P* TRAVEL 2
         ADVANCE     2              TRAVEL TO MACHINE A
         MARK        1PL
         QUEUE       WAIT1
         SEIZE       MACHA
         BPUTPIC     FILE=ATF,LINES=3,AC1,XID1
TIME *.****
PLACE P* AT 9 37
SET M1 COLOR F2
         DEPART      WAIT1
         ADVANCE     18,6
         BPUTPIC     FILE=ATF,LINES=5,AC1,XID1,XID1,XID1
TIME *.****
PLACE P* ON P2
SET P* TRAVEL 2
SET P* CLASS PART2
SET M1 COLOR F7
         ADVANCE     .3
         RELEASE     MACHA
         ADVANCE     1.7            TRAVEL TO MACHINE B
         MARK        2PL
         TABULATE    1               A TO B
         BPUTPIC     FILE=ATF,LINES=2,AC1,TB1
TIME *.****
WRITE M1 ***.**
         QUEUE       WAIT2
         SEIZE       MACHB
         BPUTPIC     FILE=ATF,LINES=3,AC1,XID1
TIME *.****
PLACE P* AT 24 37
SET M2 COLOR F2
         DEPART      WAIT2
         ADVANCE     RVEXPO(1,17)
         BPUTPIC     FILE=ATF,LINES=5,AC1,XID1,XID1,XID1
TIME *.****
PLACE P* ON P3
SET P* CLASS PART3
SET M2 COLOR F7
SET P* TRAVEL 2
         ADVANCE     .3
         RELEASE     MACHB
         ADVANCE     1.7            TRAVEL TO MACHINE C
         MARK        3PL
         TABULATE    2              A TO C
         TABULATE    6              B TO C
         BPUTPIC     FILE=ATF,LINES=3,AC1,TB2,TB6
TIME *.****
WRITE M2 ***.**
WRITE M6 ***.**
         QUEUE       WAIT3
         SEIZE       MACHC
         BPUTPIC     FILE=ATF,LINES=3,AC1,XID1
```

FIGURE 21.14 (Continued)
Program CHAP21F.GPS to animate CHAP21E.GPS. *(Continued)*

```
TIME *.****
PLACE P* AT 38 37
SET M3 COLOR F2
          DEPART       WAIT3
          ADVANCE      RVNORM(1,17,2.5)
          BPUTPIC      FILE=ATF,LINES=5,AC1,XID1,XID1,XID1
TIME *.****
PLACE P* ON P4
SET P* CLASS PART4
SET M3 COLOR F7
SET P* TRAVEL 2
          ADVANCE     .3
          RELEASE      MACHC
          ADVANCE      1.7           TRAVEL TO MACHINE D
          MARK         4PL
          TABULATE     3              A TO D
          TABULATE     7              B TO D
          TABULATE     10             C TO D
          BPUTPIC      FILE=ATF,LINES=4,AC1,TB3,TB7,TB10
TIME *.****
WRITE M3 ***.**
WRITE M7 ***.**
WRITE M10 ***.**
          QUEUE        WAIT4
          SEIZE        MACHD
          BPUTPIC      FILE=ATF,LINES=3,AC1,XID1
TIME *.****
PLACE P* AT 52 37
SET M4 COLOR F2
          DEPART       WAIT4
          ADVANCE      RVEXPO(1,16)
          BPUTPIC      FILE=ATF,LINES=5,AC1,XID1,XID1,XID1
TIME *.****
PLACE P* ON P5
SET P* CLASS PART5
SET M4 COLOR F7
SET P* TRAVEL 2
          ADVANCE      .3
          RELEASE      MACHD
          ADVANCE      1.7             TRAVEL TO MACHINE E
          MARK         5PL
          TABULATE     4              A TO E
          TABULATE     8              B TO E
          TABULATE     11             C TO E
          TABULATE     13             D TO E
          BPUTPIC      FILE=ATF,LINES=5,AC1,TB4,TB8,TB11,TB13
TIME *.****
WRITE M4 ***.**
WRITE M8 ***.**
WRITE M11 ***.**
WRITE M13 ***.**
          QUEUE        WAIT5
          SEIZE        MACHE
          BPUTPIC      FILE=ATF,LINES=3,AC1,XID1
```

FIGURE 21.14 (Continued)
Program CHAP21F.GPS to animate CHAP21E.GPS.

(Continued)

```
TIME *.****
PLACE P* at 56 37
SET M5 COLOR F2
          DEPART       WAIT5
          ADVANCE      RVNORM(1,19,2)
          BPUTPIC      FILE=ATF,LINES=5,AC1,XID1,XID1,XID1
TIME *.****
PLACE P* ON P6
SET P* CLASS PART6
SET M5 COLOR F7
SET P* TRAVEL 2
          ADVANCE      .3
          RELEASE      MACHE
          ADVANCE      1.7             PART FINISHED
          TABULATE     5             A TO OUT
          TABULATE     9             B TO OUT
          TABULATE     12            C TO OUT
          TABULATE     14            D TO OUT
          TABULATE     15            E TO OUT
          BPUTPIC      FILE=ATF,LINES=7,AC1,TB5,TB9,TB12,TB14,TB15,XID1
TIME *.****
WRITE M5 ***.**
WRITE M9 ***.**
WRITE M12 ***.**
WRITE M14 ***.**
WRITE M15 ***.**
DESTROY P*
          TERMINATE
          GENERATE     3600
          TERMINATE    1
          START        1
          PUTSTRING    ('  ')
          PUTSTRING    ('  ')
          PUTPIC       LINES=12,TB1,TB6,TB10,TB13,_
                       TB2,TB7,TB11,TB14,_
                       TB3,TB8,TB12,_
                       TB4,TB9,_
                       TB5,TB15,_
                       QA(WAIT1),QA(WAIT2),QA(WAIT3),QA(WAIT4),QA(WAIT5)
              RESULTS OF SIMULATION  (MEAN TIMES)
       A TO B   ***.**   B TO C   ***.**   C TO D   ***.**  D TO E   ***.**
       A TO C   ***.**   B TO D   ***.**   C TO E   ***.**  D TO OUT ***.**
       A TO D   ***.**   B TO E   ***.**   C TO OUT ***.**
       A TO E   ***.**   B TO OUT ***.**
       A TO OUT ***.**                                       E TO OUT ***.**
              AVERAGE QUEUES AT EACH MACHINE
        MACH  A   *.***
        MACH  B   *.***
        MACH  C   *.***
        MACH  D   *.***
        MACH  E   *.***
          PUTPIC       FILE=ATF,LINES=2,AC1
TIME *.****
END
          END
```

FIGURE 21.14 (Continued)
Program CHAP21F.GPS to animate CHAP21E.GPS.

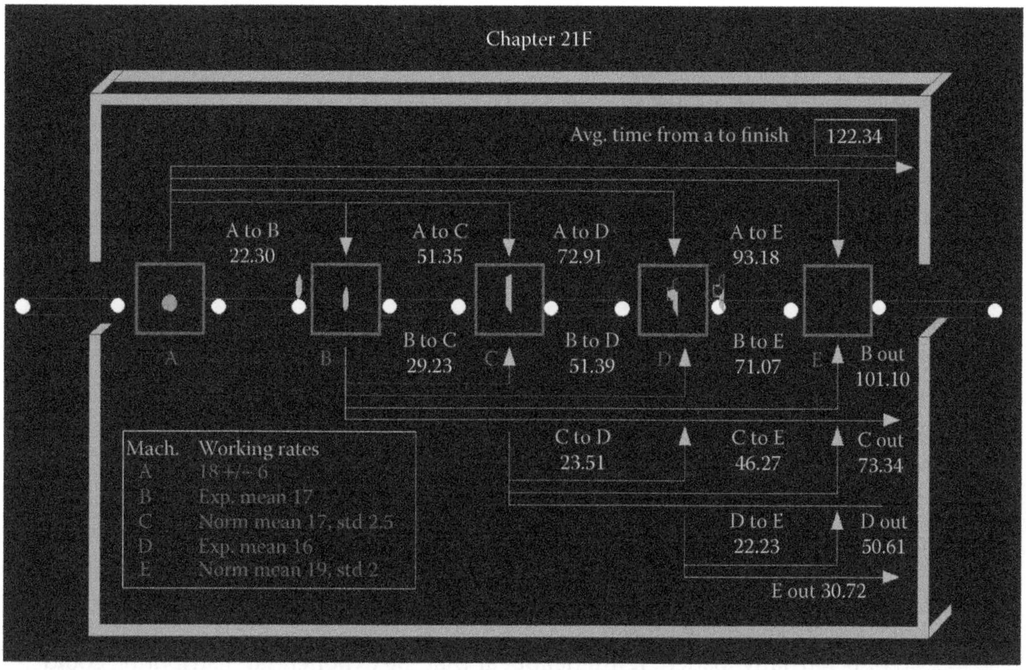

FIGURE 21.15
Animation of factory system in Example 21.4—program is CHAP21F.GPS.

21.10 Exercises

21.1. What do the following lines of code make tables of? Assume that all the GPSS/H entities have been defined.

```
(a) FIRST       TABLE        Q(WAIT),0,1,10
                ........
                TABULATE     FIRST
(b) SECOND      TABLE        FR(MACH1),250,25,25
                ........ .
                TABULATE     SECOND
(c) THIRD       TABLE        R(TROLLY),0,1,20
                ........ .
                TABULATE     THIRD
(d) FOURTH      TABLE        MP4PL,20,2,44
                ........ .
                TABULATE     FOURTH
```

21.2. Write the GPSS/H program that has a single transaction cycling as follows:

It uses a single machine that takes a time of (6,1).

It travels to a second machine in a time of (4,.5).

It takes a time of (5,.8) on this machine.

It then returns to the first machine in a time of (3,.3).

This process is repeated for 100 cycles. Determine the average cycle time for 100 cycles and the utilization of the two machines.

The notation (*a*,*b*) means a time sampled from the normal distribution with a mean of *a* and a standard deviation of *b*.

21.3. Refer to Exercise 21.2. Determine the cycle times and machine utilizations if 2, then 3, and finally five transactions are introduced to the cycle.

21.4. Redo Exercise 21.2 but now with the distributions changed from normal distributions to exponential (Poisson) distributions with means as given. How does this change the results? Run for 1, 2, 3, 4, and 5 transactions.

21.5. In many mining operations, the distributions of the times for various operations are not symmetrical and a skewed distribution is needed. This is often the case for loading a truck when the shovel or loader needs multiple passes depending on the material near it. Sometimes, it needs to move or reposition itself while a truck is ready to load or is partially loaded. It might appear that the exponential distribution could be used in this case. However, the exponential distribution tends to have a long tail that would distort the actual loading times. One possibility is to use an Erlang distribution. One such Erlang distribution is derived from the exponential distribution as follows. Suppose that there is an exponential distribution with a mean μ. Then, an Erlang distribution of order *m* is made by sampling from *m*-consecutive exponential distributions. For example, if we have the GPSS/H line:

```
ADVANCE   RVEXPO(1,10)
```

An Erlang distribution of order 3 would be as follows:

```
ADVANCE   RVEXPO(1,3.333)
ADVANCE   RVEXPO(1,3.333)
ADVANCE   RVEXPO(1,3.333)
```

The result is still an exponential distribution with a mean of 10, but the long tail has been cut off. Redo Exercise 21.4 with Erlang distributions of order 3.

21.6. Modify Example 21.1 as follows:

Have another queue to keep track of the people from the time they enter the shop until they leave. Make a table of the number of people in this queue. Make a table of the time people are in the store.

22

LOGIC Switches and the GATE Block

Longwall mining for coal mine.

22.1 LOGIC Switches and the GATE Block

22.1.1 LOGIC Switches

Consider the following Blocks:

```
        (main program)
        TEST E        &LOCK,0
        .........     ........
        .........     ........
        (secondary program)
        GENERATE      ,,,1
BACK    ADVANCE       RVEXPO(1,125)
        BLET          &LOCK = 1
```

```
ADVANCE      RVEXPO(1,12.5)
BLET         &LOCK = 0
TRANSFER     ,BACK
```

When the program begins, the value of the ampervariable, `&LOCK` is 0 (as are all ampervariables unless the `INITIAL` statement is used to specify initial values), so the transactions that enter the `TEST` Block will pass through to the next sequential Block. The transaction created in the `GENERATE` Block is put on the FEC chain for a time given by sampling from the exponential distribution with a mean of 125. After this transaction is returned to the CEC, the value of the ampervariable `&LOCK` is set equal to 1 (any other value could have been selected for the ampervariable for this example). The transaction is then put on the FEC for a time given by sampling from the exponential distribution with a mean of 12.5. Now, any transactions that enter the `TEST` Block will be delayed until the value of `&LOCK` again becomes equal to 0. This delay might represent a traffic light turning red, a break in a factory for lunch, a breakdown of an assembly line, and so on.

Rather than use ampervariables and `TEST` Blocks in this manner, there is a better way to handle such conditions in GPSS/H. This is using switches that are either *on* or *off*. The *on* condition is known as *set* and the *off* as *reset*. These switches are turned on and off by the `LOGIC` Block. Its form is as follows:

```
LOGIC (R)      (name or number of switch)
```

where `R` is a logic relationship and is either:

- `S` for set
- `R` for reset
- `I` for invert

When a transaction enters a `LOGIC` Block, the effect is as follows:

1. If the logic relationship is `R`, the switch is put in a reset position.
2. If the logic relationship is `S`, the switch is put in a set position.
3. If the logic relationship is `I`, the switch is inverted, that is, if it was set, it becomes reset and vice versa.

Examples of these might be as follows:

```
LOGIC S      HALT
LOGIC R      1
LOGIC I      WAIT
```

Notice that logic switches can be named either symbolically or numerically. The usual rules apply for selecting names or numbers for these switches.

When a program begins, all switches are in a reset position. It is possible to have the switches in a set position by means of the initial statement. A form of it is as follows:

```
INITIAL LS(switch1)/LS(switch2)/LS1
```

An example of it is as follows:

```
INITIAL    LS(HALT)/LS(PATH)/LS1
```

Alternately, in place of the parenthesis, one can use a single dollar sign ($), although this is rarely done. For example, the above line of code could have been written as follows:

```
INITIAL    LS$HALT/LS$PATH/LS1
```

If you have multiple switches all given by numbers, there is a shorthand way to have them in a set position:

```
INITIAL    LS1-5/LS9-12
```

This would put logic switches 1 through 5 in a set position and also switches 9 through 12 in a set position.

Some sample program code might be as follows:

```
          GENERATE     ,,,1               DUMMY TRANSACTION
UPTOP     ADVANCE      RVEXPO(1,200)      MACHINE WORKING
          LOGIC S      STOPIT             MACHINE DOWN
          ADVANCE      RVNORM(1,20,3.5)   MACHINE BEING FIXED
          LOGIC R      STOPIT             MACHINE FIXED
          TRANSFER     ,UPTOP             BACK TO WORK
```

The above code might be used to represent a machine that works for a certain time and then is shut down when repairs and/or maintenance is performed. The time between breakdowns is given by sampling from the exponential distribution with a mean of 200. When it is down, it is fixed or otherwise maintained. This takes a time that is normally distributed with a mean of 20 and a standard deviation of 3.5. Notice how the dummy transaction keeps looping in the program segment and alternately sets the logic switch from set to reset. Although it would not be as clear, it would have been all right to have the LOGIC Blocks as follows:

```
LOGIC I    STOPIT
```

22.1.2 The GATE Block

Logic switches generally need another Block in the main program in order to be of any use. This Block is often the GATE Block. Logic switches can be used in other Blocks such as the TEST Block, as we shall see later. This Block works as its name implies, much like a gate in the path of the transactions. When the gate is open, transactions pass through; and when it is closed, they will either wait until the gate becomes open or be routed elsewhere.

There are two forms of the GATE Block. The first is refusal mode. The general form is as follows:

```
GATE R     (logic switch)
```

The logical relation R is either LS or LR. LS is short for *logic switch set* and LR for *logic switch reset*. An example of this might be as follows:

```
GATE LS       HALT
```

When a transaction arrives at this Block, it tests to see if the logic switch HALT is in a set position. If so, it moves to the next sequential Block. If the logic switch HALT is in a reset position, the transaction remains in the previous Block. It also remains on the CEC. However, an internal flag is turned to the *on* position and the transaction is not scanned again until such time as the switch is turned to an *off* position. This happens when the LOGIC Block is entered by another transaction. If a machine is to be periodically shut down, one might use a GATE Block as follows:

```
QUEUE         WAIT
GATE LR       STOPIT
SEIZE         MACH1
DEPART        WAIT
ADVANCE       RVEXPO(1,25)
RELEASE       MACH1
```

Whenever the logic switch STOPIT is in a set position, the transaction is kept in the Block QUEUE **WAIT**.

GATE Blocks can also be used with facilities and storages as follows:

```
GATE R (label)
```

where:

R can be one of the following:

- U test for facility in use
- NU test for facility not in use
- SF test for storage full
- SNF test for storage not full
- SE test for storage empty
- SNE test for storage not empty

Some examples are as follows:

(a) GATE U MACH1

(b) GATE SNE TUGS

(c) GATE NU SHOP

(d) GATE SF BERTH

In (a), the transaction will be held in the previous Block if the facility MACH1 is *not* in use. In (b), the transaction is delayed if the storage TUGS is in any situation other than empty. Thus, if the storage of TUGS is 4 and one is being used, the transaction will be held up. In (c), the transaction is delayed if the facility SHOP is being used. Finally, for (d), the transaction is held up if the storage BERTH is not full.

22.1.2.1 The GATE Block in Conditional Transfer Mode

When a GATE Block has a second operand, this is the label of a Block. If the GATE is closed, the transaction will be transferred to the Block with this label. Thus,

```
GATE LR HALTIT,AWAY
```

will send the transaction to the Block AWAY whenever the logic switch HALTIT is in a set position.

Example 22.1: Another Small Mine Operator's Problem

Consider a small mine operator's problem. He has five trucks that haul excavated ore away to a crusher. There is a single shovel to load the trucks. The trucks travel in a cycle: load, haul, dump, and return. All times for these activities are found to be normally distributed. Loading times have a mean of 2.5 minutes and a standard deviation of .35 minutes. Travel to the dump has a mean of 7.5 minutes and a standard deviation of 1.26 minutes. Dumping takes an average of 1.75 minutes with a standard deviation of .2 minutes and returning to the shovel takes an average of 5.6 minutes and a standard deviation of 1.1 minutes. Only one truck can be loaded at a time, but there is no such restriction on dumping. The various times for the trucks are the same regardless of the truck.

The trucks periodically break down and/or must be serviced. Each has a different reliability. By calling the trucks 1, 2, 3, 4, and 5, it has been found that the time between failures for the trucks follows the exponential distribution. The time to repair a truck follows the normal distribution. Table 22.1 gives these times:

The downtimes are means for the exponential distribution, and the repair times are mean and standard deviation for the normal distribution. The mine works three shifts, 365 days a year. Each shift has 450 actual minutes of production. Even though the breakdowns of the trucks can be at any place in the system, it is sufficient to have the trucks tested for breakdowns after they dump.

The mine owner is interested in how his mine operates with the data as given.

Solution

The program is written using half-word parameter 1 to store a number from 1 to 5 to represent the five different trucks. When a truck dumps, it checks one of five different GATE Blocks depending on what type of truck it is. If a truck is down, the corresponding LOGIC switch will be in a set position and it will be held up at the GATE Block. There are five program segments to alternately put the logic switches in set and reset positions. The program is CHAP22A.GPS and its listing is given in Figure 22.1. The results of running the program are shown in Figure 22.2.

The animation of this example is important for the mining engineer as so many surface mines are similar in that they have trucks that load and cycle. The program to do the animation is given next, and then the development will be presented in great detail. The program is given first in Figure 22.3. The animation will look as shown in Figure 22.4.

TABLE 22.1

Downtimes and Repair Times for the Trucks

Truck No.	Time between Failure	Repair Time Dist.
1	400	(20, 3.5)
2	425	(35, 6.9)
3	550	(55.5, 12)
4	345	(40, 7)
5	300	(50, 8)

```
          SIMULATE
***************************
*   PROGRAM CHAP22A.GPS    *
*   MODELING DOWN TIMES    *
*   FOR TRUCKS IN A MINE   *
***************************
 TRUCKS   GENERATE      ,,,5     5 TRUCKS IN THE MINE
          ASSIGN        1,N(TRUCKS),PH  ASSIGN TRUCKS NUMBERS
 BACKUP   QUEUE         WAIT
          SEIZE         SHOVEL
          DEPART        WAIT
          ADVANCE       RVNORM(1,2.5,.35)   LOAD
          RELEASE       SHOVEL
          ADVANCE       RVNORM(1,7.5,1.26) HAUL
          ADVANCE       RVNORM(1,1.75,.2)   DUMP
          GATE LR       PH1
          ADVANCE       RVNORM(1,5.6,1.1)   RETURN
  DUMPED  TRANSFER      ,BACKUP   RETURN TO SHOVEL
          GENERATE      ,,,1
  BACKA   ADVANCE       RVEXPO(1,400)    TRUCK WORKING
          LOGIC S       1
          ADVANCE       RVNORM(1,20,3.5)
          LOGIC R       1
          TRANSFER      ,BACKA
          GENERATE      ,,,1
  BACKB   ADVANCE       RVEXPO(1,425)    TRUCK WORKING
          LOGIC S       2
          ADVANCE       RVNORM(1,35,6.9)
          LOGIC R       2
          TRANSFER      ,BACKB
          GENERATE      ,,,1
  BACKC   ADVANCE       RVEXPO(1,550)    TRUCK WORKING
          LOGIC S       3
          ADVANCE       RVNORM(1,55.5,12)
          LOGIC R       3
          TRANSFER      ,BACKC
          GENERATE      ,,,1
  BACKD   ADVANCE       RVEXPO(1,345)    TRUCK WORKING
          LOGIC S       4
          ADVANCE       RVNORM(1,40,7)
          LOGIC R       4
          TRANSFER      ,BACKD
          GENERATE      ,,,1
  BACKE   ADVANCE       RVEXPO(1,300)    TRUCK WORKING
          LOGIC S       5
          ADVANCE       RVNORM(1,50,8)
          LOGIC R       5
          TRANSFER      ,BACKE
          GENERATE      450*3*365  SIMULATE FOR A YEAR
          TERMINATE     1
          START         1
          PUTPIC        LINES=3,N(TRUCKS),FR(SHOVEL)/10.,N(DUMPED)/365.
      TRUCKS WORKING IN THE MINE  **
      UTIL. OF SHOVEL            **.**%
      LOADS DUMPED EACH DAY     ***.**
          END
```

FIGURE 22.1
Listing of program CHAP22A.GPS.

```
TRUCKS WORKING IN THE MINE   5
UTIL. OF SHOVEL            66.01%
LOADS DUMPED EACH DAY     356.64
```

FIGURE 22.2
Results from program CHAP22A.GPS.

```
        SIMULATE
***************************
*   PROGRAM CHAP22B.GPS   *
*   ANIMATION OF CHAP22A  *
***************************
  ATF     FILEDEF     'CHAP22B.ATF'
  WHERE   FUNCTION    PH1,L5
 1,BLOCKA/2,BLOCKB/3,BLOCKC/4,BLOCKD/5,BLOCKE
          REAL        &X,&Y,&Z,&R,&S,&T  DEFINE VARIABLES TO BE USED
          INTEGER     &LOADS
 TRUCKS   GENERATE    3,,0,5     5 TRUCKS IN THE MINE
          BPUTPIC     FILE=ATF,LINES=5,AC1,XID1,XID1,XID1,N(TRUCKS)
TIME *.****
CREATE TRUCKU1 T*
PLACE T* ON P7
SET T* TRAVEL 1
WRITE M3 *
          ADVANCE     1
          ASSIGN      1,N(TRUCKS),PH  ASSIGN A NUMBER TO EACH TRUCK
 BACKUP   QUEUE       WAIT
          SEIZE       SHOVEL
          DEPART      WAIT
          ADVANCE     .42       SPOT
          SEIZE       DUMMY
          BPUTPIC     FILE=ATF,LINES=3,AC1,XID1,XID1
TIME *.****
PLACE T* AT -33 -2
SET T* CLASS TRUCKU
          ADVANCE     RVNORM(1,2.5,.35)  LOAD
          RELEASE     DUMMY
          ADVANCE     .25     LEAVE SHOVEL
          RELEASE     SHOVEL
          BLET        &X=RVNORM(1,2.89,.29)
          BPUTPIC     FILE=ATF,LINES=5,AC1,XID1,XID1,XID1,&X,FR(SHOVEL)/10.
TIME *.****
PLACE T* ON P1
SET T* CLASS TRUCKF
SET T* TRAVEL **.**
WRITE M1 **.**%
          ADVANCE     &X
          BLET        &Y=RVNORM(1,2.75,.28)
          BPUTPIC     FILE=ATF,LINES=3,AC1,XID1,XID1,&Y
TIME *.****
PLACE T* ON P2
SET T* TRAVEL **.**
          ADVANCE     &Y
          BLET        &Z=RVNORM(1,1.84,.18)
          BPUTPIC     FILE=ATF,LINES=3,AC1,XID1,XID1,&Z
TIME *.****
PLACE T* ON P3
SET T* TRAVEL **.**
          ADVANCE     &Z
          SEIZE       DUMP
          BPUTPIC     FILE=ATF,LINES=2,AC1,XID1
TIME *.****
```

FIGURE 22.3
Listing of program CHAP22B.GPS. *(Continued)*

```
PLACE T* AT 47.3 -2.25
          ADVANCE     RVNORM(1,1.75,.2)   DUMP
          BLET        &LOADS=&LOADS+1
 DUMPED   RELEASE     DUMP
          BPUTPIC     FILE=ATF,LINES=2,AC1,&LOADS
TIME *.****
WRITE M2 ***
          TRANSFER    ,FN(WHERE)
  BACKH   BLET        &R=RVNORM(1,1.51,.15)
          BPUTPIC     FILE=ATF,LINES=3,AC1,XID1,XID1,&R
TIME *.****
PLACE T* ON P4
SET T* TRAVEL **.**
          ADVANCE     &R
          BLET        &S=RVNORM(1,1.72,.17)
          BPUTPIC     FILE=ATF,LINES=3,AC1,XID1,XID1,&S
TIME *.****
PLACE T* ON P5
SET T* TRAVEL **.**
          ADVANCE     &S
          BLET        &T=RVNORM(1,2.36,.24)
          BPUTPIC     FILE=ATF,LINES=3,AC1,XID1,XID1,&T
TIME *.****
PLACE T* ON P6
SET T* TRAVEL **.**
          ADVANCE     &T
          TRANSFER    ,BACKUP   RETURN TO SHOVEL
          GENERATE    ,,,1
  BACKA   ADVANCE     RVEXPO(1,400)    TRUCK WORKING
          LOGIC S     1
          ADVANCE     RVNORM(1,20,3.5)
          LOGIC R     1
          TRANSFER    ,BACKA
          GENERATE    ,,,1
  BACKB   ADVANCE     RVEXPO(1,425)    TRUCK WORKING
          LOGIC S     1
          ADVANCE     RVNORM(1,35,6.9)
          LOGIC R     1
          TRANSFER    ,BACKB
          GENERATE    ,,,1
  BACKC   ADVANCE     RVEXPO(1,550)    TRUCK WORKING
          LOGIC S     1
          ADVANCE     RVNORM(1,55.5,12)
          LOGIC R     1
          TRANSFER    ,BACKC
          GENERATE    ,,,1
  BACKD   ADVANCE     RVEXPO(1,345)    TRUCK WORKING
          LOGIC S     1
          ADVANCE     RVNORM(1,40,7)
          LOGIC R     1
          TRANSFER    ,BACKD
          GENERATE    ,,,1
  BACKE   ADVANCE     RVEXPO(1,300)    TRUCK WORKING
          LOGIC S     1
          ADVANCE     RVNORM(1,50,8)
          LOGIC R     1
          TRANSFER    ,BACKE
```

FIGURE 22.3 (Continued)
Listing of program CHAP22B.GPS.

(Continued)

```
 BLOCKA   GATE LS      1,BACKH
          BPUTPIC     FILE=ATF,LINES=3,AC1,XID1,XID1
TIME *.****
PLACE T* AT 44.5 .82
SET T* CLASS TRUCKD
          GATE LR      1
          BPUTPIC      FILE=ATF,LINES=2,AC1,XID1
TIME *.****
SET T* CLASS TRUCKU
          TRANSFER    ,BACKH
 BLOCKB   GATE LS     2,BACKH
          BPUTPIC     FILE=ATF,LINES=3,AC1,XID1,XID1
TIME *.****
PLACE T* AT 47.5 0.82
SET T* CLASS TRUCKD
          GATE LR      2
          BPUTPIC      FILE=ATF,LINES=2,AC1,XID1
TIME *.****
SET T* CLASS TRUCKU
          TRANSFER    ,BACKH
 BLOCKC   GATE LS      3,BACKH
          BPUTPIC     FILE=ATF,LINES=3,AC1,XID1,XID1
TIME *.****
PLACE T* AT 49.5 .82
SET T* CLASS TRUCKD
          GATE LR      3
          BPUTPIC      FILE=ATF,LINES=2,AC1,XID1
TIME *.****
SET T* CLASS TRUCKU
          TRANSFER    ,BACKH
 BLOCKD   GATE LS      4,BACKH
          BPUTPIC     FILE=ATF,LINES=3,AC1,XID1,XID1
TIME *.****
PLACE T* AT 44.5 2.82
SET T* CLASS TRUCKD
          GATE LR      4
          BPUTPIC      FILE=ATF,LINES=2,AC1,XID1
TIME *.****
SET T* CLASS TRUCKU
          TRANSFER    ,BACKH
 BLOCKE   GATE LS     5,BACKH
          BPUTPIC     FILE=ATF,LINES=3,AC1,XID1,XID1
TIME *.****
PLACE T* AT 47.5 2.82
SET T* CLASS TRUCKD
          GATE LR      5
          BPUTPIC      FILE=ATF,LINES=2,AC1,XID1
TIME *.****
SET T* CLASS TRUCKU
          TRANSFER    ,BACKH
          GENERATE     ,,,1     SHOVEL MOVEMENT SECTION
  WAIT    TEST E       F(SHOVEL),1     IS SHOVEL MOVING?
  AGAIN   BPUTPIC      FILE=ATF,LINES=2,AC1
TIME *.****
ROTATE BB -45 STEP 3 TIME .5
          ADVANCE      .5
          BPUTPIC      FILE=ATF,LINES=2,AC1
```

FIGURE 22.3 (Continued)
Listing of program CHAP22B.GPS.

(Continued)

```
TIME *.****
ROTATE BB 45 STEP 3 TIME .5
          ADVANCE     .5
          TEST E      F(SHOVEL),1,WAIT
          TRANSFER    ,AGAIN
          GENERATE    450*3
          TERMINATE   1
          START       1
          PUTPIC      LINES=3,N(TRUCKS),FR(SHOVEL)/10.,N(DUMPED)/1.
      TRUCKS WORKING IN THE MINE  **
      UTIL. OF SHOVEL            **.**%
      LOADS DUMPED EACH DAY     ***.**
          END
```

FIGURE 22.3 (Continued)
Listing of program CHAP22B.GPS.

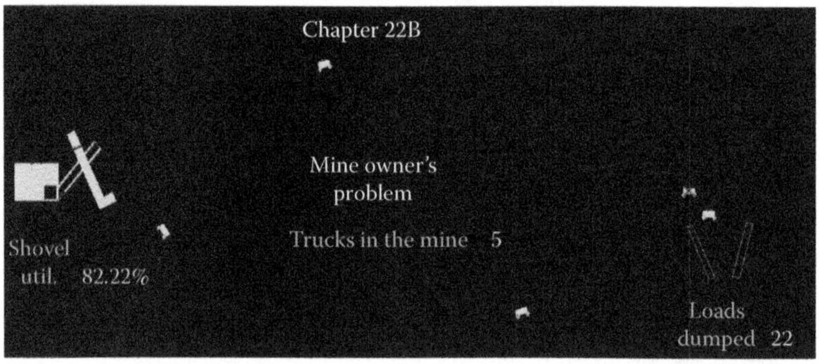

FIGURE 22.4
Screenshot of animation from program CHAP22B.GPS.

STEP-BY-STEP DEVELOPMENT OF ANIMATION FOR CHAP22B.GPS

The development of the animation for CHAP22B.GPS is presented next. In this case, the first step is to begin with the GPSS/H program CHAP22A.GPS. The initial changes are to add the line:

```
ATF FILEDEF   'CHAP22B.ATF'
```

The simulation was for 365 days. This can be changed so that the .ATF file is not large. Next the animation is begun.

The initial layout consists of the travel paths, the dump, and the shovel. The boom on the shovel will rotate, so it will be a layout object. The classes are as follows:

The loaded trucks, TRUCKF
The unloaded trucks, TRUCKU1
The unloaded trucks upside down, TRUCKU
The trucks broken down, TRUCKD

The reason for the unloaded trucks to be created for the animation both right side up and upside down is that they will be coming from the dump from the right of the screen to the left. When a truck is placed on a path going from the left to the right, it will be on top of the path; but coming back it will be on the bottom, or upside down unless different classes of trucks are created.

There are to be several paths. The paths to and from the dump are broken into three separate paths each. This is so that there will not be any encroachment of the trucks. In an actual mining situation, the modeller needs to be careful that the various travel times are of the same order of magnitude. Thus, if the loading time is, say, 1 minute and the travel time to the crusher is 20 minutes, one should break the travel times into segments. In this case, the mean travel time to the dump was 7.5 minutes and the mean time to return was 5.6 minutes. Using the relative lengths of the three paths, the following mean times for travel to the dump and return were determined:

To Dump	Mean Time
Path P1	2.89
Path P2	2.75
Path P3	1.84

Return	Mean Time
Path P4	1.51
Path P5	1.72
Path P6	2.56

After the five trucks are created for the animation, a problem arises as to where they should enter the system. One could have them immediately at the shovel at time 0, but it was decided to have them placed on an artificial path, P7, a short distance from the shovel and then have them travel to the shovel. The code for this is as shown in Figure 22.5.

The trucks are created spaced out by 3 minutes. They are given a travel time on path P7 of 1 minute. Each truck is assigned a number from 1 to 5 as it reaches the end of path P7. The PUTPIC after the START can also be used to add the statement END to the animation file. The first of many CHAP22B.ATF files can be created and the animation checked.

The animation code for the trucks being loaded and travelling to the dump is added next. This will show the trucks leaving the shovel and travelling on the three paths P1, P2, and P3, to the shovel. It is important to check that the trucks were given a clearance and were made directional when they were created. In addition, the paths should have been made as accumulating. If either of these were not done, it will become apparent now (Figure 22.6).

Notice that three variables are used to determine the times to travel each of the paths to the dump.

It is assumed that the spotting time is 25 seconds and the time for a loaded truck to leave the shovel is 15 seconds. If the RELEASE SHOVEL Block was placed immediately after the truck was finished loading, the animation would show a waiting truck

```
 TRUCKS   GENERATE    3,,0,5     5 TRUCKS IN THE MINE
          BPUTPIC     FILE=ATF,LINES=5,AC1,XID1,XID1,XID1,N(TRUCKS)
TIME *.****
CREATE TRUCKU T*
PLACE T* ON P7
SET T* TRAVEL 1
WRITE M3 *
          ADVANCE     1
          ASSIGN      1,N(TRUCKS),PH  ASSIGN TRUCKS NUMBERS
```

FIGURE 22.5
Code for start of program—trucks are created and sent to the shovel.

```
 BACKUP    QUEUE       WAIT
           SEIZE       SHOVEL
           DEPART      WAIT
           ADVANCE     .42       SPOT
           SEIZE       DUMMY
           BPUTPIC     FILE=ATF,LINES=3,AC1,XID1,XID1
TIME *.****
PLACE T* AT -33 -2
SET T* CLASS TRUCKU1
           ADVANCE     RVNORM(1,2.5,.35)  LOAD
           RELEASE     DUMMY
           ADVANCE     .25     LEAVE SHOVEL
           RELEASE     SHOVEL
           BLET        &X=RVNORM(1,2.89,.29)
           BPUTPIC     FILE=ATF,LINES=5,AC1,XID1,XID1,XID1,&X,FR(SHOVEL)/10.
TIME *.****
PLACE T* ON P1
SET T* CLASS TRUCKF
SET T* TRAVEL **.**
WRITE M1 **.**%
           ADVANCE     &X
           BLET        &Y=RVNORM(1,2.75,.28)
           BPUTPIC     FILE=ATF,LINES=3,AC1,XID1,XID1,&Y
TIME *.****
PLACE T* ON P2
SET T* TRAVEL **.**
           ADVANCE     &Y
           BLET        &Z=RVNORM(1,1.84,.18)
           BPUTPIC     FILE=ATF,LINES=3,AC1,XID1,XID1,&Z
TIME *.****
PLACE T* ON P3
SET T* TRAVEL **.**
           ADVANCE     &Z
           SEIZE       DUMP
           BPUTPIC     FILE=ATF,LINES=2,AC1,XID1
TIME *.****
PLACE T* AT 47.3 -2.25
           ADVANCE     RVNORM(1,1.75,.2)  DUMP
           BLET        &LOADS=&LOADS+1
 DUMPED    RELEASE     DUMP
           BPUTPIC     FILE=ATF,LINES=2,AC1,&LOADS
TIME *.****
WRITE M2 ***
```

FIGURE 22.6
Trucks are loaded and sent to the dump.

immediately replace it. For an actual mine, the modeller may want to have the waiting, empty truck move on a short path to the shovel.

Notice that there is a dummy facility that is SEIZEd when the shovel is loading a truck. This is used for the motion of the shovel. The segment for this can be added now. It is shown in Figure 22.7.

When the shovel is loading a truck, the facility DUMMY is being used (F(DUMMY) = 1). This causes the boom to rotate forward 45° and then back 45°. A check is made to see if the truck is still being loaded; if so, another swing of the boom is made; if not, the shovel waits until another truck is spotted and ready to be loaded.

When the trucks finish dumping, they need to be checked to see if they need repairs. This was easily done in the program CHAP22A.GPS with a single GATE Block. However, the animation needs to check each truck separately. A TRANSFER FN Block routes each of the five trucks to separate segments where each is checked. The code for this is shown in Figure 22.8.

The code in Figure 22.8 uses the five segments for the reliability for each truck (Figure 22.9).

```
            GENERATE    ,,,1    SHOVEL MOVEMENT SECTION
  WAIT      TEST E      F(DUMMY),1    IS SHOVEL MOVING?
  AGAIN     BPUTPIC     FILE=ATF,LINES=2,AC1
TIME *.****
ROTATE BB -45 STEP 3 TIME .5
            ADVANCE     .5
            BPUTPIC     FILE=ATF,LINES=2,AC1
TIME *.****
ROTATE BB 45 STEP 3 TIME .5
            ADVANCE     .5
            TEST E      F(DUMMY),1,WAIT
            TRANSFER    ,AGAIN
```

FIGURE 22.7
Segment to move the boom on the shovel.

```
 WHERE    FUNCTION     PH1,L5
 1,BLOCKA/2,BLOCKB/3,BLOCKC/4,BLOCKD/5,BLOCKE
..............
.............
BLOCKA    GATE LS      1,BACKH
           BPUTPIC     FILE=ATF,LINES=3,AC1,XID1,XID1
TIME *.****
PLACE T* AT 44.5 .82
SET T* CLASS TRUCKD
           GATE LR      1
           TRANSFER    ,BACKH
 BLOCKB    GATE LS     2,BACKH
           BPUTPIC     FILE=ATF,LINES=3,AC1,XID1,XID1
TIME *.****
PLACE T* AT 47.5 0.82
SET T* CLASS TRUCKD
           GATE LR      2
           TRANSFER    ,BACKH
 BLOCKC    GATE LS      3,BACKH
           BPUTPIC     FILE=ATF,LINES=3,AC1,XID1,XID1
TIME *.****
PLACE T* AT 49.5 .82
SET T* CLASS TRUCKD
           GATE LR      3
           TRANSFER    ,BACKH
 BLOCKD    GATE LS      4,BACKH
           BPUTPIC     FILE=ATF,LINES=3,AC1,XID1,XID1
TIME *.****
PLACE T* AT 44.5 2.82
SET T* CLASS TRUCKD
           GATE LR      4
           TRANSFER    ,BACKH
 BLOCKE    GATE LS     5,BACKH
           BPUTPIC     FILE=ATF,LINES=3,AC1,XID1,XID1
TIME *.****
PLACE T* AT 47.5 2.82
SET T* CLASS TRUCKD
           GATE LR      5
           TRANSFER    ,BACKH
```

FIGURE 22.8
Code to check each truck to see if it is down.

```
         GENERATE     ,,,1
BACKA    ADVANCE      RVEXPO(1,400)    TRUCK WORKING
         LOGIC S      1
         ADVANCE      RVNORM(1,20,3.5)
         LOGIC R      1
         TRANSFER     ,BACKA
         GENERATE     ,,,1
BACKB    ADVANCE      RVEXPO(1,425)    TRUCK WORKING
         LOGIC S      1
         ADVANCE      RVNORM(1,35,6.9)
         LOGIC R      1
         TRANSFER     ,BACKB
         GENERATE     ,,,1
BACKC    ADVANCE      RVEXPO(1,550)    TRUCK WORKING
         LOGIC S      1
         ADVANCE      RVNORM(1,55.5,12)
         LOGIC R      1
         TRANSFER     ,BACKC
         GENERATE     ,,,1
BACKD    ADVANCE      RVEXPO(1,345)    TRUCK WORKING
         LOGIC S      1
         ADVANCE      RVNORM(1,40,7)
         LOGIC R      1
         TRANSFER     ,BACKD
         GENERATE     ,,,1
BACKE    ADVANCE      RVEXPO(1,300)    TRUCK WORKING
         LOGIC S      1
         ADVANCE      RVNORM(1,50,8)
         LOGIC R      1
         TRANSFER     ,BACKE
```

FIGURE 22.9
Code for reliability of the trucks.

```
 BACKH    BLET          &R=RVNORM(1,1.51,.15)
          BPUTPIC       FILE=ATF,LINES=4,AC1,XID1,XID1,&R,XID1
TIME *.****
PLACE T* ON P4
SET T* TRAVEL **.**
SET T* CLASS TRUCKU
          ADVANCE       &R
          BLET          &S=RVNORM(1,1.72,.17)
          BPUTPIC       FILE=ATF,LINES=3,AC1,XID1,XID1,&S
TIME *,****
PLACE T* ON P5
SET T* TRAVEL **.**
          ADVANCE       &S
          BLET          &T=RVNORM(1,2.36,.24)
          BPUTPIC       FILE=ATF,LINES=3,AC1,XID1,XID1,&T
TIME *.****
PLACE T* ON P6
SET T* TRAVEL **.**
          ADVANCE       &T
          TRANSFER      ,BACKUP    RETURN TO SHOVEL
```

FIGURE 22.10
Code for empty trucks returning to the shovel.

Each truck is routed to its own GATE where it checks to see if the GATE is *open* and whether it is to be delayed for repairs. If a truck is down, it is shown as a truck with a large red cross on top of it and it is placed above the dump. When it is repaired, it can return to the shovel.

Once the above code is working, the code for the trucks to return to the shovel is added. This is straightforward (Figure 22.10).

Notice that three variables, `&R`, `&S`, and `&T` are introduced for determining the travel times on the three segments from the dump to the shovel.

After the animation program is working with all the above segments together, it is time to add the messages. These are as follows:

M1—utilization of the shovel
M2—loads dumped
M3—number of trucks working in the mine

The above may seem to be a long way to go about doing the animation but following this approach will save time for most programs. As the animation develops, the modeller appreciates the importance of adding animation to any real mining operation.

22.2 Another Form of the TRANSFER Block

In Chapter 9, two forms of the TRANSFER Block were discussed, that is, the unconditional TRANSFER and the conditional TRANSFER. In Chapter 15, the TRANSFER BOTH and TRANSFER ALL Blocks were introduced. Chapter 19 saw the TRANSFER Function Block presented. The first two of these are by far the most commonly used. However, there is still one other form of the TRANSFER Block to be presented. This is the TRANSFER PICK MODE.

This form of the TRANSFER Block will select a Block to transfer the transaction at random from a number of possible Blocks. The transaction will unconditionally go to this Block. Each of the Blocks to be selected will have the same probability of being selected. Thus, if there are three Blocks, each will be selected with probability .3333; for four Blocks, the probability is .250; and so on.

The form of the Block is as follows:

```
       TRANSFER       PICK,LOCNA,LOCNB
LOCNA  .........
       .........
       .........
LOCNB  .........
```

LOCNA and LOCNB are Block labels. The word PICK is in operand position A. LOCNA must be at a location before LOCNB. Each Block between LOCNA and LOCNB is considered in the range of the transfer. This restriction means that, in general, *only* TERMINATE *and* TRANSFER *Blocks are in the range* LOCNA *and* LOCNB. Consider the following code:

```
       TRANSFER   PICK,FIRST,LAST
FIRST  TRANSFER   ,MACH1
       TRANSFER   ,MACH2
       TRANSFER   ,MACH3
       TRANSFER   ,MACH4
LAST   TRANSFER   ,MACH5
```

A transaction will be transferred to one of the Blocks labelled MACH1, MACH2, MACH3, MACH4, and MACH5 with equal probability.

Example 22.2: Program to Illustrate the `TRANSFER PICK` Block

What will the following program (Figure 22.11) do?

The program will generate a transaction every 1 time unit. The transaction will be put on the FEC for 1 time unit. When it returns to the CEC, it will then be transferred to one of the six `TERMINATE` Blocks with equal probability (Figure 22.12).

Each of the `TERMINATE` Blocks between `FIRST` and `LAST` can be expected to be entered approximately 166 times.

The animation for this example is given by CHAP22D.GPS. The program listing is given by Figure 22.13.

A screenshot of the animation is given by Figure 22.14.

```
          SIMULATE
******************************
*   PROGRAM CHAP22C.GPS       *
*   ILLUSTRATES THE TRANSFER  *
*   PICK BLOCK                *
******************************
          GENERATE    1
          TRANSFER    PICK,FIRST,LAST
 FIRST    TERMINATE
 SECOND   TERMINATE
 THIRD    TERMINATE
 FOURTH   TERMINATE
 FIFTH    TERMINATE
 LAST     TERMINATE
          GENERATE    1001
          TERMINATE   1
          START       1
          PUTPIC      LINES=6,N(FIRST),N(SECOND),N(THIRD),_
                      N(FOURTH),N(FIFTH),N(LAST)
  TRANSACTIONS AT FIRST    ***
  TRANSACTIONS AT SECOND   ***
  TRANSACTIONS AT THIRD    ***
  TRANSACTIONS AT FOURTH   ***
  TRANSACTIONS AT FIFTH    ***
  TRANSACTIONS AT LAST     ***
          END
```

FIGURE 22.11
Program CHAP22C.GPS to illustrate `TRANSFER PICK` Block.

```
TRANSACTIONS AT FIRST    177
TRANSACTIONS AT SECOND   164
TRANSACTIONS AT THIRD    146
TRANSACTIONS AT FOURTH   176
TRANSACTIONS AT FIFTH    171
TRANSACTIONS AT LAST     166
```

FIGURE 22.12
Output from program CHAP22C.GPS.

```
          SIMULATE
******************************
*  PROGRAM CHAP22D.GPS        *
*  ANIMATION OF EXAMPLE 22.2 *
******************************
          INTEGER     &A,&B,&C,&D,&E,&F
          REAL        &TOTAL
 ATF      FILEDEF     'CHAP22D.ATF'
          GENERATE    1
          BLET        &TOTAL=&TOTAL+1
          BPUTPIC     FILE=ATF,LINES=2,AC1,&TOTAL
TIME *.****
WRITE MT ***
          TRANSFER    PICK,FIRST,LAST
 FIRST    TRANSFER    ,AWAY1
 SECOND   TRANSFER    ,AWAY2
 THIRD    TRANSFER    ,AWAY3
 FOURTH   TRANSFER    ,AWAY4
 FIFTH    TRANSFER    ,AWAY5
 LAST     TRANSFER    ,AWAY6
 AWAY1    BLET        &A=&A+1
          BPUTPIC     FILE=ATF,LINES=3,AC1,&A,(&A/&TOTAL)*100.
TIME *.****
WRITE M1 ***
WRITE M1A **.**%
          TERMINATE
 AWAY2    BLET        &B=&B+1
          BPUTPIC     FILE=ATF,LINES=3,AC1,&B,(&B/&TOTAL)*100.
TIME *.****
WRITE M2 ***
WRITE M2A **.**%
          TERMINATE
 AWAY3    BLET        &C=&C+1
          BPUTPIC     FILE=ATF,LINES=3,AC1,&C,(&C/&TOTAL)*100.
TIME *.****
WRITE M3 ***
WRITE M3A **.**%
          TERMINATE
 AWAY4    BLET        &D=&D+1
          BPUTPIC     FILE=ATF,LINES=3,AC1,&D,(&D/&TOTAL)*100
TIME *.****
WRITE M4 ***
WRITE M4A **.**%
          TERMINATE
 AWAY5    BLET        &E=&E+1
          BPUTPIC     FILE=ATF,LINES=3,AC1,&E,(&E/&TOTAL)*100
TIME *.****
WRITE M5 ***
WRITE M5A **.**%
          TERMINATE
 AWAY6    BLET        &F=&F+1
          BPUTPIC     FILE=ATF,LINES=3,AC1,&F,(&F/&TOTAL)*100
TIME *.****
WRITE M6 ***
WRITE M6A **.**%
```

FIGURE 22.13
Listing of program CHAP22D.GPS. *(Continued)*

```
                TERMINATE
                GENERATE     1001
                TERMINATE    1
                START        1
                PUTPIC       LINES=6,N(FIRST),N(SECOND),N(THIRD),_
                             N(FOURTH),N(FIFTH),N(LAST)
        TRANSACTIONS AT FIRST    ***
        TRANSACTIONS AT SECOND   ***
        TRANSACTIONS AT THIRD    ***
        TRANSACTIONS AT FOURTH   ***
        TRANSACTIONS AT FIFTH    ***
        TRANSACTIONS AT LAST     ***
                PUTPIC       FILE=ATF,LINES=2,AC1
      TIME *.****
      END
                END
```

FIGURE 22.13 (Continued)
Listing of program CHAP22D.GPS.

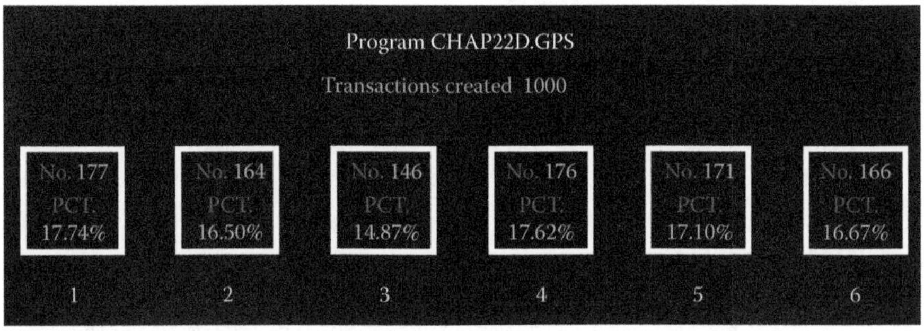

FIGURE 22.14
Screenshot of the animation for CHAP22D.GPS.

22.3 Exercises

22.1. The program written for Example 22.1 can be compacted by the use of two functions for the segment with the LOGIC switches. One function will assign a working time to each truck and the other a repair time. Write the code to do this.

22.2. Explain what each of the program lines does. Assume that all variables or entities have been defined.

```
(a)  GENERATE    ,,,1
BACK ADVANCE     RVEXPO(1,100)
     LOGIC S     SWITCH1
     ADVANCE     RVEXPO(1,10)
     LOGIC R     SWITCH1
     TRANSFER    ,BACK

(b)  GENERATE    100
     TEST E      &X,1,AWAY
     LOGIC S     FIRST
     ADVANCE     10
```

```
        LOGIC R     FIRST
        TERMINATE
AWAY    LOGIC S     SECOND
        ADVANCE     5
        LOGIC R     SECOND
        TERMINATE

(c)     GENERATE    RVNORM(1,100,10)
        LOGIC S     THIRD
        TRANSFER    .5,,NEXT
        ADVANCE     10
        LOGIC R     THIRD
        TERMINATE
NEXT    ADVANCE     5
        LOGIC R     THIRD
        TERMINATE
```

22.3. Trucks come to a crusher in a mine every 3 ± 1 minutes. Only one truck can dump at a time, which takes $1 \pm .75$ minutes. The crusher works for 50 minutes, (normally distributed, standard deviation 15% of the mean) and then is down for 10 minutes (exponentially distributed). Trucks have to wait while the crusher is down. Simulate this using a `GATE` Block. Simulate for 100 shifts of 480 minutes each.

22.4. Add the code to Exercise 22.3 to output the maximum queue, make a histogram of the trucks in the `QUEUE` and the average trucks in the queue.

22.5. Animate Exercise 22.3.

22.6. Add the necessary code for the program developed for Exercise 22.3 so that, when the crusher is down, the trucks are routed to a second crusher. The trucks travel in $1 \pm .4$ minutes and dump in 2 minutes. They then leave the system. The program should give the loads dumped at each crusher per shift. Assume that the second or spare crusher is never down.

22.7. Add the necessary animation to the one created for Exercise 22.5 to reflect the changes according to Exercise 22.6.

23

Miscellaneous Statements

Underground continuous miner.

23.1 DO LOOPS, IF, GOTO, and LET Statements

23.1.1 The GPSS/H DO LOOP

GPSS/H has DO LOOPS that can be used to greatly shorten the code for control statements. The form is quite similar to that found in other programming languages and is

```
DO integer ampervariable = lower limit,upper limit,increment
...........
...........
ENDDO
```

The increment is optional. Here is how the DO LOOP works. The integer ampervariable is set equal to its lower limit and the statements from the DO statement down to the ENDDO are executed. The integer ampervariable is then increased by the increment. If the increment is missing (it often is), the value is assumed to be 1 by default. The statements are then executed up to the ENDDO. This is continued until the value of the ampervariable is greater than the upper limit. Thus, the lines of code:

```
        DO          &I = 2,10
        CLEAR
        RMULT       9
WORKS   GENERATE    ,,,&I
        START       1
        ENDDO
```

would run the program first with `&I = 2` and then with `&I = 3`, and so on, up to and including `&I = 10` for a total of nine times. There would be nine reports written. The random number sampling would start in position 9.

It is not possible to decrement an ampervariable in a `DO LOOP`.

The lines of code:

```
DO             &K = 4,10,2
CLEAR
RMULT          777
LET            &VALUE = &K+3
START          1
ENDDO
```

would run a program first for values of `&K = 4` and `&VALUE = 7`. The second time, the values would be `&K = 6` (increment is 2) and `&VALUE = 9`, and so on, up to `&K = 10` and `&VALUE = 13`.

For example, consider the short program that uses the above code. The program is given by CHAP23A.GPS (Figure 23.1).

The output from the program is shown in Figure 23.2.

It is possible to have nested `DO` loops. If so, each must have its own `ENDDO` statement.

```
                  SIMULATE
***************************
*   PROGRAM CHAP23A.GPS   *
*   ILLUSTRATION          *
*   OF THE DO LOOP        *
***************************
                  INTEGER      &VALUE,&I
                  GENERATE     1
                  ADVANCE      1.5,.3
                  BPUTPIC      &VALUE
 VALUE IS **
                  TERMINATE    1
                  DO           &I=4,10,2
                  RMULT        777
                  LET          &VALUE=&I+2
                  START        1
                  ENDDO
                  END
```

FIGURE 23.1
Listing of program CHAP23A.GPS.

```
VALUE IS  6
VALUE IS  8
VALUE IS 10
VALUE IS 12
```

FIGURE 23.2
Output from program CHAP23A.GPS.

```
        DO           &J = 2,6
        CLEAR
        RMULT        54321
TRUCKS  GENERATE     ,,,&J
        DO           &I = 1,3
        STORAGE      S(TUGS),&I
        START        1
        ENDDO
        ENDDO
```

The value of `&J` is first set equal to 2. The Block

```
TRUCKS GENERATE      ,,,&J
```

would become

```
TRUCKS GENERATE      ,,,2
```

next, the value of I is 1 and the `STORAGE` is as follows:

```
STORAGE        S(TUGS),1
```

The program is run for these values. The value of `&I` is next incremented to 2 and the program run with the `STORAGE` of `TUGS` equal to 2 (`&J` remains equal to 2). Thus, the main program will be executed 15 times (`&J = 2,3,4,5,6` and `&I = 1,2,3`).

Inventory problems are common for the mining engineer. This is especially so for mines in remote areas where supplies are not readily available. The next example has many parallels for mining.

Example 23.1*: Inventory Example

The owner of a garage stocks snow tyres for sale during the winter months. He places one order at the end of summer and cannot receive any more tyres if the demand is greater than the supply. Tyres cost \$20 + \$25*(number of tyres ordered). The \$20 is a fixed cost no matter how many tyres are ordered. The tyres sell for \$45 each. If any tyres are left over, there is a penalty of \$5 per tyre in *holding* costs. If a person wants to buy a tyre and it is no longer in stock, the \$20 profit is considered as a loss. From past records, the owner feels that the demand for tyres is given by the following probability distribution.

Determine the amount of tyres the garage owner should order to maximize his expected profit. Simulate for 200 winters.

Solution

The way to do the simulation is to first assume a supply of a reasonable amount of tyres. Suppose this amount is 120. Then, using the data in Table 23.1, a demand is simulated by means of Monte Carlo simulation. Suppose this is 100. This means that the store made a profit of

$$(\$100 * \$45) - \$20 - (\$120 * \$25) - \$5 * (\$120 - \$100) = \$1380$$

If the demand had been 130, the profit would have been:

$$(\$120 * \$45) - \$20 - (\$120 * 25) - \$20 * (\$130 - \$120) = \$2180$$

This is done for a large number of possible demands, say 200. The expected profit is then the average of the simulated ones. This is then taken to be the expected profit (or loss)

* This example is modified from one found in "Introduction to Operations Research Models," Cooper, Bhat, and LeBlanc, W. B. Sanders, Philadelphia, PA, 1977, p. 344.

TABLE 23.1

Distribution of Demand for Tyres

Demand	Probability
100	.03
105	.05
110	.10
115	.15
120	.18
125	.14
130	.12
135	.10
140	.08
145	.03
150	.02

for the given supply amount. The program assumes supply amounts of from 90 to 150 in increments of 5, that is, amounts of 90, 95, 100,...,150.

GPSS/H is ideal for simulating such inventory problems. The program to do the simulation is given in Figure 23.3.

The use of the `DO LOOP` means that the program needs to be run only one time. The `CLEAR` statement clears out the previous stored results and resets the internal clock.

```
           SIMULATE
*****************************
*   PROGRAM CHAP23B.GPS       *
*   INVENTORY EXAMPLE         *
*****************************
           INTEGER     &I,&STOCK
           REAL        &PROFIT,&RESULT
  DEMAND   FUNCTION    RN1,D11
.03,100/.08,105/.18,110/.33,115/.51,120/.65,125/.77,130
.87,135/.95,140/.98,145/1,150
           GENERATE    ,,,1    DUMMY TRANSACTION
  BACK1    ASSIGN      1,FN(DEMAND),PH      DETERMINE DEMAND
           TEST GE     &STOCK,PH1,DOWN1     ENOUGH SUPPLIED?
           BLET        &PROFIT=&PROFIT+PH1*45.-(20.+25.*&STOCK)-_
                       5.*(&STOCK-PH1)
           ADVANCE     1        ONE YEAR PASSES
           TRANSFER    ,BACK1
 DOWN1     BLET        &PROFIT=&PROFIT+&STOCK*45.-(20.+25.*&STOCK)-_
                       20.*(PH1-&STOCK)
           ADVANCE     1        ONE YEAR PASSES
           TRANSFER    ,BACK1
           DO          &I=90,150,5     BEGIN DO LOOP
           CLEAR
           RMULT       7777
           LET         &STOCK=&I
           GENERATE    200
           BLET        &RESULT=&PROFIT/200.
           TERMINATE   1
           START       1
           PUTPIC      LINES=3,&STOCK,&RESULT
        RESULTS OF SIMULATION
        NUMBER OF ITEMS IN STOCK WAS ***
        EXPECTED PROFIT              ****.**
           LET         &PROFIT=0
           ENDDO
           END
```

FIGURE 23.3
Listing of program CHAP23B.GPS.

```
RESULTS OF SIMULATION
NUMBER OF ITEMS IN STOCK WAS  90
EXPECTED PROFIT              1119.00
RESULTS OF SIMULATION
NUMBER OF ITEMS IN STOCK WAS  95
EXPECTED PROFIT              1319.00
RESULTS OF SIMULATION
NUMBER OF ITEMS IN STOCK WAS 100
EXPECTED PROFIT              1519.00
RESULTS OF SIMULATION
NUMBER OF ITEMS IN STOCK WAS 105
EXPECTED PROFIT              1705.00
RESULTS OF SIMULATION
NUMBER OF ITEMS IN STOCK WAS 110
EXPECTED PROFIT              1875.25
RESULTS OF SIMULATION
NUMBER OF ITEMS IN STOCK WAS 115
EXPECTED PROFIT              2003.50
RESULTS OF SIMULATION
NUMBER OF ITEMS IN STOCK WAS 120
EXPECTED PROFIT              2084.50
RESULTS OF SIMULATION
NUMBER OF ITEMS IN STOCK WAS 125
EXPECTED PROFIT              2107.75
RESULTS OF SIMULATION
NUMBER OF ITEMS IN STOCK WAS 130
EXPECTED PROFIT              2082.00
RESULTS OF SIMULATION
NUMBER OF ITEMS IN STOCK WAS 135
EXPECTED PROFIT              2010.75
RESULTS OF SIMULATION
NUMBER OF ITEMS IN STOCK WAS 140
EXPECTED PROFIT              1908.00
RESULTS OF SIMULATION
NUMBER OF ITEMS IN STOCK WAS 145
EXPECTED PROFIT              1775.50
RESULTS OF SIMULATION
NUMBER OF ITEMS IN STOCK WAS 150
EXPECTED PROFIT              1632.50
```

FIGURE 23.4
Results from program CHAP23B.GPS.

Since ampervariables are not cleared, it is necessary to have the ampervariable `&PROFIT` reset to zero (Figure 23.4).

From the output, it is found that the optimum number of tyres to order is 125, which will result in an expected profit of $2107.75 However, if 120 or 130 are ordered, the expected profit is only slightly less. If desired, the program can be rerun with the supply amounts taken as being from 120 to 130 in increments of 1. This is left for an exercise.

23.1.2 The GETLIST Statement

Once ampervariables are defined, they can be read into the program by means of the `GETLIST` statement. This causes the program execution to stop until the ampervariables as specified in the `GETLIST` specification are read from the user's input device. A prompt appears on the screen until the data is input. The general form of this is

```
label GETLIST (FILE = filename),(list of ampervariables)
```

The label is optional. For `FILE = filename` the common one is `GUSER`, which is used for interactive screen input. If more than one ampervariable is read in by a single `GETLIST` statement, they *must be separated by blanks not commas*. For example, if values of 2, 3, and 7 are to be assigned to `&I`, `&J`, and `&K`, respectively, the data must be input as follows:

2 3 7

If a `GETLIST` statement asks for more than one value to be read in and less than this number is input, the prompt remains on the screen until all the data is input. *The* `FILE = GUSER` *specification can be omitted* as GPSS/H assumes this is the file by default.

Example 23.2: Modification of Example 23.1

Go back to Example 23.1 and add the necessary code so that the number of years to simulate for is a variable. Also, have the supply amounts and the increment as variables.

Solution
The changes are made and the results are program CHAP23C.GPS. A listing of it is given next in Figure 23.5.

23.1.3 The IF, GOTO, and HERE Statements in GPSS/H

It is possible to have `IF` statements in GPSS/H. These can be used in the control statements to check on the input data or after the program has executed to prompt the user to rerun the program often with different data. The form of the `IF` statement is

```
label   IF SNA1,R,SNA2
            ..........
            ..........
            ENDIF
```

The relational operator, `R`, is one of the following:

'E'	Equal
'NE'	Not equal
'LE'	Less than or equal
'L'	Less than
'G'	Greater than
'GE'	Greater than or equal
'L'	Less than

If the condition is true, the group of statements after it are executed. Some examples of `IF` statements are as follows:

```
(a) IF          &X'E'20
                 LET    &Y = 200
      ENDIF

(b) IF          &ANS'E''Y'
                 LET    &Z = 1
      ENDIF

(c) IF          W(TRUCKS)'LE'5
                 LET    &KKK = 0
                 LET    &JJJ = 1
      ENDIF
```

```
          SIMULATE
****************************
*  PROGRAM CHAP23C.GPS      *
*  INVENTORY EXAMPLE        *
*  NOW WITH MODIFICATIONS   *
****************************
          INTEGER     &I,&STOCK,&YEARS,&SUPSTART,&FINISH,&INC
          PUTSTRING   ('  ')
          PUTSTRING   ('  HOW MANY YEARS TO SIMULATE FOR?')
          PUTSTRING   ('  ')
          GETLIST     &YEARS
          PUTSTRING   ('  ')
          PUTSTRING   (' WHAT SUPPLY AMOUNT DO YOU WANT TO START WITH?')
          PUTSTRING   ('  ')
          GETLIST     &SUPSTART
          PUTSTRING   ('  ')
          PUTSTRING   (' WHAT SUPPLY AMOUNT DO YOU WANT TO FINISH WITH?')
          PUTSRING    ('  ')
          GETLIST     &FINISH
          PUTSTRING   ('  ')
          PUTSTRING   ('  WHAT INCREMENT TO USE?')
          PUTSTRING   ('  ')
          GETLIST      &INC
          PUTPIC      LINES=5,&YEARS,&SUPSTART,&FINISH,&INC
     DATA USED IN SIMULATION - DATA IN THOUSANDS
   YEARS TO SIMULATE FOR      ****
   SUPPLY AMT. TO START       ****
   SUPPLY AMT. TO FINISH      ****
   INCREMENT                     *
          REAL        &PROFIT,&RESULT
 DEMAND   FUNCTION    RN1,D11
.03,100/.08,105/.18,110/.33,115/.51,120/.65,125/.77,130
.87,135/.95,140/.98,145/1,150
          GENERATE    ,,,1   DUMMY TRANSACTION
 BACK1    ASSIGN      1,FN(DEMAND),PH     DETERMINE DEMAND
          TEST GE     &STOCK,PH1,DOWN1    ENOUGH SUPPLIED?
          BLET        &PROFIT=&PROFIT+PH1*45.-(20.+25.*&STOCK)-_
                      5.*(&STOCK-PH1)
          ADVANCE     1       ONE YEAR PASSES
          TRANSFER    ,BACK1
DOWN1     BLET        &PROFIT=&PROFIT+&STOCK*45.-(20.+25.*&STOCK)-_
                      20.*(PH1-&STOCK)
          ADVANCE     1       ONE YEAR PASSES
          TRANSFER    ,BACK1
          DO         &I=&SUPSTART,&FINISH,&INC    BEGIN DO LOOP
          CLEAR
          RMULT       7777
          LET         &STOCK=&I
          GENERATE    &YEARS
          BLET        &RESULT=&PROFIT/&YEARS
          TERMINATE   1
          START       1
          PUTPIC      LINES=3,&STOCK,&RESULT
        RESULTS OF SIMLATION
        NUMBER OF ITEMS IN STOCK WAS ***
        EXPECTED PROFIT             ****.**
          LET         &PROFIT=0
          ENDDO
          END
```

FIGURE 23.5
Listing of program CHAP23C.GPS.

It is possible to have ELSEIF and ELSE statements such as are found in other programming languages. Their form would be as follows:

```
label  IF      condition 1
       ....
       ....
       ELSEIF  condition 2
       ....

       ....
       ELSE    condition 3
       ....
       ....
       ENDIF
```

An example of these might be as follows:

```
IF        &I'E'2
LET       &J = 0
ELSEIF    &I'E'3
LET       &J = 1
ELSE
LET       &J = 2
ENDIF
```

Often, the GOTO is used with the IF statement. This is simply

```
GOTO label
```

For example, one might have code such as

```
PUTSTRING     (' THE DATA YOU TYPED IN IS AS FOLLOWS')
PUTPIC        LINES = 3,&I,&J,&SPEED
    THE VALUE OF I = ***
    THE VALUE OF J = ***
    THE SPEED IS   = ***.**
PUTSTRING     (' ')
PUTSTRING     (' ARE YOU HAPPY WITH THESE?')
PUTSTRING     (' RESPOND: 1 = YES; 0 = NO')
GETLIST       &L
IF            &L'E'0
GOTO          TRYAGAIN
ENDIF
```

The effect of the above is to prompt the user to examine the data already input. If the data is not satisfactory, control returns to where the data was originally input.

It is possible to have control return to a target of the GOTO that is a dummy statement known as the HERE statement. The form of this is

```
label HERE
```

Thus, in the previous example, there might have been a statement:

```
TRYAGAIN HERE
```

This statement is analogous to the Fortran CONTINUE statement.

Example 23.3: Classical Inventory Problem

A classical inventory problem is one where goods are continually being depleted. These might be an item being sold in a store or, for a mine, items being used in the mine such as spare parts for machines. As stocks shrink, there is a point where an order is placed to replenish the supply. This is called the reorder point (ROP). The quantity to be reordered is the reorder quantity (ROQ). There will normally be a time delay before the order arrives. It costs money to have extra items available, although not nearly as much if a part is required and not available. One such inventory problem will be presented next in the form of sales of products.

Suppose that there initially are 100 units available at the start of the sales period. Each day sales vary according to the uniform distribution 10 ± 3. When stocks fall below the ROP (to be determined) an order for more items is placed. This ROQ is 100. The time to fill the order is from 6 to 12 days uniformly distributed. The penalty for not having an item in stock is \$5.54. Having inventory incurs a slight penalty of \$.123 per item per day. Determine the optimum ROP. Stocks are replenished only at the end of the day.

Solution

The program to simulate this is given by CHAP23D.GPS. Its listing is given in Figure 23.6.

It is instructive to examine this code in detail.

```
GENERATE      1,,,,1          START OF THE DAY
BLET          &DMD = 7*FRN1+7     DAILY DEMAND
TEST GE       &STOCK,&DMD,TRBL        CAN THE SUPPLY MEET DEMAND?
```

A worker comes to work each day. He has a higher priority than the transaction that determines if an order is to be placed for more stock. He determines what the demand for the day is to be by sampling the function 10 ± 3. This is given by the following line of code:

```
BLET       &DMD = 7*FRN1+7 DAILY DEMAND
```

When FRN1 is 0, this returns a 7 and when it is .999999, it returns a 13 since &DMD is an integer. If there is enough stock on hand, then the following lines are executed:

```
BLET             &STOCK = &STOCK-&DMD
BLET             &ONHAND = &ONHAND+&STOCK     HOW MUCH ON HAND
TERMINATE        1    NO MORE WORK FOR THE DAY
```

The stock on hand is reduced by the projected demand and the total stock on hand is incremented. The day is over.

If there is not enough stock on hand to meet the demand, the following lines of code are executed (Figure 23.7).

```
          SIMULATE
**********************************
*  PROGRAM CHAP23D.GPS           *
*  CLASSIC INVENTORY EXAMPLE     *
**********************************
          REAL        &ONHAND    RUNNING TOTAL OF STOCK ON HAND
          REAL        &COST3,&COST4,&COST5
          INTEGER     &DMD,&STOCK,&LOST,&ROP,&ROQ
*    &DMP = DAILY DEMAND
*    &STOCK =  HOW MUCH YOU HAVE ON HAND EACH DAY (START WITH 100)
*    &LOST  =  HOW MUCH YOU LOST IN SALES EACH DAY
*    &ROP   =  WHEN YOU MAKE A REORDER - WILL CHANGE
*    &ROQ   =  HOW MUCH YOU REORDER - CAN VARY
          LET         &STOCK=100     START WITH 100 SQUIDGETS
          LET         &ROQ=100       AMOUNT TO REORDER EACH TIME
          PUTSTRING   ('  ')
          PUTSTRING   ('  ')
          PUTSTRING   ('  INPUT THE INITIAL REORDER POINT')
          PUTSTRING   ('  ')
          GETLIST     &ROP
          PUTSTRING   ('  ')
          PUTSTRING   ('  WHAT IS THE COST TO HOLD ITEM IN STOCK EACH DAY')
          PUTSTRING   ('  ')
          REAL        &COST1,&COST2
*    COST1 IS COST TO HAVE ITEM IN STOCK AT END OF DAY
*    COST2 IS COST IF ITEM IS OUT OF STOCK
          GETLIST     &COST1
          PUTSTRING   ('  ')
          PUTSTRING   (' WHAT IS THE COST IF ITEM OUT OF STOCK?')
          PUTSTRING   ('  ')
          GETLIST     &COST2
          GENERATE    1,,,,1      START OF THE DAY
          BLET        &DMD=7*FRN1+7     DAILY DEMAND
          TEST GE     &STOCK,&DMD,TRBL   CAN THE SUPPLY MEET DEMAND?
          BLET        &STOCK=&STOCK-&DMD
          BLET        &ONHAND=&ONHAND+&STOCK     HOW MUCH ON HAND
          TERMINATE   1   NO MORE WORK FOR THE DAY
 TRBL     BLET        &LOST=&LOST+(&DMD-&STOCK)
          BLET        &STOCK=0    NO MORE STOCK
          TERMINATE   1
          GENERATE    ,,,1
 BACK     TEST LE     &STOCK,&ROP    TIME TO MAKE NEW ORDER?
          ADVANCE     4*FRN1+6       PLACE ORDER
          BLET        &STOCK=&STOCK+&ROQ
          TRANSFER    ,BACK
          START       10000   SIMULATE FOR 10000 DAYS
          LET         &COST3=(&ONHAND/10000.)*&COST1
          LET         &COST4=(&LOST/10000.)*&COST2
          LET         &COST5=&COST3+&COST4
          PUTPIC      LINES=4,&ONHAND,&COST3,&COST4,&COST5
   ONHAND *********
   COST3  *****.**
   COST4  *****.**
   COST5  *****.**
          PUTPIC      LINES=3,&ROP,&ONHAND/10000.,&LOST,&COST5
 ROP =  *** AVG. STOCK ON HAND = *******.**  EACH DAY
            AMOUNT LOST    =   *****
            TOTAL COST     =   *****.**
           END
```

FIGURE 23.6
Listing of program CHAP23D.GPS.

```
TRBL     BLET        &LOST=&LOST+(&DMD-&STOCK)
         BLET        &STOCK=0     NO MORE STOCK
         TERMINATE   1            NO MORE WORK FOR THE DAY
```

FIGURE 23.7
Sample output for stock on hand and stock demand from program CHAP23D.GPS.

The amount of stock lost is incremented. The amount of stock on hand is zeroed.

Another segment of the program checks on the inventory amount. These are the lines are as follows:

```
         GENERATE     ,,,1
BACK     TEST LE      &STOCK,&ROP    TIME TO MAKE NEW ORDER?
         ADVANCE      4*FRN1+6       PLACE ORDER
         BLET         &STOCK=&STOCK+&ROQ
         TRANSFER     ,BACK
```

When the stock gets at or below the reorder point, an order is placed for new stock. It will be delayed by 8 ± 2 days. After this, the stock supply is replenished by the amount ordered.

Sample output from the program is shown in Figure 23.8:

```
ROP =    85 AVG. STOCK ON HAND =        51.82  EACH DAY
            AMOUNT LOST    =    2966
            TOTAL COST     =       7.99
```

FIGURE 23.8
Sample output from program CHAP23D.GPS.

Example 23.4: Iron Ore Supply Problem

One of the best mine inventory examples dates back many years. The following example is taken from the textbook, *Analysis for Production and Operations Management* by Bowman and Fetter (1961),* page 338. It was proposed to be solved by the authors two ways:

1. A technique known as *incremental analysis*
2. Monte Carlo simulation (by hand, of course!)

The problem is stated next. Simulation will be used to solve it as incremental analysis cannot solve the general case that GPSS/H can easily do. The interested reader is referred to their text for their method of solution by incremental analysis.

"Consider the Cleveland Steel Company's problem: It must store iron ore from Minnesota for use during the winter while the lakes are frozen and lake traffic is not possible. The amount of iron ore necessary for the winter's needs is uncertain for two reasons: (1) the season of ice is a variable, and (2) the amount of iron ore consumed per unit time is also a variable.

If too much iron ore has been stored during the winter, unnecessary storage costs would have been incurred. If too little iron has been stored during the winter, it is necessary to bring the needed ore in by train, a substantially more expensive operation than to bring by boat (Figure 23.9)."

* Bowman, E. H., and Fetter, R. B., *Analysis for Production and Operations Management*, Richard D. Irwin, ed., Homewood, IL, 1961.

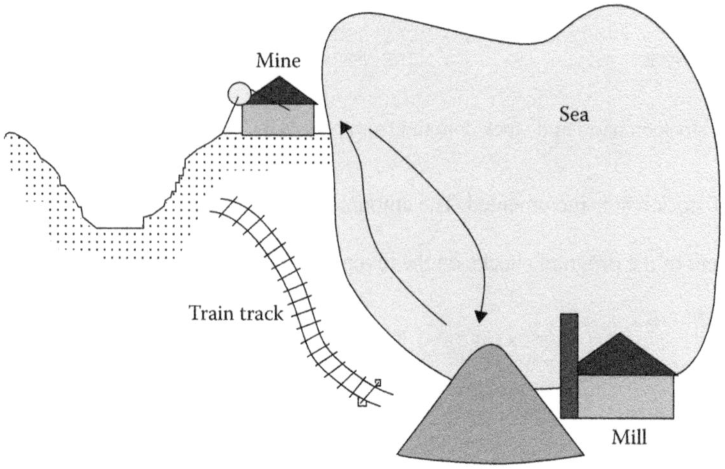

FIGURE 23.9
Sketch of iron ore company's problem.

The following is the data given by Bowman and Fetter:

s = transport cost for a ton of ore by ship = \$3
t = transport a ton of ore by train = \$10
h = holding a ton of ore for the whole winter = \$1

Probability distribution for number of weeks in the ice season:

Probability	0%	8%	47%	31%	11%	2%	1%	0%
Weeks	11	12	13	14	15	16	17	18

Probability distribution for the average ore demand (in tons) per week during the winter season:

Probability	0%	4%	16%	42%	21%	9%	4%	2%	1%	1%
Demand	7800	8000	8200	8400	8600	8800	9000	9200	9400	9600

Solution
The solution to this is obtained by the program CHAP23E.GPS. Different supply amounts are selected and the simulation is run for a great many years. In order to have precision, the number of years to simulate for was taken to be 20,000.* The program listing is given in Figure 23.10.

The results of running the program for different supply amounts yields results as given in Table 23.2.

For Table 23.2, it is seen that the amount of ore to supply for the ice season is 117,000 tons. However, the solution is not very sensitive to changes in the supply amount. Exercise 23.10 involves a change to the example.

The animation for this example, as for most inventory examples, is done mainly to verify that the program is correct. Considering that the program CHAP23E.GPS was run for 20,000 years, the .ATF file would be of considerable length. Program CHAP23F.GPS gives the animation. Its listing is shown in Figure 23.11.

* In the textbook by Bowmen and Fetter, the number of years to simulate by hand was around 50 years.

```
*****************************************
*    PROGRAM CHAP23E.GPS                *
*    MILL SUPPLY PROBLEM                *
*    FROM BOWMAN AND FETTER             *
*    ANALYSIS FOR PRODUCTION MANAGEMENT *
*    RICHARD IRWIN, PUB. HOMEWOOD, IL 1961 *
*    PROBLEM 6, CHAPTER 10, PAGE 338    *
*****************************************
          SIMULATE
          INTEGER     &I,&YEARS
          REAL        &SHIPCOST,&STORECST,&TRAINCST,&SUPPLY,_
                      &DEMAND,&COST
          CHAR*1      &ANS
          DO          &I=1,25
          PUTSTRING   ('  ')
          ENDDO
          PUTSTRING   ('      ORE TO SUPPLY TO A MILL FOR')
          PUTSTRING   ('      THE WINTER MONTHS             ')
          PUTSTRING   ('  ')
 ALLNEW   PUTSTRING   ('   INPUT THE COST TO SEND BY SHIP')
          PUTSTRING   ('  ')
          GETLIST     &SHIPCOST
          PUTSTRING   ('  ')
          PUTSTRING   ('   INPUT THE COST TO STORE A TON')
          PUTSTRING   ('  ')
          GETLIST     &STORECST
          PUTSTRING   ('  ')
          PUTSTRING   ('   INPUT THE COST TO SEND BY TRAIN')
          PUTSTRING   ('  ')
          GETLIST     &TRAINCST
          PUTSTRING   ('  ')
          PUTSTRING   ('  HOW MANY YEARS DO YOU WANT TO SIMULATE FOR?')
          PUTSTRING   ('  ')
          GETLIST     &YEARS
          PUTSTRING   ('  ')
 NEWSUP   PUTSTRING   ('    HOW MANY TONS DO YOU WANT TO USE FOR')
          PUTSTRING   ('    THE SUPPLY AMOUNT?')
          PUTSTRING   ('  ')
          GETLIST     &SUPPLY
          PUTSTRING   ('  ')
          PUTSTRING   ('   SIMULATION IN PROGRESS....')
          PUTSTRING   ('  ')
 WEEKS    FUNCTION    RN1,D6
0.08,12/.55,13/.86,14/.97,15/.99,16/1,17
 TONS     FUNCTION    RN1,D9
0.04,8000/.2,8200/.62,8400/.83,8600/.92,8800/.96,9000/.98,9200
0.99,9400/1,9600
          GENERATE    ,,,1,1      DUMMY TRANSACTION
 UPTOP    BLET        &DEMAND=FN(WEEKS)*FN(TONS)  DETERMINE DEMAND
          TEST L      &DEMAND,&SUPPLY,EEEE1      IS THERE ENOUGH ORE?
          BLET        &COST=&COST+&SHIPCOST*&SUPPLY+&STORECST*_
                      (&SUPPLY-&DEMAND)
          ADVANCE     1           GO TO NEXT YEAR
          TRANSFER    ,UPTOP      BACK FOR ANOTHER YEAR
 EEEE1    BLET        &COST=&COST+&SHIPCOST*&SUPPLY+&TRAINCST*_
```

FIGURE 23.10
Listing of program CHAP23E.GPS. *(Continued)*

```
                    (&DEMAND-&SUPPLY)
         ADVANCE      1           GO TO NEXT YEAR
         TRANSFER     ,UPTOP      BACK FOR ANOTHER YEAR
         GENERATE     &YEARS,,,,1       SIMULATE FOR YEARS
         BLET         &COST=&COST/&YEARS   DETERMINE AVERAGE COST
         TERMINATE    1           END OF SIMULATION
         START        1
         DO           &I=1,25
         PUTSTRING    ('  ')
         ENDDO
         PUTPIC       LINES=8,&YEARS,&SHIPCOST,&STORECST,&TRAINCST,_
&SUPPLY,&COST
              ______________________________________
             |  YEAR TO SIMULATE FOR       ******    |
             |  COST TO SEND BY SHIP          ***.** |
             |  COST TO STORE A TON           ***.** |
             |  COST TO SEND BY TRAIN         ***.** |
             |  FOR A SUPPLY OF          ********    |
             |  THE EXPECTED COST IS     ********    |
              ______________________________________
         CLEAR
         LET         &COST=0
         PUTSTRING    ('  ')
         PUTSTRING    ('  ')
         PUTSTRING    ('  DO YOU WANT TO GO BACK AND DO THIS')
PUTSTRING    ('  EXAMPLE?  (Y/N)')
         GETLIST      &ANS
         IF           (&ANS'E''Y')OR(&ANS'E''y')
         PUTSTRING    ('  ')
         PUTSTRING    ('  DO YOU WANT TO CHANGE THE SUPPLY AMOUNTS? (Y/N)')
PUTSTRING    ('  ')
         GETLIST      &ANS
         IF           (&ANS'E''Y')OR(&ANS'E''y')
         PUTSTRING    ('  ')
         GOTO         NEWSUP
         ENDIF
         PUTSTRING    ('  ')
         GOTO         ALLNEW
         ENDIF
         PUTSTRING    ('  ')
         PUTSTRING    ('  SIMULATION OVER....')
          END
```

FIGURE 23.10 (Continued)
Listing of program CHAP23E.GPS.

TABLE 23.2
Results of Simulation from Program CHAP23E.GPS

Supply Amount	Expected Cost ($)
113,000	386,592
114,000	384,397
115,000	382,903
116,000	382,944
117,000	381,716
118,000	381,867
119,000	383,127
120,000	384,173
121,000	385,445
122,000	387,664

```
********************************************
*     PROGRAM CHAP23F.GPS                  *
*     ANIMATION OF CHAP23E.GPS             *
********************************************
          SIMULATE
ATF       FILEDEF          'CHAP23F.ATF'
          INTEGER          &I,&YEARS,&RCOUNT
          LET              &RCOUNT = 1
          REAL             &SHIPCOST,&STORECST,&TRAINCST,&SUPPLY,_
          &DEMAND,         &COST
          CHAR*1           &ANS
          DO               &I = 1,25
          PUTSTRING        (' ')
          ENDDO
          PUTSTRING        ('       ORE TO SUPPLY TO A MILL FOR')
          PUTSTRING        ('       THE WINTER MONTHS       ')
          PUTSTRING        (' ')
ALLNEW    PUTSTRING        (' INPUT THE COST TO SEND BY SHIP')
          PUTSTRING        (' ')
          GETLIST          &SHIPCOST
          PUTSTRING        (' ')
          PUTSTRING        (' INPUT THE COST TO STORE A TON')
          PUTSTRING        (' ')
          GETLIST          &STORECST
          PUTSTRING        (' ')
          PUTSTRING        (' INPUT THE COST TO SEND BY TRAIN')
          PUTSTRING        (' ')
          GETLIST          &TRAINCST
          PUTSTRING        (' ')
          PUTSTRING        (' HOW MANY YEARS DO YOU WANT TO SIMULATE FOR?')
          PUTSTRING        (' ')
          GETLIST          &YEARS
          PUTSTRING        (' ')
NEWSUP    PUTSTRING        (' HOW MANY TONS DO YOU WANT TO USE FOR')
          PUTSTRING        (' THE SUPPLY AMOUNT?')
          PUTSTRING        (' ')
          GETLIST          &SUPPLY
          PUTSTRING        (' ')
          PUTSTRING        (' SIMULATION IN PROGRESS....')
          PUTSTRING        (' ')
WEEKS     FUNCTION         RN1,D6
0.08,12/.55,13/.86,14/.97,15/.99,16/1,17
  TONS    FUNCTION         RN1,D9
0.04,8000/.2,8200/.62,8400/.83,8600/.92,8800/.96,9000/.98,9200
0.99,9400/1,9600
          GENERATE         ,,,1,1       DUMMY TRANSACTION
          BPUTPIC          FILE = ATF,LINES = 5,AC1,XID1,XID1,XID1,XID1
TIME *.****
```

FIGURE 23.11
Listing of program CHAP23F.GPS. *(Continued)*

```
CREATE TE T*
PLACE T* ON PT1
CREATE BF B*
PLACE B* ON PB1
  UPTOP   BLET          &DEMAND = FN(WEEKS)*FN(TONS) DETERMINE DEMAND
          BPUTPIC       FILE = ATF,LINES = 8,AC1,N(UPTOP),&SUPPLY,&DEMAND,_
                        &SUPPLY-&DEMAND,XID1,XID1,XID1
TIME *.****
WRITE M1 ****
WRITE M2 *******
WRITE M3 *******
WRITE M4 ******
PLACE B* ON PB1
SET B* CLASS BF
SET B* TRAVEL.2
          ADVANCE       .3
          BPUTPIC       FILE = ATF,LINES = 4,AC1,XID1,XID1,XID1
TIME *.****
SET B* CLASS BE
PLACE B* ON PB2
SET B* TRAVEL.2
          ADVANCE       .2
          TEST L        &DEMAND,&SUPPLY,EEEE1 IS THERE ENOUGH ORE?
          BLET          &COST = &COST+&SHIPCOST*&SUPPLY+&STORECST*_
                        (&SUPPLY-&DEMAND)
          REAL          &COSTA
          BLET          &COSTA = &COST/&RCOUNT
          BLET          &RCOUNT = &RCOUNT+1
          BPUTPIC       FILE = ATF,LINES = 3,AC1,&COST,&COSTA
TIME *.****
WRITE M5 ******
WRITE M6 ******
          ADVANCE       .5
          TRANSFER      ,UPTOP       BACK FOR ANOTHER YEAR
EEEE1     BLET          &COST = &COST+&SHIPCOST*&SUPPLY+&TRAINCST*_
                        (&DEMAND-&SUPPLY)
          BLET          &COSTA = &COST/&RCOUNT
          BLET          &RCOUNT = &RCOUNT+1
          BPUTPIC       FILE = ATF,LINES = 3,AC1,&COST,&COSTA
TIME *.****
WRITE M5 ******
WRITE M6 ******
          BPUTPIC       FILE = ATF,LINES = 4,AC1,XID1,XID1,XID1
TIME *.****
SET T* CLASS TF
PLACE T* ON PT1
SET T* TRAVEL.2
          ADVANCE       .2
          BPUTPIC       FILE = ATF,LINES = 4,AC1,XID1,XID1,XID1
TIME *.****
SET T* CLASS TE
PLACE T* ON PT2
SET T* TRAVEL.2
          ADVANCE       .3
          BPUTPIC       FILE = ATF,LINES = 3,AC1,XID1
TIME *.****
```

FIGURE 23.11 (Continued)
Listing of program CHAP23F.GPS.

(Continued)

```
PLACE T* ON PT1
SET PATH PT1 SPEED 0
          TRANSFER      ,UPTOP  BACK FOR ANOTHER YEAR
          GENERATE      ,,,1
   BACK   ADVANCE       .2      WAIT FOR SHIP TO COME
          BPUTPIC       FILE = ATF,LINES = 3,AC1
TIME *.****
CREATE PILE1 PILE
PLACE PILE AT 50 -50
          ADVANCE       .1
          BPUTPIC       FILE = ATF,LINES = 2,AC1
TIME *.****
SET PILE CLASS PILE2
          ADVANCE       .1
          BPUTPIC       FILE = ATF,LINES = 2,AC1
TIME *.****
SET PILE CLASS PILE3
          ADVANCE       .1
          BPUTPIC       FILE = ATF,LINES = 2,AC1
TIME *.****
SET PILE CLASS PILE4
          ADVANCE       .1
          BPUTPIC       FILE = ATF,LINES = 2,AC1
TIME *.****
SET PILE CLASS PILE5
          ADVANCE       .1
          BPUTPIC       FILE = ATF,LINES = 2,AC1
TIME *.****
SET PILE CLASS PILE6
          ADVANCE       .1
          BPUTPIC       FILE = ATF,LINES = 2,AC1
TIME *.****
SET PILE CLASS PILE7
          ADVANCE       .1
          BPUTPIC       FILE = ATF,LINES = 2,AC1
TIME *.****
SET PILE CLASS PILE8
          ADVANCE       .1
          BPUTPIC       FILE = ATF,LINES = 2,AC1
TIME *.****
DESTROY PILE
          TRANSFER ,BACK
********************************
* SMOKE STACK SEGMENT    *
********************************
          GENERATE      1,.25,0
          BPUTPIC       FILE = ATF,LINES = 4,AC1,XID1,XID1,XID1
TIME *.****
CREATE SMOKE1 S*
PLACE S* ON PSM1
SET S* TRAVEL 1
          ADVANCE       .5
          BPUTPIC       FILE = ATF,LINES = 2,AC1,XID1,XID1
TIME *.****
SET S* CLASS SMOKE2
          ADVANCE       .3
          BPUTPIC       FILE = ATF,LINES = 2,AC1,XID1,XID1
TIME *.****
```

FIGURE 23.11 (Continued)
Listing of program CHAP23F.GPS.

(Continued)

```
SET S* CLASS SMOKE3
          ADVANCE        .2
          BPUTPIC        FILE = ATF,LINES = 2,AC1,XID1
TIME *.****
DESTROY S*
          TERMINATE
          GENERATE       1,.5,.25
  BPUTPIC        FILE = ATF,LINES = 4,AC1,XID1,XID1,XID1
TIME *.****
CREATE SMOKE1 S*
PLACE S* ON PSM2
SET S* TRAVEL 1
          ADVANCE        .5
          BPUTPIC        FILE = ATF,LINES = 2,AC1,XID1,XID1
TIME *.****
SET S* CLASS SMOKE2
          ADVANCE        .3
          BPUTPIC        FILE = ATF,LINES = 2,AC1,XID1,XID1
TIME *.****
SET S* CLASS SMOKE3
          ADVANCE        .2
          BPUTPIC        FILE = ATF,LINES = 2,AC1,XID1
TIME *.****
DESTROY S*
          TERMINATE
          GENERATE       .9,.3,.1
          BPUTPIC        FILE = ATF,LINES = 4,AC1,XID1,XID1,XID1
TIME *.****
CREATE SMOKE1 S*
PLACE S* ON PSM3
SET S* TRAVEL 1
          ADVANCE        .5
          BPUTPIC        FILE = ATF,LINES = 2,AC1,XID1,XID1
TIME *.****
SET S* CLASS SMOKE2
          ADVANCE        .3
          BPUTPIC        FILE = ATF,LINES = 2,AC1,XID1,XID1
TIME *.****
SET S* CLASS SMOKE3
          ADVANCE        .2
          BPUTPIC        FILE = ATF,LINES = 2,AC1,XID1
TIME *.****
DESTROY S*
          TERMINATE
          GENERATE       1,.1,.2
          BPUTPIC        FILE = ATF,LINES = 4,AC1,XID1,XID1,XID1
TIME *.****
CREATE SMOKE1 S*
PLACE S* ON PSM4
SET S* TRAVEL 1
          ADVANCE        .5
          BPUTPIC        FILE = ATF,LINES = 2,AC1,XID1,XID1
TIME *.****
SET S* CLASS SMOKE2
          ADVANCE        .3
          BPUTPIC        FILE = ATF,LINES = 2,AC1,XID1,XID1
TIME *.****
```

FIGURE 23.11 (Continued)
Listing of program CHAP23F.GPS.

(Continued)

```
SET S* CLASS SMOKE3
          ADVANCE          .2
          BPUTPIC          FILE = ATF,LINES = 2,AC1,XID1
TIME *.****
DESTROY S*
          TERMINATE
          GENERATE         &YEARS,,,,1          SIMULATE FOR YEARS
          BLET             &COST = &COST/&YEARS   DETERMINE AVERAGE COST
          TERMINATE        1       END OF SIMULATION
          START            1
          PUTPIC           FILE = ATF,LINES = 2,AC1
TIME *.****
END
          DO               &I = 1,25
          PUTSTRING        (' ')
          ENDDO
          PUTPIC           LINES = 8,&YEARS,&SHIPCOST,&STORECST,&TRAINCST,_
                           &SUPPLY,&COST
                |- - - - - - - - - - - - - - - - - - - - - - - - - -|
                | YEARS TO SIMULATE FOR       ******                |
                | COST TO SEND BY SHIP            ***.**            |
                | COST TO STORE A TON             ***.**            |
                | COST TO SEND BY TRAIN           ***.**            |
                | FOR A SUPPLY OF           ********                |
                | THE EXPECTED COST IS      ********                |
                |- - - - - - - - - - - - - - - - - - - - - - - - - -|
          CLEAR
          BLET             &COST = 0
          PUTSTRING        (' ')
          PUTSTRING        (' ')
          PUTSTRING        (' DO YOU WANT TO GO BACK AND DO THIS')
          PUTSTRING        (' EXAMPLE? (Y/N)')
          GETLIST          &ANS
          IF               (&ANS'E''Y')OR(&ANS'E''y')
          PUTSTRING        (' ')
          PUTSTRING        (' DO YOU WANT TO CHANGE ONLY THE SUPPLY AMOUNTS?
(Y/N)')
          PUTSTRING        (' ')
          GETLIST          &ANS
          IF               (&ANS'E''Y')OR(&ANS'E''y')
          PUTSTRING        (' ')
          GOTO             NEWSUP
          ENDIF
          PUTSTRING        (' ')
          GOTO             ALLNEW
          ENDIF
          PUTSTRING        (' ')
          PUTSTRING        (' SIMULATION OVER....')
          END
```

FIGURE 23.11 (Continued)
Listing of program CHAP23F.GPS.

23.2 Exercises

23.1. In Example 23.1, suppose the modeller has put the line of code:

```
BLET &VALUE = &I+2
```

instead of

```
LET &VALUE = &I+2
```

What would the output have been and why?

23.2. Modify the program given for Example 23.2 to determine the exact amount of tyres to order for the winter to the nearest integer.

23.3. State exactly what the following statements do:

```
        SIMULATE
        CHAR*1          &ANS
        INTEGER         &TK789,&TK793
AGAIN   PUTSTRING       (' ')
        PUTSTRING       (' INPUT THE NUMBER OF 789 TRUCKS')
        PUTSTRING       (' ')
        GETLIST         &TK789
        PUTSTRING       (' ')
        PUTSTRING       (' INPUT THE NUMBER OF 793 TRUCKS')
        PUTSTRING       (' ')
        GETLIST         &TK793
        PUTSTRING       (' ')
        PUTPIC          LINES = 3,&TK789,&TK793
  THE DATA YOU JUST KEYED IN IS:
NUMBER OF 789 TRUCKS **
NUMBER OF 793 TRUCKS **
        PUTSTRING       (' ARE YOU HAPPY WITH THESE? (Y/N)')
        PUTSTRING       (' ')
        GETLIST         &ANS
        IF     (&ANS'E''Y')OR(&ANS'E''y')
          GOTO NEXT
          ENDIF
          GOTO AGAIN
NEXT    PUTSTRING       (' ')
        PUTSTRING       (' DATA OK SO FAR...')
```

23.4. A variation of Example 23.1 is known as *The Newsboy's Problem*. This is to determine how many newspapers a newsboy should purchase for sale on a street corner. The newsboy goes to the newspaper office and orders papers each day and then sells them on a street corner. If he cannot sell all the ones he orders, he loses money on the unsold papers. If the demand is greater than his supply, he cannot sell any more papers. A simple variation of this problem is the following: Suppose the newsboy pays \$.30 for each paper. He sells them for \$.55. The probability of sales for each day is given in Table 23.3.

TABLE 23.3

Newsboy's Sales Record

Newspapers	Probability of Demand (Sales)
50	.10
55	.20
60	.40
65	.25
70	.05

For sake of simplicity, assume all sales are in units of 5. How many papers should the newsboy stock each day to maximize his expected profit?

23.5. Suppose that the newsboy in Exercise 23.5 can run quickly to a news stand and purchase extra papers for $.40 each if the demand is greater than his original supply. How does this affect the answer to the exercise?

23.6. Even though there is not much demand for unsold newspapers, assume that the newsboy is somehow able to receive $.20 for each unsold newspaper at the end of his sales day. How does this affect the result in Exercise 23.5?

24

The SELECT *and* COUNT *Blocks*

Hydraulic loader.

24.1 The SELECT Block

GPSS/H has several powerful Blocks that can be used to search several entities and test their attributes for a specific condition when a transaction enters the Block. When the condition is met, a record of this is placed in one of the transaction's parameters. Such scans are common in real-life situations. A mining truck approaches several shovels and will check for the one with the shortest queue of waiting trucks. The Block to do this scan and test at the same time is the SELECT Block. When a transaction enters this Block, a scan of selected entity members is made. When one of these scanned members is found that satisfies some stated condition, the scan is terminated. Examples of where such a Block might be used to model real-life situations are as follows:

1. A person enters a bank that has five tellers each with a queue of people waiting for service. The person will look at each queue and determine which is the shortest and then join that one.
2. A part comes along an assembly line where three machines can work on it. The part will be sent to the first machine that is not in use. If all three machines are in use, the part is sent to another section of the plant.

In order to do this scanning and testing, the `SELECT` Block needs to know what entities to scan, what the test is, where to put the result of the scan, and what to do if the scan is not successful. As a result, the `SELECT` Block can have up to six operands. As such, it is the most complicated Block since the `GENERATE` Block. One general form of the `SELECT` Block is:

```
SELECT ℛ    A,B,C,D,E,(F)
```

ℛ is a relational operator that is one of the following:

- `G` greater than
- `GE` greater than or equal to
- `E` equal to
- `NE` not equal to
- `LE` less than or equal
- `L` less than

`A` is the parameter number into which the first entity number that satisfies the test is to be placed. Thus, if queues numbered from 5 to 8 are scanned and queue number 7 satisfies the condition, then number 7 is placed in the transaction's parameter given by A.

`B` is the smallest number of the entities to be scanned.

`C` is the largest number of the entities to be scanned.

`D` is what the test is (see examples below).

`E` is the family name of the SNA to be used in the scan. This might be `F`, `FR`, `PH`, `S`, `SR`, `Q`, and so on.

`F` (optional) is a Block label where the transaction is transferred to if the scan is not successful.

The `SELECT` Block is best understood by considering examples of its use.

(a) `SELECT E    3PF,1,5,2,Q`

When a transaction enters this Block, a scan will be made of the queues from 1 to 5. These will be tested to see if any has a queue length of 2. If so, the number of the queue will be placed in the transaction's parameter #3. If no queue from 1 to 5 has a queue length equal to 2, the transaction moves to the next sequential Block. If all the queue lengths are zero except at queues 4 and 5 where they are both 2, the number 4 is placed in the transaction's full-word parameter number 3.

(b) `SELECT G    5PH,3,7,250,FR`

A scan is made of facilities numbered 3 to 7 starting with facility 3 and going up to facility 7. Once one is found to have a fractional utilization greater than .250 (recall that the fractional utilization is expressed in parts per thousand), the scan is stopped, and the facility number is placed in the transaction's half-word parameter number 5. If no facility from 3 to 7 has a fractional utilization greater than .250, the transaction moves to the next sequential Block.

(c) `SELECT LE    10PH,1,3,1,R,AWAY`

Storages numbered from 1 to 3 are scanned. If one has a remaining storage of less than or equal to 1, the number of it is copied in the transaction's 10th half-word parameter. If no storage in the scan satisfies the criteria, the transaction is routed to the Block with the label AWAY.

(d) `SELECT NE 4PH,7,12,0,PH`

The scan is of the transaction's half-word parameters numbered from 7 to 12. Once one of these is found to be not equal to zero, the number of it is copied in the transaction's 4th half-word parameter.

(e) `SELECT E   FN(ONE),PH3,PH4,3,Q,DOWN`

The parameter to place the result of the scan (if successful) is given by reference to the function ONE. Suppose it is 2. The scan will be to see if any of the scanned queues has a length of 3. The queues to be scanned will depend on the values of the transaction's 3rd and 4th half-word parameters. Suppose these are 5 and 9, respectively. Then queues 5 through 9 are scanned and, if any has a length of 3, the number of it is placed in the transaction's parameter number 2. If none of the queues is equal to 3, the transaction is routed to the Block with the label DOWN.

Example 24.1: Parts on Assembly Line

Parts come along an assembly line to be worked on by one of three identical machines. The parts arrive at a rate of one every 8 ± 4.5 minutes. The machines finish a part in 20 ± 5 minutes. If a machine is free, the part is worked on by that machine. But if all three machines are busy, the part is sent away to another part of the factory. Determine the utilization of the three machines and how many parts are sent away for 20 shifts of 480 minutes each.

Solution

The program to do the simulation is given in Figure 24.1.

```
           SIMULATE
****************************
*  PROGRAM CHAP24A.GPS       *
*  EXAMPLE OF SELECT BLOCK   *
****************************
 PARTS     GENERATE     8,4.5  PARTS COME ALONG
           SELECT E     2,1,3,0,F,AWAY   CHECK TO SEE IF MACHINE FREE
           SEIZE        PH2
           ADVANCE      20,5
           RELEASE      PH2
           TERMINATE
  AWAY     TERMINATE
           GENERATE     480*20
           TERMINATE    1
           START        1
           PUTPIC       LINES=6,N(PARTS),FR1/10.,FR2/10.,FR3/10.,N(AWAY)
          RESULTS OF SIMULATION
  PARTS TO ARRIVE              ****
  UTIL. OF MACHINE 1            ***.**%
  UTIL. OF MACHINE 2            ***.**%
  UTIL. OF MACHINE 3           ***.**%
  PARTS TO BE TURNED AWAY ****
           END
```

FIGURE 24.1
Listing of program CHAP24A.GPS.

```
              RESULTS OF SIMULATION
PARTS TO ARRIVE              1198
UTIL. OF MACHINE 1              81.56%
UTIL. OF MACHINE 2              75.14%
UTIL. OF MACHINE 3            62.48%
PARTS TO BE TURNED AWAY   151
```

FIGURE 24.2
Results of program CHAP24A.GPS.

The output from the program is as shown in Figure 24.2.

For the 20 shifts, 1198 parts came to the three machines. Because all three machines were busy, 151 were turned away. The utilization of the three machines was 81.6% for machine 1, 75.1% for machine 2, and 62.5% for machine 3. The reason for the decrease in utilization is because, when a new part arrives at the machines, the scan is to see if any one of the three machines is free. The scan stops when one is found to be free and the scanning always starts at the first machine.

This problem could also have been solved using the `TRANSFER ALL` Block.

Notice that, in this example, the parameter type was not specified. As long as you only have half-word parameters, this is all right. But if the transaction had different types of parameters, it is necessary to specify the type in the `SELECT` Block. For example,

```
GENERATE        20,6.5,,,,2PH,10PF
........        .................
........        .................
SELECT E        2,1,5,1,Q
```

would give an error. The correct form of the `SELECT` Block would have to be

```
SELECT E        2PH,1,5,1,Q (or 2PF)
```

The animation will take considerably more lines of code, since each of the three machines needs to be animated separately. In addition, time for the parts to enter the system and then leave will have to be added. Times for these were arbitrarily chosen. For an actual situation, these times would have to be measured. The program to do the animation is CHAP24B.GPS. Its listing in given in Figure 24.3. The animation is shown in Figure 24.4.

```
          SIMULATE
****************************
*  PROGRAM CHAP24B.GPS      *
*  ANIMATION OF CHAP24A.GPS *
****************************
          INTEGER      &PENTER,&ALLDONE,&TAWAY
 ATF      FILEDEF      'CHAP24B.ATF'
 WHERE    FUNCTION     PH2,L3
1,MACH1/2,MACH2/3,MACH3
 PARTS    GENERATE     8,4.5  PARTS COME ALONG
          BLET         &PENTER=&PENTER+1
          BPUTPIC      FILE=ATF,LINES=5,AC1,XID1,XID1,XID1,&PENTER
TIME *.****
CREATE PART P*
PLACE P* ON P1
SET P* TRAVEL   1
WRITE M6 ****
```

FIGURE 24.3
Listing of program CHAP24B.GPS. *(Continued)*

```
          ADVANCE     1
          SELECT E    2,1,3,0,F,AWAY   CHECK TO SEE IF MACHINE FREE
          TRANSFER    ,FN(WHERE)
 MACH1    SEIZE       1
          BPUTPIPC    FILE=ATF,LINES=3,AC1,XID1
TIME *.****
PLACE P* AT -10 19
SET M1 COLOR RED
          ADVANCE     20,5
          RELEASE     1
          BPUTPIC     FILE=ATF,LINES=5,AC1,XID1,XID1,FR1/10.
TIME *.****
PLACE P* ON P3
SET P* TRAVEL .5
SET M1 COLOR GREEN
WRITE M1 ***.**%
          ADVANCE     .5
          BLET        &ALLDONE=&ALLDONE+1
          BPUTPIC     FILE=ATF,LINES=3,AC1,XID1,&ALLDONE
TIME *.****
DESTROY P*
WRITE M4 ****
          TERMINATE
MACH2     SEIZE       2
          BPUTPIPC    FILE=ATF,LINES=3,AC1,XID1
TIME *.****
PLACE P* AT -10 6
SET M2 COLOR RED
          ADVANCE     20,5
          RELEASE     2
          BPUTPIC     FILE=ATF,LINES=5,AC1,XID1,XID1,FR2/10.
TIME *.****
PLACE P* ON P3
SET P* TRAVEL .5
SET M2 COLOR GREEN
WRITE M2 ***.**%
          ADVANCE     .5
          BLET        &ALLDONE=&ALLDONE+1
          BPUTPIC     FILE=ATF,LINES=3,AC1,XID1,&ALLDONE
TIME *.****
DESTROY P*
WRITE M4 ****
          TERMINATE
 MACH3    SEIZE       3
          BPUTPIPC    FILE=ATF,LINES=3,AC1,XID1
TIME *.****
PLACE P* AT -10 -5
SET M3 COLOR RED
          ADVANCE     20,5
          RELEASE     3
          BPUTPIC     FILE=ATF,LINES=5,AC1,XID1,XID1,FR3/10.
TIME *.****
PLACE P* ON P3
SET P* TRAVEL .5
SET M3 COLOR GREEN
WRITE M3 ***.**%
          ADVANCE     .5
          BLET        &ALLDONE=&ALLDONE+1
          BPUTPIC     FILE=ATF,LINES=3,AC1,XID1,&ALLDONE
TIME *.****
```

FIGURE 24.3 (Continued)
Listing of program CHAP24B.GPS.

(Continued)

```
DESTROY P*
WRITE M4 ****
          TERMINATE
  AWAY    BPUTPIC      FILE=ATF,LINES=3,AC1,XID1,XID1
TIME *.****
PLACE P* ON P2
SET P* TRAVEL 1
          ADVANCE      1
          BLET         &TAWAY=&TAWAY+1
          BPUTPIC      FILE=ATF,LINES=3,AC1,XID1,&TAWAY
TIME *.****
DESTROY P*
WRITE M5 ****
          TERMINATE
          GENERATE     480
          TERMINATE    1
          START        1
          PUTPIC       FILE=ATF,LINES=2,AC1
TIME *.****
END
          PUTPIC       LINES=6,N(PARTS),FR1/10.,FR2/10.,FR3/10.,N(AWAY)
         RESULTS OF SIMULATION
  PARTS TO ARRIVE              ****
  UTIL. OF MACHINE 1            ***.**%
  UTIL. OF MACHINE 2            ***.**%
  UTIL. OF MACHINE 3           ***.**%
  PARTS TO BE TURNED AWAY ****
          END
```

FIGURE 24.3 (Continued)
Listing of program CHAP24B.GPS.

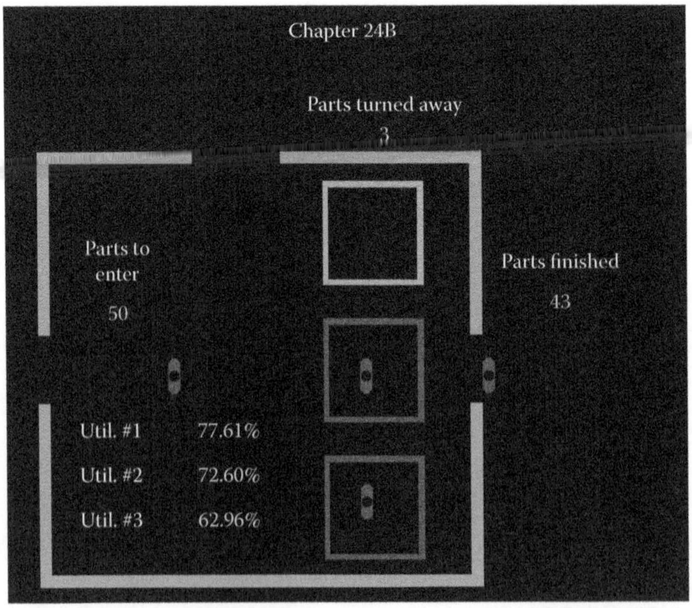

FIGURE 24.4
Screenshot of animation from CHAP24B.GPS.

24.2 The COUNT Block

The COUNT Block resembles the SELECT Block in that, when a transaction enters it, it triggers a scan of specified entities. The result of the scan is placed in a specified parameter. In the case of the SELECT Block, once the scan finds an entity to satisfy the given test, the scan is over. The COUNT Block counts the number of the entities that satisfy the criteria and places the number counted that satisfy the test criteria into the transaction's specified parameter. One general form of the COUNT Block is:

```
COUNT ℛ        A,B,C,D,E
```

$\mathcal{R}$ is one of the relational operators used in the SELECT Block. Thus, it must be one of the following: G, GE, E, NE, L, or LE.

A, B, C, D, and E have the same meanings as the operands in the SELECT Block.

Some examples of this Block are as follows:

```
(a) COUNT E    1PH,1,4,0,Q
(b) COUNT G    3PH,3,6,250,FR
```

In (a), a count is made of all the queues from 1 to 4 that have lengths 0. This number is placed in the transaction's half-word parameter number 1.

In (b), a count is made of the facilities from 3 to 6 that have fractional utilization greater than 250. The number of these facilities is placed in the transaction's half-word parameter number 3.

Since the result of the entities in a COUNT Block will always be a number, there cannot be any F operand.

24.3 Other Forms of the SELECT and COUNT Blocks

24.3.1 The SELECT Block in MIN/MAX Mode

The SELECT Block can be used to scan a group of entities and determine which one has a maximum (or minimum) value. The general form of this is as follows:

```
SELECT MIN (or MAX)    A,B,C,,E
```

The word MIN (or MAX) must be in the auxiliary operator (aux op) position, which is one position away from the SELECT Block. The A, B, C, and E operands are the same as for the regular SELECT Block described previously. There is no D operand.

When a transaction enters this form of SELECT Block, a scan is made of the entities specified by the B and C operands. The processor selects the minimum (or maximum) from the entity class and places that number into the transaction's parameter number as specified by the A operand.

```
(a) SELECT MIN    3,1,4,,FR
(b) SELECT MAX    1,3,7,,Q
(c) SELECT MIN    5,1,3,,R
```

In (a), facilities from 1 to 4 are scanned and the processor will place the number of the one that has the least fractional utilization into the transaction's parameter number 3. Thus, if facility 1 had an FR of 350, facility 2 of 599, facility 3 of 500, and facility 4 of 222, the number 4 would be in the transaction's fourth parameter.

In (b), queues 3 to 7 are scanned. The number that has the greatest length is placed in the transaction's parameter number 1. In case of a tie, that is, if both queue 4 and queue 7 had equal lengths of 5 and this was the maximum, the number 4 is placed in parameter 1. This is also the case if SELECT MIN is used.

In (c), storages from 1 to 3 are scanned and the one with the greatest remaining storage is placed in the transaction's parameter number 1.

24.3.2 Use with Logic Switches

A SELECT or a COUNT Block can be used with logic switches. The general form is

```
SELECT LS (or LR)    A,B,C,,,(F)
```

Operands D and E are omitted and F is optional.

The scan is made of logic switches from B (minimum value) to C (maximum value). As soon as the processor encounters one that satisfies the test (either LS for set or LR for reset), the scan is finished, and the number of the logic switch is placed in the transaction's parameter number given by operand A. The F operand is a Block label where the transaction is routed to if no logic switch satisfies the scan criteria.

The COUNT Block in this mode is

```
COUNT LS (or LR)     A,B,C
```

The only difference from the SELECT Block is that the operand F is not used. In this case, a scan is made of the logic switches from B to C and the number of logic switches in a set (or reset) condition is placed in parameter number as given by operand A. For example, suppose a transaction entered the following two sequential Blocks in a program.

```
SELECT LS            3PH,2,6
COUNT LR             4PH,2,6
```

and suppose that the logic switches 2 through 6 were as follows:

LS2 is reset

LS3 is reset

LS4 is set

LS5 is reset

LS6 is reset

The transaction would have a 4 in parameter 3 and also a 4 in parameter 4.

24.3.3 Use with Facilities and Storages

The form of the SELECT Block is

```
SELECT (aux op)      A,B,C,,,(F)
```

where (aux op) can be one of the following:

Aux Op	Meaning
`U`	Facility in use
`NU`	Facility not in use
`SE`	Storage empty
`SNE`	Storage not empty
`SF`	Storage full
`SNF`	Storage not full

There are no `D` or `E` operands and the `F` operand is optional. The meaning of the operands is the same as before.

The general form of the `COUNT` Block in this mode is

```
COUNT (aux op)   A,B,C
```

A scan is done of the entities and the number of them satisfying the criteria is placed in the transaction's parameter number as specified by the `A` operand. Several examples of these Blocks are as follows:

(a) `SELECT NU 1PH,3,6`

(b) `COUNT NU 2PH,3,6`

(c) `SELECT SE 3PH,1,7,,,AWAY`

(d) `COUNT SF 4PH,1,7`

In (a), the scan is of facilities 3 to 6. Once one is found to be not in use, the scan is finished and the number of the facility is placed in the transaction's parameter number 1.

In (b), a count is made of the facilities from 3 to 6 that are not in use. This number is placed in parameter number 3.

In (c), a scan of storages from 1 to 7 is made to determine if any are empty. The number of the first one to satisfy the criteria is placed in parameter 3 and the scan is stopped. If no storage is found to satisfy the criteria, the transaction is sent to the Block with the label `AWAY`.

In (d), a count is made of the full storages from 1 to 7. This number is placed in parameter 4.

Example 24.2: Quickline versus Individual Queues

Did you ever wonder why banks, post offices, airline ticket agents, and other places where multiple servers are used now have customers wait in an individual queue rather than forming separate queues at each teller or agent? The single queue system is known as a *quickline* system. This example will illustrate why a quickline system is better than individual queues.

First, consider the case of individual queues. Suppose customers arrive in a store that has six clerks behind desks to serve the customers. Customers come in and first see if any clerk is free. If so, the customer will go to that desk. If all the clerks are busy, the person will go to the back of the shortest queue. Once in a queue at a desk, no queue jumping is allowed.

Customers arrive in a Poisson stream, with an interarrival time of 10 seconds. A customer will transact business as given in Table 24.1.

TABLE 24.1

Business Data for Customers

Business	Percentage of Time	Time Taken
Type 1	28	(15, 2)
Type 2	17	(32, 5)
Type 3	30	(55, 8)
Type 4	25	(65, 10)

The time is sampled from the normal distribution so that the figures in the table are (mean and standard deviation). This system is modelled with both individual queues and the quickline. The store works 10 hours straight.

Solution

The program to model the quickline is quite easy to write using the clerks as storages. The listing of it is as shown in Figure 24.5. The results of the simulation are given in Figure 24.6.

This shows that, at one time, there actually were 19 people in the queue waiting for a free clerk. The average number of people in the queue was 0.72. The clerks were busy 73.5% of the time.

The program for the individual queues is given as CHAP24D.GPS (Figure 24.7).

A single global queue, `WAIT`, is used for all the customers as they enter the system. Each server has its own queue. The results of the program are as shown in Figure 24.8.

A comparison of the two systems reveals that the quickline system has an average of .72 customers in the system waiting for service while the individual queue system

```
          SIMULATE
************************
*  PROGRAM CHAP24C.GPS  *
*  CLASSIC EXAMPLE OF   *
*  QUEUES IN A STORE    *
*  MODEL WITH QUICKLINE *
************************
 MEAN      FUNCTION     RN1,D4
.28,15/.4,32/.7,55/1,65
 STDDEV    FUNCTION     PH1,D4
15,2/32,5/50,8/60,10
          STORAGE      S(TELLER),6
          GENERATE     RVEXPO(1,10)  CUSTOMERS ARRIVE
          ASSIGN       1,FN(MEAN),PH DETERMINE MEAN TIME FOR BUSINESS
          ASSIGN       2,FN(STDDEV),PH  DET. STD. DEV. FOR BUSINESS
          QUEUE        QLINE     FORM SINGLE QUEUE
          ENTER        TELLER    IS A TELLER FREE?
          DEPART       QLINE     LEAVE THE QUEUE
          ADVANCE      RVNORM(1,PH1,PH2)     DO BUSINESS
          LEAVE        TELLER    FREE THE TELLER
          TERMINATE             LEAVE THE BANK
          GENERATE     3600*10*100
          TERMINATE    1
          START        1
          PUTPIC       LINES=4,QM(QLINE),QA(QLINE),SR(TELLER)/10.
       RESULTS OF SIMULATION
   MAX. IN QUEUE AT ANY ONE TIME ***
   AVG. IN QUEUE                    *.**
   UTIL OF TELLERS                ***.**%
           END
```

FIGURE 24.5
Listing of program CHAP24C.GPS.

```
        RESULTS OF SIMULATION
    MAX. IN QUEUE AT ANY ONE ITME  19
    AVG. IN QUEUE                   0.72
    UTIL OF TELLERS                73.45%
```

FIGURE 24.6
Output from program CHAP24C.GPS.

```
          SIMULATE
**************************
*   PROGRAM CHAP24D.GPS    *
*   CLASSIC EXAMPLE OF     *
*   QUEUES IN A STORE      *
*   MODEL WITH INDIVIDUAL *
*   QUEUES                 *
**************************
 MEAN      FUNCTION     RN1,D4
.28,15/.4,32/.7,55/1,65
 STDDEV    FUNCTION     PH1,D4
15,2/32,5/50,8/60,10
 WAIT      EQU          10,Q
           GENERATE     RVEXPO(1,10)  CUSTOMERS ARRIVE
           ASSIGN       3,FN(MEAN),PH DETERMINE MEAN TIME FOR BUSINESS
           ASSIGN       4,FN(STDDEV),PH  DET. STD. DEV. FOR BUSINESS
           QUEUE        WAIT
           SELECT E     1PH,1,6,0,F,AWAY   SEE IF A CLERK IS FREE
 BACK      QUEUE        PH1
           SEIZE        PH1
           DEPART       PH1
           DEPART       WAIT
           ADVANCE      RVNORM(1,PH3,PH4)
           RELEASE      PH1
           TERMINATE                  LEAVE THE STORE
 AWAY      SELECT MIN   1PH,1,6,,Q    SELECT MINIMUM QUEUE
           TRANSFER     ,BACK
           GENERATE     3600*10*100
           TERMINATE    1
           START        1
           PUTPIC       LINES=9,QM(WAIT),QA(WAIT),FR1/10.,FR2/10.,_
                        FR3/10.,FR4/10.,FR5/10.,FR6/10.
       RESULTS OF SIMULATON
  MAX. IN SHOP AT ANY ONE TIME ***
  AVG. IN SHOP                  **.**
  UTIL OF TELLER NO. 1         ***.**%
  UTIL OF TELLER NO. 2         ***.**%
  UTIL OF TELLER NO. 3         ***.**%
  UTIL OF TELLER NO. 4         ***.**%
  UTIL OF TELLER NO. 5         ***.**%
  UTIL OF TELLER NO. 6         ***.**%
          END
```

FIGURE 24.7
Listing of program CHAP24D.GPS.

has 1.07 customers waiting for service. At one time, the individual queue system has 29 customers waiting for a free clerk. The quickline has a maximum number of 19. In the individual queue system, the servers were busy from 90.24% to 51.79%. This is because an arriving customer will first scan all the clerks and always go to the first numerical clerk if more than one clerk is free.

The animation is given for the case of the quickline. It is program CHAP24E.GPS. The listing is as shown in Figure 24.9.

```
          RESULTS OF SIMULATON
      MAX. IN SHOP AT ANY ONE TIME  29
      AVG. IN SHOP                   1.07
      UTIL OF TELLER NO. 1          90.24%
      UTIL OF TELLER NO. 2          85.32%
      UTIL OF TELLER NO. 3          79.01%
      UTIL OF TELLER NO. 4          71.32%
      UTIL OF TELLER NO. 5          62.26%
      UTIL OF TELLER NO. 6          51.79%
```

FIGURE 24.8
Results of program CHAP24D.GPS.

```
          SIMULATE
*****************************
*  PROGRAM CHAP24E.GPS      *
*  ANIMATION OF CHAP24C.GPS *
*  CLASSIC EXAMPLE OF       *
*  QUEUES IN A STORE        *
*  MODEL WITH QUICKLINE     *
*************************
 MEAN     FUNCTON      RN1,D4
.28,15/.4,32/.7,55/1,65
 STDDEV   FUNCTON      PH1,D4
15,2/32,5/50,8/60,10
          INTEGER      &COMEIN,&GOOUT
  ATF     FILEDEF      'CHAP24E.ATF'
          GENERATE     RVEXPO(1,10)  CUSTOMERS ARRIVE
          BLET         &COMEIN=&COMEIN+1
          ASSIGN       1,FN(MEAN),PH DETERMINE MEAN TIME FOR BUSINESS
          ASSIGN       2,FN(STDDEV),PH  DET. STD. DEV. FOR BUSINESS
          BPUTPIC      FILE=ATF,LINES=5,AC1,XID1,XID1,XID1,&COMEIN
TIME *.****
CREATE PERSON P*
PLACE P* ON P1
SET P* TRAVEL 4
WRITE M8 ***
          ADVANCE      4
          QUEUE        QLINE    FORM SINGLE QUEUE
          TRANSFER     ALL,FIRST,LAST1,10
 FIRST    SEIZE        WORK1
          DEPART       QLINE
          BPUTPIC      FILE=ATF,LINES=4,AC1,XID1,QA(QLINE)
TIME *.****
PLACE P* AT -24 23
SET W1 COLOR RED
WRITE M7 *.**
          ADVANCE      RVNORM(1,PH1,PH2)
          RELEASE      WORK1
          BPUTPIC      FILE=ATF,LINES=5,AC1,XID1,XID1,FR(WORK1)/10.
TIME *.****
PLACE P* ON P2
SET P* TRAVEL 3.5
SET W1 COLOR GREEN
WRITE M1 **.**%
          ADVANCE      3.5
          BLET         &GOOUT=&GOOUT+1
          BPUTPIC      FILE=ATF,LINES=3,AC1,XID1,&GOOUT
```

FIGURE 24.9
Listing of program CHAP24E.GPS. *(Continued)*

```
TIME *.****
DESTROY P*
WRITE M9 ***
          TERMINATE
          SEIZE        WORK2
          DEPART       QLINE
          BPUTPIC      FILE=ATF,LINES=4,AC1,XID1,QA(LINE)
TIME *.****
PLACE P* AT -24 15
SET W2 COLOR RED
WRITE M7 *.**
          ADVANCE      RVNORM(1,PH1,PH2)
          RELEASE      WORK2
          BPUTPIC      FILE=ATF,LINES=5,AC1,XID1,XID1,FR(WORK2)/10.
TIME *.****
PLACE P* ON P2
SET P* TRAVEL 3.5
SET W2 COLOR GREEN
WRITE M2 **.**%
          ADVANCE      3.5
          BLET         &GOOUT=&GOOUT+1
          BPUTPIC      FILE=ATF,LINES=3,AC1,XID1,&GOOUT
TIME *.****
DESTROY P*
WRITE M9 ***
          TERMINATE
          SEIZE        WORK3
          DEPART       QLINE
          BPUTPIC      FILE=ATF,LINES=4,AC1,XID1,QA(QLINE)
TIME *.****
PLACE P* AT -24 8
SET W3 COLOR RED
WRITE M7 *.**
          ADVANCE      RVNORM(1,PH1,PH2)
          RELEASE      WORK3
          BPUTPIC      FILE=ATF,LINES=5,AC1,XID1,XID1,FR(WORK3)/10.
TIME *.****
PLACE P* ON P2
SET P* TRAVEL 3.5
SET W3 COLOR GREEN
WRITE M3 **.**%
          ADVANCE      3.5
          BLET         &GOOUT=&GOOUT+1
          BPUTPIC      FILE=ATF,LINES=3,AC1,XID1,&GOOUT
TIME *.****
DESTROY P*
WRITE M9 ***
          TERMINATE
          SEIZE        WORK4
          DEPART       QLINE
          BPUTPIC      FILE=ATF,LINES=4,AC1,XID1,QA(QLINE)
TIME *.****
PLACE P* AT -24 -1
SET W4 COLOR RED
WRITE M7 *.**
          ADVANCE      RVNORM(1,PH1,PH2)
          RELEASE      WORK4
          BPUTPIC      FILE=ATF,LINES=5,AC1,XID1,XID1,FR(WORK4)/10.
TIME *.****
```

FIGURE 24.9 (Continued)
Listing of program CHAP24E.GPS.

(Continued)

```
PLACE P* ON P2
SET P* TRAVEL 3.5
SET W4 COLOR GREEN
WRITE M4 **.**%
          ADVANCE      3.5
          BLET         &GOOUT=&GOOUT+1
          BPUTPIC      FILE=ATF,LINES=3,AC1,XID1,&GOOUT
TIME *.****
DESTROY P*
WRITE M9 ***
          TERMINATE
          SEIZE        WORK5
          DEPART       QLINE
          BPUTPIC      FILE=ATF,LINES=4,AC1,XID1,QA(QLINE)
TIME *.****
PLACE P* AT -24 -9
SET W5 COLOR RED
WRITE M7 *.**
          ADVANCE      RVNORM(1,PH1,PH2)
          RELEASE      WORK5
          BPUTPIC      FILE=ATF,LINES=5,AC1,XID1,XID1,FR(WORK5)/10.
TIME *.****
PLACE P* ON P2
SET P* TRAVEL 3.5
SET W5 COLOR GREEN
WRITE M5 **.**%
          ADVANCE      3.5
          BLET         &GOOUT=&GOOUT+1
          BPUTPIC      FILE=ATF,LINES=3,AC1,XID1,&GOOUT
TIME *.****
DESTROY P*
WRITE M9 ***
          TERMINATE
 LAST     SEIZE        WORK6
          DEPART       QLINE
          BPUTPIC      FILE=ATF,LINES=4,AC1,XID1,QA(QLINE)
TIME *.****
PLACE P* AT -24 -17
SET W6 COLOR RED
WRITE M7 *.**
          ADVANCE      RVNORM(1,PH1,PH2)
          RELEASE      WORK6
          BPUTPIC      FILE=ATF,LINES=5,AC1,XID1,XID1,FR(WORK6)/10.
TIME *.****
PLACE P* ON P2
SET P* TRAVEL 3.5
SET W6 COLOR GREEN
WRITE M6 **.**%
          ADVANCE      3.5
          BLET         &GOOUT=&GOOUT+1
          BPUTPIC      FILE=ATF,LINES=3,AC1,XID1,&GOOUT
TIME *.****
DESTROY P*
WRITE M9 ***
          TERMINATE
          GENERATE     3600*10
          TERMINATE    1
          START        1
          PUTPIC
```

FIGURE 24.9 (Continued)
Listing of program CHAP24E.GPS.

(Continued)

```
LINES=9,QM(QLINE),QA(QLINE),FR(WORK1)/10.,FR(WORK2)/10.,_
                     FR(WORK3)/10.,FR(WORK4)/10.,FR(WORK5)/10.,FR(WORK6)/10.
       RESULTS OF SIMULATON
  MAX. IN QUEUE AT ANY ONE TIME ***
  AVG. IN QUEUE                   *.**
  UTIL OF SERVER 1              ***.**%
  UTIL OF SERVER 2              ***.**%
  UTIL OF SERVER 3              ***.**%
  UTIL OF SERVER 4              ***.**%
  UTIL OF SERVER 5              ***.**%
  UTIL OF SERVER 6              ***.**%
          PUTPIC     FILE=ATF,LINES=2,AC1
TIME *.****
END
          END
```

FIGURE 24.9 (Continued)
Listing of program CHAP24E.GPS.

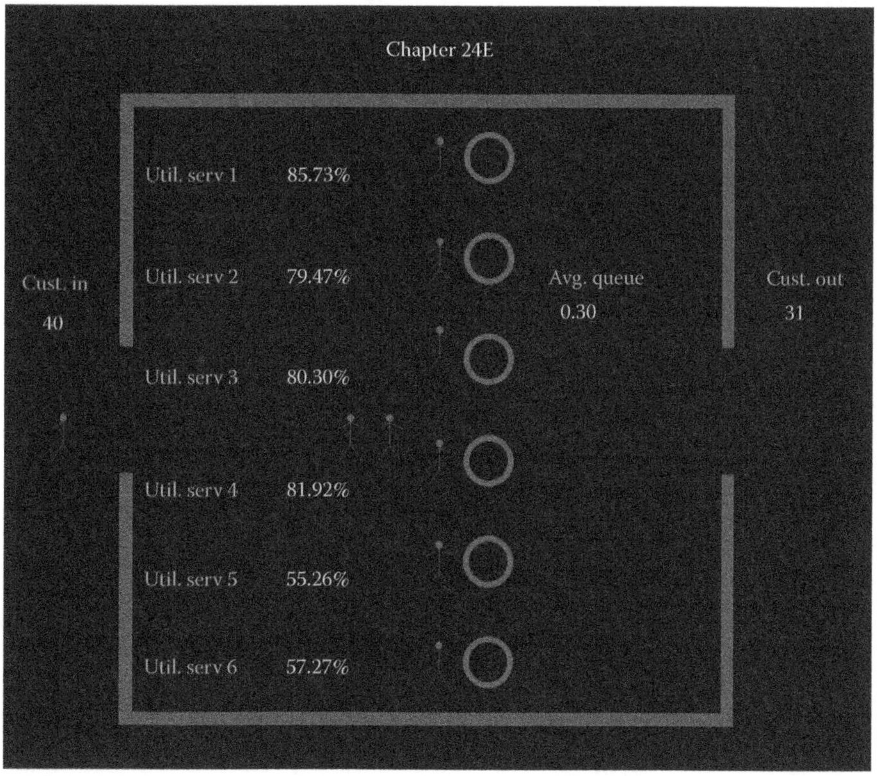

FIGURE 24.10
Screenshot of animation from program CHAP24E.GPS.

The animation program is now considerably longer than the original program, CHAP24C.GPS. This is because each of the six servers needs to be modelled separately for the animation. The animation looks as shown in Figure 24.10.

Example 24.3: An Example Using the COUNT Block

The COUNT Block is not as common as the SELECT Block in mining simulations. One example of its use is presented here. Suppose that there are five loaders in a mine.

TABLE 24.2

Reliabilities of Machines

Loader No.	Working	Down
1	100 ± 50	10 ± 5
2	200 ± 100	20 ± 10
3	300 ± 150	30 ± 15
4	400 ± 200	40 ± 20
5	500 ± 250	50 ± 25

Trucks arrive to be loaded every 5 ± 2 minutes. Each loader is identical and can load a truck in 20 ± 5 minutes. The truck will use the first loader if it is free. If the first loader is not available, the truck will attempt to use the second loader and then the third up to the fifth. If no loader is available, the truck will wait until one becomes available. The reliabilities of the loaders are given in Table 24.2.

The program to simulate this system is CHAP24F and is given in Figure 24.11. It uses both the COUNT and SELECT Blocks.

```
          SIMULATE
******************************
*  PROGRAM CHAP24F.GPS        *
*  EXAMPLE OF COUNT BLOCK     *
******************************
COME      GENERATE     5,2    TRUCK ARRIVES
          QUEUE        WAIT
BACKA     COUNT E      2PH,1,5,0,F   IS ANY SHOVEL FREE?
          TEST NE      PH2,0,AWAY    YES, AT LEAST ONE
          SELECT E     1PH,1,5,0,F   WHICH ONE IS FREE?
          SEIZE        PH1
          SEIZE        PH1+5
          DEPART       WAIT          LEAVE THE QUEUE
          ADVANCE      20,5          LOAD
          RELEASE      PH1
          RELEASE      PH1+5
          TERMINATE                  TRUCK LEAVES
*****************************
*  SHOVEL RELIABILITIES      *
*****************************
          GENERATE     ,,,1,1
 FIRST    ADVANCE      100,50
          SEIZE        1
          ADVANCE      10,5
          RELEASE      1
          TRANSFER     ,FIRST
          GENERATE     ,,,1,1
 SECOND   ADVANCE      200,100
          SEIZE        2
          ADVANCE      20,10
          RELEASE      2
          TRANSFER     ,SECOND
          GENERATE     ,,,1,1
 THIRD    ADVANCE      300,150
          SEIZE        3
          ADVANCE      30,15
          RELEASE      3
          TRANSFER     ,THIRD
          GENERATE     ,,,1,1
```

FIGURE 24.11
Listing of program CHAP24F.GPS.

(Continued)

```
FOURTH  ADVANCE     400,200
        SEIZE       4
        ADVANCE     40,20
        RELEASE     4
        TRANSFER    ,FOURTH
        GENERATE    ,,,1,1
FIFTH   ADVANCE     500,250
        SEIZE       5
        ADVANCE     50,25
        RELEASE     5
        TRANSFER    ,FIFTH
AWAY    ADVANCE     1
        TRANSFER    ,BACKA
        GENERATE    480*10000
        TERMINATE   1
        START       1
        PUTPIC
LINES=7,N(COME)/10000.,QA(WAIT),FR6/10.,FR7/10.,FR8/10.,_
                    FR9/10.,FR10/10.
   TRUCKS TO ARRIVE PER SHIFT  ***.**
   AVG. QUEUE                    *.**
   UTIL. OF SHOVEL 1            **.**%
   UTIL. OF SHOVEL 2            **.**%
   UTIL. OF SHOVEL 3            **.**%
   UTIL. OF SHOVEL 4            **.**%
   UTIL. OF SHOVEL 5            **.**%
        END
```

FIGURE 24.11 (Continued)
Listing of program CHAP24F.GPS.

The first COUNT Block is COUNT 2PH,1,5,0,F. This scans the five facilities (loaders) to count the number that are free. If none are free, the count is 0. The next Block is

```
TEST NE   PH2,0,AWAY      YES, AT LEAST ONE
```

If no loader is free, control passes to the Block with the label AWAY. This waits a minute and then sends the transaction back to see if any loader is now free. Eventually, a loader will be free and the TEST Block will be true and the transaction can pass through to the next Block:

```
SELECT E  1PH,1,5,0,F    WHICH ONE IS FREE?
```

This determines which of the five loaders is free and allows the truck to move to the first free loader. The results are shown in Figure 24.12.

Since the arrival rate has a mean time of 5 minutes and there are five loaders that can load in a mean time of 20 minutes, the results appear reasonable.

The animation is given by program CHAP24G.GPS. A screenshot is shown in Figure 24.13.

```
TRUCKS TO ARRIVE PER SHIFT   95.98
AVG. QUEUE                    0.44
UTIL. OF SHOVEL 1            85.17%
UTIL. OF SHOVEL 2            83.20%
UTIL. OF SHOVEL 3            81.06%
UTIL. OF SHOVEL 4            78.18%
UTIL. OF SHOVEL 5            72.31%
```

FIGURE 24.12
Results of program 24F.GPS.

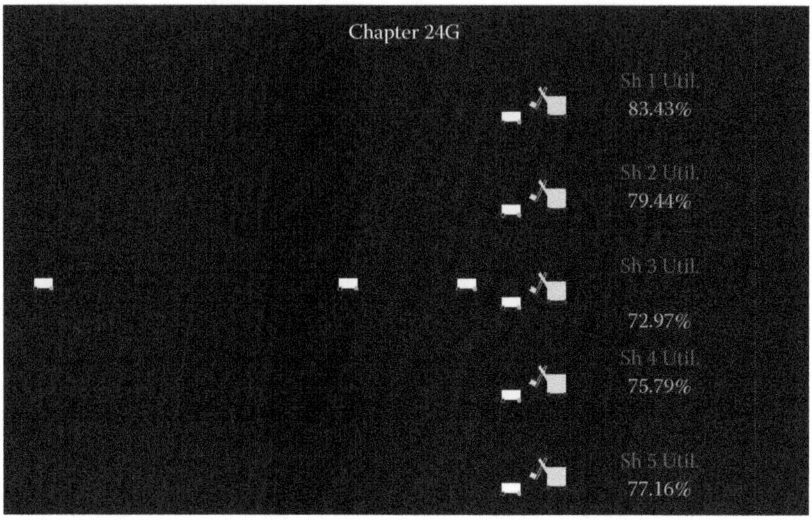

FIGURE 24.13
Screenshot of animation from CHAP24G.GPS.

24.4 Exercises

24.1. State what each of the following Blocks will do:

```
(a) SELECT E    2PH,4,6,8,S
(b) SELECT E    1PH,1,3,0,Q
(c) SELECT NE   4PH,1,3,0,Q,AWAY
(d) SELECT GE   3PH,2,5,500,FR
(e) SELECT NE   12PH,1,4,0,R,DDDD
(f) COUNT E     2PH,1,3,0,Q
(g) COUNT LE    1PH,2,6,200,F
(h) SELECT MAX  5PH,1,3,,R
(i) SELECT MIN  3PH,2,4,,FR
(j) SELECT E    1PH,PH2,PH3,PH4,Q
(k) SELECT E    3PH,PH1,PH2,4,S,AWAY
```

24.2. The service times for Example 24.2 were all assumed to be normally distributed. Suppose that the service times were now all exponential with the same means as before. For example, the service time of (15, 2) was given by `RVNORM(1,15,2)`. Now it will be given by `RVEXPO(1,15)`. Determine the difference in the two systems.

24.3. The animation for Example 24.3 can be made a bit more illuminating. For example, when the trucks leave the path they arrive on for loading by a loader, they immediately *pop* in front of the first loader that is free. Add paths to each loader and have the trucks *spot* or move to each loader. Add the code to the simulation program to count the trucks that have been loaded and add this to the animation.

TABLE 24.3

Reliabilities for Six Machines

Machine	Working	Down
1	70 ± 20	12 ± 4
2	Exp. mean 60	Exp. mean 8
3	50 ± 10	8 ± 3
4	55 ± 12	10 ± 5
5	Normal mean 50, standard deviation 6	Normal mean 5, standard deviation 75
6	Exp. mean 25	Exp. mean 1.5

24.4. An engineer has six machines in his mine. It is essential that at least one is always working. The engineer has the following data relating to the reliability of the machines and their downtimes. This is given in Table 24.3.

The engineer wants to simulate the system for the next 10 years, checking at 5 minute intervals to count the number of machines that are down. Write the program to do this simulation. How many times were all machines down?

24.5. The engineer in Exercise 24.4 can purchase special equipment for each of the machines in his mine so that the mean downtimes are halved. Thus, machine 1 will now be down for a time of 6 ± 4; machine 2 will be down for a time given by the exponential distribution with a mean of 4; machine 3 has a downtime given by 4 ± 3; machine 4 has a downtime given by 5 ± 5; machine 5 has a downtime given by the normal distribution with a mean of 2.5 and a standard deviation of .5; machine 6 for a time given by the exponential distribution with a mean of .75. Will there now be any time when all six machines might be down?

25

Variables and Expressions

Ore pass system for mine.

25.1 Variables and Expressions

25.1.1 Arithmetic in GPSS/H (Review)

We have been doing arithmetic in GPSS/H when it was needed without commenting on it. This is because arithmetic is allowed in the operands of the various Blocks and is done in a logical manner. Thus, when we have a Block such as

```
ADVANCE 2*100
```

it was not necessary to point out that the time on the FEC for the transaction was 200. However, it is now necessary to formally go through the steps done in performing arithmetic because there are certain cautions which must be observed. These have to do with the original integer nature of calculations in GPSS.

As introduced in Chapter 13, arithmetic is accomplished in GPSS/H using SNAs together with various arithmetic operations. These operations are repeated here:

- \+ addition
- − subtraction (both unary and binary subtraction)
- / division
- * multiplication
- @ modular division

An arithmetic expression can be placed in an operand or referenced in a manner similar to the way that functions are referenced. This might be done in GPSS/H, for example, when a particular expression is referenced multiple times and, therefore, can save considerable time in writing the code. Referencing of expressions is done by defining the expression using either a `VARIABLE` or an `FVARIABLE` expression. `VARIABLE` is used when the result is to be an integer. `FVARIABLE` is used when the result is a real number. The form is

```
(label) VARIABLE (expression)
```

or

```
(label) FVARIABLE (expression)
```

Numbers can be used for variable labels. Therefore,

```
TOMMY VARIABLE       3PH+FR(MACH)/200-Q(WAIT)
   1 VARIABLE        &FIRST&SECOND+QM(NEXT)
CALC FVARIABLE       3+Q(LAST)*F(MACH)-R(TUGS)
   16 FVARIABLE       45.34*FR(MACH1)/10
```

are possible ways to define variables `TOMMY, 1, CALC,` and `16.` Notice that no spaces are allowed in expressions where, in other programming languages such as Fortran, blank spaces are recommended for clarity. A blank space in this case will terminate the expression.

Referencing variables is done by

```
V(name) (or, V$name)
```

When an expression is referenced, it is evaluated and the result is returned. Evaluation of an expression follows the usual rules found in other programming languages. This means from left to right with multiplication, division, and modular division having precedence over addition and subtraction. Parentheses are used for grouping and clarity. Whatever expression is innermost in nested parentheses is done first. Inner parentheses have preference over outer parentheses.

If the variable expression is defined by a `VARIABLE` statement, only integer calculations are done. All division is integer division; that is, the result is truncated. The variable

```
JERRY    VARIABLE    3/2+1
```

would return a value of 2 when referenced by `V(JERRY)`.

With `FVARIABLES`, the expression is evaluated by doing floating-point calculations. If necessary, non-decimal values are converted to floating-point values. So, if the variable `JERRY` above had been defined as

```
JERRY    FVARIABLE   3/2+1
```

the value returned would have been 2.5.

If an expression is used in an operand, then, in general, integer calculations are performed *unless* floating-point results are indicated. Also, GPSS/H has two built-in functions to handle arithmetic calculations where one wants to specify either fixed-point or floating-point calculations. These are `FIX` for fixed-point conversion and `FLT` for floating-point conversion. Once you specify `FLT` in an expression for a single SNA, the whole expression is evaluated as though it was to be done using floating-point calculations. Consider the following examples:

```
(a)   ADVANCE   3/2+1
(b)   ADVANCE   FLT(3)/2+1
```

In (a), the delay time is 2.0, but in (b), the delay time is 2.5.

25.1.2 The PRINT Block

It is possible to have statistical information sent to the report while the program is being run. This statistical information will be the SNAs associated with a particular entity at the time the information is sent to the report. This is done using a `PRINT` Block. When a transaction enters this Block, all the statistics associated with the specified entity (or entities) are sent to the report file. The form of the `PRINT` Block is

```
PRINT A,B,C,D
```

`A` is the lower limit of the entity range (often omitted)

`B` is the upper limit of the entity range (often omitted)

`C` is the family name of the entity to be printed out

`D` used to be a printer directive but is now ignored—it will not be used here

If `A` and `B` are omitted, all the statistics for the entity class are printed out.
Some examples of the `PRINT` Block are as follows:

```
(a)      PRINT  1,3,S
(b)      PRINT  ,,Q
(c)      PRINT  3,7,F
```

In (a), the storages 1 to 3 are printed out.

In (b), all the queue statistics are printed out.

In (c), statistics for facilities 3 to 7 are printed out.

If labels are used for an entity, as is the common case, it is not possible to have only selected ones printed out.

Since a `PRINT` Block will add output to the normal GPSS/H report every time a transaction passes through it, caution must be taken in using it. It is mostly used for debugging purposes and, then, only when the program is run for a limited time.

Some other possible entities that can be used in the PRINT Block are as follows:

AMP	All ampervariables
B	Current and total Block execution counts
C	Absolute and relative clock values
F	Facilities
LG	All logic switches that are in a set position
N	Current and total Block execution counts
Q	Queue statistics
RN	Random number stream
S	Storage statistics
T	Table statistics
W	Current and total Block execution counts

Notice that if B, N, or W is used, the statistics sent to the report are identical.

Example 25.1: A PRINT Block Example

A simple example will be presented here to show how one might use the PRINT Block. As mentioned, it is primarily used for debugging purposes.

A mining truck comes to a station every 10 ± 4 minutes. Only one truck can be serviced at this station. It takes 9 ± 5 minutes for the service. The trucks then move to a second station taking 1 minute where service takes 11 ± 5 minutes for service. Trucks then leave the system. Simulate for one shift of 480 minutes and print out the statistics of the queues every 120 minutes.

The program for this is easily written. It is CHAP25A.GPS. A listing is as shown in Figure 25.1.

When the program is run, the list file will now contain the statistics for the two queues for times 120, 240, and 360. Selected portions of the .LIS file are as follows:

```
QUEUE  MAXIMUM  AVERAGE  TOTAL   ZERO    PERCENT  AVERAGE    CURRENT
       CONTENTS CONTENTS ENTRIES ENTRIES ZEROS    TIME/UNIT  CONTENTS
WAIT1  1        0.099    12      6       50.0     0.993      0
WAIT2  3        1.482    11      1       9.1      16.173     3

QUEUE  MAXIMUM  AVERAGE  TOTAL   ZERO    PERCENT  AVERAGE    CURRENT
       CONTENTS CONTENTS ENTRIES ENTRIES ZEROS    TIME/UNIT  CONTENTS
WAIT1  1        0.056    23      15      65.2     0.581      0
WAIT2  4        2.058    22      1       4.5      22.455     3

QUEUE  MAXIMUM  AVERAGE  TOTAL   ZERO    PERCENT  AVERAGE    CURRENT
       CONTENTS CONTENTS ENTRIES ENTRIES ZEROS    TIME/UNIT  CONTENTS
WAIT1  2        0.164    35      17      48.6     1.691      0
WAIT2  5        2.525    34      1       2.9      26.733     4
```

As expected, the second station has a build up of trucks waiting for service. After 120 minutes, the maximum queue is 3 and the current queue also 3. After 240 minutes, the same results were obtained. After 360 minutes the queue was 4.

```
          SIMULATE
**************************************
*  PROGRAM 25A.GPS   SHORT EXAMPLE   *
*  TO ILLUSTRATE THE PRINT BLOCK     *
**************************************
          GENERATE     10,4    TRUCKS COME ALONG
          QUEUE        WAIT1   FIRST QUEUE
          SEIZE        MACH1   FIRST SERVICE
          DEPART       WAIT1
          ADVANCE      9,5
          RELEASE      MACH1
          ADVANCE      1      TRAVEL TO SECOND SERVICE
          QUEUE        WAIT2
          SEIZE        MACH2
          DEPART       WAIT2
          ADVANCE      11,5
          RELEASE      MACH2
          TERMINATE           LEAVE THE SYSTEM
          GENERATE     480
          TERMINATE    1
          GENERATE     120
          PRINT        ,,Q
          TERMINATE
          START        1
          END
```

FIGURE 25.1
Listing of program CHAP25A.GPS.

25.1.3 Boolean Variables

The TEST Block and the GATE Block have been used to allow the programmer to control the flow of transactions through the program Blocks. When the transactions enter a TEST Block or a GATE Block, depending on the type, the transaction may be delayed until some condition is true. It may pass through to the next sequential Block, or it may be routed to another part of the program. There may be more than one reason for these types of delays. These types of delay situations can be modelled quite easily by the use of Boolean variables. These allow the programmer to specify user-supplied logic conditions to control the flow of transactions through a system. As such, they can be used to model very complex situations that require many different conditions to be satisfied. For example, a plane attempting a landing at a distant airport might need to meet the following conditions: Is the airport open? Is the runway clear? Is there room in the hangar for the particular type of plane? A truck in the mine may have the following conditions: Have there been 10,000 tons mined in Area B, and is the mine working?

Two types of variables have been introduced: fixed-point and floating-point variables. A third type of variable used in GPSS/H is known as a Boolean variable. This is a variable that is defined by the programmer. It will have only one of two values. These are either 0 or 1. Just as with other variables, the Boolean variable will have an associated expression (called a *logical expression*) that is evaluated. The value of the expression will be either 1 (true) or 0 (false).

Boolean expressions are made up of SNAs or entities connected by one or more of the following:

1. Relational operators
2. Boolean operators
3. Logical operators

These will be covered next.

25.1.4 Relational (Comparison) Operators

These were introduced with the TEST and SELECT Blocks. For convenience, they will be repeated here. When they are used in the TEST Block or the SELECT Block, they are used alone; but in Boolean expressions, they must have single apostrophes on either side:

Relational Operator	Meaning
'G'	Greater than
'GE'	Greater than or equal
'E'	Equal
'NE'	Not equal
'LE'	Less than or equal
'L'	Less than

Some Boolean expressions illustrating the above are as follows:

(a) Q(TOM)'E'Q(BILL)
(b) &FIRST'L'&SECOND
(c) FR(MACHA)'G'250

In (a), if the length of the queue TOM is equal to the length of the queue BILL, the expression is true; otherwise, the value is 0.

In (b), the ampervariable &FIRST must be less than the ampervariable &SECOND in order for the expression to be equal to 1.

In (c), the utilization of the facility MACHA must be greater than 250 for the expression to be true (recall that the utilization of a facility is expressed in parts per thousand).

GPSS/H also allows the following symbols to be used as alternatives.

Operator	Equivalent
'G'	>
'GE'	>=
'L'	<
'LE'	<=
'E'	=
'NE'	!=

Thus,

```
Q(TOM)'E'Q(BILL)
```

could have been written as follows:

```
Q(TOM) = Q(BILL)
```

Both forms will be used in this book.

25.1.5 Boolean Operators

The real power of Boolean variables comes from using Boolean operators to connect relational operators. There are three Boolean operators in GPSS/H. These are AND, OR, and NOT.

These are used by inserting these actual words between expressions that are enclosed in parentheses. The value of each is identical to similar operators in languages such as Fortran. Thus, the Boolean operator AND returns a true result only when the value of the expressions on both sides of it are true. Thus,

```
(true)AND(true)          is true or 1
(false)AND(true)         is false or 0
(true)AND(false)         is false or 0
(false)AND(false)        is false or 0
```

The operator OR returns a true result if *either* or both of the values of the expressions is true. Thus,

```
(true)OR(true)           is true or 1
(false)OR(true)          is true or 1
(true)OR(false)          is true or 1
(false)OR(false)         is false or 0
```

NOT inverts the value of an expression. Thus,

```
NOT(true)                is false or 0
NOT(false)               is true or 1
```

Some examples are as follows:
Assumptions

```
Q(TOM) = 3
Q(BILL) = 2
&X = 5
&Y = 6
PH1 = -4
```

Boolean expressions:

(a) `(Q(TOM)'LE'4)AND(PH1'G'-5)`
(b) `(Q(BILL)'E'2)OR&X'E'6)`
(c) `(&Y'LE'&X)AND(PH1'G'0)`
(d) `(Q(TOM)'G'2)AND(NOT(&Y'E'6))`

(a) and (b) are true (value 1) but (c) and (d) are false (value 0).

Alternate symbols that can be used for OR and AND are + for OR and * for AND. Since these symbols are also used for arithmetic operations, this means that one cannot do addition or multiplication in Boolean expressions. Thus, if one wishes to do any addition or multiplication in a Boolean expression, one has to do this in another variable and reference it in the Boolean expression. For example,

```
                     ONE   VARIABLE    &X+&Y-PH2
(Boolean expression) (Q(TOM)> = 2)OR(V(ONE) = 0)
```

The Boolean expression will be true if the queue at TOM is equal to or greater than 2 or if the variable ONE is equal to 0.

The choice of + and * for OR and AND is also confusing, because it is so easy to look at the plus sign and mentally associate it with AND. However unfortunate this is, it is a part of the GPSS/H language. This is a carryover from the early versions of GPSS. There is no symbol that can be used for NOT as this was not a feature of the early versions of GPSS.

The previous examples could have been written as:

(a) `(Q(TOM)'LE'4)*(PH1'G'-5)`
(b) `(Q(BILL)'E'2)+&X = 6)`
(c) `(X(SECOND)'LE'X(FIRST))*(PH1'G'0)`
(e) `(Q(TOM)'G'2)*(NOT(X(SECOND)'E'6))`

Since it is much more logical to write out AND, OR, or NOT, this will be the practice in this book.

Many of the parentheses used above are not needed as we shall shortly learn.

25.1.6 Logical Operators

It is possible to use logical operators to reference various entities in GPSS/H. A logical operator will check on the status of an entity condition. If this check is true, the value is 1, otherwise 0. The logical operators are SNAs. Some of them are as follows:

Logical Operator	Condition Referenced
FU(name) (or F(name))	Is the facility in use?
FNU(name)	Is the facility not in use?
SE(name)	Is the storage empty?
SNE(name)	Is the storage not empty?
SF(name)	Is the storage full?
LS(name)	Is the logic switch set?
LR(name)	Is the logic switch reset?

The above entities referenced by the logical operator could have been numbers, in which case the reference parentheses are optional. GPSS/H also supports the older reference of these logical operators using a single dollar ($) sign. Thus,

```
LS$FIRST      and    LS(FIRST)
```

are the same.

Some examples of these operators are

(a) SNF(TUGS)
(b) LR(STOP12)
(c) F(MACH1)
(d) LS6

In (a), if the storage of TUGS is not full, the value is 1.

In (b), if the logic switch STOP12 is reset, the value is 1.

In (c), if the facility MACH1 is in use, the value is 1.

In (d), the logic switch 6 is referenced. If it is set, the value is 1.

By combining relational operators, Boolean operators and logical operators, a great many complex situations can be modelled easily. For example, consider the following:

```
(LR(GOIN))AND(Q(WAIT)'LE'3)AND(FNU(MACH2))
```

In order for this to be true, the following conditions must all be true:

1. The logic switch GOIN must be reset.
2. The queue at WAIT must be less than or equal to 3.
3. MACH2 must be free.

25.1.7 Referencing Boolean Variables

Boolean variables are defined and referenced in much the same way as other variables. The general form is

```
(name) BVARIABLE (expression)
```

It is possible to have a Boolean variable with a number for a label. One references the Boolean variable by BV(name) or BVn, where (name) is the label or *n* the number. Thus, one might have something like

```
TEST E     BV(STOPIT),1
```

When a transaction arrives at this TEST Block, the value of the Boolean variable STOPIT is determined. Unless it has the value 1, the transaction cannot move to the next sequential Block. Instead it is kept on the CEC and scanned again when a rescan is made.

It is also possible to reference Boolean variables using the single dollar sign ($). Thus, the example above could have been written as follows:

```
TEST E     BV$STOPIT,1
```

25.1.8 Rules for Evaluation of Boolean Expressions

The rules for evaluation of Boolean expressions are as follows:

1. Logical and relational operators have preference in a left to right order.
2. The operators AND, OR, and NOT are evaluated in a left to right order.

Parentheses can (and should) be used for clarity. When used, whatever is in the innermost parenthesis is evaluated first.

Example 25.2: Simulation of a Port for Hauling Ore

Many mining operations are located where the ore has to be taken by train to a port for loading and shipping to a market. This is the case for iron ore mines in Western Australia. At a port, two types of ships arrive for loading and hauling the ore to market. They arrive at an interval time of 23 ± 8 hours. 40% are small ships and the rest are large ships. The small ships take 35 ± 10 hours to load, and the large ships load in 45 ± 12 hours. There are two tug boats available for berthing and unberthing the ships. The small ships require one tug boat, but the large ships require both of the tugs to berth and unberth. The time to berth and unberth a small ship is 1 ± .15 hour. The time for a large ship is 2 ± .3 hours. The port is periodically shut for one reason or another, such as maintenance or storms. These downtimes occur at random but according to the exponential distribution with mean of 50 hours. The port is open again in a time given by the normal distribution with a mean of 5 hours and a standard deviation of .8. When the port is down, no ships can enter or leave. If a ship is berthing or unberthing when the port is down, it can continue to berth or unberth.

The company that owns the port is considering adding a third berth. It costs $150 for each hour a ship is at the harbour either waiting in the queue at the harbour entrance or loading. Once the ship is free from the berth, there is no charge. The cost of a new berth is $2.5 million. The port will serve the mine for at least another 10 years. Should the mine add a new berth?

Solution

The program to model this example uses four Boolean variables: `SMALL`, `LARGE`, `SMGO`, and `LRGO`. `SMALL` is true if there is a berth free, a tug free, and the port is open. `LARGE` is true for the same conditions but requires two tugs to be free. `SMGO` is to allow the small ships to leave the harbour, so it requires that the harbour be open and a single tug free. `LRGO` requires both that the harbour is open and that two tugs are available. When a ship enters the harbour, the time of entry is marked in its first real parameter. If it is a small ship, it is routed to the segment where it waits for the Boolean variable `SMALL` to be true. If it is a large ship, it is routed to the segment where the Boolean variable `LARGE` is true (Figure 25.2).

The output from the program is shown in Figure 25.3.

The harbour is used 93.4% of the time. The tugs are not used much, but that is understandable since they are used only for berthing and unberthing the ships, which take only around 1 or 2 hours. The annual cost of the ships in the harbour is $3,986,500.

To determine whether a third berth should be built, it is necessary to rerun program CHAP25B but with the storage for the berths increased from 2 to 3. This is the only change needed, so the program listing will not be given. The results of making this change are as shown in Figure 25.4.

The changes in the results are that the berths are now used only 62.58% of the time, which is to be expected. The cost has been reduced to $2,487,306 per year. This is approximately one and a half million dollars and, since the port will be there for at least another 10 years, would seem to justify the addition of the third berth.

The animation is given by program CHAP25C.GPS. It is more complex than the program to do the simulation, as the original GPSS/H program uses `ENTER` Blocks. These are modelled by the `TRANSFER BOTH` Block. A listing of the program is given in Figure 25.5. A screenshot of the animation is given in Figure 25.6.

```
          SIMULATE
******************************
*  PROGRAM CHAP25B.GPS       *
*  IRON ORE PORT EXAMPLE     *
******************************
 TYPE       FUNCTION   RN1,D2
.4,1/1,2
 WHERE      FUNCTION   PH1,L2
1,BLOCKA/2,BLOCKB
            REAL       &COST1,&COST2,&TOTCOST
 SMTAB      TABLE      MP1PL,20,1,20
 LATAB      TABLE      MP1PL,25,1,20
            STORAGE    S(TUGS),2/S(BERTHS),2
  SMALL     BVARIABLE  (R(BERTHS)'GE'1)AND(R(TUGS)'GE'1)AND(LR(PORT))
  LARGE     BVARIABLE  (R(BERTHS)'GE'1)AND(R(TUGS)'E'2)AND(LR(PORT))
  SMGO      BVARIABLE  (R(TUGS)'GE'1)AND(LR(PORT))
  LRGGO     BVARIABLE  (R(TUGS)'E'2)AND(LR(PORT))
            GENERATE   23,8,,,,1PH,1PL  SHIPS ARRIVE
            MARK       1PL              RECORD TIME SHIP ENTERS
            ASSIGN     1,FN(TYPE),PH    WHAT TYPE IS IT?
            QUEUE      WAIT             JOIN QUEUE
            TRANSFER   ,FN(WHERE)       PLACE IN PROPER QUEUE TO WAIT
 BLOCKA     TEST E     BV(SMALL),1      CAN SMALL BOAT ENTER HARBOR?
            DEPART     WAIT
            ENTER      TUGS   USE A TUG BOAT   USE A TUG
            ENTER      BERTHS  USE A BERTH     RESERVE A BERTH
            ADVANCE    1,.15   BERTH           TUG BERTHS THE SHIP
            LEAVE      TUGS                    FREE THE TUG
            ADVANCE    35,10   LOAD            LOAD WITH IRON
            TEST E     BV(SMGO),1              CAN SHIP LEAVE?
            TABULATE   SMTAB                   RECORD TIME IN HARBOR
            ENTER      TUGS                    USE A TUG
            LEAVE      BERTHS                  FREE A BERTH
            ADVANCE    1,.15                   LEAVE HARBOR
            LEAVE      TUGS
            TERMINATE
 BLOCKB     TEST E     BV(LARGE),1     CAN LARGE SHIP ENTER HARBOR?
            DEPART     WAIT
            ENTER      TUGS,2                 USE TWO TUGS
            ENTER      BERTHS                 RESERVE A BERTH
            ADVANCE    2,.3                   TUGS BERTH SHIP
            LEAVE      TUGS,2                 FREE THE TUGS
            ADVANCE    45,12                  LOAD WITH IRON ORE
            TEST E     BV(LRGGO),1            CAN SHIP LEAVE?
            TABULATE   LATAB                  RECORD TIME IN HARBOR
            ENTER      TUGS,2                 USE TUGS
            LEAVE      BERTHS                 FREE A BERTH
            ADVANCE    2,.3                   LEAVE HARBOR
            LEAVE      TUGS,2
            TERMINATE
***************************
*  PORT CLOSURE SEGMENT   *
***************************
            GENERATE   ,,,1   DUMMY TRANSACTION
  BACK      ADVANCE    RVEXPO(1,50)
```

FIGURE 25.2
Listing of program CHAP25B.GPS. *(Continued)*

```
    LOGIC S     PORT
    ADVANCE     RVNORM(1,5,.8)
    LOGIC R     PORT
    TRANSFER    ,BACK
    GENERATE    24*365*10
    TERMINATE   1
    START       1
    LET         &COST1=N(BLOCKA)/10.*150*TB(SMTAB)
    LET         &COST2=N(BLOCKB)/10.*150*TB(LATAB)
    LET         &TOTCOST=&COST1+&COST2
    PUTPIC      LINES=7,N(BLOCKA)/10.,N(BLOCKB)/10,SR(BERTHS)/10.,_
                SR(TUGS)/10.,TB(SMTAB),TB(LATAB),&TOTCOST
  SMALL SHIPS TO ENTER HARBOR EACH YEAR    ***
  LARGE SHIPS TO ENTER HARBOR EACH YEAR    ***
  UTIL. OF HARBOR                          ***.**%
  UTIL. OF TUGS                            ***.**%
  AVG. TIME IN THE HARBOR FOR SMALL SHIPS   **.**
  AVG. TIME IN THE HARBOR FOR LARGE SHIPS   **.**
  ANNUAL COST FOR SHIPS                 ********
    END
```

FIGURE 25.2 (Continued)
Listing of program CHAP25B.GPS.

```
SMALL SHIPS TO ENTER HARBOR EACH YEAR    149
LARGE SHIPS TO ENTER HARBOR EACH YEAR    230
UTIL. OF HARBOR                           93.40%
UTIL. OF TUGS                             12.27%
AVG. TIME IN THE HARBOR FOR SMALL SHIPS   57.98
AVG. TIME IN THE HARBOR FOR LARGE SHIPS   77.61
ANNUAL COST FOR SHIPS                   3986500
```

FIGURE 25.3
Output from program CHAP25B.GPS.

```
SMALL SHIPS TO ENTER HARBOR EACH YEAR    148
LARGE SHIPS TO ENTER HARBOR EACH YEAR    232
UTIL. OF HARBOR                           62.58%
UTIL. OF TUGS                             12.30%
AVG. TIME IN THE HARBOR FOR SMALL SHIPS   36.76
AVG. TIME IN THE HARBOR FOR LARGE SHIPS   47.92
ANNUAL COST FOR SHIPS                   2487306
```

FIGURE 25.4
Output from program CHAP25B after the number of berths is changed to 3.

```
           SIMULATE
******************************
*  PROGRAM CHAP25C.GPS       *
*  ANIMATION FOR CHAP25B.GPS  *
******************************
 TYPE      FUNCTION   RN1,D2
.4,1/1,2
 WHERE     FUNCTION   PH1,L2
1,BLOCKA/2,BLOCKB
           RMULT      12345
           REAL       &X,&Y
           INTEGER    &SM,&LG
  ATF      FILEDEF    'CHAP25C.ATF'
           REAL       &COST1,&COST2,&TOTCOST
 SMTAB     TABLE      MP1PL,25,1,50
 LATAB     TABLE      MP1PL,35,1,50
           STORAGE    S(TUGS),2/S(BERTHS),2
  SMALL    BVARIABLE  (R(BERTHS)'GE'1)AND(R(TUGS)'GE'1)AND(LR(PORT))
  LARGE    BVARIABLE  (R(BERTHS)'GE'1)AND(R(TUGS)'E'2)AND(LR(PORT))
  SMGO     BVARIABLE  (R(TUGS)'GE'1)AND(LR(PORT))
  LRGGO    BVARIABLE  (R(TUGS)'E'2)AND(LR(PORT))
           PUTPIC     FILE=ATF,LINES=2,AC1
TIME *.****
WRITE M1 Port Open
           GENERATE   23,8,0,,,1PH,1PL  SHIPS ARRIVE
           MARK       1PL              RECORD TIME SHIP ENTERS
           ASSIGN     1,FN(TYPE),PH    WHAT TYPE IS IT?
           QUEUE      WAIT             JOIN QUEUE
           TRANSFER   ,FN(WHERE)       PLACE IN PROPER QUEUE TO WAIT
 BLOCKA    BLET       &SM=&SM+1
           BPUTPIC    FILE=ATF,LINES=5,AC1,XID1,XID1,XID1,&SM
TIME *.****
CREATE SMALL S*
PLACE S* ON P1
SET S* TRAVEL .25
WRITE M2 ***
           ADVANCE    .25
           TEST E     BV(SMALL),1      CAN SMALL BOAT ENTER HARBOR?
           TRANSFER   BOTH,,BERTH2
           SEIZE      BERTH1
           DEPART     WAIT
           ENTER      TUGS   USE A TUG BOAT
           ENTER      BERTHS  USE A BERTH
           BLET       &X=.3*FRN1+.85
           BPUTPIC    FILE=ATF,LINES=4,AC1,XID1,XID1,XID1,&X
TIME *.****
PLACE S* ON P2
SET S* CLASS SMALLT
SET S* TRAVEL **.**
           ADVANCE    &X
           LEAVE      TUGS
           BPUTPIC    FILE=ATF,LINES=3,AC1,XID1,XID1
TIME *.****
PLACE S* AT 30 12
SET S* CLASS SMALL
           ADVANCE    35,10   LOAD            LOAD WITH IRON
```

FIGURE 25.5
Listing of program CHAP25C.GPS. *(Continued)*

```
              TEST E      BV(SMGO),1                CAN SHIP LEAVE?
              TABULATE    SMTAB                     RECORD TIME IN HARBOR
              BPUTPIC     FILE=ATF,LINES=3,AC1,MP1PL,TB(SMTAB)
TIME *.****
WRITE M4 ***.**
WRITE M5 ***.**
              ENTER       TUGS                      USE A TUG
              LEAVE       BERTHS                    FREE A BERTH
              RELEASE     BERTH1
              BLET        &X=.85*FRN1+.3
              BPUTPIC     FILE=ATF,LINES=4,AC1,XID1,XID1,XID1,&X
TIME *.****
PLACE S* ON P4
SET S* CLASS SMALLT
SET S* TRAVEL **.**
              ADVANCE     &X
              LEAVE       TUGS
              BPUTPIC     FILE=ATF,LINES=2,AC1,XID1
TIME *.****
DESTROY S*
              TERMINATE
 BERTH2       SEIZE       BERTH2
              DEPART      WAIT
              ENTER       TUGS                      USE A TUG
              ENTER       BERTHS                    USE A BERTH
              BLET        &X=.3*FRN1+.85
              BPUTPIC     FILE=ATF,LINES=4,AC1,XID1,XID1,XID1,&X
TIME *.****
PLACE S* ON P3
SET S* CLASS SMALLT
SET S* TRAVEL **.**
              ADVANCE     &X
              LEAVE       TUGS
              BPUTPIC     FILE=ATF,LINES=3,AC1,XID1,XID1
TIME *.****
PLACE S* AT   24 -1
SET S* CLASS SMALL
              ADVANCE     35,10   LOAD              LOAD WITH IRON
              TEST E      BV(SMGO),1                CAN SHIP LEAVE?
              TABULATE    SMTAB                     RECORD TIME IN HARBOR
              BPUTPIC     FILE=ATF,LINES=3,AC1,MP1PL,TB(SMTAB)
TIME *.****
WRITE M4 ***.**
WRITE M5 ***.**
              ENTER       TUGS                      USE A TUG
              LEAVE       BERTHS                    FREE A BERTH
              RELEASE     BERTH2
              BLET        &X=.85*FRN1+.3
              BPUTPIC     FILE=ATF,LINES=4,AC1,XID1,XID1,XID1,&X
TIME *.****
PLACE S* ON P5
SET S* CLASS SMALLT
SET S* TRAVEL **.**
              ADVANCE     &X
              LEAVE       TUGS
              BPUTPIC     FILE=ATF,LINES=2,AC1,XID1
TIME *.****
```

FIGURE 25.5 (Continued)
Listing of program CHAP25C.GPS.

(Continued)

```
DESTROY S*
            TERMINATE
 BLOCKB     BLET        &LG=&LG+1
            BPUTPIC     FILE=ATF,LINES=5,AC1,XID1,XID1,XID1,&LG
TIME *.****
CREATE LARGE L*
PLACE L* ON P1
SET L* TRAVEL .25
WRITE M3 ***
            ADVANCE     .25
            TEST E      BV(LARGE),1     CAN LARGE SHIP ENTER HARBOR?
            TRANSFER    BOTH,,BERTHX
            SEIZE       BERTH1
            DEPART      WAIT
            ENTER       TUGS,2                  USE TWO TUGS
            ENTER       BERTHS                  RESERVE A BERTH
            BLET        &Y=.6*FRN1+1.7
            BPUTPIC     FILE=ATF,LINES=4,AC1,XID1,XID1,XID1,&Y
TIME *.****
PLACE L* ON P2
SET L* CLASS LARGET
SET L* TRAVEL **.**
            ADVANCE     &Y
            LEAVE       TUGS,2                  FREE THE TUGS
            BPUTPIC     FILE=ATF,LINES=3,AC1,XID1,XID1
TIME *.****
PLACE L* AT  30 12
SET L* CLASS LARGE
            ADVANCE     45,12                   LOAD WITH IRON ORE
            TEST E      BV(LRGGO),1             CAN SHIP LEAVE?
            TABULATE    LATAB                   RECORD TIME IN HARBOR
            BPUTPIC     FILE=ATF,LINES=3,AC1,MP1PL,TB(LATAB)
TIME *.****
WRITE M6 ***.**
WRITE M7 ***.**
            ENTER       TUGS,2                  USE TUGS
            LEAVE       BERTHS                  FREE A BERTH
            RELEASE     BERTH1                 RELEASE THE BERTH
            BLET        &Y=.6*FRN1+1.7
            BPUTPIC     FILE=ATF,LINES=4,AC1,XID1,XID1,XID1,&Y
TIME *.****
PLACE L* ON P4
SET L* CLASS LARGET
SET L* TRAVEL **.**
            ADVANCE     &Y
            LEAVE       TUGS,2
            BPUTPIC     FILE=ATF,LINES=2,AC1,XID1
TIME *.****
DESTROY L*
            TERMINATE
BERTHX      SEIZE       BERTH2
            DEPART      WAIT
            ENTER       TUGS,2                  USE TWO TUGS
            ENTER       BERTHS                  RESERVE A BERTH
            BLET        &Y=.6*FRN1+1.7
            BPUTPIC     FILE=ATF,LINES=4,AC1,XID1,XID1,XID1,&Y
TIME *.****
```

FIGURE 25.5 (Continued)
Listing of program CHAP25C.GPS.

(Continued)

```
PLACE L* ON P3
SET L* CLASS LARGET
SET L* TRAVEL **.**
          ADVANCE    &Y
          LEAVE      TUGS,2                  FREE THE TUGS
          BPUTPIC    FILE=ATF,LINES=3,AC1,XID1,XID1
TIME *.****
PLACE L* AT   24 -1
SET L* CLASS LARGE
          ADVANCE    45,12                   LOAD WITH IRON ORE
          TEST E     BV(LRGGO),1             CAN SHIP LEAVE?
          TABULATE   LATAB                   RECORD TIME IN HARBOR
          BPUTPIC    FILE=ATF,LINES=3,AC1,MP1PL,TB(LATAB)
TIME *.****
WRITE M6 ***.**
WRITE M7 ***.**
          ENTER      TUGS,2                  USE TUGS
          LEAVE      BERTHS                  FREE A BERTH
          RELEASE    BERTH2
          BLET       &Y=.6*FRN1+1.7
          BPUTPIC    FILE=ATF,LINES=4,AC1,XID1,XID1,XID1,&Y
TIME *.****
PLACE L* ON P5
SET L* CLASS LARGET
SET L* TRAVEL **.**
          ADVANCE    &Y
          LEAVE      TUGS,2
          BPUTPIC    FILE=ATF,LINES=2,AC1,XID1
TIME *.****
DESTROY L*
          TERMINATE
****************************
*  PORT CLOSURE SEGMENT    *
****************************
          GENERATE   ,,,1   DUMMY TRANSACTION
  BACK    ADVANCE    RVEXPO(1,50)
          LOGIC S    PORT
          ADVANCE    RVNORM(1,5,.8)
          LOGIC R    PORT
          TRANSFER   ,BACK
          GENERATE   24*365
          TERMINATE  1
          START      1
          BPUTPIC    FILE=ATF,LINES=2,AC1
TIME *.****
END
          LET        &COST1=N(BLOCKA)*15*TB(SMTAB)
          LET        &COST2=N(BLOCKB)*15*TB(LATAB)
          LET        &TOTCOST=&COST1+&COST2
          PUTPIC     LINES=7,N(BLOCKA)/10.,N(BLOCKB)/10,SR(BERTHS)/10.,_
                     SR(TUGS)/10.,TB(SMTAB),TB(LATAB),&TOTCOST
        SMALL SHIPS TO ENTER HARBOR EACH YEAR   ***
        LARGE SHIPS TO ENTER HARBPR EACH YEAR   ***
        UTIL. OF HARBOR                         ***.**%
        UTIL. OF TUGS                           ***.**%
        AVG. TIME IN THE HARBOR FOR SMALL SHIPS  **.**
        AVG. TIME IN THE HARBOR FOR LARGE SHIPS  **.**
        ANNUAL COST FOR SHIPS                  ********
          END
```

FIGURE 25.5 (Continued)
Listing of program CHAP25C.GPS.

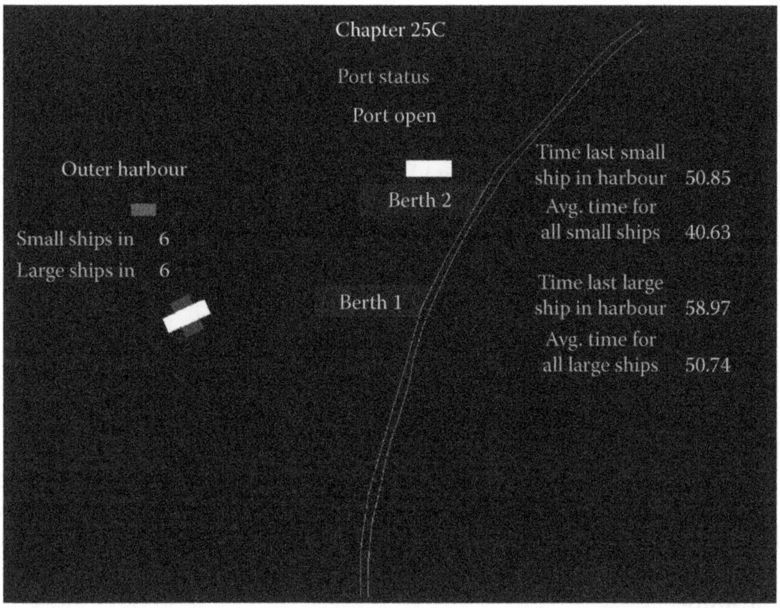

FIGURE 25.6
Screenshot of animation for program CHAP25C.GPS.

25.2 Exercises

25.1. State how each of the following expressions would be evaluated:

```
(a)  (Q(FIRST)'LE'1)AND(&X'NE'0)
(b)  (R(TUGS)'E'3)OR(S(BIGSHIP)'E'1)
(c)  (&X'LE'&Z)AND(&Y'E'3)AND(&A'NE'0)
(d)  (AC1'GE'480)OR(&LOADS'GE'58)
(e)  (FR(MACH1)'LE'250)AND(FR(MACH2)'GE'500)OR(LS(SWITCH1))
```

25.2. What will the value of the following expressions be? Assume the following:

```
F(MACH1) = 1, Q(WAIT) = 3, FR(MACH1) = 500, &X = 499.99, R(TUGS = 3,
   S(TUGS) = 6
```

```
(a) FIRST VARIABLE 9/4+F(MACH1)-Q(WAIT)
(b) SECOND VARIABLE FR(MACH1)-499.99+R(TUGS)/2
(c) THIRD FVARIABLE 4.321*FR(MACH1)/10.
(d) FOURTH FVARIABLE Q(WAIT)/2-&X
(e) FIFTH FVARIABLE S(TUGS)*&X-Q(WAIT)/4.0
(f) SIXTH VARIABLE 4.321*FR(MACH1)/10.
(g) SEVENTH VARIABLE Q(WAIT)/2-&X
(h) EIGHTH VARIABLE S(TUGS)*&X-Q(WAIT)/4.0
```

25.3. A small mine has four trucks and a single shovel. Only one truck can be loaded or dump at a time. It takes 2 ± 1 minutes to load a truck. Travel to the crusher takes 14 ± 5 minutes and dumping into the crusher takes $1 \pm .1$ minutes. The return to the shovel is done in 12 ± 4 minutes. The mine owner wants to have *exactly* 50 loads dumped from this pit each shift, so after a truck dumps, it checks to see how many loads have been dumped. Once 50 loads are dumped, the trucks are diverted to another area in the mine. In addition, the mine owner does not want to have the pit worked for more than 400 minutes. Build a model to simulate this mine to determine what comes first, the 50 loads or 400 minutes of work.

25.4. For Example 25.2, a faster loader can be purchased for a cost of half a million dollars. This will result in a small ship being loaded in 28 ± 8 hours and a large ship in 37 ± 9 hours. Is this expenditure justified?

25.5. Change Example 25.2 so that the cost associated with the time the ships are in the harbour can be an input variable. Suppose that the cost of the ships' delay is decreasing. Determine the cost such that the difference between the two berth harbour and the three berth harbour will be a million dollars per year. Assume that the cost for the three berths does not change but remains at $150 per hour.

26

The BUFFER Block

Front end loader (FEL) used in a mine.

26.1 The BUFFER Block and the SPLIT Block

26.1.1 The BUFFER Block

In order to fully appreciate the way a GPSS/H program works, it is important always to understand the way transactions are moved on the various chains. Recall that, once a transaction is moved by the processor, it will be moved forward as far as possible until one of three things happens to it:

1. It is terminated.
2. It is put on the FEC.
3. It is blocked.

Once certain Blocks are executed, a re-scan of the CEC is initiated. Some of these Blocks are as follows:

```
SEIZE/RELEASE
ENTER/LEAVE
LOGIC
PRIORITY
```

When one of the above happens, the processor will move the transaction as far as it can and then do a re-scan of the CEC. So far, this is always what we wanted to have happen. Sometimes, one will want a re-scan before one of the Blocks that causes an automatic re-scan is executed. The `BUFFER` Block is used when a re-scan of the CEC is desired. The Block is simply as follows:

```
BUFFER
```

There are no operands and no SNAs associated with it. When a transaction enters the Block, a re-scan of the CEC is made. If there are transactions waiting to be moved, the one(s) with higher priority will then be moved. How it works can best be understood by considering a short example.

Example 26.1: Waiting at a Tool Crib

Consider a tool crib in a mine. A worker needs a tool and comes to the crib that has a single worker. The person has a filled out slip that he or she hands to the worker. The worker then takes the tool request and fetches the tool. Workers arrive according to the normal distribution with a mean of 10 and a standard deviation of 1. The worker goes to the tool crib and finds the tool and returns to the worker. The time to obtain the tool, do the paper work, and return is normally distributed with a mean of 9 and a standard deviation of .9. Determine how many workers are in the system for a typical day and how busy the worker is.

Solution

Consider the program CHAP26X.GPS, for example. It might appear to be all right but the results are not correct as will be shown in Figure 26.1. The output is shown in Figure 26.2.

The results show that the miners arrived and waited for the whole shift! Not a single one was able to obtain the tool. The reason is that, when the worker in the tool crib opened the `GATE` for the miner to wait for his tool, the `GATE` was immediately closed and the miner still not able to pass through. The `BUFFER` Block will ensure that this does not happen. When the worker transaction enters this Block, a re-scan of the CEC is done. The miner transaction is at the following Block becomes:

```
GATE LS    TOOLA
```

The transaction now has a priority of 10, although any priority greater than zero would have sufficed. Thus, the miner can pass through the Block and will wait for the next `GATE` Block to be *opened* for him.

```
          SIMULATE
*****************************
*  PROGRAM CHAP26X.GPS       *
*  THIS PROGRAM WILL NOT     *
*  SIMULATE EXAMPLE 26.1     *
*****************************
          GENERATE    RVNORM(1,10,1)   WORKER ARRIVES
          QUEUE       WAIT      JOIN QUEUE
          GATE LS     TOOLA     WAIT FOR TOOL CRIB OPERATOR
          GATE LS     TOOLB     REQUEST TAKEN, WAIT FOR IT TO BE FILLED
          DEPART      WAIT      LEAVE THE QUEUE
 DONE     TERMINATE             BACK TO THE MINE
          GENERATE    ,,,1
 WAIT     TEST G      Q(WAIT),0   ANY WORKER THERE?
          SEIZE       WORKER
          LOGIC S     TOOLA       OPEN THE GATE
          LOGIC R     TOOLA       CLOSE THE GATE
          ADVANCE     RVNORM(1,9,.9) OBTAIN THE TOOL
          LOGIC S     TOOLB       OPEN THE GATE
          LOGIC R     TOOLB       CLOSE THE GATE
          RELEASE     WORKER
          TRANSFER    ,WAIT       WAIT FOR ANOTHER MINER
          GENERATE    480
          TERMINATE   1
          START       1
          PUTPIC      LINES=4,QA(WAIT),FR(WORKER)/10.,N(DONE)
      RESULTS OF SIMULATION
   AVG. QUEUE OF WORKERS      **.**
   UTIL. OF WORKER            **.**%
   MINERS WITH TOOLS          **.**
          END
```

FIGURE 26.1
Listing of incorrect program CHAP26X.GPS.

```
   RESULTS OF SIMULATION
AVG. QUEUE OF WORKERS      23.79
UTIL. OF WORKER            96.12%
MINERS WITH TOOLS           0.00
```

FIGURE 26.2
Output for program CHAP26X.GPS.

The correct program is CHAP26A.GPS. Its listing is given in Figure 26.3.

The program is now run for 10 shifts. If this had been done for the previous program CHAP26X.GPS, an error OUT OF COMMON would have occurred.

The results of the simulation are shown in Figure 26.4.

These results are now correct. In a shift of 480 minutes, 48 miners were expected to arrive and the simulation results show that 47.5 did indeed receive the tools. The tool crib worker was busy 89.36% of the time.

The animation is given by CHAP26B.GPS. Its listing is shown in Figure 26.5. The animation is shown in Figure 26.6.

The animation has various times added to make it more realistic. Thus, the results of the simulation from running CHAP26B.GPS will differ slightly from the original program CHAP26A.GPS.

```
          SIMULATE
******************************
*  PROGRAM CHAP26A.GPS       *
*  EXAMPLE OF BUFFER BLOCK   *
******************************
          GENERATE    RVNORM(1,10,1),,,,10   MINER ARRIVES
          QUEUE       WAIT       JOIN QUEUE
          GATE LS     TOOLA      WAIT FOR TOOL CRIB OPERATOR
          GATE LS     TOOLB      REQUEST TAKEN, WAIT FOR IT TO BE FILLED
          DEPART      WAIT       LEAVE THE QUEUE
 DONE     TERMINATE              BACK TO THE MINE
          GENERATE    ,,,1
 WAIT     TEST G      Q(WAIT),0   ANY WORKER THERE?
          SEIZE       WORKER
          LOGIC S     TOOLA        OPEN THE GATE
          BUFFER
          LOGIC R     TOOLA        CLOSE THE GATE
          ADVANCE     RVNORM(1,9,.9) OBTAIN THE TOOL
          LOGIC S     TOOLB        OPEN THE GATE
          BUFFER
          LOGIC R     TOOLB        CLOSE THE GATE
          RELEASE     WORKER
          TRANSFER    ,WAIT        WAIT FOR ANOTHER WORKER
          GENERATE    480*10
          TERMINATE   1
          START       1
          PUTPIC      LINES=4,QA(WAIT),FR(WORKER)/10.,N(DONE)/10.
      RESULTS OF SIMULATION
   AVG. QUEUE OF MINERS      **.**
   UTIL. OF WORKER           **.**%
   MINERS WITH TOOLS         **.**   PER SHIFT
          END
```

FIGURE 26.3
Listing of correct program CHAP26A.GPS.

```
   RESULTS OF SIMULATION
AVG. QUEUE OF MINERS      0.93
UTIL. OF WORKER          89.36%
MINERS WITH TOOLS        47.50   PER SHIFT
```

FIGURE 26.4
Output from program CHAP26A.GPS.

```
          SIMULATE
******************************
*  PROGRAM CHAP26B.GPS       *
*  ANIMATION OF CHAP26A.GPS  *
******************************
  ATF     FILEDEF     'CHAP26B.ATF'
          REAL        &X
          INTEGER     &COUNT
          GENERATE    RVNORM(1,10,1),,,,10   MINER ARRIVES
          BPUTPIC     FILE=ATF,LINES=4,AC1,XID1,XID1,XID1
TIME *.****
CREATE MINER M*
PLACE M* ON P1
SET M* TRAVEL 2.5
          ADVANCE     2.5
          QUEUE       WAIT       JOIN QUEUE
          GATE LS     TOOLA      WAIT FOR TOOL CRIB OPERATOR
          ADVANCE     1
          BPUTPIC     FILE=ATF,LINES=2,AC1,XID1
```

FIGURE 26.5
Listing of program CHAP26B.GPS. *(Continued)*

```
TIME *.****
PLACE M* at 0 0
          GATE LS     TOOLB      REQUEST TAKEN, WAIT FOR IT TO BE FILI
          DEPART      WAIT       LEAVE THE QUEUE
          BPUTPIC     FILE=ATF,LINES=4,AC1,XID1,XID1,XID1
TIME *.****
PLACE M* ON P3
SET M* CLASS MINER2
SET M* TRAVEL 2.5
          ADVANCE     2.5
          BPUTPIC     FILE=ATF,LINES=2,AC1,XID1
TIME *.****
DESTROY M*
 DONE     TERMINATE              BACK TO THE MINE
          GENERATE    ,,,1
          BPUTPIC     FILE=ATF,LINES=3,AC1,XID1,XID1
TIME *.****
CREATE WORKER W*
PLACE W* AT -10 5
 WAIT     TEST G      Q(WAIT),0    ANY WORKER THERE?
          SEIZE       WORKER
          LOGIC S     TOOLA        OPEN THE GATE
          BUFFER
          LOGIC R     TOOLA        CLOSE THE GATE
          BLET        &X=RVNORM(1,9,.9)
          BPUTPIC     FILE=ATF,LINES=3,AC1,XID1,XID1,&X
TIME *.****
PLACE W* ON P2
SET W* TRAVEL **.**
          ADVANCE     &X
          BPUTPIC     FILE=ATF,LINES=3,AC1
TIME *.****
CREATE MESSAGE MESS
PLACE MESS AT 3.2 13.75
          ADVANCE     2
          BPUTPIC     FILE=ATF,LINES=2,AC1
TIME *.****
DESTROY MESS
          LOGIC S     TOOLB        OPEN THE GATE
          BUFFER
          LOGIC R     TOOLB        CLOSE THE GATE
          RELEASE     WORKER
          BLET        &COUNT=&COUNT+1
          BPUTPIC     FILE=ATF,LINES=4,AC1,XID1,&COUNT,FR(WORKER)/10.
TIME *.****
PLACE W* AT -10 5
WRITE M11 ***
WRITE M10 **.**%
          TRANSFER    ,WAIT        WAIT FOR ANOTHER WORKER
          GENERATE    480*10
          TERMINATE   1
          START       1
          PUTPIC      LINES=4,QA(WAIT),FR(WORKER)/10.,N(DONE)/10.
      RESULTS OF SIMULATION
   AVG. QUEUE OF MINERS      **.**
   UTIL. OF WORKER           **.**%
   MINERS WITH TOOLS         **.**   PER SHIFT
          PUTPIC      FILE=ATF,LINES=2,AC1
TIME *.****
END
          END
```

FIGURE 26.5 (Continued)
Listing of program CHAP26B.GPS.

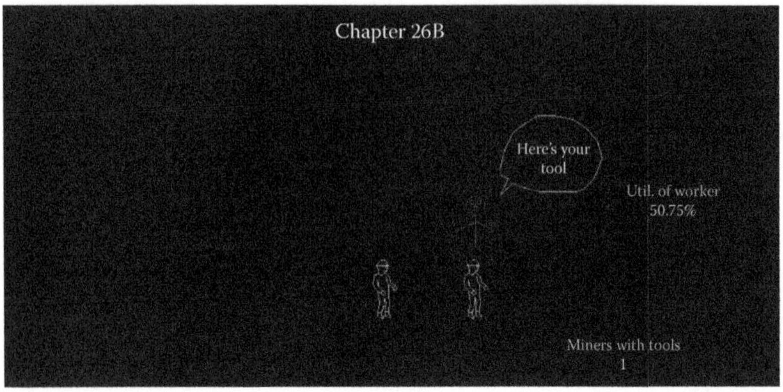

FIGURE 26.6
Screenshot of animation from program CHAP26B.GPS.

26.1.2 The SPLIT Block

Transactions have been placed in our models by the GENERATE Block. In fact, this is the *only* way to create original transactions. However, once a transaction is in a model, it is possible to make clones of the original transactions. These clones normally will be identical to the original transactions, although they can be made to differ. As far as being identical to the original transactions, the clones will always be identical in the priority level and the time of entry (their mark time). This latter point is worth noting. If the original transaction entered the model at time 2050, and, at time 3500, a new transaction was cloned, the clone has a mark time of 2050, *not* 3500. The clones normally will have the same number and type of parameters as the original, but it is here that the clones can be made to differ. How to do this will be discussed below.

The Block that creates these clones is the SPLIT Block. The following is the form to create identical transactions:

```
SPLIT  n,(label)
```

where
n is the number of clones to create
label is the Block label the cloned transactions are routed to

When a transaction enters a SPLIT Block, n identical transactions are created and leave the Block one at a time (incrementing the Block count as they leave). These are all routed to the Block whose label is specified in the B operand of the SPLIT Block. The original transaction is *not* routed to this Block but goes to the next sequential Block. In fact, this original transaction is moved before the clones are. Some examples of the SPLIT Block are as follows:

```
(a) SPLIT    1,DOWN1
(b) SPLIT    10,UPTOP
```

In (a), one new transaction is created and sent to the Block with the label DOWN1. In (b), 10 new transactions are created and sent to the Block with the label UPTOP. In both cases, the original transactions are routed to the next sequential Block.

Often, it is desired to have the original transaction and the clones routed to the same Block. This can be done by making the next sequential Block the one where the clones are sent, that is,

```
       SPLIT    3,NEXT1
NEXT1  (next Block)
```

Here the original and the three clones are sent to the same Block.

These SPLIT Blocks can be very handy in programming problems where a single unit comes along and several things have to be done on different parts of the unit simultaneously. For example, suppose a truck comes to the repair shop for several different operations. Once the truck is positioned on a rack, all operations commence. Suppose that the truck needs three separate operations. Thus, one person may make an adjustment to the front, another to the rear, and a third bolts on a part. All three of these operations are being done at the same time. The model for this is done in GPSS/H using the SPLIT Block. The truck transaction makes two clones, and the clones are worked on separately, as well as the original transaction (the truck) being worked on. Only when all three operations (the original and the two clones) are finished can the truck be moved. How to do this will be covered later.

Example 26.2: An Overhead Crane Model

In a mineral processing plant, ore buckets come along a conveyor belt line every 4 ± 1.2 minutes. The belt moves the buckets in 4 minutes to the end where they are to be moved to the next process by an overhead crane. This crane is used to lift them and carry them to another section where they will be worked on further. It takes $2 \pm .8$ minutes for the crane to load and transport the buckets. The crane must then return to the original position. This takes 1.6 minutes.

Solution

The program to model this is given by CHAP26C.GPS (Figure 26.7).

```
          SIMULATE
*****************************
*  PROGRAM CHAP26C.GPS      *
*  EXAMPLE OF SPLIT BLOCK   *
*****************************
          GENERATE    4,1.2    ORE COMES ALONG
          ADVANCE     4        MOVES ON BELT
          QUEUE       WAIT     LEAVES BELT, STACKS UP
          SEIZE       CRANE    IS CRANE AVAILABLE?
          DEPART      WAIT     YES, USE CRANE
          ADVANCE     2,.8     MOVE TO NEXT STATION
          SPLIT       1,AWAY
          ADVANCE     1.6      CRANE RETURNS
          RELEASE     CRANE    FREE CRANE
          TERMINATE
  AWAY    ADVANCE     5        NEXT PROCESS
          TERMINATE
          GENERATE    60*8
          TERMINATE   1
          START       1
          PUTPIC      FR(CRANE)/10.
  CRANE WAS BUSY ***.**% OF THE TIME
          END
```

FIGURE 26.7
Listing of program CHAP26C.GPS.

The program yields the following results:

```
CRANE WAS BUSY   84.84% OF THE TIME
```

The program is relatively straightforward. The ore buckets come along and wait for the crane. If it is free, the crane takes a bucket to the next process. Once the bucket is at this process, the crane must return to the conveyor belt. It needs to do things. Hence, the `SPLIT` Block. One transaction is used to represent the crane moving back to the belt. This is the `ADVANCE 1.6` Block. After this time delay, the crane is released and can pick up another ore bucket. The other transaction represents the bucket that has been transported to the next process. The bucket is to move as soon as it is brought to the next conveyor belt.

The animation program is worthy of study, as it involves some animation techniques not used previously. The animation program is given by CHAP26D.GPS. A listing of the program is given in Figure 26.8.

The animation uses three classes. These are `BUCKET`, `CRANE`, and `CRANEB` (Figure 26.9).

```
          SIMULATE
*****************************
*  PROGRAM CHAP27D.GPS      *
*  ANIMATION OF CHAP26C.GPS *
*****************************
 ATF      FILEDEF     'CHAP27B.ATF'
          REAL        &X
          GENERATE    4,1.2   ORE BUCKET COMES ALONG  BELT
          BPUTPIC     FILE=ATF,LINES=4,AC1,XID1,XID1,XID1
TIME *.****
CREATE BUCKET B*
PLACE B* ON P1
SET B* TRAVEL 4
          ADVANCE     4
          QUEUE       WAIT     END OF BELT
          SEIZE       CRANE    IS CRANE AVAILABLE?
          DEPART      WAIT     YES, USE CRANE
          BLET        &X=1.6*FRN1+1.2
          BPUTPIC     FILE=ATF,LINES=5,AC1,XID1,XID1,XID1,&X
TIME *.****
SET B* CLASS CRANEB
PLACE C1 AT 1000 1000
PLACE B* ON P3
SET B* TRAVEL **.**
          ADVANCE     &X
          SPLIT       1,AWAY
          BPUTPIC     FILE=ATF,LINES=4,AC1,XID1,XID1,XID1
TIME *.****
SET B* CLASS CRANE
PLACE B* ON P4
SET B* TRAVEL 1.6
          ADVANCE     1.6      RETURN TO BELT
          BPUTPIC     FILE=ATF,LINES=3,AC1,XID1
TIME *.****
DESTROY B*
PLACE C1 AT 0 19
          RELEASE     CRANE   FREE CRANE FOR NEXT BUCKET
          BPUTPIC     FILE=ATF,LINES=2,AC1,FR(CRANE)/10.
TIME *.****
WRITE M1 **.**%
          TERMINATE
  AWAY    BPUTPIC      FILE=ATF,LINES=4,AC1,XID1,XID1,XID1
```

FIGURE 26.8
Listing of program CHAP26D.GPS.

(Continued)

```
TIME *.****
CREATE BUCKET BB*
PLACE BB* ON P2
SET BB* TRAVEL 5
          ADVANCE      5        NEXT PROCESS
          BPUTPIC      FILE=ATF,LINES=2,AC1,
TIME *.****
DESTROY BB*
          TERMINATE
          GENERATE     60*8
          TERMINATE    1
          START        1
          PUTPIC       FR(CRANE)/10.
  CRANE WAS BUSY ***.**% OF THE TIME
          PUTPIC        FILE=ATF,LINES=2,AC1
TIME *.****
END
           END
```

FIGURE 26.8 (Continued)
Listing of program CHAP26D.GPS.

FIGURE 26.9
Three classes used in animation: `BUCKET`, `CRANE`, and `CRANEB`.

When the program starts, a layout object named C1 is shown where the crane rests waiting for an ore bucket. This can be seen by going to Draw Mode and clicking on the crane. When an ore bucket, `BUCKET`, arrives at the end of the first conveyor belt, the crane has to pick it up and move it to the next belt. Two things happen simultaneously in the animation. The `BUCKET` is turned into the class, `CRANEB`, and the layout object is moved off the screen to position (1000, 1000). This is an arbitrarily chosen position far from what is seen on the screen. The ore bucket is now moved to the second conveyor belt. At this point, two things are to happen simultaneously in the animation. The ore bucket is to move along the second conveyor belt and the crane returns to the initial position. The crane movement to the left is done by the following code:

```
      BPUTPIC      FILE = ATF,LINES = 4,AC1,XID1,XID1,XID1
TIME *.****
SET B* CLASS CRANE
PLACE B* ON P4
```

```
SET B* TRAVEL 1.6
     ADVANCE          1.6      RETURN TO BELT
     BPUTPIC          FILE = ATF,LINES = 3,AC1,XID1
TIME *.****
DESTROY B*
PLACE C1 AT 0 19
```

First, the crane is given the class CRANE so it shows as the crane without the bucket. It moves back to position on path P4. There it is DESTROYed, and the original layout object C1 is taken off the screen and placed back in the original position.

The animation code for the bucket has to be examined also. One cannot use the old XID1s for the bucket on the second belt. The code for this is as follows:

```
 AWAY     BPUTPIC              FILE = ATF,LINES = 4,AC1,XID1,XID1,XID1
TIME *.****
CREATE BUCKET BB*
PLACE BB* ON P2
SET BB* TRAVEL 5
          ADVANCE              5        NEXT PROCESS
          BPUTPIC              FILE = ATF,LINES = 2,AC1,XID1
TIME *.****
DESTROY BB*
          TERMINATE
```

A new object needs to be created to represent the ore bucket. This is called BB. It is placed on path P2 and moved to its end where it is destroyed. The animation is given by program CHAP26D.GPS. A screenshot is shown in Figure 26.10.

This example is worth studying, as an animation often must be made up of many different layout objects. This is often the case where the SPLIT Block is used.

Example 26.3: A Port Problem

A small iron ore mine is located near the ocean. It loads the ore into containers at the mine and transports these by trucks to the loading barges at a dock, where the containers are loaded onto a large ship. The ship can hold from 95 to 105 such containers. When a ship docks, there are always enough containers available. A ship comes once a month. It is scheduled to arrive on the first of the month but often is delayed. The delay can be anywhere from 0 to 2 days uniformly distributed. When a ship arrives, it first positions itself, which takes 15 minutes. During this time, two tug boats position themselves at

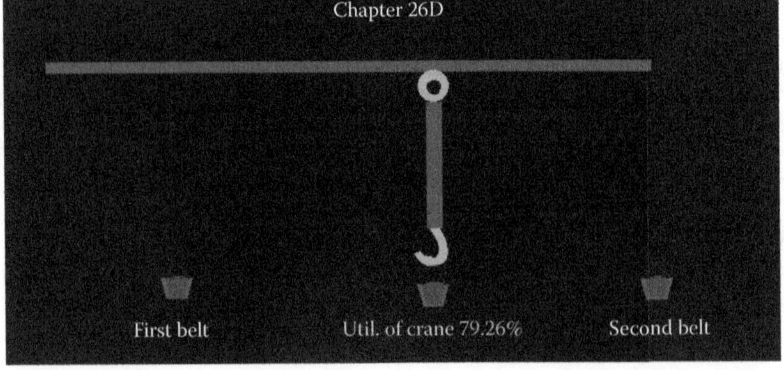

FIGURE 26.10
Screenshot from program CHAP26D.GPS.

the ship to guide it to the dock. This takes 30 minutes. The ship then needs further positioning before a crane can commence loading the containers. This takes 10 minutes. Once positioned, the crane begins to load the containers. The crane carries a loaded container to the ship and deposits it in a time of 1.98 ± .18 minutes. It returns to pick up another container in 1.5 ± .12 minutes. Determine the expected loading time for the ship at the dock. Neglect the time for it to be positioned by the tug boats.

The program to do this simulation is CHAP26E.GPS. Its listing is shown in Figure 26.11.

The output from the program will depend on the random number stream used. One possible output is as shown in Figure 26.12.

The animation is given by CHAP26F.GPS. The listing is given in Figure 26.13. Again, it should be studied closely to see how layout objects have been used to make the

```
          SIMULATE
***********************************************
*    PROGRAM CHAP26E.GPS                      *
*    PORT EXAMPLE                             *
***********************************************
***********************************************
*                                             *
*  DEFINE VARIABLES TO BE USED NEXT           *
***********************************************
          INTEGER     &RANSEED    SETS THE RANDOM NUMBER SEED
          INTEGER     &K          DUMMY FOR USE IN DO LOOP
          INTEGER     &LOADS      COUNTER FOR LOADS ON BARGE
          INTEGER     &LOADED
****************************************
*  NEXT MAKES A CLEAN SCREEN           *
****************************************
          DO          &K=1,23
            PUTSTRING ('  ')
            ENDDO
          PUTSTRING   ('  ')
          PUTSTRING   ('  INPUT A NUMBER TO SEED THE RANDOM NUMBER STREAM')
          PUTSTRING   ('  ')
          GETLIST     &RANSEED
          RMULT       &RANSEED
          REAL        &Y,&Z,&TIME
 SWING1   FUNCTION    RN1,C2
0,.030/1,.036
 SWING2   FUNCTION    RN1,C2
0,.023/1,.027
 AMT      FUNCTION    RN1,C2
0,95/1,106
******************************************
*   SHIP ARRIVES                         *
******************************************
          GENERATE    ,,,1,10,1PH,2PL     SHIP ARRIVES
          ADVANCE     24,24               DELAY TIME FOR SHIP
          BLET        &LOADS=FN(AMT)     HOW BIG IS THE SHIP?
          ADVANCE     .25     TAKES 15 MINUTES TO TRAVEL TO DOCK
          ADVANCE     .5      30 MINUTES FOR TUGS TO DOCK BOAT
          ADVANCE     .16667      10 MINUTES FOR POSITIONING
          MARK        1PL      MAKE RECORD OF TIME LOADING STARTS
          SPLIT       1,UNLOAD
          TEST E      &LOADED,&LOADS    IS SHIP LOADED?
          BLET        &TIME=MP1PL       TIME SHIP LOADING
          BPUTPIC     LINES=2,&LOADS,&TIME
       NUMBER OF LOADS  ***
       TIME TO LOAD      ***.**
          TERMINATE   1
```

FIGURE 26.11
Listing of program CHAP26E.GPS. *(Continued)*

```
*****************************************
*  SEGMENT FOR LOADING SHIP             *
*****************************************
 UNLOAD   SEIZE       LOADING         SEIZE THE LOADER
 BACK     BLET        &Y=FN(SWING1)   SWING OF BOOM WITH
          BLET        &Z=FN(SWING2)   SWING OF BOOM BACK
          ADVANCE     &Y
          ADVANCE     &Z
          BLET        &LOADED=&LOADED+1    COUNT LOADS
          TEST E      &LOADED,&LOADS,BACK DONE LOADING?
          RELEASE     LOADING
          TERMINATE
*******************************************************
*    END SUBPROGRAM FOR UNLOADING                     *
*                                                     *
*******************************************************
          START       1
          END
```

FIGURE 26.11 (Continued)
Listing of program CHAP26E.GPS.

```
NUMBER OF LOADS  104
TIME TO LOAD       6.03
```

FIGURE 26.12
One possible output from program CHAP26E.GPS.

```
          SIMULATE
**********************************************
*   PROGRAM CHAP26F.GPS                      *
*   ANIMATION FOR PORT EXAMPLE               *
**********************************************
*
*
***********************************************
*                                             *
*  DEFINE VARIABLES TO BE USED NEXT           *
***********************************************
 ATF      FILEDEF     'CHAP26F.ATF'
          INTEGER     &RANSEED    SETS THE RANDOM NUMBER SEED
          INTEGER     &K        DUMMY FOR USE IN DO LOOP
          INTEGER     &LOADED,&LOADS
          REAL        &TIMEIN
***************************************
*  NEXT MAKES A CLEAN SCREEN          *
***************************************
          DO          &K=1,23
            PUTSTRING ('  ')
            ENDDO
          PUTSTRING   ('  ')
          PUTSTRING   ('  INPUT A NUMBER TO SEED THE RANDOM NUMBER STREAM')
          PUTSTRING   ('  ')
          GETLIST     &RANSEED
          RMULT       &RANSEED
          REAL        &Y,&Z
  SWING1  FUNCTION    RN1,C2
0,.030/1,.036
```

FIGURE 26.13
Listing of program CHAP26F.GPS. *(Continued)*

```
 SWING2   FUNCTION    RN1,C2
0,.023/1,.027
 AMT      FUNCTION    RN1,C2
0,95/1,106
******************************************
*   SHIP ARRIVES                         *
******************************************
          GENERATE    ,,,1,10,1PH,2PL     SHIP ARRIVES
          BLET        &LOADS=FN(AMT)     HOW BIG IS THE SHIP?
          MARK        1PL       MAKE RECORD OF TIME OF ARRIVAL
          BPUTPIC     FILE=ATF,LINES=5,AC1,XID1,XID1,XID1,&LOADS
TIME *.****
CREATE SHIP P*
PLACE P* ON P1
SET P* TRAVEL .25
WRITE M3 ***
          ADVANCE     .25    TAKES 15 MINUTES TO TRAVEL TO DOCK
          BPUTPIC     FILE=ATF,LINES=5,AC1
TIME *.****
PLACE T3 ON P3
PLACE T2 ON P2
SET T3 TRAVEL .15
SET T2 TRAVEL .15
          ADVANCE      .15
          BPUTPIC      FILE=ATF,LINES=6,AC1,XID1,XID1,XID1
TIME *.****
SET T3 COLOR BACKDROP
SET T2 COLOR BACKDROP
SET P* CLASS SHIPD
PLACE P* ON P4
SET P* TRAVEL .5
          ADVANCE     .5
          BPUTPIC     FILE=ATF,LINES=3,AC1,XID1
TIME *.****
SET P* CLASS SHIP
SET ANC COLOR F21
          ADVANCE     .16667
          SPLIT       1,UNLOAD
          TEST E      &LOADED,&LOADS
          BLET        &TIMEIN=MP1PL
          BPUTPIC     FILE=ATF,LINES=6,AC1,XID1,XID1,XID1,&TIMEIN
TIME *.****
SET P* CLASS SHIPD
PLACE P* ON P6B
SET P* TRAVEL  .1
SET ANC COLOR BACKDROP
WRITE M2 **.**
          ADVANCE     .1
          BPUTPIC     FILE=ATF,LINES=4,AC1,XID1,XID1,XID1
TIME *.****
SET P* CLASS SHIP
PLACE P* ON P6C
SET P* TRAVEL .15
          ADVANCE     .15
          BPUTPIC     FILE=ATF,LINES=2,AC1,XID1
TIME *.****
DESTROY P*
          TERMINATE   1          SIMULATION OVER
```

FIGURE 26.13 (Continued)
Listing of program CHAP26F.GPS.

(Continued)

```
*****************************************
*  SUBROUTINE FOR UNLOADING BARGE       *
*****************************************
 UNLOAD    ADVANCE     0        BEGIN UNLOADING
           SEIZE       LOADING         SEIZE THE LOADER
 BACK      BLET        &Y=FN(SWING1)
           BLET        &Z=FN(SWING2)
           BPUTPIC     FILE=ATF,LINES=2,AC1,&Y
TIME *.****
ROTATE BB 150 TIME *.****
           ADVANCE     &Y
           BPUTPIC     FILE=ATF,LINES=2,AC1,&Z
TIME *.****
ROTATE BB -150 TIME *.****
           ADVANCE     &Z
           BLET        &LOADED=&LOADED+1
           TEST G      &LOADED,&LOADS,BACK1
           TRANSFER    ,AWAY
BACK1      BPUTPIC     FILE=ATF,LINES=2,AC1,&LOADED
TIME *.****
WRITE M1 ***
           TRANSFER    ,BACK
 AWAY      RELEASE     LOADING
           TERMINATE
*******************************************************
*    END SUBROUTINE FOR UNLOADING                     *
*
*******************************************************
***********************************
*   CLOCK SEGMENT
***********************************
           GENERATE    ,,,1,150,12PH,12PL   1 MASTER CLOCK TRANSACTION
           BPUTPIC     FILE=ATF,LINES=2,AC1
TIME *.****
ROTATE HH SPEED -30 STEP 6
           INTEGER     &HOUR
  QQQQ     BPUTPIC     FILE=ATF,LINES=2,AC1,&HOUR
TIME *.****
WRITE M200 **
           ADVANCE     1
           BLET        &HOUR=&HOUR+1
           TRANSFER    ,QQQQ
           START       1
           PUTPIC      FILE=ATF,LINES=2,AC1
TIME *.****
END
             END
```

FIGURE 26.13 (Continued)
Listing of program CHAP26F.GPS.

animation match the reality of the example. There are different classes to represent the different stages of the ship. These are as shown in Figure 26.14.

When the ship approaches the port, it is `SHIP`. When the tugs attach to the ship, it becomes Class `SHIPD`. The Class `ANCHOR` is a layout object, `ANC`, that is originally coloured Backdrop so it is invisible. Once the ship is docked, its colour is changed to a dull pink to show up. When the ship is loaded, its colour changes back to Backdrop.

A screenshot of the animation is given in Figure 26.15. Notice that a clock segment has been added. The stored program CLOCK.GPS can be modified easily to be included with most of the animations. In this case, only the hour hand is used.

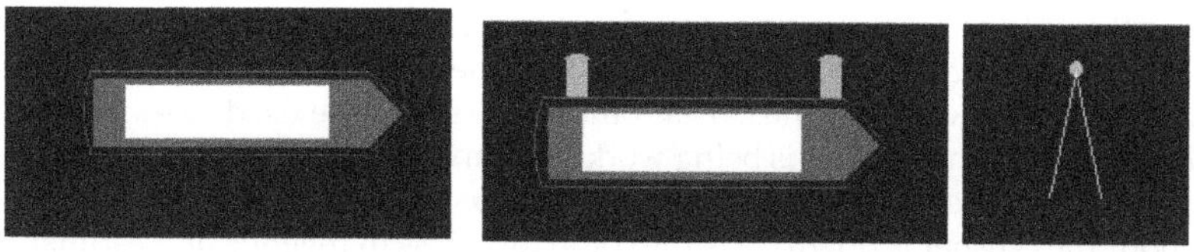

FIGURE 26.14
Three of the classes used in animation program CHAP26F.GPS. These are SHIP, SHIPD, and ANCHOR.

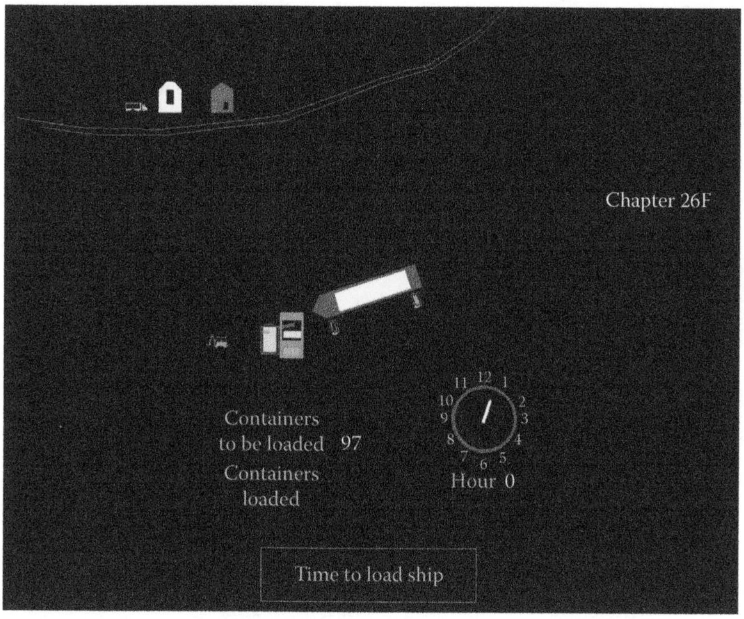

FIGURE 26.15
Screenshot of animation from program CHAP26F.GPS.

Example 26.4: A Spare Parts Problem

Inventory problems are standard in any study of operations research. Most are easily solved using simulation. The following example is one that illustrates how GPSS/H handles such problems.

A crusher in a mine uses a part that is subject to periodic failure. Whenever the part fails, the crusher is immediately turned off. The failed part is then removed and a good spare part is installed if one is available, or as soon as one becomes available, and the machine is turned on again. Failed parts can be repaired and used over and over.

The lifetime of a part is normally distributed with a mean of 375 hours and a standard deviation of 80 hours. It takes 4 ± 1 hours to remove a failed part from the machine. The variation in time includes the time for the miner to come to the crusher, remove the broken part, and take it to the repair shop. The time required to install a replacement part is 6 ± 1.5 hours, which includes all miscellaneous times such as taking the part to the crusher. Repair time for a failed part is normally distributed with a mean and a standard deviation of 18 and 1.4 hours, respectively.

A single miner is responsible for removing a failed part from the machine and installing a replacement part. There is a repair shop where the failed parts are fixed so they can be used again. The repair shop has other broken parts to fix. These other items arrive in a Poisson stream with a mean interarrival time of 12 hours. Their service time is 9.5 ± 4 hours. These other items have a higher priority than the failed parts used in the crusher. The reason for this is because, if the crusher is down, ore can be stockpiled and later dumped into the crusher by a front end loader, whereas the other failed items are deemed essential to the mine and mill working. Thus, these other parts should be fixed as soon as possible. Even so, if a part for the crusher is being worked on and another failed item arrives at the repair shop, the crusher part will continue to be worked on until it is repaired.

Each hour the crusher is down costs the company a loss in revenue. It is estimated that this is $155 per hour. Each spare part for the crusher costs the company $175 per week to keep in stock. Determine how many spares to have. Assume that there is no other time needed to take a broken part to the shop and that the miner will immediately begin installation of the spare part if one is available.

Solution

The program to do the simulation is CHAP26G.GPS and is shown in Figure 26.16. A SPLIT Block is needed when the crusher breaks down, as two things are desired to be done at the same time (namely, the broken part is to be repaired and a spare part, if available, installed).

```
                SIMULATE
********************************
*   PROGRAM CHAP26G.GPS        *
*   SPARE PARTS PROGRAM        *
********************************
                INTEGER      &SPARES,&I,&NOSPARE
                REAL         &COST
                GENERATE     ,,,1         ONE MINER TO LOOK AFTER CRUSHER
       WAIT     SEIZE        CRUSHER      CRUSHER WORKS
                ADVANCE      RVNORM(1,375,80)   CRUSHER IS WORKING
                RELEASE      CRUSHER      CRUSHER IS DOWN
                ADVANCE      4,1          MINER REMOVES PART
                SPLIT        1,REPAIR
                SEIZE        RSHOP
                ADVANCE      RVNORM(1,18,1.4)    REPAIR PART
                BLET         &SPARES=&SPARES+1
                RELEASE      RSHOP
                TERMINATE                 PART IS FIXED
     REPAIR     TEST G       &SPARES,0 IS A SPARE AVAILABLE?
                BLET         &SPARES=&SPARES-1    TAKE A SPARE
                ADVANCE      6,1.5
                TRANSFER     ,WAIT        READY TO GO AGAIN
                GENERATE     RVEXPO(1,12),,,,10    OTHER PARTS
                QUEUE        SHOP
                SEIZE        RSHOP
                DEPART       SHOP
                ADVANCE      9.5,4        FIX OTHER ITEMS
                RELEASE      RSHOP
                TERMINATE
                GENERATE     24*365*100    SIMULATE FOR 100 YEARS
                TERMINATE    1
                DO           &I=1,4
                CLEAR
                LET          &SPARES=&I-1
                LET          &NOSPARE=&SPARES
                START        1
                LET          &COST=(1-FR(CRUSHER)/1000.)*155
                LET          &COST=&COST*24*7
                LET          &COST=&COST+&NOSPARE*175
                PUTPIC       LINES=2,&NOSPARE,FR(CRUSHER)/10.,&COST
       NO. SPARES **       UTIL. OF CRUSHER **.**%
       TOTAL COST     *****.**
                ENDDO
                END
```

FIGURE 26.16
Listing of program CHAP26G.GPS.

The program is run four times using a DO LOOP. Notice that a new variable, &NOSPARE, is used to keep track of the number of spares for the system. The reason is that the original number of spares, &SPARES, changes as the program is being run. The output from the program is shown in Figure 26.17.

As can be seen, the optimum number of spare parts is 2.

The animation is quite revealing—not so much for the crusher but for the repair shop. The original problem was to determine the number of spare parts for the crusher. It was stated that the other parts being repaired are more important than the crusher, so these were given a higher priority for being repaired. The animation is given by CHAP26H.GPS. Its listing is given in Figure 26.18

A screenshot from the animation is shown in Figure 26.19.

Notice that the repair shop has a large queue of other items to be repaired before the failed part from the crusher can be repaired. In this example, the crusher is still working as one of the two spares has been installed. It is left as an exercise, to determine how the system (crusher plus repair shop) might be improved if more than one item can be repaired at a time.

```
NO. SPARES  0    UTIL. OF CRUSHER 76.26%
TOTAL COST    6182.74
NO. SPARES  1    UTIL. OF CRUSHER 94.72%
TOTAL COST    1551.12
NO. SPARES  2    UTIL. OF CRUSHER 96.62%
TOTAL COST    1229.99
NO. SPARES  3    UTIL. OF CRUSHER 97.11%
TOTAL COST    1278.67
```

FIGURE 26.17
Output from program CHAP26G.GPS.

```
         SIMULATE
********************************
*  ANIMATION FOR PROGRAM       *
*  CHAP26G.GPS                 *
*  SPARE PARTS PROGRAM         *
********************************
 ATF     FILEDEF     'CHAP26.ATF'
         INTEGER     &SPARES,&NOSPARE
         LET         &SPARES=2
         PUTPIC      FILE=ATF,LINES=3,AC1,&SPARES
TIME *.****
WRITE M1 *
SET S3 COLOR BACKDROP
         LET         &NOSPARE=&SPARES
         REAL        &COST
         GENERATE    ,,,1     ONE MINER TO LOOK AFTER CRUSHER
   WAIT  SEIZE       CRUSHER   CRUSHER WORKS
         ADVANCE     RVNORM(1,375,80)  CRUSHER IS WORKING
         RELEASE     CRUSHER    CRUSHER IS DOWN
         LOGIC S     WORK       SET SWITCH
         BPUTPIC     FILE=ATF,LINES=2,AC1
TIME *.****
SET PL CLASS CRDOWN
         ADVANCE     4,1           MECHANIC REMOVES PART
         SPLIT       1,REPAIR
         BPUTPIC     FILE=ATF,LINES=2,AC1
TIME *.****
SET S3 COLOR F2
         SEIZE       RSHOP
         BPUTPIC     FILE=ATF,LINES=2,AC1
```

FIGURE 26.18
Listing of program CHAP26H.GPS. *(Continued)*

```
TIME *.****
PLACE S3 AT 50 15
          ADVANCE      RVNORM(1,18,1.4)    REPAIR PART
          BLET         &SPARES=&SPARES+1    UPDATE SPARES
          RELEASE      RSHOP
          BPUTPIC      FILE=ATF,LINES=3,AC1
TIME *.****
PLACE S3 AT 50 7
SET S3 COLOR BACKDROP
          TERMINATE                   PART IS FIXED
 REPAIR   TEST G       &SPARES,0        IS A SPARE AVAILABLE?
          BLET         &SPARES=&SPARES-1    TAKE A SPARE
          BPUTPIC      FILE=ATF,LINES=2,AC1,&SPARES
TIME *.****
WRITE M1 *
          ADVANCE      6,1.5
          LOGIC R      WORK
          BPUTPIC      FILE=ATF,LINES=2,AC1
TIME *.****
SET PL CLASS PLUNGER
          TRANSFER     ,WAIT     READY TO GO AGAIN
          GENERATE     RVEXPO(1,12),,,,10    OTHER PARTS
          BPUTPIC      FILE=ATF,LINES=4,AC1,XID1,XID1,XID1
TIME *.****
CREATE OTHERS O*
PLACE O* ON P1
SET O* TRAVEL 6
          ADVANCE      6
          QUEUE        SHOP
          SEIZE        RSHOP
          DEPART       SHOP
          BPUTPIC      FILE=ATF,LINES=2,AC1,XID1
TIME *.****
PLACE O* AT 50 15
          ADVANCE      9.5,4     FIX OTHER ITEMS
          BPUTPIC      FILE=ATF,LINES=4,AC1,XID1,XID1,XID1
TIME *.****
SET O* CLASS OTHERSD
PLACE O* ON P2
SET O* TRAVEL 2
          ADVANCE      .5
          RELEASE      RSHOP
          ADVANCE      1.5
          BPUTPIC      FILE=ATF,LINES=2,AC1,XID1
TIME *.****
DESTROY O*
          TERMINATE
          GENERATE       ,,,1
          BPUTPIC        FILE=ATF,LINES=2,AC1
TIME *.****
ROTATE PL 3 STEP 4 TIME .2
  BACK    GATE LR        WORK
          ADVANCE        .2
          BPUTPIC        FILE=ATF,LINES=2,AC1
TIME *.****
ROTATE PL -6 STEP 4 TIME .2
          ADVANCE        .2
          BPUTPIC        FILE=ATF,LINES=2,AC1
TIME *.****
ROTATE PL 6 STEP 4 TIME .2
          ADVANCE        .2
          TRANSFER       ,BACK
```

FIGURE 26.18 (Continued)
Listing of program CHAP26H.GPS.

(Continued)

```
              GENERATE     24*365   SIMULATION TIME FOR ANIMATION
              TERMINATE    1
              START        1
              PUTPIC       FILE=ATF,LINES=2,AC1
    TIME *.****
    END
              LET          &COST=(1-FR(CRUSHER)/1000.)*155
              LET          &COST=&COST*24*7
              LET          &COST=&COST+&NOSPARE*175
              PUTPIC       LINES=2,&NOSPARE,FR(CRUSHER)/10.,&COST
       NO. SPARES **     UTIL. OF CRUSHER **.**%
       TOTAL COST    *****.**
              END
```

FIGURE 26.18 (Continued)
Listing of program CHAP26H.GPS.

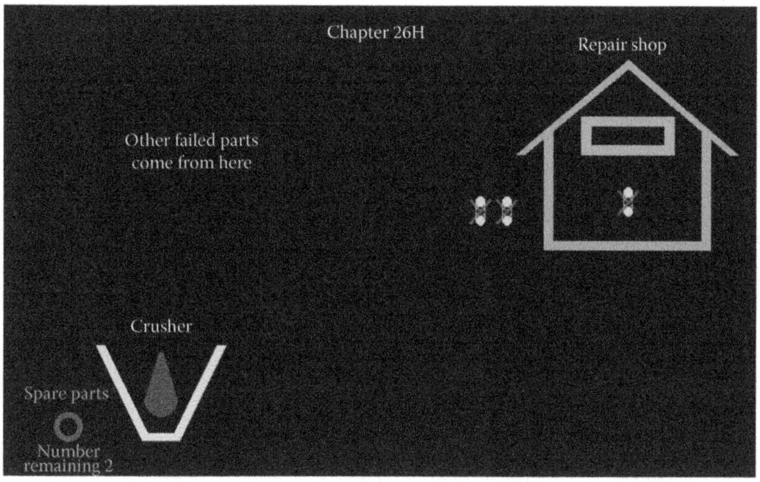

FIGURE 26.19
Screenshot of animation from program CHAP26H.GPS.

This example illustrates the importance of having animation—often the mining engineer is concerned with one aspect of the mine and uses a simulation model to assist in his or her decision. However, the completed model then points up other possibilities to examine.

Example 26.5: Modelling Conveyor Belts

Nearly every large mining operation will have a conveyor belt or system of conveyor belts to transport the ore from the pit to the mill. They are easily modelled by use of the SPLIT Block. One way to do this is given in this example. A simple model is used to illustrate conveyor belts. Trucks come from a mine every 20 ± 4 minutes. They travel to a crusher in 8 minutes. At the crusher, they take 1.5 minutes to dump. The crusher also takes 1.5 minutes to crush the load, which then empties to a conveyor belt. The belt carries the crushed ore to another section where it is removed from the belt. Once trucks finish dumping, they return to the mine in 6 minutes.

This example shows how to model a conveyor belt. This example is best illustrated by the animation, so this is the only program that will be given. The program to make the animation file is CHAP26I.GPS. Its listing is given in Figure 26.20.

It is instructive to examine the program closely. When the truck arrives at the crusher with its ore, it dumps in 1.5 minutes. This is the Block.

```
           SIMULATE
***********************************
*  PROGRAM CHAP27I.GPS            *
*  THIS USES THE SPLIT BLOCK TO   *
*  ILLUSTRATE HOW ONE MIGHT MODEL *
*  A CONVEYOR BELT                *
***********************************
  ATF      FILEDEF      'CHAP27G.ATF'
           INTEGER      &LOADS
           GENERATE     20,4      TRUCK COME ALONG
           BPUTPIC      FILE=ATF,LINES=4,AC1,XID1,XID1,XID1
TIME *.****
CREATE TRUCKF T*
PLACE T* ON P3
SET T* TRAVEL 8
           ADVANCE      8         TRAVELS TO CRUSHER
           BPUTPIC      FILE=ATF,LINES=2,AC1,XID1
TIME *.****
SET T* CLASS TRUCKD
           ADVANCE      1.5       DUMPS
           LOGIC S      CRUSHER  SETS CRUSHER IN MOTION
           BLET         &LOADS=&LOADS+1   COUNT LOADS
           BPUTPIC      FILE=ATF,LINES=2,AC1,&LOADS
TIME *.****
WRITE M1 ***
           SPLIT        23,NEXT   READY TO DUMP ORE INTO THE CRUSHER
           BPUTPIC      FILE=ATF,LINES=4,AC1,XID1,XID1,XID1
TIME *.****
SET T* CLASS TRUCKE
PLACE T* ON P4
SET T* TRAVEL 6
           ADVANCE      6          TRUCK RETURNS TO MINE
           BPUTPIC      FILE=ATF,LINES=2,AC1,XID1
TIME *.****
DESTROY T*
           TERMINATE
  NEXT     ADVANCE      1.5        WAIT WHILE TRUCK DUMPS INTO CRUSHER
           SEIZE        BELT
           BPUTPIC      FILE=ATF,LINES=4,AC1,XID1,XID1,XID1
TIME *.****
CREATE ORE O*
PLACE O* ON P1
SET O* TRAVEL  8
           ADVANCE      .2
           RELEASE      BELT
           ADVANCE      7.8
           BPUTPIC      FILE=ATF,LINES=3,AC1,XID1,XID1
TIME *.****
PLACE O* ON P2
SET O* TRAVEL .6
           ADVANCE      .6
           BPUTPIC      FILE=ATF,LINES=2,AC1,XID1
TIME *.****
DESTROY O*
           TERMINATE
           GENERATE     ,,,1
 WAIT      GATE LS      CRUSHER   WAIT FOR ORE TO BE DUMPED
           ADVANCE      1.5        TRUCK DUMPS INTO CRUSHER
           BPUTPIC      FILE=ATF,LINES=2,AC1
```

FIGURE 26.20
Listing of program CHAP26I.GPS.

(Continued)

```
TIME *.****
PLACE CR ON P5
          ADVANCE      5
          LOGIC R      CRUSHER
          BPUTPIC      FILE=ATF,LINES=2,AC1
TIME *.****
PLACE CR AT -22 -8
          TRANSFER     ,WAIT
          GENERATE     ,,,1        DUMMY TRANSACTION
  WHEEL   TEST E       FS(BELT),1 IS BELT TO MOVE?
          BPUTPIC      FILE=ATF,LINES=3,AC1
TIME *.****
ROTATE W1 SPEED -60
ROTATE W2 SPEED -60
 AWAY     ADVANCE      .2
          TEST E       FS(BELT),0,AWAY
          ADVANCE      6
          BPUTPIC      FILE=ATF,LINES=3,AC1
TIME *.****
ROTATE W1 TO 0
ROTATE W2 TO 0
          TRANSFER     ,WHEEL
          GENERATE     450
          TERMINATE    1
          START        1
          PUTPIC       FILE=ATF,LINES=2,AC1
TIME *.****
END
          END
```

FIGURE 26.20 (Continued)
Listing of program CHAP26I.GPS.

```
ADVANCE     1.5
```

Next, it makes 23 clones via the Block:

```
SPLIT     23,NEXT
```

The parent transaction is the truck, and it immediately starts back to the mine via the next few Blocks. The 23 clones are sent to the Block `NEXT`. The number of these clones was only for the animation—there might have been more or less depending on the model.

```
NEXT     ADVANCE    1.5 WAIT WHILE TRUCK DUMPS INTO CRUSHER
         SEIZE      BELT
         BPUTPIC    FILE = ATF,LINES = 4,AC1,XID1,XID1,XID1
TIME *.****
CREATE ORE O*
PLACE O* ON P1
SET O* TRAVEL 8
         ADVANCE    .2
         RELEASE    BELT
         ADVANCE    7.8
```

The 23 clones all immediately are routed to the Block `NEXT`.

```
NEXT   ADVANCE     1.5
```

This allows the crusher to begin crushing, which takes 1.5 minutes. Next, the first of the clones enters a `SEIZE` Block, which is a facility named `BELT`. A unit of ore is created and placed on the belt and sent to travel to its end in 8 time units. After .2 time units, the facility `BELT` is released to allow the second clone to make a second unit of ore that is placed on the

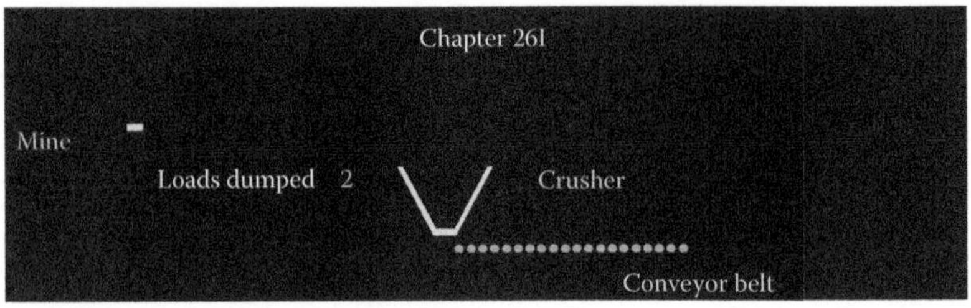

FIGURE 26.21
Screenshot of animation from program CHAP26I.GPS.

conveyor belt. This is repeated until all 23 of the clones are placed on the belt. The animation shows the ore spaced out on the belt. Figure 26.21 gives a screenshot of the animation.

26.1.3 The SPLIT Block Serialization Option

Often, the SPLIT Block is used to create transactions that are to be routed to different Blocks. For example, there might be four clones to be made and these are all to be sent to different Blocks. One way to do this is to use several SPLIT Blocks. For example, consider the code that creates three clones and routes them all to the NEXT Block.

```
        GENERATE      ,,,1
        SPLIT         3,NEXT
        ..........
NEXT    ............
```

If it is desired to route the transactions to different Blocks, this can be accomplished by having the following code:

```
        GENERATE      ,,,1
        SPLIT         1,NEXT
        ..........
NEXT    SPLIT         1,NEXT2
        ..........
NEXT2   SPLIT         1,NEXT3
        .........
NEXT3   ........
```

Each clone will have the same number of parameters as the parent. Each parameter will have the same value(s) as in the original parent transaction. Each will have a different XID1. There is a way to have each clone different and that is given by the Block:

```
SPLIT  n,(label),(parameter number)
```

where:
- n is the number of clones
- (label) is the label to send the clones to
- (parameter number) is the parameter to begin the serialization

The serialization is sequential starting with the value already in the parent's parameter. For example, consider the following Block:

```
SPLIT  3,NEXT,1PH
```

Three clones will be created and sent to the Block with the label NEXT. Assume that the original number in the parent's first half-word parameter was 5. The values placed in the transaction's half-word parameter 1 will be as follows.

```
Parent          6
First clone     7
Second clone    8
Third clone     9
```

To illustrate this, consider the following short program called CHAP26J.GPS as shown in Figure 26.22.

A single transaction is created at time 0. Its first half-word parameter is given the value 5. Next, three clones are created by the SPLIT Block and routed to the Block NEXT. Serialization is done in the transaction's first half-word parameter. The original value was 5, so the parent has the new value of 6 and each of the clones will have the values 7, 8, and 9 in their first half-word parameter. The transactions' XID1 will be 1, 2, 3, and 4. When the program is run, the output is shown in Figure 26.23.

Suppose one wanted to route the three clones in program CHAP26J.GPS to different Blocks and have the serialization option. This can be done, but one needs to be aware of what will happen. For example, consider the program made by modifying CHAP26J.GPS to route the clones to different Blocks. This is program CHAP26K.GPS. Before running the program or looking at the output, it is instructive to try to guess what the results will be (Figure 26.24). The results of running the program are shown in Figure 26.25.

The first clone has its value of half-word parameter equal to 2 when it enters the Block with the label NEXT. The SPLIT Block increases this by 1, so it is now 3. A similar thing happens for the second clone to give its first half-word parameter a value of 5.

```
          SIMULATE
*************************
*   PROGRAM CHAP26J.GPS   *
*   ILLUSTRATION OF       *
*   SPLIT BLOCK WITH      *
*   SERIALIZATION         *
*************************
          GENERATE      ,,,1  ONE TRANSACTION
          ASSIGN        1,5,PH
          SPLIT         3,NEXT,1PH
          BPUTPIC       AC1,XID1,PH1
 TIME *  XID1 *  PH1 *
          TERMINATE     1
 NEXT     ADVANCE       5
          BPUTPIC       AC1,XID1,PH1
 TIME *  XID1 *  PH1 *
          TERMINATE     1
          START         4
          END
```

FIGURE 26.22
Listing of program CHAP27J.GPS.

```
TIME 0  XID1 1  PH1 6
TIME 5  XID1 2  PH1 7
TIME 5  XID1 3  PH1 8
TIME 5  XID1 4  PH1 9
```

FIGURE 26.23
Output from program CHAP26J.GPS.

```
            SIMULATE
            GENERATE     ,,,1
            SPLIT        1,NEXT,1PH
            BPUTPIC      AC1,XID1,PH1
TIME **       XID1 IS *   PH1 IS *
            TERMINATE
NEXT        ADVANCE      5
            SPLIT        1,NEXT1,1PH
            BPUTPIC      AC1,XID1,PH1
TIME **       XID1 IS *   PH1 IS *
            TERMINATE
NEXT1       SPLIT        1,NEXT2,1PH
            BPUTPIC      AC1,XID1,PH1
TIME **       XID1 IS *   PH1 IS *
            TERMINATE
NEXT2         BPUTPIC    AC1,XID1,PH1
TIME **       XID1 IS *   PH1 IS *
            TERMINATE    1
            START        1
            END
```

FIGURE 26.24
Listing of program CHAP26K.GPS.

```
TIME  0   XID1 IS 1  PH1 IS 1
TIME  5   XID1 IS 2  PH1 IS 3
TIME  5   XID1 IS 3  PH1 IS 5
TIME  5   XID1 IS 4  PH1 IS 6
```

FIGURE 26.25
Results of running program CHAP26K.GPS.

26.2 Exercises

26.1. In Example 26.1, the miners came to the tool crib according to the normal distribution. Suppose that they now arrive in a Poisson stream with a mean of 10. What is the average queue? What is the average time for a miner at the tool crib? This includes both waiting for the worker and obtaining the tool.

26.2. Modify program CHAP26C.GPS so that the output shows what percentage of the time the crane is moving a bucket of ore and what percentage of time it is returning to the belt area.

26.3. Add the following to the animation in Example 26.2: A clock to show the time and a message to show the number of buckets carried to the second conveyor belt.

26.4. Change the times in program CHAP26E.GPS for Example 26.3 as follows:

	Original Time	New Time
Ship arriving:	15 minutes	15 ± 5 minutes
Tugs dock ship	30	30 ± 2.8 minutes
Misc. time	10 minutes	10 ± 4 minutes

26.5. Example 26.4 has the repair shop able to repair only one item at a time. Suppose a second repair mechanic was added. Change program CHAP26G.GPS to reflect how much more efficient this would be.

27

The ASSEMBLY Block

One of the many statues carved from salt in the salt mine in Wielitzka, Poland (see also Chapter 7).

27.1 Assembly Sets

All transactions belong to different groups known as *assembly sets*. When a transaction is created via the `GENERATE` Block, it is assigned its own assembly set. Once a transaction is assigned its own assembly set, it remains there until it leaves the system. These sets are not numbered or named, so a programmer cannot refer to them. Only after a transaction leaves the system and re-enters later can it be assigned a different assembly set. When only a single transaction is in an assembly set, there is not much of interest with the set. It is when there is more than one transaction belonging to a particular assembly set that it is of interest to the programmer.

When a transaction enters a `SPLIT` Block, the cloned transactions belong to the same assembly set as the original. Even if these cloned transactions themselves enter a `SPLIT` Block, the newly cloned transactions belong to the same assembly set. Several Blocks are used with the concept of assembly sets. The first is the `ASSEMBLE` Block.

27.2 The ASSEMBLE Block

The ASSEMBLE Block acts in a manner opposite to the SPLIT Block. The SPLIT Block clones new transactions into the system and the ASSEMBLE Block removes them. The form of it is as follows:

```
ASSEMBLE  n
```

n is the number of transactions to be removed from the system minus 1. When a transaction enters the ASSEMBLE Block, it is delayed until other members of its assembly set also arrive in the Block where each is removed. Only when the counter, as given by the operand n is reached, is the original transaction allowed to move to the next sequential Block. It is immaterial if the first transaction to arrive at the ASSEMBLE Block is the original uncloned (parent) transaction or not. The first transaction in the Block is the one that is allowed to move on. The rest are removed from the simulation just as if they had entered a TERMINATE Block.

For example, when a transaction enters the Block

```
ASSEMBLE  2
```

it will be delayed until one other transaction from its assembly set also enters the ASSEMBLE Block and is subsequently destroyed. The first transaction to enter the Block can then be moved to the next sequential Block. A program might have a set of Blocks such as

```
SPLIT           2,DOWNA
                ......      ......
                ......      ......
      DOWNA     ......      ......
                ......      ......
                ASSEMBLE  3
```

The SPLIT Block creates two clones when the original transaction enters it. Later, two of the transactions in the same assembly set are removed from the system. The transaction that goes to the sequential Block after the ASSEMBLE Block is not necessarily the original transaction.

It is possible to have the same ASSEMBLE Block working on more than one assembly set. For a given assembly set, it is possible to have assembling operations being done at more than one ASSEMBLE Block. Thus,

```
         SPLIT      6,DOWN1
         ......     .......
DOWN1    ......     .......
         ......     .......
         TRANSFER   ,FN(AWAY)
BLOCKA   ASSEMBLE   2
         .......    .......
         .......    .......
BLOCKB   ASSEMBLE   4
         .......    .......
         .......    .......
```

The Block `TRANSFER ,FN(AWAY)` might send some of the cloned transactions to the Block labelled `BLOCKA`, whereas others might go to the Block labelled `BLOCKB`.

Example 27.1: Assembly Line Example

A single worker is positioned at the end of an assembly line. The assembly line brings a quantity of items that are stacked up in front of him. There are always enough of these items for the worker. He needs to fill a case with these and then put the case on a conveyor belt for further movement. Each case can hold six of these items. The worker needs to do the following:

1. Take an empty box from a stack and bring it to a bench in front of the assembly line.
2. Pick up an item, inspect it, and place a label on the item.
3. Place an inspection tag on the item and put it in the box.
4. When there are six of these items in the box, close the box, and place it back on the conveyor belt.

These operations take the following times:

1. 1 ± .25 minutes
2. Mean of 2, standard deviation of .25 minutes (normally distribution)
3. Mean of 2 minutes, exponentially distributed
4. .5 ± .1 minutes

Solution

This example is solved by using both `SPLIT` and `ASSEMBLE` Blocks. The worker is represented by a single transaction. The worker begins by entering a `SPLIT` Block and starts to work. He does this by making himself: (`SEIZE HIMSELF`). The parent clone waits until he has filled a case. When this is done, the Block `RELEASE HIMSELF` is executed, and the worker will begin filling a new case. The program for this is CHAP27A.GPS and the listing is given in Figure 27.1.

The program is written so that the user can run it multiple times starting at different positions in the random number streams. The program will show that the expected output is around 12 cases per shift.

Example 27.2: Trucks in a Mine Needing Service

A mine has two service bays for trucks. One is for trucks that need only routine servicing, and the other bay is for trucks needing major service. The bays are different and not interchangeable. Trucks that arrive needing routine services will have required the following operations:

1. Obtain petrol
2. Have the driver perform routine checks
3. Oil and grease
4. Miscellaneous

Once a truck enters this bay, these four items are performed simultaneously as there are always enough miners available. Only one truck can be serviced at a time. The other service bay can also handle only one truck at a time. Trucks enter the service area according to the normal distribution with a mean of 25 minutes and a standard

```
            SIMULATE
**********************************
*  PROGRAM CHAP27A.GPS             *
*  EXAMPLE OF THE ASSEMBLY BLOCK   *
**********************************
            INTEGER     &CASES
            CHAR*1      &ANS
 AGAIN      PUTSTRING   ('  ')
            GENERATE    ,,,1          ONE WORKER
  BACKUP    SEIZE       PERSON
            SPLIT       1,BACKUP
            ADVANCE     1,.25
            ADVANCE     RVNORM(1,2,.25)
            ADVANCE     RVEXPO(1,2)
            ADVANCE     .5,.1
            RELEASE     PERSON
            ASSEMBLE    6
            BLET        &CASES=&CASES+1
            TERMINATE
            GENERATE    400
            TERMINATE   1
            START       1
            PUTPIC      &CASES
  CASES FINISHED **
            PUTSTRING   ('  ')
            PUTSTRING   ('  DO AGAIN (Y/N)?')
            GETLIST     &ANS
            IF   (&ANS'E''Y')OR(&ANS'E''y')
            LET   &CASES=0
            GOTO AGAIN
            ENDIF
            END
```

FIGURE 27.1
Listing of program CHAP27A.GPS.

deviation of 4 minutes. 10% of these trucks need major service and are routed to this particular service bay. The rest need service according to the following:

Type of Service	Time
Petrol	17 ± 6 minutes
Driver checks	Exponential mean 15 minutes
Oil and grease	Normal distribution, mean of 18 minutes, and standard deviation of 3 minutes
Miscellaneous	17 ± 10 minutes

Major service takes a time given by the normal distribution with a mean of 180 minutes and a standard deviation of 25 minutes.

The mine manager is concerned that the repair facility is taking too long either for routine maintenance or for the major repairs. He needs data to support his request for an additional service bay or perhaps two. He would like to have a record of how long trucks take for each type of service. He wants this time to include time for service and time waiting for the service bay.

The mine works 24 hours a day and trucks continually arrive at the service bay. The simulation should be done for 100 days and the results given.

The program to do this simulation is CHAP27B.GPS. The listing of this program is given in Figure 27.2.

Some aspects of the coding will be discussed next.

```
          SIMULATE
**********************************
*  PROGRAM CHAP27B.GPS           *
*  EXAMPLE OF THE ASSEMBLE BLOCK *
**********************************
 TIMEIN   TABLE       MP1PL,15,1,30
 TIMEB    TABLE       MP1PL,70,2,30
 TIMEW    TABLE       MP2PL,20,1,30
 MYOUT    FILEDEF     'CHAP27B.OUT'
          REAL        &DAYS
          PUTSTRING   ('  ')
          PUTSTRING   ('  HOW MANY DAYS TO SIMULATE FOR?')
          PUTSTRING   ('  ')
          GETLIST     &DAYS
 TRUCKS   GENERATE    RVNORM(1,25,4),,,,,2PL   TRUCKS ARRIVE
          TRANSFER   .1,,AWAY     10% MAJOR REPAIRS
 TRUCKA   MARK        1PL          TIME TRUCK ARRIVES AT BAY
          QUEUE       WAITR
          GATE LR     OKIN         IS BAY BEING USED?
          DEPART      WAITR
          MARK        2PL
          LOGIC S     OKIN         NO, SHUT GATE UNTIL TRUCK FINISHED
          PRIORITY    10
          SPLIT       1,NEXT
          ADVANCE     17,6    FILL WITH PETROL
          TRANSFER    ,FINISH
 NEXT     SPLIT       1,NEXT2
          ADVANCE     RVEXPO(1,15)   CHECKS
          TRANSFER    ,FINISH
 NEXT2    SPLIT       1,NEXT3
          ADVANCE     RVNORM(1,18,3)  OIL AND GREASE
          TRANSFER    ,FINISH
 NEXT3    ADVANCE     17,10          SERVICE OTHER PARTS
 FINISH   ASSEMBLE    4           WAIT UNTIL ALL 4 JOBS DONE
          TABULATE    TIMEIN      TABLE OF TOTAL TIMES FOR ROUTINE JOBS
          TABULATE    TIMEW       TABLE OF WORK TIME FOR ROUTINE JOBS
          LOGIC R     OKIN        OPEN GATE FOR NEXT TRUCK
          TERMINATE
 AWAY     MARK        1PL         TIME OTHER TRUCKS ARRIVE FOR SERVICE
          QUEUE       WAIT
          SEIZE       BAY
          DEPART      WAIT
          ADVANCE     RVNORM(1,180,25)
          RELEASE     BAY
          TABULATE    TIMEB     RECORD OF TIMES AT BAY
          TERMINATE
          GENERATE    480*&DAYS*3
          TERMINATE   1
          START       1
          PUTPIC      LINES=10,FILE=MYOUT,&DAYS,N(TRUCKS)/&DAYS,_
                      N(TRUCKA)/&DAYS,_
                      N(AWAY)/&DAYS,TB(TIMEIN),TB(TIMEW),TB(TIMEB),_
                      FR(BAY)/10.,QA(WAITR),QA(WAIT)
             SIMULATION FOR   **** DAYS
   TRUCKS TO ENTER SERVICE AREA/DAY            ***.**
```

FIGURE 27.2
Listing of program CHAP27B.GPS. *(Continued)*

```
          ADVANCE     .2   .2,.1
          RELEASE     WORKER2
          BPUTPIC     FILE=ATF,LINES=4,AC1,FR(WORKER2)/10.
TIME *.****
SET W2 COLOR F6
MOVE W2 .1 22.05 -12.7
WRITE M2 **.**%
          TRANSFER    ,BACK
          GENERATE    24*1000
          TERMINATE   1
          START       1
          PUTPIC      LINES=3,FR(WORKER1)/10.,FR(WORKER2)/10.,QA(WAIT)
   UTIL. OF WORKER 1  **.**%
   UTIL. OF WORKER 2  **.**%
   AVG. TRUCKS WAITING FOR SERVICE  *.**
          PUTPIC     FILE=ATF,LINES=2,AC1
TIME *.****
END
          END
```

FIGURE 27.2 (Continued)
Listing of program CHAP27B.GPS.

It is desired to keep a record of the times for the different trucks at each bay. The trucks requiring longer service are sent to the second bay. Since only one truck can be serviced at this bay, SEIZE/RELEASE Blocks are used. This cannot be done for the trucks requiring the four different services. Instead, a GATE Block is used for these trucks. When a truck needs these four services, it encounters the following Block:

```
GATE LR          OKIN
```

If no truck is in the bay, the switch is in a reset position and a truck can pass through for service. The truck then leaves the GATE Block and immediately sets the switch OKIN to a set position

```
LOGIC S          OKIN
```

so that no other truck can be serviced until the switch is put in a reset position. Once all four services are complete, the truck leaves the ASSEMBLE 4 Block and sets the logic switch OKIN to a reset position.

```
LOGIC R          OKIN
```

The output from the program is given in Figure 27.3.

The animation of this is worth studying as it is not as straightforward as it might seem. The program to do this is given by CHAP27C.GPS. A listing is given in Figure 27.4.

The animation is an example of how one has to be creative to illustrate the correctness of the simulation program. A screenshot of the animation is shown in Figure 27.5.

```
                SIMULATION FOR     500 DAYS
TRUCKS TO ENTER SERVICE AREA/DAY                57.62
TRUCKS TO NEED ROUTINE SERVICE/DAY              51.97
TRUCKS NEEDING MAJOR SERVICE/DAY                 5.65
AVG. TIME IN SYSTEM FOR ROUTINE SERVICE         50.54
AVG. TIME IN BAY FOR ROUTINE SERVICE            24.88
AVG. TIME IN SYSTEM FOR MAJOR SERVICE          396.01
UTIL. OF MAJOR SERVICE REPAIR UNIT              70.54%
AVG. NO. OF TRUCKS WAITING FOR ROUTINE SERVICE 0.926
AVG. NO. OF TRUCKS WAITING FOR MAJOR SERVICE   0.848
```

FIGURE 27.3
Output from program CHAP27B.GPS.

```
          SIMULATE
**********************************
*  PROGRAM CHAP27C.GPS           *
*  ANIMATION OF CHAP27B.GPS      *
**********************************
 TIMEIN   TABLE       MP1PL,15,1,30
 TIMEB    TABLE       MP1PL,70,2,30
 TIMEW    TABLE       MP2PL,20,1,30
 ATF      FILEDEF     'CHAP27C.ATF'
          REAL        &DAYS
          PUTSTRING   ('  ')
          PUTSTRING   ('  HOW MANY DAYS TO SIMULATE FOR?')
          PUTSTRING   ('  ')
          GETLIST     &DAYS
          PUTPIC      FILE=ATF,LINES=2,AC1
TIME *.****
SET TT COLOR BACKDROP
 TRUCKS   GENERATE    RVNORM(1,25,4),,0,,,1PH,2PL   TRUCKS ARRIVE
          BPUTPIC     FILE=ATF,LINES=4,AC1,XID1,XID1,XID1
TIME *.****
CREATE TRUCK T*
PLACE T* ON P1
SET T* TRAVEL 5
          ADVANCE     5
          TRANSFER    .1,,AWAY     10% MAJOR REPAIRS
 TRUCKA   MARK        1PL        TIME TRUCK ARRIVES AT BAY
          QUEUE       WAITR
          GATE LR     OKIN       IS BAY BEING USED?
          DEPART      WAITR
          MARK        2PL
          LOGIC S     OKIN       NO, SHUT GATE UNTIL TRUCK FINISHED
          BPUTPIC     FILE=ATF,LINES=4,AC1,XID1,PL2
TIME *.****
SET TT COLOR RED
DESTROY T*
WRITE M5 ***.**
          SPLIT       1,NEXT
          BPUTPIC     FILE=ATF,LINES=2,AC1
TIME *.****
WRITE M1 UNDERWAY
          ADVANCE     17,6    FILL WITH PETROL
          BPUTPIC     FILE=ATF,LINES=2,AC1
TIME *.****
WRITE M1 FINISHED
          TRANSFER    ,FINISH
 NEXT     SPLIT       1,NEXT2
          BPUTPIC     FILE=ATF,LINES=2,AC1
TIME *.****
WRITE M2 UNDERWAY
          ADVANCE     RVEXPO(1,15)   CHECK FOR OIL LEAKS
          BPUTPIC     FILE=ATF,LINES=2,AC1
TIME *.****
WRITE M2 FINISHED
          TRANSFER    ,FINISH
 NEXT2    SPLIT       1,NEXT3
          BPUTPIC     FILE=ATF,LINES=2,AC1
```

FIGURE 27.4
Listing of program CHAP27C.GPS. *(Continued)*

```
TIME *.****
WRITE M3 UNDERWAY
          ADVANCE     RVNORM(1,18,3)  DRIVER DOES ROUTINE CHECKS
          BPUTPIC     FILE=ATF,LINES=2,AC1
TIME *.****
WRITE M3 FINISHED
          TRANSFER    ,FINISH
 NEXT3    BPUTPIC     FILE=ATF,LINES=2,AC1
TIME *.****
WRITE M4 UNDERWAY
          ADVANCE     17,10         SERVICE OTHER PARTS
          BPUTPIC     FILE=ATF,LINES=2,AC1
TIME *.****
WRITE M4 FINISHED
 FINISH   ASSEMBLE    4           WAIT UNTIL ALL 4 JOBS DONE
          BPUTPIC     FILE=ATF,LINES=2,AC1,XID1
TIME *.****
CREATE TRUCK T*
          TABULATE    TIMEIN      TABLE OF TOTAL TIMES FOR ROUTINE JOBS
          TABULATE    TIMEW       TABLE OF WORK TIME FOR ROUTINE JOBS
          BPUTPIC     FILE=ATF,LINES=12,AC1,AC1,MP2PL,TB(TIMEW),XID1,XID1
TIME *.****
WRITE M1
WRITE M2
WRITE M3
WRITE M4
WRITE M5
WRITE M6  **.**
WRITE M7 ***.**
WRITE M8  **.**
SET TT COLOR BACKDROP
PLACE T* ON P2
SET T* TRAVEL 4
          ADVANCE     .2
          LOGIC R     OKIN        OPEN GATE FOR NEXT TRUCK
          ADVANCE     3.8
          BPUTPIC     FILE=ATF,LINES=2,AC1,XID1
TIME *.****
DESTROY T*
          TERMINATE
 AWAY     MARK        1PL         TIME OTHER TRUCKS ARRIVE FOR SERVICE
          BPUTPIC     FILE=ATF,LINES=2,AC1,XID1
TIME *.****
PLACE T* ON P4
          QUEUE       WAIT
          SEIZE       BAY
          DEPART      WAIT
          MARK        2PL
          BPUTPIC     FILE=ATF,LINES=3,AC1,XID1,PL2
TIME *.****
PLACE T* AT 0 -13
WRITE M9 ****.**
          ADVANCE     RVNORM(1,180,25)
          RELEASE     BAY
          TABULATE    TIMEB      RECORD OF TIMES AT BAY
          BPUTPIC     FILE=ATF,LINES=6,AC1,XID1,XID1,AC1,MP2PL,TB(TIMEB)
TIME *.****
```

FIGURE 27.4 (Continued)
Listing of program CHAP27C.GPS.

(Continued)

```
PLACE T* ON P3
SET T* TRAVEL 4
WRITE M10 ****.**
WRITE M11  ***.**
WRITE M12  ***.**
          ADVANCE      4
          BPUTPIC      FILE=ATF,LINES=2,AC1,XID1
TIME *.****
DESTROY T*
          TERMINATE
          GENERATE     480*3*&DAYS
          TERMINATE    1
          START        1
          PUTPIC       FILE=ATF,LINES=2,AC1
TIME *.****
END
          PUTPIC       LINES=10,&DAYS,N(TRUCKS)/&DAYS,N(TRUCKA)/&DAYS,_
                       N(AWAY)/&DAYS,_
                       TB(TIMEIN),TB(TIMEW),TB(TIMEB),FR(BAY)/10.,_
                       QA(WAITR),QA(WAIT)
               SIMULATION FOR  *** DAYS
   TRUCKS TO ENTER SERVICE AREA                          ***.**
   TRUCKS TO NEED ROUTINE SERVICE                        ***.**
   TRUCKS NEEDING MAJOR SERVICE                          ***.**
   AVG. TIME IN SYSTEM FOR ROUTINE SERVICE                **.**
   AVG. TIME IN BAY FOR ROUTINE SERVICE                   **.**
   AVG. TIME IN SYSTEM FOR MAJOR SERVICE                 ***.**
   UTIL. OF MAJOR SERVICE REPAIR UNIT                     **.**%
   AVG. NO. OF TRUCKS WAITING FOR ROUTINE SERVICE *.***
   AVG. NO. OF TRUCKS WAITING FOR MAJOR SERVICE    *.***
          END
```

FIGURE 27.4 (Continued)
Listing of program CHAP27C.GPS.

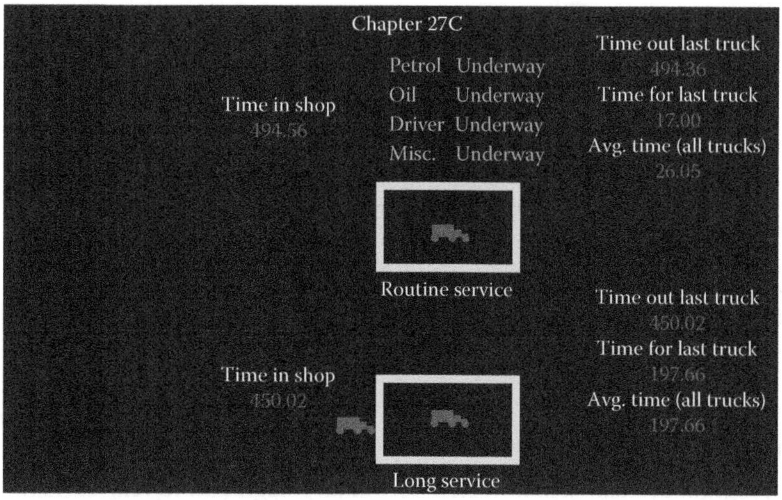

FIGURE 27.5
Screenshot of animation from program CHAP27C.GPS.

The single class is the TRUCK. There is a layout object called TT created from this class that is positioned in the centre of the service bay for trucks with the four services. This bay is labelled *routine service* in Figure 27.5. When the animation starts, the layout object is given the backdrop colour so it disappears from view. The trucks approach the service bays on a path P2. 10% are routed to the lower service bay. This is labelled *long service* in the layout. A short path is created in front of this where the trucks briefly travel before being placed in the centre of the bay. This was done so that, if a truck was in for long service and another one arrived at the end of path P1, it would be moved to the front of the service bay. The other trucks requiring normal service would continue to queue at the end of the path. Figure 27.5 shows the situation with one truck in each bay. A second truck is waiting at the lower bay for service and another truck is approaching the service area on path P1 (not shown).

When a truck can leave path P2 and enter the service bay for routine service, the layout object is given the colour RED, which is what the trucks are. The truck on path P2 is *destroyed* and removed from the animation.

```
BPUTPIC      FILE = ATF,LINES = 4,AC1,XID1,PL2
TIME *.****
SET TT COLOR RED
DESTROY T*
WRITE M5 ***.**
```

The truck transaction then enters the various SPLIT Blocks for the four services.

```
      BPUTPIC     FILE = ATF,LINES = 2,AC1,XID1
TIME *.****
CREATE TRUCK T*
      TABULATE    TIMEIN     TABLE OF TOTAL TIMES FOR ROUTINE JOBS
      TABULATE    TIMEW      TABLE OF WORK TIME FOR ROUTINE JOBS
      BPUTPIC     FILE = ATF,LINES = 12,AC1,AC1,MP2PL,TB(TIMEW),XID1,XID1
TIME *.****
WRITE M1
WRITE M2
WRITE M3
WRITE M4
WRITE M5
WRITE M6   **.**
WRITE M7  ***.**
WRITE M8   **.**
SET TT COLOR BACKDROP
PLACE T* ON P2
SET T* TRAVEL 4
      ADVANCE       .2
      LOGIC R       OKIN      OPEN GATE FOR NEXT TRUCK
      ADVANCE       3.8
```

A new truck transaction is created for the animation. This is placed on path P2 to leave the system. The layout object, TT, is set equal to the backdrop colour so it becomes invisible. The slight delay of .2 time units is given to allow the truck to leave the service bay before a waiting one replaces it.

Example 27.3: A PERT Diagram for a Mine

PERT (**P**rogram **E**valuation and **R**eview **T**echnique) diagrams are very useful in planning mining construction projects. Such diagrams indicate what jobs have to be done, the statistical distribution of the time to do each job, and, in many cases, the number of workers for each job. The diagram is set out to indicate the sequence of the jobs. For example, consider a new mine that is about to commence construction. The initial pit

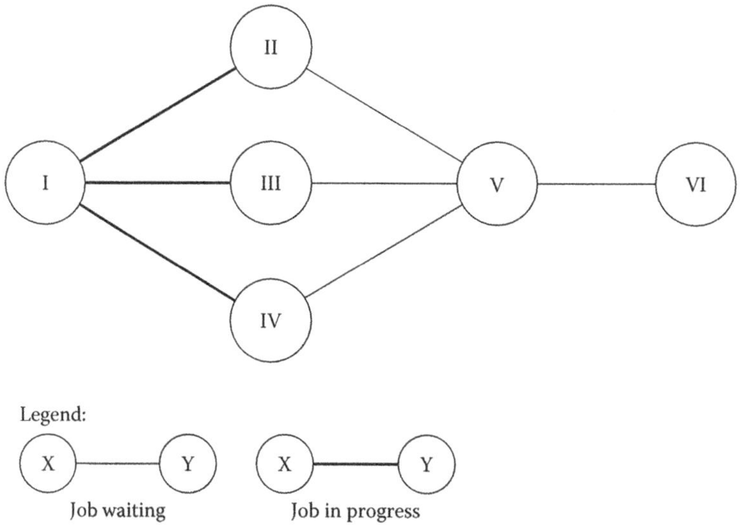

FIGURE 27.6
PERT diagram for a mine construction.

can be started while supplies for construction of the building are ordered. Other equipment is also ordered. Once the building supplies arrive, construction can begin. For an actual mine, the PERT diagram can be extremely complex. Figure 27.6 gives a simple PERT diagram for a mine construction project.

This project has only seven jobs. The assumption is made that there are always enough workers available so that each job can be started when it is scheduled. The mine starts construction at node 1. At the start, jobs from I to II, from I to III, and from I to IV are started simultaneously. Once the job from I to II is completed, job from II to V can commence. Similarly, the jobs from III to V and from IV to V cannot start until the jobs from I to III and from I to IV are done, respectively. The job at V cannot begin until all three of these initial jobs are completed. Once this final job is done, the project is complete. The various times for each job are given in Table 27.1.

If one used only averages, the expected time to do the project would be 90 time units. This is obtained from taking the maximum of the first three jobs (max(20, 25, 22)) and adding this (25) to the maximum of the next three jobs which is 10 (max(30, 35, 33)) and the mean of the last job of 30. The program to simulate this example is CHAP27D.GPS (Figure 27.7).

The output from the program is given next in Figure 27.8.

TABLE 27.1
Times to Do Different Jobs

Job	Time
I–II	20 ± 10
I–III	25 ± 15
I–IV	22 ± 12
II–V	30 ± 10
III–V	35 ± 15
IV–V	33 ± 10
V–VI	30 ± 10

```
 *********************************
*  PROGRAM CHAP27D.GPS            *
*  PERT DIAGRAM THERE ARE AN      *
*  INFINITE NUMBER OF WORKERS     *
*  FOR EACH JOB                   *
**********************************
         SIMULATE
         INTEGER     &I,&COUNT,&C,&AMT
         REAL        &SUMTIME,&MIN,&MAX,&T
         LET         &AMT=500        PROVIDE AMPLE NUMBER OF WORKERS
         LET         &MIN=100000     WANT TO DETERMINE MIN. TIME FOR JOB
         LET         &MAX=0          AND MAX. TIME FOR JOB
         DO          &I=1,25         CLEAR SCREEN
         PUTSTRING   ('  ')
         ENDDO
         PUTSTRING   ('  HOW MANY SIMULATED JOBS TO DO?')
         PUTSTRING   ('  ')
         GETLIST     &C
         PUTSTRING   ('  ')
         PUTSTRING   ('  SIMULATION IN PROGRESS....')
         PUTSTRING   ('  ')
         STORAGE     S(WORKER),&AMT      MANY WORKERS
 TIMES   TABLE       MP1PL,80,5,20
         GENERATE    ,,,1,,1PL           DO ONE PROJECT
 BACKUP  BLET        &COUNT=&COUNT+1
         BUFFER
         MARK        1PL
         SPLIT       1,AWAY1
         ENTER       WORKER
         ADVANCE     20,10   DO JOB 1 - 2
         LEAVE       WORKER
         ENTER       WORKER
         ADVANCE     30,10   DO JOB 2 - 5
         LEAVE       WORKER
         TRANSFER    ,AWAY2   WAIT FOR OTHER JOBS TO FINISH
 AWAY1   SPLIT       1,AWAY3
         ENTER       WORKER   DO 1 TO 3
         ADVANCE     25,15
         LEAVE       WORKER
         ENTER       WORKER
         ADVANCE     35,15     DO JOB 3 - 5
         LEAVE       WORKER
         TRANSFER    ,AWAY2   WAIT FOR OTHER JOBS TO FINISH
 AWAY3   ENTER       WORKER
         ADVANCE     22,12     DO JOB 1 - 4
         LEAVE       WORKER
         ENTER       WORKER
         ADVANCE     33,10     DO JOB 4 - 5
         LEAVE       WORKER
  AWAY2  ASSEMBLE    3       WAIT FOR THREE JOBS TO FINISH
         ENTER       WORKER
         ADVANCE     30,10  DO JOB 5 - 6
         LEAVE       WORKER
         BLET        &SUMTIME=&SUMTIME+MP1PL
         TABULATE    TIMES
```

FIGURE 27.7
Listing of program CHAP27D.GPS. *(Continued)*

```
          BLET        &T=MP1PL
          TEST G      &T,&MAX,AWAY10
          BLET        &MAX=&T
AWAY10    TEST L      &T,&MIN,AWAY11
          BLET        &MIN=&T
AWAY11    SPLIT       1,BACKUP
          TERMINATE
          GENERATE    ,,,1,1
          TEST G      &COUNT,&C
          TERMINATE   1
          START       1
          DO         &I=1,25
          PUTSTRING  ('   ')
          ENDDO
          PUTPIC     LINES=8,&C,&SUMTIME/&C,&MIN,&MAX,TD(TIMES)
      ####################################################
      ##             <<<< PERT DIAGRAM >>>>              ##
      ##   NUMBER OF JOBS DONE                ******     ##
      ##   AVERAGE TIME TO DO EACH JOB         ****.**   ##
      ##   MINIMUM TIME TO DO A JOB             ***.**   ##
      ##   MAXIMUM TIME TO DO A JOB             ***.**   ##
      ##   STD. DEV. OF ALL JOB TIMES           ***.**   ##
      ####################################################
          PUTSTRING    ('   ')
          PUTSTRING    ('   SIMULATION OVER....')
           END
```

FIGURE 27.7 (Continued)
Listing of program CHAP27D.GPS.

```
####################################################
##             <<<< PERT DIAGRAM >>>>              ##
##   NUMBER OF JOBS DONE               10000       ##
##   AVERAGE TIME TO DO EACH JOB          94.86    ##
##   MINIMUM TIME TO DO A JOB             62.20    ##
##   MAXIMUM TIME TO DO A JOB            128.29    ##
##   STD. DEV. OF ALL JOB TIMES           10.35    ##
####################################################
```

FIGURE 27.8
Output from program CHAP27D.GPS.

The expected average time for a project was 90 time units. The simulated average was not far from this as it is 94.86. The maximum time do a project was 128.29 time units and the minimum was 62.20.

The program will shortly be modified and run again but this time with a limited number of total workers and each job will have a required number of workers. First, the animation will be given. The animation program is CHAP27E.GPS. Its listing is given in Figure 27.9. The animation is shown in Figure 27.10.

The animation gives the average time for each job as it is completed. It also gives the minimum and maximum times for any job. As each job is being done, the path turns from dull grey to white as do the corresponding nodes.

```
********************************
*  PROGRAM CHAP27E.GPS          *
*  ANIMATION OF CHAP27D.GPS     *
********************************
          SIMULATE
  ATF     FILEDEF     'CHAP27E.ATF'
          INTEGER      &I,&COUNT,&C,&AMT,&X
          LET          &X=1
          REAL         &SUMTIME,&MIN,&MAX,&T
          LET          &AMT=500        PROVIDE AMPLE NUMBER OF WORKERS
          LET          &MIN=100000     WANT TO DETERMINE MIN. TIME FOR JOB
          LET          &MAX=0          AND MAX. TIME FOR JOB
          DO           &I=1,25         CLEAN SCREEN
          PUTSTRING    ('  ')
          ENDDO
          PUTSTRING    ('  HOW MANY SIMULATED JOBS TO DO?')
          PUTSTRING    ('  ')
          GETLIST      &C
          PUTSTRING    ('  ')
          PUTSTRING    ('  SIMULATION IN PROGRESS....')
          PUTSTRING    ('  ')
          STORAGE      S(WORKER),&AMT      MANY WORKERS
 TIMES    TABLE        MP1PL,80,5,20
          GENERATE     ,,,1,,1PL           DO ONE PROJECT
 BACKUP   BLET         &COUNT=&COUNT+1
          BPUTPIC      FILE=ATF,LINES=2,AC1,&COUNT
TIME *.****
WRITE M4 ***
          BUFFER
          MARK         1PL
          SPLIT        1,AWAY1
          ENTER        WORKER
          BPUTPIC      FILE=ATF,LINES=6,AC1
TIME *.****
SET I12 COLOR F4
SET b1 COLOR F3
SET nn1 COLOR F3
SET b2 COLOR F3
SET nn2 COLOR F3
          ADVANCE      20,10   DO JOB 1 - 2
          BPUTPIC      FILE=ATF,LINES=6,AC1
TIME *.****
SET I12 COLOR F26
SET b1 COLOR F26
SET nn1 COLOR F26
SET b2 COLOR F26
SET nn2 COLOR F26
          LEAVE        WORKER
          ENTER        WORKER
          BPUTPIC      FILE=ATF,LINES=6,AC1
TIME *.****
SET 125 COLOR F4
SET b2 COLOR F3
SET nn2 COLOR F3
SET b5 COLOR F3
```

FIGURE 27.9
Listing of program CHAP27E.GPS. *(Continued)*

```
SET nn5 COLOR F3

          ADVANCE     30,10   DO JOB 2 - 5
          BPUTPIC     FILE=ATF,LINES=6,AC1
TIME *.****
SET 125 COLOR F26
SET b2 COLOR F26
SET nn2 COLOR F26
SET b5 COLOR F26
SET nn5 COLOR F26
          LEAVE       WORKER
          TRANSFER    ,AWAY2   WAIT FOR OTHER JOBS TO FINISH
 AWAY1    SPLIT       1,AWAY3
          ENTER       WORKER   DO 1 TO 3
          BPUTPIC     FILE=ATF,LINES=6,AC1
TIME *.****
SET I13 COLOR F4
SET b1 COLOR F3
SET nn1 COLOR F3
SET b3 COLOR F3
SET nn3 COLOR F3

          ADVANCE     25,15
          BPUTPIC     FILE=ATF,LINES=6,AC1
TIME *.****
SET I13 COLOR F26
SET b1 COLOR F26
SET nn1 COLOR F26
SET b3 COLOR F26
SET nn3 COLOR F26
          LEAVE       WORKER
          ENTER       WORKER
          BPUTPIC     FILE=ATF,LINES=6,AC1
TIME *.****
SET I35 COLOR F4
SET b3 COLOR F3
SET nn3 COLOR F3
SET b5 COLOR F3
SET nn5 COLOR F3

          ADVANCE     35,15     DO JOB 3 - 5
          BPUTPIC     FILE=ATF,LINES=6,AC1
TIME *.****
SET I35 COLOR F26
SET b3 COLOR F26
SET nn3 COLOR F26
SET b5 COLOR F26
SET nn5 COLOR F26
          LEAVE       WORKER
          TRANSFER    ,AWAY2   WAIT FOR OTHER JOBS TO FINISH
 AWAY3    ENTER       WORKER
          BPUTPIC     FILE=ATF,LINES=6,AC1
TIME *.****
SET I14 COLOR F4
SET b1 COLOR F3
SET nn4 COLOR F3
SET b4 COLOR F3
```

FIGURE 27.9 (Continued)
Listing of program CHAP27E.GPS.

(Continued)

```
SET nn4 COLOR F3

          ADVANCE     22,12     DO JOB 1 - 4
          BPUTPIC     FILE=ATF,LINES=6,AC1
TIME *.****
SET I14 COLOR F26
SET b1 COLOR F26
SET nn4 COLOR F26
SET b4 COLOR F26
SET nn4 COLOR F26
          LEAVE       WORKER
          ENTER       WORKER
          BPUTPIC     FILE=ATF,LINES=6,AC1
TIME *.****
SET I45 COLOR F4
SET b4 COLOR F3
SET nn4 COLOR F3
SET b5 COLOR F3
SET nn5 COLOR F3
          ADVANCE     33,10     DO JOB 4 - 5
          BPUTPIC     FILE=ATF,LINES=6,AC1
TIME *.****
SET I45 COLOR F26
SET b4 COLOR F26
SET nn4 COLOR F26
SET b5 COLOR F26
SET nn5 COLOR F26
          LEAVE       WORKER
  AWAY2   ASSEMBLE    3      WAIT FOR THREE JOBS TO FINISH
          ENTER       WORKER
          BPUTPIC     FILE=ATF,LINES=6,AC1
TIME *.****
SET I56 COLOR F4
SET b5 COLOR F3
SET nn5 COLOR F3
SET b6 COLOR F3
SET nn6 COLOR F3
          ADVANCE     30,10  DO JOB 5 - 6
          BPUTPIC     FILE=ATF,LINES=6,AC1
TIME *.****
SET I56 COLOR F26
SET b5 COLOR F26
SET nn5 COLOR F26
SET b6 COLOR F26
SET nn6 COLOR F26
          LEAVE       WORKER
          BLET        &SUMTIME=&SUMTIME+MP1PL
          TABULATE    TIMES
          BPUTPIC     FILE=ATF,LINES=4,AC1,TB(TIMES),MP1PL,&COUNT
TIME *.****
WRITE M2 ***.**
WRITE M5 ***.**
WRITE M1 ***
          BLET        &T=MP1PL
          TEST G      &T,&MAX,AWAY10
          BLET        &MAX=&T
  AWAY10  TEST L      &T,&MIN,AWAY11
```

FIGURE 27.9 (Continued)
Listing of program CHAP27E.GPS.

(Continued)

```
             BLET          &MIN=&T
    AWAY11   SPLIT         1,BACKUP
             BPUTPIC       FILE=ATF,LINES=3,AC1,&MIN,&MAX
  TIME *.****
  WRITE M10 ***.**
  WRITE M11 ***.**
             TERMINATE
             GENERATE      ,,,1     DUMMY TRANSACTION
  WAIT       TEST NE       &COUNT,&X
             BPUTPIC       FILE=ATF,LINES=2,AC1
  TIME *.****
  WRITE M3  "just finished"
             BLET          &X=&COUNT
             ADVANCE       5
             BPUTPIC       FILE=ATF,LINES=2,AC1
  TIME *.****
  WRITE M3
             TRANSFER       ,WAIT
             GENERATE      ,,,1,1
             TEST G        &COUNT,&C
             TERMINATE     1
             START         1
             DO            &I=1,25
             PUTSTRING     ('  ')
             ENDDO
             PUTPIC        LINES=8,&C,&SUMTIME/&C,&MIN,&MAX,TD(TIMES)
          #####################################################
          ##             <<<< PERT DIAGRAM >>>>               ##
          ##   NUMBER OF JOBS DONE              ******        ##
          ##   AVERAGE TIME TO DO EACH JOB        ****.**     ##
          ##   MINIMUM TIME TO DO A JOB            ***.**     ##
          ##   MAXIMUM TIME TO DO A JOB            ***.**     ##
          ##   STD. DEV. OF ALL JOB TIMES          ***.**     ##
          #####################################################
             PUTSTRING     ('  ')
             PUTSTRING     ('  SIMULATION OVER....')
             PUTPIC        FILE=ATF,LINES=2,AC1
  TIME *.****
  END
             END
```

FIGURE 27.9 (Continued)
Listing of program CHAP27E.GPS.

Example 27.4: Modification to Example 27.3

The assumption made in solving Example 27.3 is that there is an infinite supply of workers available to do each job as it has to be done. This is rarely the case. For example, suppose that there are only 20 workers available for the mine. Suppose the workers needed for each job are as given in Table 27.2.

Thus, when a project starts, all three jobs starting at node 1 cannot be done simultaneously since a total of 21 workers are needed (8 + 7 + 6) but only 20 are available. Therefore, one of the three jobs cannot be done until enough workers are available.

The changes to the program are easily made. Only a few lines need be changed. The program is CHAP27F.GPS. Its listing is given in Figure 27.11.

The changes that were made are discussed next.

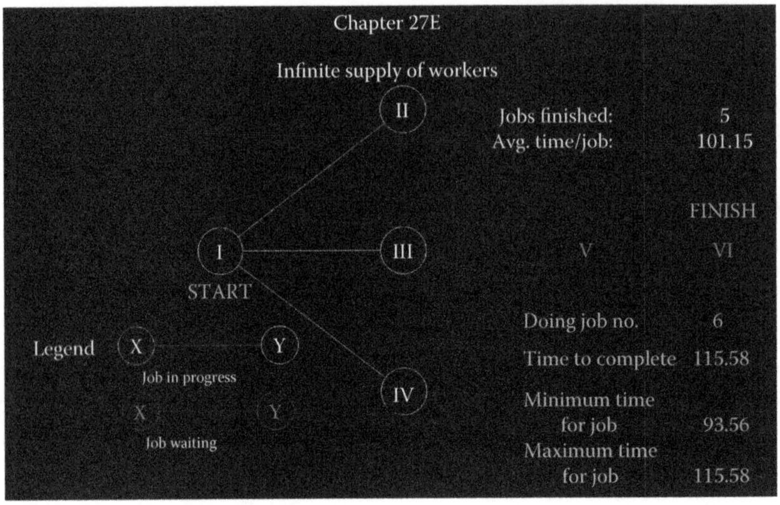

FIGURE 27.10
Screenshot of animation from program CHAP27E.GPS.

TABLE 27.2
Number of Workers Needed for Each Job

Job	Workers Needed
1–2	8
1–3	7
1–4	6
2–5	10
3–5	6
4–5	9
5–6	15

The line of code that specifies the number of workers as being large needs to be removed or placed before the lines that ask the user to input the number of workers. The old line is

```
LET        &AMT = 500      PROVIDE AMPLE NUMBER OF WORKERS
```

The new code is

```
PUTSTRING     (' HOW MANY WORKERS ARE IN THE MINE?')
PUTSTRING     (' ')
GETLIST       &AMT
PUTSTRING     (' ')
```

```
*******************************
*  PROGRAM CHAP27F.GPS        *
*  PERT DIAGRAM               *
*  THERE ARE A FINITE NUMBER  *
*  OF WORKERS                 *
*  FOR EACH JOB.              *
*******************************
          SIMULATE
          INTEGER     &I,&COUNT,&C,&AMT
          REAL        &SUMTIME,&MIN,&MAX,&T
          LET         &MIN=100000    WANT TO DETERMINE MIN. TIME FOR JOB
          LET         &MAX=0         AND MAX. TIME FOR JOB
          DO          &I=1,25        CLEAR SCREEN
          PUTSTRING   ('  ')
          ENDDO
          PUTSTRING   ('  HOW MANY SIMULATED JOBS TO DO?')
          PUTSTRING   ('  ')
          GETLIST     &C
          PUTSTRING   ('   ')
          PUTSTRING   ('  HOW MANY WORKERS ARE IN THE MINE?')
          PUTSTRING   ('  ')
          GETLIST     &AMT
          PUTSTRING   ('  ')
          PUTSTRING   ('  SIMULATION IN PROGRESS....')
          PUTSTRING   ('  ')
          STORAGE     S(WORKER),&AMT      MANY WORKERS
 TIMES    TABLE       MP1PL,80,5,20
          GENERATE    ,,,1,,1PL           DO ONE PROJECT
 BACKUP   BLET        &COUNT=&COUNT+1
          BUFFER
          MARK        1PL
          SPLIT       1,AWAY1
          ENTER       WORKER,8
          ADVANCE     20,10   DO JOB 1 - 2
          LEAVE       WORKER,8
          ENTER       WORKER,10
          ADVANCE     30,10   DO JOB 2 - 5
          LEAVE       WORKER,10
          TRANSFER    ,AWAY2   WAIT FOR OTHER JOBS TO FINISH
 AWAY1    SPLIT       1,AWAY3
          ENTER       WORKER,7   DO 1 TO 3
          ADVANCE     25,15
          LEAVE       WORKER,7
          ENTER       WORKER,6
          ADVANCE     35,15     DO JOB 3 - 5
          LEAVE       WORKER,6
          TRANSFER    ,AWAY2   WAIT FOR OTHER JOBS TO FINISH
 AWAY3    ENTER       WORKER,6
          ADVANCE     22,12    DO JOB 1 - 4
          LEAVE       WORKER,6
          ENTER       WORKER,9
          ADVANCE     33,10    DO JOB 4 - 5
          LEAVE       WORKER,9
 AWAY2    ASSEMBLE    3      WAIT FOR THREE JOBS TO FINISH
          ENTER       WORKER,15
```

FIGURE 27.11
Listing of program CHAP27F.GPS.

(Continued)

```
          ADVANCE     30,10  DO JOB 5 - 6
          LEAVE       WORKER,15
          BLET        &SUMTIME=&SUMTIME+MP1PL
          TABULATE    TIMES
          BLET        &T=MP1PL
          TEST G      &T,&MAX,AWAY10
          BLET        &MAX=&T
  AWAY10  TEST L      &T,&MIN,AWAY11
          BLET        &MIN=&T
  AWAY11  SPLIT       1,BACKUP
          TERMINATE
          GENERATE    ,,,1,1
          TEST G      &COUNT,&C
          TERMINATE   1
          START       1
          DO         &I=1,25
          PUTSTRING  ('  ')
          ENDDO
          PUTPIC     LINES=10,&C,&SUMTIME/&C,&MIN,&MAX,TD(TIMES),_
                     &AMT,SR(WORKER)/10.
      #####################################################
      ##            <<<< PERT DIAGRAM >>>>                ##
      ##  NUMBER OF JOBS DONE                  ******     ##
      ##  AVERAGE TIME TO DO EACH JOB           ****.**   ##
      ##  MINIMUM TIME TO DO A JOB               ***.**   ##
      ##  MAXIMUM TIME TO DO A JOB               ***.**   ##
      ##  STD. DEV. OF ALL JOB TIMES             ***.**   ##
      ##  NUMBER OF WORKERS IN THE MINE           **      ##
      ##  UTIL. OF THE WORKERS                    **.**%  ##
      #####################################################
          PUTSTRING    ('  ')
          PUTSTRING    ('  SIMULATION OVER....')
          END
```

FIGURE 27.11 (Continued)
Listing of program CHAP27F.GPS.

This just inputs the number of workers available in the mine. Each ENTER Block needs to now have the number of workers needed. For example, the modified ENTER Block for the job from node 1 to node 2 requires eight workers be available. Thus, this becomes

```
ENTER      WORKER,8
ADVANCE    20,10 DO JOB 1 - 2
LEAVE      WORKER,8
```

Similar changes are needed for all the other ENTER/LEAVE Blocks. The program can further be modified to give the utilization of the workers. The output from the program is shown in Figure 27.12.

With only 20 workers in the mine, the average time to do the project increases from 94.86 to 126.45. The standard deviation of the time to do each project also increases from 10.35 to 15.54.

```
###################################################
##            <<<< PERT DIAGRAM >>>>             ##
##   NUMBER OF JOBS DONE              10000      ##
##   AVERAGE TIME TO DO EACH JOB       126.45    ##
##   MINIMUM TIME TO DO A JOB           77.74    ##
##   MAXIMUM TIME TO DO A JOB          174.47    ##
##   STD. DEV. OF ALL JOB TIMES         15.54    ##
##   NUMBER OF WORKERS IN THE MINE      20       ##
##   UTIL. OF THE WORKERS               68.22%   ##
###################################################
```

FIGURE 27.12
Output from running program CHAP27F.GPS.

27.3 Exercises

27.1. Add the code to the program CHAP27A.GPS for Example 27.1 so that each time the program is re-run, different random numbers are used. This is done using the `RMULT` statement with a variable. The user is asked to input a number for each run.

27.2. Refer to Example 27.2. The mining engineer wants to have data to decide if a second repair bay is needed for trucks requiring a longer service. Modify program CHAP27B.GPS to provide this data.

27.3. Suppose that the times for each job in Exercise 27.2 were given by the normal distribution with the means as given with the standard deviation of 15% of the mean. How does this change the expected average times for the project?

27.4. In Exercise 27.2, assume that the various times are now given by the exponential distribution with mean as before. Determine the expected time in the system for each type of service.

27.5. In Example 27.4, the assumption was that 20 workers were available. Suppose 25 were available. How does the result change? What if there were 30 workers available?

28

MATCH, GATHER, and PREEMPT Block

Hydraulic loader and truck.

28.1 MATCH, GATHER, and PREEMPT Blocks

28.1.1 The MATCH Block

In Chapter 27, the ASSEMBLE Block was introduced. Its use was to wait until a specified number of transactions belonging to the same assembly set entered it. Nothing was done until the number specified by its operand had entered. Then a number one less than this were destroyed and one transaction moved to the next sequential Block. There are two other Blocks closely associated with the ASSEMBLE Block that will be introduced here. The first is the MATCH Block. This Block has another MATCH Block, known as *the conjugate* Block in another segment of the program. This conjugate Block is also a MATCH Block. This Block has as its operand the label of the original Block. The original MATCH Block has as its label the operand of the other MATCH Block. One can think of the two MATCH Blocks as pointing at each other. For example, one might have

```
        (segment 1)
BLOCKA  MATCH          BLOCKB
        ......... .
        ......... .
        (segment 2)
BLOCKB  MATCH          BLOCKA
```

The way the MATCH Block works is as follows. When a member of an assembly set arrives at the conjugate Block it checks to see if another transaction from the same assembly set is at its conjugate Block. If so, both transactions move to their next sequential Blocks. If not, the arriving transaction waits until a transaction of the same assembly set arrives at the conjugate Block.

Typical code might be as follows:

```
        SPLIT   1,AWAY
        ......... .
        ......... .
        ......... .
BLOCKA  MATCH   BLOCKB
        ......... .
        ......... .
AWAY    ......... .
        ......... .
BLOCKB  MATCH   BLOCKB
        ...........
```

The next example gives an illustration of how one might use a MATCH Block for a mining simulation.

Example 28.1: Trucks Come for Major Repairs

Trucks in a mine arrive at a single repair bay where they need to have a part overhauled. They arrive in a Poisson stream every 5.25 hours. One worker takes the damaged part out, whereas a second prepares the replacement part. This preparation takes place simultaneously with the removal. Removal of the damaged part takes a time given by the normal distribution with a mean of 3.75 hours and a standard deviation. of .45 hours. Preparation of the new part takes a time given by the normal distribution with a mean of 4 hours and a standard deviation of .4 hours. Once both of the jobs are done, the first worker readies the truck for the new part whereas the other worker does minor adjustments to the part. The first worker takes .25 ± .15 hours whereas the second worker takes .2 ± .1 hours to do the adjustments. Once both of these jobs are done, the truck can leave the service bay. When a truck arrives at the service bay, work cannot begin until both workers are free.

The program to simulate this is given by CHAP28A.GPS. Its listing is in Figure 28.1. Once a truck arrives, it joins a queue. It then checks to see if both workers are free. If so, it enters a SPLIT Block and a clone is created. This clone represents the second worker. The cloned transaction is routed to the Block with the label AWAY. Once either worker is finished, he needs to wait until the other worker is finished. The MATCH Block does this. Once the match is made, both workers continue to finish their work. Finally, an ASSEMBLE Block is used to allow the truck to leave the system.

The animation of CHAP28A.GPS is given by the program CHAP28B.GPS. Since the basic time unit is an hour, the animation speed was slowed to .3. This was to allow the viewer to see the two workers moving to work on the different trucks. The listing of the program is given in Figure 28.2.

```
          SIMULATE
*****************************
*  CHAP28A.GPS   EXAMPLE OF  *
*  THE MATCH BLOCK           *
*****************************
 GOIN     BVARIABLE    (F(WORKER1)'E'0)AND(F(WORKER2)'E'0)
          GENERATE     RVEXPO(1,5.25)    TRUCKS ARRIVE FOR SERVICE
          QUEUE        WAIT
          TEST E       BV(GOIN),1
          DEPART       WAIT
          SPLIT        1,AWAY        OTHER WORKER BEGINS WORK
          SEIZE        WORKER1       FIRST WORKER
          ADVANCE      RVNORM(1,3.75,.45)    TAKES ENGINE PART OUT
  WORK2   MATCH        WORK1         WAIT UNTIL OTHER MECHANIC DONE
          ADVANCE      .25,.15       FIRST WORKER DOES A BIT MORE
          RELEASE      WORKER1       FREE FIRST WORKER
  BACK    ASSEMBLE     2
          TERMINATE
  AWAY    SEIZE        WORKER2
          ADVANCE      RVNORM(1,4.,.4)
  WORK1   MATCH        WORK2
          ADVANCE      .2,.1
          RELEASE      WORKER2
          TRANSFER     ,BACK
          GENERATE     24*1000
          TERMINATE    1
          START        1
          PUTPIC       LINES=3,FR(WORKER1)/10.,FR(WORKER2)/10.,QA(WAIT)
   UTIL. OF WORKER 1   **.**%
   UTIL. OF WORKER 2   **.**%
   AVG. TRUCKS WAITING FOR SERVICE   *.**
          END
```

FIGURE 28.1
Listing of program CHAP28A.GPS.

```
          SIMULATE
*****************************
*  CHAP28B.GPS  ANIMATION    *
*  OF CHAP28A.GPS            *
*****************************
 GOIN     BVARIABLE    (F(WORKER1)'E'0)AND(F(WORKER2)'E'0)
 ATF      FILEDEF      'CHAP28B.ATF'
TRUCK     GENERATE     RVEXPO(1,5.25),,0,,,10PH,10PL    TRUCKS ARRIVE
          ASSIGN       1,N(TRUCK),PH
          BPUTPIC      FILE=ATF,LINES=4,AC1,PH1,PH1,PH1
 TIME *.****
 CREATE TRUCK T*
 PLACE T* ON P1
 SET T* TRAVEL .25
          ADVANCE      .25
          QUEUE        WAIT
          TEST E       BV(GOIN),1
          DEPART       WAIT
          SPLIT        1,AWAY        OTHER WORKER BEGINS WORK
          SEIZE        WORKER1       FIRST WORKER
          BPUTPIC      FILE=ATF,LINES=5,AC1,PH1
TIME *.****
SET W1 COLOR F2
```

FIGURE 28.2
Listing of program CHAP28B.GPS.

(Continued)

```
PLACE W1 ON P3
SET W1 TRAVEL .1
PLACE T* AT 14.25 -2.1
          ADVANCE     .1
          ADVANCE     RVNORM(1,3.75,.45)     TAKES ENGINE PART OUT
  WORK2   MATCH       WORK1          WAIT UNTIL OTHER MECHANIC DONE
          ADVANCE     .25  .25,.15        FIRST WORKER DOES A BIT MORE
          RELEASE     WORKER1        FREE FIRST WORKER
          BPUTPIC     FILE=ATF,LINES=5,AC1,FR(WORKER1)/10.
TIME *.****
SET W1 COLOR F6
PLACE W1 ON P5
SET W1 TRAVEL .1
WRITE M1 **.**%
          ADVANCE     .1
   BACK   ASSEMBLE    2
          BPUTPIC     FILE=ATF,LINES=3,AC1,PH1,PH1
TIME *.****
PLACE T* ON P2
SET T* TRAVEL .2
          ADVANCE     .2
          BPUTPIC     FILE=ATF,LINES=2,AC1,PH1
TIME *.****
DESTROY T*
          TERMINATE
  AWAY    SEIZE       WORKER2
          BPUTPIC     FILE=ATF,LINES=3,AC1
TIME *.****
SET W2 COLOR F2
MOVE W2 .1 14.25 -8.56
          ADVANCE     RVNORM(1,4.,.4)
  WORK1   MATCH       WORK2
          ADVANCE     .2   .2,.1
          RELEASE     WORKER2
          BPUTPIC     FILE=ATF,LINES=4,AC1,FR(WORKER2)/10.
TIME *.****
SET W2 COLOR F6
MOVE W2 .1 22.05 -12.7
WRITE M2 **.**%
          TRANSFER    ,BACK
          GENERATE    24*1000
          TERMINATE   1
          START       1
          PUTPIC      LINES=3,FR(WORKER1)/10.,FR(WORKER2)/10.,QA(WAIT)
   UTIL. OF WORKER 1  **.**%
   UTIL. OF WORKER 2  **.**%
   AVG. TRUCKS WAITING FOR SERVICE  *.**
          PUTPIC     FILE=ATF,LINES=2,AC1
TIME *.****
END
          END
```

FIGURE 28.2 (Continued)
Listing of program CHAP28B.GPS.

Notice that the truck transaction's XID1 was not used for the animation but, rather, the number of each truck was placed in the first half-word parameter. This was necessary due to the `SPLIT` Block.

A screenshot of the animation is given in Figure 28.3.

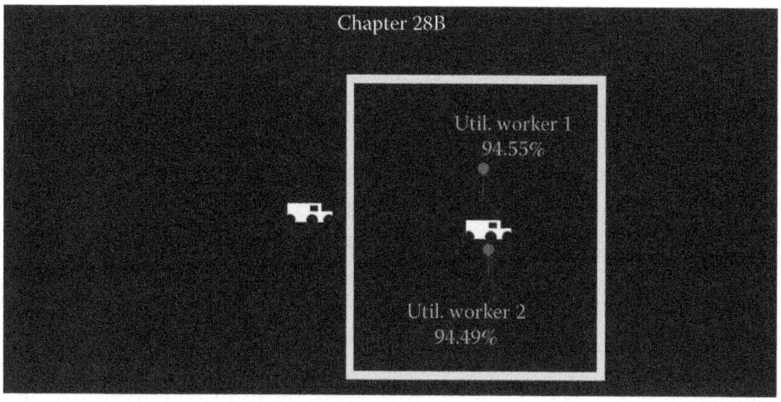

FIGURE 28.3
Animation for Example 28.1. This is program CHAP28B.GPS.

28.1.2 The GATHER Block

The GATHER Block acts much like the ASSEMBLE Block encountered in the previous chapter. It has a single operand so the general form is as follows:

```
GATHER    n
```

where *n* is a positive integer. The purpose of the Block is to hold up transactions of the same family, that is, those created by a SPLIT Block until *n* have been collected. It then allows all *n* transactions to proceed to the next Block. Unlike the ASSEMBLE Block, this Block has little or no application in mining. Instead, when one wants to delay a process until a specified number of units is reached, one uses the TEST Block.

28.1.3 PREEMPT Block

When a transaction has a facility seized, arriving transactions have to wait until the facility is released. This is normally what is desired. When the facility is released, waiting transactions will use it according to which had arrived first. The exception to this is when a waiting transaction has a higher priority than one that might have arrived ahead of it. It is possible for a transaction to actually interrupt a transaction that has seized a facility and take its place. This is done by the use of a PREEMPT Block. This Block has the general form:

```
PREEMPT       (facility name),PR
```

The first operand is the name of the facility to be pre-empted. The second must be PR and will be explained in the next section. It can be and often is omitted. When a transaction arrives at a facility that is currently being used by a transaction the PREEMPT Block will actually remove the transaction that has seized the facility and let the pre-empting transaction take its place. The pre-empted transaction will have time to finish with the SEIZE Block. So it is set aside on what is call the *interrupt chain*. The amount of time needed to finish is tagged with the transaction and, when it again can seize the facility, it continues to use the facility until this time has expired. The classic example of this is for a repair shop that repairs vehicles as they arrive. However, when certain special vehicles come for repairs, it is necessary to stop work on the current vehicle and immediately commence work on the arriving vehicle. For example, suppose a mining truck is being repaired. The time to do the repair was determined to be 15.80 minutes. Suppose that it has seized the repair facility and it has been on the FEC for 10.5 minutes when the mine manager's car comes for immediate repairs. The vehicle being repaired will be set aside just as in real life, and the mine manager's car worked on. When this car is repaired, the original truck will again be worked on for the remaining 5.3 minutes. The utilization of the facility will reflect the time that the original transaction actually seized the facility plus the time it was on the interrupt chain.

When a transaction is finished with the pre-emption, the Block to return the facility to its original transaction is

```
RETURN     (facility name)
```

If the second operand in the PREEMPT Block is omitted, pre-emption will always take place no matter what the priority level of the transactions.

The following example will illustrate how the PREEMPT Block works.

Example 28.2: Study of a Repair Shop for a Mine

A small mine has a repair shop for simple repairs. Two types of trucks arrive. They both arrive according to the uniform distribution, 12 ± 4 minutes .35% are type 1 that are repaired in a time given by the normal distribution with a mean of 12.5 and a standard deviation of 1. Type 2 trucks are repaired according to the exponential distribution with a mean of 10.8. Type 2 trucks are more important that type 1 trucks so, if there is queue, the type 2 trucks are placed at the head of the queue. When a truck comes for service, it takes 4 minutes to travel to the repair shop and 4 minutes to return regardless of the truck type. The mine manager first wants a study of the repair system to see how busy it is. He plans to purchase new equipment that will speed up the repairs. Thus, type 1 trucks will be repaired according to the normal distribution but now with a mean of 8.2 and a standard deviation of 1; type 2 trucks will be repaired according to the exponential distribution but now with a mean of 6.6. Because of the decrease in repair times, the manager now wants other vehicles to also be repaired at the same shop. These other vehicles arrive according to the uniform distribution, 24 ± 8, and take a repair time given by the normal distribution with a mean of 6.5 and a standard deviation of 1. These other vehicles are critical to the operation of the mine so that when they arrive at the shop, all work stops on current vehicles at the shop and the arriving vehicle is worked on.

Solution

The initial part of this example will be simulated first. The program to do the simulation is CHAP28C.GPS. Its listing is given in Figure 28.4. The results of the simulation are shown in Figure 28.5.

As can be seen, the repair shop is operating 95% of the time. The animation is given by program CHAP28D.GPS. Its listing is given in Figure 28.6. A screenshot of the animation produced by this program is shown in Figure 28.7.

When type 3 trucks are added to the system the PREEMPT Block is added. The program to provide the animation is CHAP28E.GPS. The listing of the program is given in Figure 28.8 and the results are provided in Figure 28.9.

```
          SIMULATE
******************************
*   PROGRAM 28C.GPS            *
*   PROGRAM TO SIMULATE FIRST *
8   PART OF EXAMPLE 28.2       *
******************************
 TYPE      FUNCTON      RN1,D2
.35,1/1,2
          GENERATE      12,4   TRUCKS ARRIVE
          ASSIGN        1,FN(TYPE),PH
          ADVANCE       4   TRAVEL TO GARAGE
          TEST E        PH1,2,NEXT
          PRIORITY      1
 NEXT     QUEUE         WAIT
          SEIZE         MECH
          DEPART        WAIT
          TEST E        PH1,1,NEXT1
          ADVANCE       RVNORM(1,12.5,1)
          RELEASE       MECH
 FIRST    TERMINATE
 NEXT1    ADVANCE       RVEXPO(1,10.8)
          RELEASE       MECH
          ADVANCE       4      RETURN TO THE MINE
SECOND    TERMINATE
          GENERATE      480*100
          TERMINATE     1
          START         1
          PUTPIC        LINES=4,FR(MECH)/10.,N(FIRST)/100.,N(SECOND)/100.
     RESULTS OF SIMULATON
     UTIL OF REPAIR SHOP  **.**%
     TYPE 1 TRUCKS REPAIRED EACH SHIFT  **.**
     TYPE 2 TRUCKS REPAIRED EACH SHIFT  **.**
          END
```

FIGURE 28.4
Listing of program CHAP28C.GPS.

```
RESULTS OF SIMULATON
UTIL OF REPAIR SHOP  95.01%
TYPE 1 TRUCKS REPAIRED EACH SHIFT  13.79
TYPE 2 TRUCKS REPAIRED EACH SHIFT  26.19
```

FIGURE 28.5
Results from program CHAP28C.GPS.

```
          SIMULATE
********************************
*  PROGRAM CHAP28D.GPS         *
*  ANIMATION OF CHAP28C.GPS    *
********************************
 ATF      FILEDEF       'CHAP28D.ATF'
 TYPE     FUNCTON       RN1,D2
.35,1/1,2
          GENERATE      12,4   TRUCKS ARRIVE
          ASSIGN        1,FN(TYPE),PH
          TEST E        PH1,2,NEXT
          PRIORITY      1
 NEXT     TEST E        PH1,1,TYPE2
          BPUTPIC       FILE=ATF,LINES=4,AC1,XID1,XID1,XID1
```

FIGURE 28.6
Listing of program CHAP28D.GPS.

(*Continued*)

```
TIME *.****
CREATE TRUCK1 T*
PLACE T* ON P1
SET T* TRAVEL 4
          ADVANCE     4
          QUEUE       WAIT
          SEIZE       MECH
          BPUTPIC     FILE=ATF,LINES=2,AC1,XID1
TIME *.****
PLACE T* AT 0 0
          DEPART      WAIT
          ADVANCE     RVNORM(1,11,1)
          BPUTPIC     FILE=ATF,LINES=3,AC1,XID1,XID1
TIME *.****
PLACE T* ON P2
SET T* TRAVEL 4
          ADVANCE     .5
          RELEASE     MECH
          ADVANCE     3.5
          BPUTPIC     FILE=ATF,LINES=2,AC1,XID1
TIME *.****
DESTROY T*
 FIRST    TERMINATE
 TYPE2    BPUTPIC     FILE=ATF,LINES=4,AC1,XID1,XID1,XID1
TIME *.****
CREATE TRUCK2 T*
PLACE T* ON P1
SET T* TRAVEL 4
          ADVANCE     4
          QUEUE       WAIT
          SEIZE       MECH
          BPUTPIC     FILE=ATF,LINES=2,AC1,XID1
TIME *.****
PLACE T* AT 0 0
          DEPART      WAIT
          ADVANCE     RVEXPO(1,8.8)
          BPUTPIC     FILE=ATF,LINES=3,AC1,XID1,XID1
TIME *.****
PLACE T* ON P2
SET T* TRAVEL 4
          ADVANCE     .5
          RELEASE     MECH
          ADVANCE     3.5
          BPUTPIC     FILE=ATF,LINES=2,AC1,XID1
TIME *.****
DESTROY T*
 SECOND   TERMINATE
          GENERATE    480*100
          TERMINATE   1
          START       1
          PUTPIC      LINES=4,FR(MECH)/10.,N(FIRST)/100.,N(SECOND)/10(
     RESULTS OF SIMULATON
     UTIL OF REPAIR SHOP  **.**%
     TYPE 1 TRUCKS REPAIRED EACH SHIFT  **.**
     TYPE 2 TRUCKS REPAIRED EACH SHIFT  **.**
          PUTPIC      FILE=ATF,LINES=2,AC1
TIME *.****
END
          END
```

FIGURE 28.6 (Continued)
Listing of program CHAP28D.GPS.

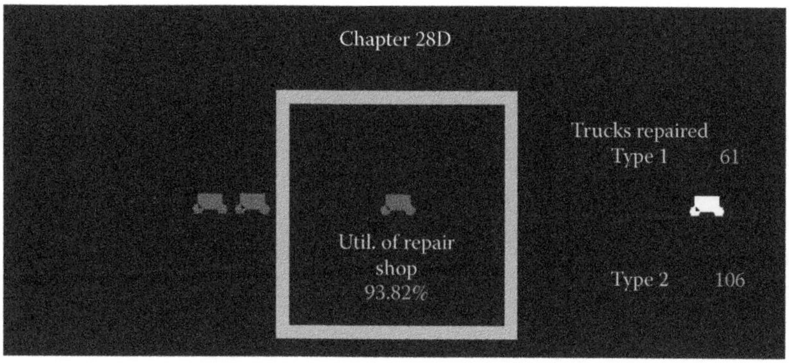

FIGURE 28.7
Screenshot of animation from program CHAP28D.GPS.

```
            SIMULATE
*******************************
*  PROGRAM CHAP28E.GPS        *
*  EXAMPLE OF PREEMPT BLOCK   *
*******************************
 ATF        FILEDEF     'CHAP28E.ATF'
            INTEGER     &COUNT1,&COUNT2,&COUNT3
 TYPE       FUNCTON     RN1,D2
.35,1/1,2
            GENERATE    12,4   TRUCKS ARRIVE
            ASSIGN      1,FN(TYPE),PH
            TEST E      PH1,2,NEXT
            PRIORITY    1
 NEXT       TEST E      PH1,1,TYPE2
            SEIZE       DUMMY
            BPUTPIC     FILE=ATF,LINES=4,AC1,XID1,XID1,XID1
TIME *.****
CREATE TRUCK1 T*
PLACE T* ON P1A
SET T* TRAVEL .667
            ADVANCE     .667
            RELEASE     DUMMY
            BPUTPIC     FILE=ATF,LINES=3,AC1,XID1,XID1
TIME *.****
PLACE T* ON P1
SET T* TRAVEL 3.333
            ADVANCE     3.333
            QUEUE       WAIT
            SEIZE       MECH
            DEPART      WAIT
            BPUTPIC     FILE=ATF,LINES=3,AC1,XID1,QA(WAIT)
TIME *.****
PLACE T* AT 0 0
WRITE M30 **.**
            ADVANCE     RVNORM(1,8.2,1)
            BPUTPIC     FILE=ATF,LINES=3,AC1,XID1,XID1
TIME *.****
PLACE T* ON P2
SET T* TRAVEL 4
            ADVANCE     .5
            RELEASE     MECH
            BLET        &COUNT1=&COUNT1+1
            BPUTPIC     FILE=ATF,LINES=3,AC1,&COUNT1,FR(MECH)/10.
```

FIGURE 28.8
Listing of program CHAP28E.GPS.

(Continued)

```
TIME *.****
WRITE M10 ***
WRITE M20 **.**%
          ADVANCE      3.5
          BPUTPIC      FILE=ATF,LINES=2,AC1,XID1
TIME *.****
DESTROY T*
 FIRST    TERMINATE
 TYPE2    SEIZE        DUMMY
          BPUTPIC      FILE=ATF,LINES=4,AC1,XID1,XID1,XID1
TIME *.****
CREATE TRUCK2 T*
PLACE T* ON P1A
SET T* TRAVEL .667
          ADVANCE      .667
          RELEASE      DUMMY
          BPUTPIC      FILE=ATF,LINES=3,AC1,XID1,XID1
TIME *.****
PLACE T* ON P1
SET T* TRAVEL 3.333
          ADVANCE      3.333
          QUEUE        WAIT
          SEIZE        MECH
          DEPART       WAIT
          BPUTPIC      FILE=ATF,LINES=3,AC1,XID1,QA(WAIT)
TIME *.****
PLACE T* AT 0 0
WRITE M30 **.**
          ADVANCE      RVEXPO(1,6.5)
          BPUTPIC      FILE=ATF,LINES=3,AC1,XID1,XID1
TIME *.****
PLACE T* ON P2
SET T* TRAVEL 4
          ADVANCE      .5
          RELEASE      MECH
          BLET         &COUNT2=&COUNT2+1
          BPUTPIC      FILE=ATF,LINES=3,AC1,&COUNT2,FR(MECH)/10.
TIME *.****
WRITE M11 ***
WRITE M20 **.**
          ADVANCE      3.5
          BPUTPIC      FILE=ATF,LINES=2,AC1,XID1
TIME *.****
DESTROY T*
 SECOND   TERMINATE
          GENERATE     24,8     OTHER TRUCKS ARRIVE
          SEIZE        DUMMY
          BPUTPIC      FILE=ATF,LINES=4,AC1,XID1,XID1,XID1
TIME *.****
CREATE TRUCK3 T*
PLACE T* ON P1A
SET T* TRAVEL .667
          ADVANCE      .667
          RELEASE      DUMMY
          BPUTPIC      FILE=ATF,LINES=3,AC1,XID1,XID1
TIME *.****
PLACE T* ON P1
SET T* TRAVEL 3.333
          ADVANCE      3.333
          QUEUE        WAIT
          TEST E       F(MECH),1,AWAY2
          BPUTPIC      FILE=ATF,LINES=2,AC1
```

FIGURE 28.8 (Continued)
Listing of program CHAP28E.GPS.

(Continued)

```
TIME *.****
WRITE M1 FACILITY BEING PREEMPTED
  AWAY2    PREEMPT      MECH
           DEPART       WAIT
           BPUTPIC      FILE=ATF,LINES=3,AC1,XID1,QA(WAIT)
TIME *.****
PLACE T* AT 0 3
WRITE M30 **.**
           ADVANCE      RVNORM(1,6.5,1)
           BPUTPIC      FILE=ATF,LINES=4,AC1,XID1,XID1
TIME *.****
PLACE T* ON P2
SET T* TRAVEL 4.5
WRITE M1
           ADVANCE      .5
           RETURN       MECH
           BLET         &COUNT3=&COUNT3+1
           BPUTPIC      FILE=ATF,LINES=3,AC1,&COUNT3,FR(MECH)/10.
TIME *.****
WRITE M12 ***
WRITE M20 **.**%
           ADVANCE      4
           BPUTPIC      FILE=ATF,LINES=2,AC1,XID1
TIME *.****
DESTROY T*
 THIRD     TERMINATE
           GENERATE     480*100
           TERMINATE    1
           START        1
           PUTPIC       LINES=5,FR(MECH)/10.,N(FIRST)/100.,_
                        N(SECOND)/100.,N(THIRD)/100.
     RESULTS OF SIMULATON
     UTIL OF REPAIR SHOP   **.**%
     TYPE 1 TRUCKS REPAIRED EACH SHIFT   **.**
     TYPE 2 TRUCKS REPAIRED EACH SHIFT   **.**
     TYPE 3 TRUCKS REPAIRED EACH SHIFT   **.**
           PUTPIC       FILE=ATF,LINES=2,AC1
TIME *.****
END
           END
```

FIGURE 28.8 (Continued)
Listing of program CHAP28E.GPS.

```
RESULTS OF SIMULATON
UTIL OF REPAIR SHOP  92.60%
TYPE 1 TRUCKS REPAIRED EACH SHIFT  13.57
TYPE 2 TRUCKS REPAIRED EACH SHIFT  26.36
TYPE 3 TRUCKS REPAIRED EACH SHIFT  19.92
```

FIGURE 28.9
Output from program CHAP28E.GPS.

This program also provides the animation. As can be seen, with the faster repair equipment, it is also possible to repair the critical vehicles and the repair shop is working 92.6% of the time. The animation is shown in Figure 28.10.

28.1.4 The PREEMPT PR Option

Is it possible to PREEMPT a facility that is already being pre-empted? The answer is yes, providing that PR operand on the PREEMPT Block is used and the transaction that is doing

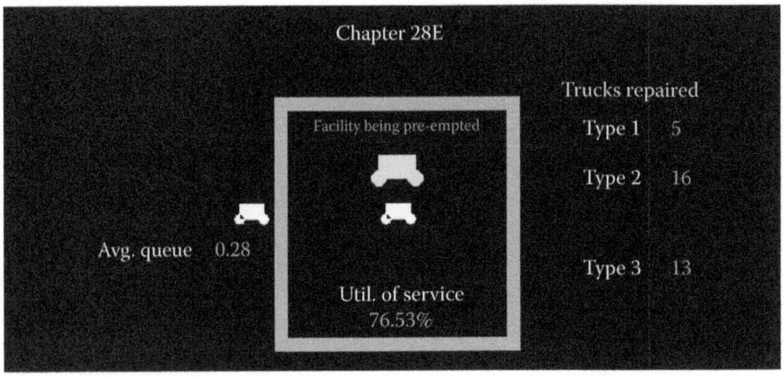

FIGURE 28.10
Screenshot of animation from program CHAP28E.GPS.

the pre-empting has a higher priority than the transaction which is already pre-empting the facility. The next example illustrates this. It is in four parts to illustrate the concept of priority as well as pre-empting.

Example 28.3: A Repair Shop for a Mine

A repair shop in a mine for minor repairs has two types of trucks that arrive for repairs. Type 1 arrive every 18 ± 4 minutes and type 2 trucks every 26 ± 8 minutes. Type 1 trucks are repaired according to the Poisson distribution with a mean repair time of 8 minutes; type 2 trucks are also repaired according to the Poisson distribution with a mean repair time of 6 minutes. The repair facility can repair only one truck at a time. The mining engineer wants simulation models for the following systems:

1. How the repair show presently works. This includes how busy the shop is; how long it takes for each type truck in the shop (both queuing and repairs), the average queue for the trucks and the maximum number of trucks at any one time. For this exercise assume that the arrival rate of the trucks is not dependent on the number of trucks at the repair shop.
2. What happens if the type 2 trucks have a higher priority than the type 1 trucks. Thus, if a type 2 truck arrives at the shop and there are trucks waiting, it will move ahead of type 1 trucks for repairs.
3. The type 2 trucks are more important than the type 1 trucks, so what happens if the type 2 trucks can pre-empt type 1 trucks if they are being repaired.
4. The repair shop now operates 24 hours a day. What will happen if it is shut from midnight to 4 a. m.?

Solution

The first part of this example is modelled by program CHAP28F.GPS. Its listing in shown in Figure 28.11.

The results from the program are shown in Figure 28.12.

These results will be compared with those from the next three programs.

The changes to the program CHAP28F.GPS to have type 2 trucks have priority are easily made and are in program CHAP28G.GPS. Its listing will not be given since the only change is to the second `GENERATE` Block. It is now as follows:

```
GENERATE 26,8,,,1,12PH,12PL OTHER TRUCKS ARRIVE
```

```
                    SIMULATE
******************************
*  PROGRAM CHAP28F.GPS       *
*  EXAMPLE OF REPAIR SHOP    *
*  TWO TYPES OF TRUCKS ARRIVE *
*  STANDARD QUEUE CRITERIA   *
******************************
 FIRST    TABLE       MP1PL,2,1,20
 SECOND   TABLE       MP1PL,2,1,20
          GENERATE    18,4,,,,12PH,12PL     TRUCKS ARRIVE
          MARK        1PL
          QUEUE       WAIT     REPAIR
          SEIZE       REPAIR
          DEPART      WAIT
          ADVANCE     RVEXPO(1,8)
          RELEASE     REPAIR
          TABULATE    FIRST
          TERMINATE
          GENERATE    26,8,,,,12PH,12PL    OTHER TRUCKS ARRIVE
          MARK        1PL
          QUEUE       WAIT
          SEIZE       REPAIR
          DEPART      WAIT
          ADVANCE     RVEXPO(1,6)
          RELEASE     REPAIR
          TABULATE    SECOND
          TERMINATE
          GENERATE    480*3*100
          TERMINATE   1
          START       1
          PUTPIC      LINES=5,FR(REPAIR)/10.,TB(FIRST),TB(SECOND),_
                      QM(WAIT),QA(WAIT)
 UTIL OF REPAIR **.**%
 AVG. TIME IN SYSTEM FOR TYPE 1 TRUCKS **.**
 AVG. TIME IN SYSTEM FOR TYPE 2 TRUCKS **.**
 MAXIMUM QUEUE AT SHOP     **
 AVG. QUEUE AT SHOP        **.**
          END
```

FIGURE 28.11
Listing of program CHAP28F.GPS.

```
UTIL OF REPAIR 67.97%
AVG. TIME IN SYSTEM FOR TYPE 1 TRUCKS 14.12
AVG. TIME IN SYSTEM FOR TYPE 2 TRUCKS 13.21
MAXIMUM QUEUE AT SHOP     11
AVG. QUEUE AT SHOP        0.61
```

FIGURE 28.12
Results from running program CHAP28F.GPS.

This shows that type 2 trucks have a priority of 1. The results of running the program are shown in Figure 28.13.

The expected time in the system for type 2 trucks has been reduced from 13.21 to 10.44. Next, the concept of pre-empt is introduced. This is given by program CHAP28H.GPS. Even though only a few lines of code have been changed, the full listing of the program is given in Figure 28.14. The results are shown in Figure 28.15.

The shop is busy roughly the same amount of time for the three cases of approximately 68% but the times in the shop for the trucks change dramatically when the concept of pre-empt is allowed. The time in the shop for type 2 trucks is now 6.10, which is approximately what the mean time for repairs is.

```
                    UTIL OF REPAIR 67.85.%
          AVG. TIME IN SYSTEM FOR TYPE 1 TRUCKS 15.61
          AVG. TIME IN SYSTEM FOR TYPE 2 TRUCKS 10.44
          MAX. QUEUE IN SHOP   10
          AVG. QUEUE AT SHOP    0.59
                   END
```

FIGURE 28.13
Results from running program CHAP28G.GPS.

```
          SIMULATE
********************************
*  PROGRAM CHAP28H.GPS          *
*  EXAMPLE CHAP28F.GPS BUT NOW *
*  WITH THE PREEMPT BLOCK       *
********************************
 FIRST    TABLE      MP1PL,2,1,20
 SECOND   TABLE      MP1PL,2,1,20
          GENERATE   18,4,,,,12PH,12PL     TRUCKS ARRIVE
          MARK       1PL
          QUEUE      WAIT      REPAIR
          SEIZE      REPAIR
          DEPART     WAIT
          ADVANCE    RVEXPO(1,8)
          RELEASE    REPAIR
          TABULATE   FIRST
          TERMINATE
          GENERATE   26,8,,,1,12PH,12PL    OTHER TRUCKS ARRIVE
          MARK       1PL
          QUEUE      WAIT
          PREEMPT    REPAIR
          DEPART     WAIT
          ADVANCE    RVEXPO(1,6)
          RETURN     REPAIR
          TABULATE   SECOND
          TERMINATE
          GENERATE   480*3*100
          TERMINATE  1
          START      1
          PUTPIC     LINES=5,FR(REPAIR)/10.,TB(FIRST),TB(SECOND),_
                     QM(WAIT),QA(WAIT)
 UTIL OF REPAIR **.**%
 AVG. TIME IN SYSTEM FOR TYPE 1 TRUCKS **.**
 AVG. TIME IN SYSTEM FOR TYPE 2 TRUCKS **.**
 MAX. QUEUE IN SHOP     **
 AVG. QUEUE IN SHOP     **.**
          END
```

FIGURE 28.14
Listing of program CHAP28H.GPS.

```
UTIL OF REPAIR 67.83.%
AVG. TIME IN SYSTEM FOR TYPE 1 TRUCKS 18.82
AVG. TIME IN SYSTEM FOR TYPE 2 TRUCKS  6.10
MAX. QUEUE IN SHOP     8
AVG. QUEUE IN SHOP    0.43
```

FIGURE 28.15
Results from running program CHAP28H.GPS.

```
            GENERATE    ,,,1,10       DUMMY TRANSACTION
BACK        ADVANCE     1200          OPEN UNTIL MIDNIGHT
            PREEMPT     REPAIR,PR     CLOSE THE SHOP
            ADVANCE     240           WAIT 4 HOURS
            RETURN      REPAIR        OPEN THE SHOP
            TRANSFER     ,BACK
```

FIGURE 28.16
Code to be added to program CHAP28I.GPS.

```
UTIL OF REPAIR 84.63%
AVG. TIME IN SYSTEM FOR TYPE 1 TRUCKS 108.63
AVG. TIME IN SYSTEM FOR TYPE 2 TRUCKS 32.88
MAX. QUEUE IN SHOP  37
AVG. QUEUE IN SHOP   6.29.
```

FIGURE 28.17
Results from running program CHAP28I.GPS.

In order to close the repair shop from midnight to 4 a. m., the concept of a pre-empt with priority option is used. A dummy transaction is created. The lines of code added are shown in Figure 28.16. The results of the program are given in Figure 28.17.

The times the trucks are in the shop are large because of the shop being closed for the 4 hours. It will be left as an exercise to modify the program CHAP28I.GPS to have the shop remain open at midnight to finish working on any truck that may be either in the shop being repaired or in the queue waiting for repairs.

28.2 Exercises

28.1. In program CHAP28B, the number of each truck was placed in the truck transaction's first half-word parameter. This was used for the animation. What would happen if program CHAP28B used the transaction's XID1 for the animation?

28.2. Modify the program given in Example 28.1 so that there will now be a clock in the animation. The clock should only show the hour hand. Also, add a counter to give the number of trucks serviced.

28.3. Program CHAP28B can be improved a bit. When a truck is waiting for service and one finishes, the animation shows the finished truck and the waiting one changing positions at the same time. Modify the program so that there is a slight delay while the finished truck is placed on the path to leave.

28.4. Consider Example 28.3 point 4. If it is midnight and the shop is closed there may be a truck being repaired and others waiting. Add the code so that the shop will now be shut until all these trucks have been repaired. Have the average time that the shop was shut down as part of the output.

28.5. The mining engineer is satisfied with the results from Example 28.3 and decides that the pre-empt case is the one she wants (point 3). She can send other trucks of type 1 to the repair shop. She will do this until the utilization is 90%. By adding more trucks, the arrival rate will be increased to 16 ± 4, then 14 ± 4, next 12 ± 4, and so on. Determine the correct arrival rate for as close to 90% utilization as possible.

29

MACROs

Head frame of old mine.

29.1 MACROs in GPSS/H Subroutines

29.1.1 MACROs

When writing the GPSS/H code for all the examples in this book so far, the lines of code were always written out as they would appear in the .LIS file. No shorthand method was given. Many times a sequence or segment of code has Blocks that are nearly the same but differ only in the labels and/or the operands. Consider an example of products on an assembly line having to stop at two stations in a row. The code might look as follows:

```
QUEUE        WAIT1 FIRST STATION
SEIZE        MACH1
DEPART       WAIT1
ADVANCE      20,8
RELEASE      MACH1
............
............
QUEUE        WAIT2 SECOND STATION
SEIZE        MACH2
DEPART       WAIT2
ADVANCE      18,5
RELEASE      MACH2
...............
...............
QUEUE        WAIT3 THIRD STATION
SEIZE        MACH3
DEPART       WAIT3
ADVANCE      19.5,6
RELEASE      MACH2
```

Code such as the above series of Blocks can be written in a compact form using the concept of GPSS/H MACROs.

Any experienced GPSS/H programmer soon learns how useful MACROs can be in writing lengthy programs. Not only do they compact lines of code, but they make it easier to ensure that the program code is correct.

The general form of a MACRO is as follows:

```
(label)      STARTMACRO
             (line of code with #A's, #B's, etc to represent variables)
             (next line of code.....)
             .............
             (last line
             ENDMACRO label)
```

The (*label*) is necessary. The statements STARTMACRO and ENDMACRO must appear as shown.

The MACRO definition consists of statements and must appear before any Blocks referencing it. It normally is placed before any programming Blocks but does not necessarily need to be placed there.

For the example of products coming to two stations as indicated in the first paragraph above, a MACRO might be defined as follows:

```
MYFIRST      STARTMACRO
             QUEUE  #A
             SEIZE  #B
             DEPART #A
             ADVANCE       #C,#D
             RELEASE       #B
             ENDMACRO
```

To invoke the MACRO, the lines of code in the program are written as:

```
MYFIRST        MACRO  WAIT1,MACH1,20,8
               .........
MYFIRST        MACRO  WAIT2,MACH2,18,5
               ...........
MYFIRST        MACRO WAIT3,MACH3,19.5,6
```

The MACRO becomes expanded during compiling, and the resulting code for the first MACRO is

```
QUEUE          WAIT1
SEIZE          MACH1
DEPART         WAIT1
ADVANCE        20,8
RELEASE        MACH1
```

In the MACRO, WAIT1 replaces #A, MACH1 replaces #B, 20 replaces #C, and 8 replaces #D. The next example shows how this MACRO is used in a program.

Example 29.1: Trucks in a Mine Come for Two Services

Trucks come to a service area every 25 ± 10 minutes. Only one truck can be serviced at a time. This service takes 20 ± 8 minutes. The truck then travels in 3 minutes to a second service area where, again, only one truck can be serviced at a time. This second service takes 18 ± 5 minutes. Simulate for 100 shifts of 480 minutes to determine the utilization of each service bay.

The program CHAP29A.GPS gives the program to simulate this using a MACRO. The listing is given next in Figure 29.1.

```
                SIMULATE
******************************
*  PROGRAM CHAP29A.GPS        *
*  EXAMPLE OF A MACRO         *
******************************
 MYFIRST  STARTMACRO
          QUEUE          #A
          SEIZE          #B
          DEPART         #A
          ADVANCE        #C,#D
          RELEASE        #B
          ENDMACRO
          GENERATE       25,10     TRUCKS COME ALONG
 MYFIRST  MACRO          WAIT1,MACH1,20,8
          ADVANCE        3         TRAVEL TO SECOND SERVICE
 MYFIRST  MACRO          WAIT2,MACH2,18,5
          TERMINATE
          GENERATE       480*100    100 SHIFTS
          TERMINATE      1
          START          1
          PUTPIC         LINES=2,FR(MACH1)/10.,FR(MACH2)/10.
 UTIL OF MACH 1 **.**%
 UTIL OF MACH 2 **.**%
          END
```

FIGURE 29.1
Listing of program CHAP29A.GPS.

The program for this example is quite short so may not illustrate the economy of using MACROs, but in a longer program, the savings in writing the code can be substantial. This is especially so when writing code for making .ATF files when so many of the program lines lend themselves to MACROs.

The output from the program is as given in Figure 29.2.

The .LIS file shows how the complier handles MACROs. A portion of this is shown in Figure 29.3.

Notice how the GPSS/H compiler expands the macros. It places the plus (+) sign before each Block.

Another reason to use MACROs is that they make a program easy to detect typing errors. For example, consider the following lines of code:

```
FIRST MACRO   STOPA,MACHA,JUMPA,PERSONA,AWAYA,PLACEA
FIRST MACRO   STOPB,MACHB,JUMPB,PERSONB,AWAYB,PLACEB
FIRST MACRO   STOPC,MACHC,JUMPC,PERSONC,AWAYC,PLACEC
```

These are very easy to look at and see if each is correct. If one had typed PLACE as PLACCE, it is quite easy to detect as the lines of code will not line up. If one did not use a macro, there would be a large number of lines of code to look at to detect any errors.

As mentioned, MACROs are often used when making .ATF files. In fact, it is a good idea to have a library of MACROs that can be imported before making the program for the .ATF file. To illustrate this, consider the following two MACROs:

```
PLACEAT              STARTMACRO
                     BPUTPIC         FILE = ATF,LINES = 2,AC1,XID1
TIME *.****
PLACE #A* AT #B #C
                     ENDMACRO
TRAVEL               STARTMACRO
                     BLET            #A = RVNORM(1,#B,.1*#B)
                     BPUTPIC         FILE = ATF,LINES = 3,AC1,XID1,XID1,#A
TIME *.****
PLACE #C* ON P#D
SET #C* TRAVEL **.**
                     ADVANCE         #A
                     ENDMACRO
```

```
UTIL OF MACH 1 80.45%
UTIL OF MACH 2 72.69%
```

FIGURE 29.2
Output from program CHAP29A.GPS.

```
       MYFIRST  MACRO      WAIT1,MACH1,20,8
 2 +            QUEUE      WAIT1
 3 +            SEIZE      MACH1
 4 +            DEPART     WAIT1
 5 +            ADVANCE    20.8
 6 +            RELEASE    MACH1
 7              ADVANCE    3           TRAVEL TO
       MYFIRST  MACRO      WAIT2,MACH2,18,5
 8 +            QUEUE      WAIT2
 9 +            SEIZE      MACH2
10 +            DEPART     WAIT2
11 +            ADVANCE    18.5
```

FIGURE 29.3
Portion of CHAP29A.LIS showing how the MACROs are expanded.

In the program, the output file for the .ATF file has the label ATF. Consider the effect of the code:

```
PLACEAT MACRO              T,20,30
TRAVEL MACRO               &X,20,T,13
```

Assume that the simulation time is 123.45 and the transaction's XID1 is 35. When the MACROs are expanded by the compiler, the code becomes:

```
          BPUTPIC          FILE = ATF,LINES = 2,AC1,XID1
TIME 123.4500
PLACE T35 AT 20 30
          BLET             &X = RVNORM(1,20,2)
          BPUTPIC          FILE = ATF,LINES = 3,AC1,XID1,XID1,&X
TIME 123.4500
PLACE T35 ON P13
SET T35 TRAVEL (the value of &X)
          ADVANCE (the value of &X)
```

MACROs given here may seem to be complex to write and understand, but their use is highly recommended. Program CHAP29B.GPS was written to illustrate the above two MACROs. It is short but instructive to carefully examine it, as the use of macros can greatly shorten not only the number of lines of code but the time to write the code. The listing of program CHAP29B is given in Figure 29.4.

A single transaction is generated to invoke the two MACROs. Two ADVANCE Blocks were added, one after the transaction is generated and one after the first MACRO is

```
          SIMULATE
********************************
*  PROGRAM CHAP29B.GPS.         *
*  ILLUSTRATION OF MACROS       *
*  SHORT PROGRAM TO ILLUSTRATE  *
*  HOW MACROS WORK.             *
********************************
 ATF      FILEDEF      'CHAP29B.ATF'
          REAL        &X
 PLACEAT  STARTMACRO
          BPUTPIC      FILE=ATF,LINES=2,AC1,XID1
TIME *.****
PLACE #A* AT #B #C
          ENDMACRO
 TRAVEL   STARTMACRO
          BLET      #A=RVNORM(1,#B,.1*#B)
          BPUTPIC   FILE=ATF,LINES=3,AC1,XID1,XID1,#A
TIME *.****
PLACE #C* ON P#D
SET #C* TRAVEL **.**
          ADVANCE   #A
          ENDMACRO
          GENERATE  ,,,1,,10PL,10PH
          ADVANCE  3,2
  PLACEAT MACRO    T,20,30
          ADVANCE  10,2.5
  TRAVEL  MACRO    &X,20,T,15
          TERMINATE   1
          START      1
          END
```

FIGURE 29.4
Listing of program CHAP29B.GPS.

```
TIME 2.0812
PLACE T1  AT 20 30
TIME 12.0780
PLACE T1  ON P15
SET T1  TRAVEL 20.44
END
```

FIGURE 29.5
Output from program CHAP29B.GPS. This is the file CHAP29B.ATF.

invoked. There will be five lines of code created in the file, CHAP29B.ATF. These lines are as shown in Figure 29.5.

In a typical mining simulation, one might have the trucks travelling on many different paths. A large surface mine might have as many as 400 path segments. Rather than have multiple lines of code for each segment, one can now have a single line of code.

29.1.2 Subroutines

It is possible to have subroutines in GPSS/H just as in other programming languages. A subroutine is where one leaves the main body of programming code to perform other programming instructions and return to the main program. While this is not as common in GPSS/H as in other languages, there is one use of a subroutine that can be quite handy. First, how a subroutine works is explained.

The subroutine is written as follows:

```
(label) (start of subroutine)
       lines of code
       ...... .
       TRANSFER (code to return to main)
```

The subroutine can come before the actual program Blocks begin (the first `GENERATE` Block) or even after the main lines of code (but it must be before the `START` statement). It normally is written before the first `GENERATE` Block with the other statements.

A subroutine is accessed by the `TRANSFER` Block:

```
TRANSFER       SBR,(label),(parameter number)
```

The (*label*) is the label where control is transferred to, namely, the first Block in the subroutine. The (*parameter number*) is a parameter where the number of the `TRANSFER` Block is stored. Recall that each Block is numbered consecutively by the compiler. This Block number is what is placed in the parameter number. For example,

```
TRANSFER       SBR,MYSUB,3PH
```

This specifies that control passes to the subroutine with the label `MYSUB` and the GPSS/H number of the `TRANSFER` Block is placed in the transaction's 3rd half-word parameter. Since this number must be an integer, one cannot use floating point parameters for subroutines.

Once control passes to the subroutine and the Blocks are executed, control must return to the main program. Care must be taken to not return to the initial `TRANSFER` Block or an infinite loop may occur. Consider the following Blocks that might have made up the subroutine, `MYSUB`:

```
MYSUB ADVANCE     10,2
      TEST E      &X,100,NEXT
      BLET        &X = 250
NEXT  TRANSFER    ,PH3+1
```

Here, the Blocks in the subroutine are executed and control returns to the Block in the main program with the number PH3 +1. This is one Block more than the one that transferred to the subroutine. Thus, if the Block where the transfer to the subroutine was Block number 123 in the main program, the subroutine returns control to Block number 124 in the main program.

The above example of a subroutine does not show any advantage in using a subroutine. One of the most common uses is provided in the next example.

In many mining simulations, once the simulation program is developed and the animation correct, it is not necessary to create an animation file every time the program is written. These animation files can become quite large and there really is no reason to create them every time. When a program is run, the operator can be asked whether an animation file is desired. For example, consider the lines of code:

```
          CHAR*1         ANS
          PUTSTRING      (' DO YOU WANT AN ANIMATION FILE')
          PUTSTRING      (' TO BE CREATRED? (Y/N)')
          PUTSTRING      (' NOTE: "YES" MUST BE CAPITAL Y!')
          PUTSTRING      (' ')
          GETLIST        &ANS
          .............

          .............
     ANIM TEST E                    &ANS,'Y',PH3+2
          TRANSFER       ,PH3+1
          .............

          .............
          (main program starts here)
```

In the main program before *every* BPUTPIC, there will be the Block:

```
TRANSFER          SBR,ANIM,3PH
```

This transfers control to the subroutine ANIM. The subroutine checks to see if there is to be an animation file created. If there is to be an animation file, control returns to the main program and the next sequential Block in the program, namely, the BPUTPIC Block will be executed; if not, control returns to the Block one beyond the BPUTPIC.

Example 29.2: Model of Mine Travel on a Segment with Optional Animation

A segment for trucks to travel in a mine is broken into four segments. These are segments A, B, C, and D. Trucks enter on the left and travel to the end of the segment where they leave the system. The travel times on the four segments is given by the normal distribution with means as shown in Table 29.1. The standard deviations are all 10% of the means.

Write the program to simulate the trucks travelling on the segment using MACROs and a subroutine that prompts the uses to create an animation file if so desired.

The program to simulate this system is given by CHAP29C.GPS. This is shown in Figure 29.6.

TABLE 29.1

Data for Travel Times for CAHP30C.GPS

Segment	Mean Travel Time
A	5.9
B	5.3
C	3.4
D	3.4

```
          SIMULATE
**********************************
*   PROGRAM CHAP29C.GPS  EXAMPLE *
*   OF BOTH MACRO AND SUBROUTINE *
**********************************
   ATF    FILEDEF      'CHAP29C.ATF'
          CHAR*1       &ANS
          REAL         &X,&Y,&Z,&W
 TIMES    TABLE        MP1PL,15,1,30
          PUTSTRING    ('  ')
          PUTSTRING    ('  ')
          PUTSTRING    ('  DO YOU WANT ANIMATION? (Y/N)')
          PUTSTRING    ('  NOTE: "YES" IS CAPITAL "Y"')
          PUTSTRING    ('  ')
          GETLIST      &ANS
   ANIM   TEST E       &ANS,'Y',PH3+2
          TRANSFER     ,PH3+1
  TRAVEL  STARTMACRO
          BLET         #A=RVNORM(1,#B,.1*#B)
          TRANSFER     SBR,ANIM,3PH
          BPUTPIC      FILE=ATF,LINES=3,AC1,XID1,XID1,#A
TIME *.****
PLACE #C* ON P#D
SET #C* TRAVEL **.**
          ADVANCE      #A
          ENDMACRO
          GENERATE     20,5,,,,10PH,10PL    TRUCKS ARRIVE AT A
          MARK         1PL                  TIME TRUCK ARRIVES
          TRANSFER     SBR,ANIM,3PH
          BPUTPIC      FILE=ATF,LINES=3,AC1,XID1,PL1
TIME *.****
CREATE TRUCK T*
WRITE M1 ***.**
 TRAVEL   MACRO       &X,5.9,T,1
 TRAVEL   MACRO       &Y,5.3,T,2
 TRAVEL   MACRO       &Z,3.4,T,3
 TRAVEL   MACRO       &W,3.4,T,4
          TABULATE    TIMES
          TRANSFER    SBR,ANIM,3PH
          BPUTPIC     FILE=ATF,LINES=5,AC1,XID1,AC1,MP1PL,TB(TIMES)
TIME *.****
DESTROY T*
WRITE M2 ***.**
WRITE M3  **.**
WRITE M4  **.**
          TERMINATE
          GENERATE   480
          TERMINATE  1
          START      1
          IF   &ANS'E''Y'
          PUTPIC      FILE=ATF,LINES=2,AC1
TIME *.****
END
          ENDIF
           END
```

FIGURE 29.6
Listing of program CHAP29C.GPS.

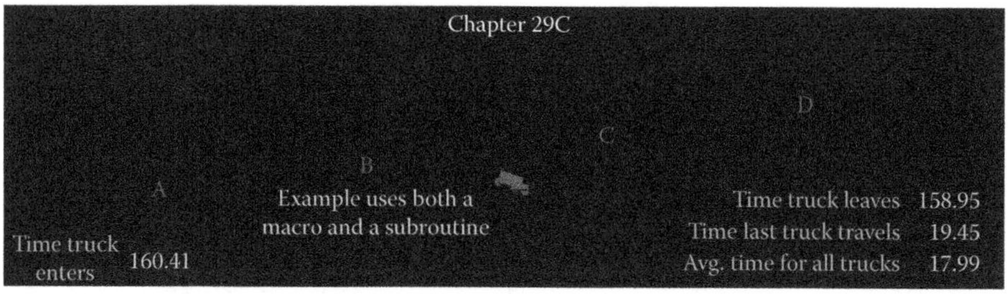

FIGURE 29.7
Screenshot of animation from program CHAP29C.GPS.

The animation is shown in Figure 29.7.

Notice that the `TRANSFER SBR` was placed before every `BPUTPIC` Block in the main program. The `PUTPIC` statement at the end of the program cannot have this as it is not a Block. Instead, an `IF` statement is used.

Example 29.3: Trucks Travelling in a Mine

A very common model for working mines are trucks as they travel along paths. The paths are often broken into segments, so that the mean travel times are of the same order of magnitude. Suppose that a long path in a mine consists of seven segments. The segment lengths are as follows:

Segment	Length
A to B	24.67
B to C	17.24
C to D	13.23
D to E	24.67
E to F	21.23
F to G	21.00
G to H	32.35

Two types of trucks enter at point A every 8 ± 2.3 minutes. 35% are type 1 trucks and the rest type 2 trucks. The mean time to travel from A to H is 30 minutes for type 1 trucks and 33 minutes for type 2 trucks. The engineer needs a model that shows the trucks travelling from A to H in a *smooth* manner. This means that the animation will show the trucks appearing to travel on each path for a time approximately proportional to its length. Each segment travel time is to be sampled from the normal distribution with mean selected proportional to the segment length and standard deviation of 8% of these.

Table 29.2 is from a spread sheet to calculate the relative times for each segment for each type of truck. The total length of all the segments is 154.39. The per cent of the length of segment from A to B is 16. Thus, for example, for truck type 1, on segment A to B, it will take 4.79 minutes (.16 times 30); for truck type 2, from F to G, it will take 4.49 minutes.

The program to simulate this is CHAP29D.GPS. The program uses a `MACRO` for travel on each segment. A listing is given in Figure 29.8. The animation is shown in Figure 29.9.

TABLE 29.2

Data from Spread Sheet Giving Relative Travel Time for Each Segment

Travel Times			**30**	**33**
Truck Type			**1**	**2**
A to B	24.67	0.16	4.79	5.27
B to C	17.24	0.11	3.35	3.68
C to D	13.23	0.09	2.57	2.83
D to E	24.67	0.16	4.79	5.27
E to F	21.23	0.14	4.13	4.54
F to G	21.00	0.14	4.08	4.49
G to H	32.35	0.21	6.29	6.91
Total	154.39			

```
          SIMULATE
******************************
*  PROGRAM CHAP29D.GPS        *
*  FURTHER EXAMPLE OF MACROS *
******************************
 ATF      FILEDEF      'CHAP29D.ATF'
     1    FUNCTION     PH1,M2
1,RVNORM(1,4.79,.08*4.79)/2,RVNORM(1,5.27,.08*5.27)
     2    FUNCTION     PH1,M2
1,RVNORM(1,3.35,.08*3.35)/2,RVNORM(1,3.68,.08*3.68)

     3    FUNCTION     PH1,M2
1,RVNORM(1,2.57,.08*2.57)/2,RVNORM(1,2.83,.08*2.83)

     4    FUNCTION     PH1,M2
1,RVNORM(1,4.79,.08*4.79)/2,RVNORM(1,5.27,.08*5.27)

     5    FUNCTION     PH1,M2
1,RVNORM(1,4.13,.08*4.13)/2,RVNORM(1,4.54,.08*4.54)

     6    FUNCTION     PH1,M2
1,RVNORM(1,4.08,.08*4.08)/2,RVNORM(1,4.49,.08*4.49)

     7    FUNCTION     PH1,M2
1,RVNORM(1,6.29,.08*6.29)/2,RVNORM(1,6.91,.08*6.91)
 TYPE     FUNCTION      RN1,D2
.35,1/1,2
          REAL         &A,&B,&C,&D,&E,&F,&G
 TRAVEL   STARTMACRO
          BLET         #A=FN#B
          BPUTPIC      FILE=ATF,LINES=3,AC1,XID1,XID1,#A
TIME *.****
PLACE T* ON P#C
SET T* TRAVEL **..**
          ADVANCE      #A
          ENDMACRO
          GENERATE     8,2.3     TRUCKS COMES ALONG
          ASSIGN       1,FN(TYPE),PH    WHAT TYPE TRUCK
          BLET         &A=FN1
          BPUTPIC      FILE=ATF,LINES=4,AC1,PH1,XID1,XID1,XID1,&A
TIME *.****
```

FIGURE 29.8
Listing of program CHAP29D.GPS.

(Continued)

```
CREATE TRUCK* T*
PLACE T* ON P1
SET T* TRAVEL **.**
          ADVANCE      &A          TRAVEL FROM A TO B
 TRAVEL   MACRO        &B,2,2
 TRAVEL   MACRO        &C,3,3
 TRAVEL   MACRO        &D,4,4
 TRAVEL   MACRO        &E,5,5
 TRAVEL   MACRO        &F,6,6
 TRAVEL   MACRO        &G,7,7
          BPUTPIC      FILE=ATF,LINES=2,AC1,XID1
TIME *.****
DESTROY T*
          TERMINATE
          GENERATE     400
          TERMINATE    1
          START        1
          PUTPIC       FILE=ATF,LINES=2,AC1
TIME *.****
END
          END
```

FIGURE 29.8 (Continued)
Listing of program CHAP29D.GPS.

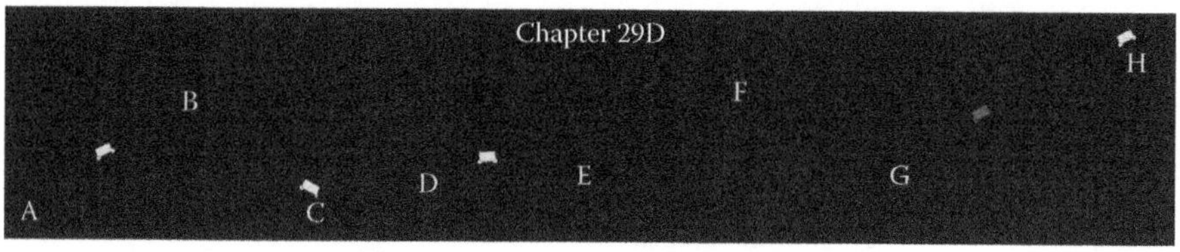

FIGURE 29.9
Screenshot of animation from CHAP29D.GPS.

Example 29.4: Another Example of a Subroutine

A single machine in an assembly line paints five different products that arrive according to the distributions:

Part #	Arrival Dist.
1	30 ± 5
2	25 ± 10
3	30 ± 11
4	40 ± 12
5	26 ± 8

All times are in minutes. Painting takes 5.5 ± 2 minutes for each part. Only one part can be painted at a time. The manager is concerned that the painting machine is working too hard and wonders what the utilization of it is. If it is working more than 92% of the time, it should be replaced.

Solution
The program to do the simulation consist s of a single subroutine and five short segments. Its listing is shown in Figure 29.10. The output from the program is given in Figure 29.11.

As can be seen, the painting machine is busy 90.97% of the time. The critical utilization is 92% so it should be replaced.

```
          SIMULATE
******************************
*   PROGRAM CHAP29D.GPS       *
*   EXAMPLE OF SUBROUTINE     *
******************************
          INTEGER     &PART1,&PART2,&PART3,&PART4,&PART5
   FIRST  QUEUE       WAIT
          SEIZE       MACH
          DEPART      WAIT
          ADVANCE     5.5,2
          RELEASE     MACH
          TRANSFER    ,PH1+1
          GENERATE    30,5
          TRANSFER    SBR,FIRST,1PH
          BLET        &PART1=&PART1+1
          TERMINATE
          GENERATE    25,10
          TRANSFER    SBR,FIRST,1PH
          BLET        &PART2=&PART2+1
          TERMINATE
          GENERATE    36,11
          TRANSFER    SBR,FIRST,1PH
          BLET        &PART3=&PART3+1
          TERMINATE
          GENERATE    40,12
          TRANSFER    SBR,FIRST,1PH
          BLET        &PART4=&PART4+1
```

FIGURE 29.10
Listing of program CHAP29F.GPS.

(Continued)

```
            TERMINATE
            GENERATE   26,8
            TRANSFER   SBR,FIRST,1PH
            BLET       &PART5=&PART5+1
            TERMINATE
            GENERATE   480*3*10
            TERMINATE  1
            START      1
            PUTPIC     LINES=6,&PART1/10.,&PART2/10.,&PART3/10.,&PART4/10.,_
                       &PART5/10.,FR(MACH)/10.
    NO. OF PARTS 1 MADE EACH DAY **.**
    NO. OF PARTS 2 MADE EACH DAY **.**
    NO. OF PARTS 3 MADE EACH DAY **.**
    NO. OF PARTS 4 MADE EACH DAY **.**
    NO. OF PARTS 5 MADE EACH DAY **.**
     MACHINE WAS BUSY **.**%
            END
```

FIGURE 29.10 (Continued)
Listing of program CHAP29F.GPS.

```
NO. OF PARTS 1 MADE EACH DAY 48.00
NO. OF PARTS 2 MADE EACH DAY 57.50
NO. OF PARTS 3 MADE EACH DAY 39.40
NO. OF PARTS 4 MADE EACH DAY 36.30
NO. OF PARTS 5 MADE EACH DAY 55.80
 MACHINE WAS BUSY 90.97%
```

FIGURE 29.11
Output from program CHAP29E.GPS.

29.2 Exercises

29.1. There is a layout CHAP29CC.LAY. This has two roads added to the layout for CHAP29C.LAY. One has the segments E—F—G—H and the other has segments I—J—K. Take the program CHAP29C.GPS and add the code to have trucks travelling on these paths. On segment E—F—G—H, they travel from right to left and on segment I—J—K, they travel from left to right. The program should be written by modifying the program CHAP29C.GPS. The travel times can be approximated from the travel time for the segments in CHAP29C.

29.2. The times given for travel in Example 29.2 are to be changed. Suppose that the total travel time for the four segments is a mean of 20 time units. The lengths of the four segments, A, B, C, and D are to be input variables. Change program CHAP29C.GPS, so that the program determines the mean travel times based on the ratios of the lengths of the four segments. For example, suppose that the segments had lengths of 10, 20, 30, and 40. The total travel time is still 20. Thus the mean time for segment A would be determined to be 1/10 of 20 or 2; for segment B, it would be 4; for segment C it would be 6; and for segment D it would be 8 (sum is 20).

29.3. Consider the program CHAP29D.GPS. The animation can be improved to show the number of each type truck to enter and then leave the system. Also, the animation should show the average travel time for each type truck as it leaves the system.

29.4. In Example 29.5, there was no priority to painting the various parts by the machine. Suppose that the parts were given priorities so that the one that took the least time had the highest priority, the one with the second fastest time, the next priority, and so on. How does this change the output?

30

A Few More PROOF Icons

Small drill for a quarry.

30.1 A Few More PROOF Icons

30.1.1 Making x–y Plots

There is an icon on the PROOF menu bar that allows the user to make x–y plots of the data in the simulation. This icon is as shown below:

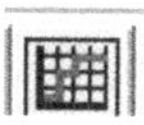

This icon is used for making the x–y plots. When it is clicked, the following on-screen menu comes up:

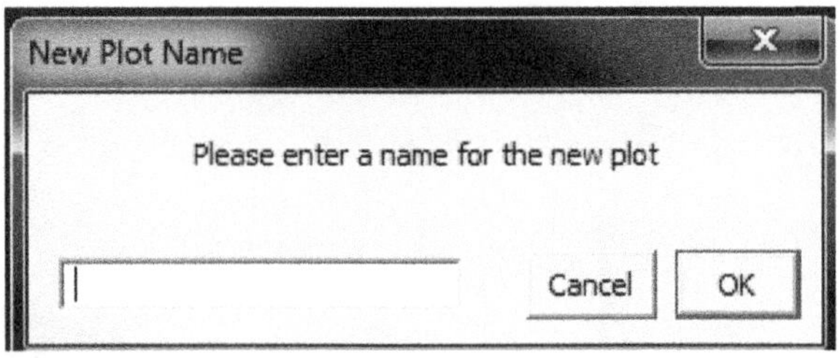

The name is case sensitive. When a name is given for the plot, another menu appears:

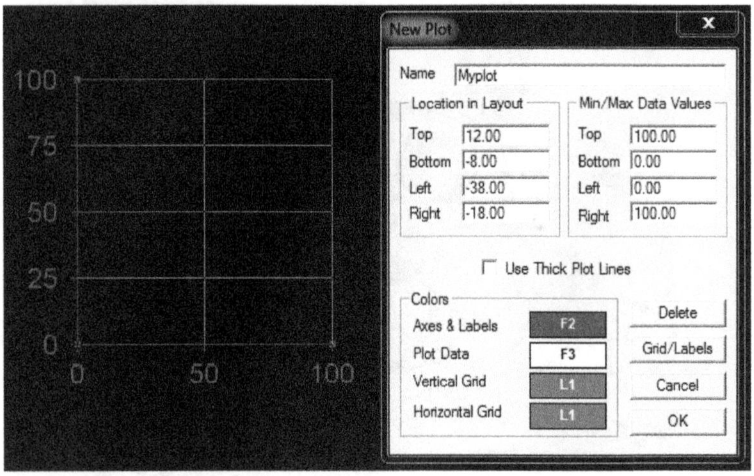

A sample plot axis is also shown. This is shown in red but this can be changed. This plot can be moved around by clicking on the origin and dragging it to the desired location. The various options for the plot are found in the on-screen menu. For example, the maximum values for the values to be plotted are given by default as 0 to 100 for both the values on the x and y axes. These are easily changed depending on the date to be plotted.

The animation command for a plot is

```
PLOT (name) xold yold xnew ynew
```

There will be a line drawn from the old coordinates (xold, yold) to the new one (xnew, ynew).

The use of a plot can best be demonstrated by means of an example.

Example 30.1: The Gold Miner's Problem (Revisited)

A small gold mine has a single loader and six trucks. There is a single crusher. Various times for the operation are as follows:

Load	2.1 ± .75
Travel to crusher	6.6 ± 2.2
Dump	1.2 ± .5
Return	4.3 ± .5

The miner wants an animation of his operation showing the production on an x–y plot.

Solution

The program to do the simulation and animation is CHAP30A.GPS. The model is simple as only the PLOT feature is to be illustrated. Exercise 30.1 will ask for more details (Figure 30.1).

```
              SIMULATE
************************************
*   PROGRAM CHAP30A.GPS            *
*   EXAMPLE OF A PLOT              *
************************************
    ATF       FILEDEF      'CHAP30A.ATF'
              REAL         &X,&Y
              REAL         &XOLD,&YOLD,&YNEW
            GENERATE     1.5,,0,6        PUT TRUCKS IN THE MINE
              BLET         &X=1.0*FRN1+3.8
              BPUTPIC      FILE=ATF,LINES=4,AC1,XID1,XID1,XID1,&X
TIME *.****
CREATE TRUCK T*
PLACE T* ON P1
SET T* TRAVEL **.**
              ADVANCE      &X
 UPTOP        QUEUE        WAIT
              SEIZE        SHOVEL
              DEPART       WAIT
              BPUTPIC      FILE=ATF,LINES=2,AC1,XID1
TIME *.****
PLACE T* AT -18.5 -12.45
              ADVANCE      2.1,.75
              BLET         &Y=2.4*FRN1+5.4
              BPUTPIC      FILE=ATF,LINES=4,AC1,XID1,XID1,XID1,&Y
TIME *.****
SET T* CLASS TRUCKL
PLACE T* ON P2
SET T* TRAVEL **.**
              ADVANCE      1.       MOVE TRUCK AWAY FROM SHOVEL
              RELEASE      SHOVEL   FREE SHOVEL
```

FIGURE 30.1
Listing of program CHAP30A.GPS.

(Continued)

```
            ADVANCE      &Y-1
            QUEUE        CRUSH
            SEIZE        CRUSHER
            DEPART       CRUSH
            BPUTPIC      FILE=ATF,LINES=2,AC1,XID1
TIME *.****
pLACE T* AT 78.15 -3.45
            ADVANCE      1.2,.5
            BLET         &X=1.0*FRN1+3.8
            BLET         &YNEW=&YNEW+1
            BPUTPIC
FILE=ATF,LINES=5,AC1,XID1,XID1,XID1,&X,&XOLD,&YOLD,AC1,&YNEW
TIME *.****
PLACE T* ON P1
set T* CLASS TRUCK
SET T* TRAVEL **.**
PLOT LOADS **.* **.*  **.*  **.*
            ADVANCE      1        LEAVE CRUSHER AREA
            RELEASE      CRUSHER
            ADVANCE      &X-1      TRAVEL BACK
            BLET         &XOLD=AC1
            BLET         &YOLD=&YNEW
            TRANSFER     ,UPTOP
            GENERATE     480
            TERMINATE    1
            START       1
            PUTPIC      FILE=ATF,LINES=2,AC1
TIME *.****
END
            END
```

FIGURE 30.1 (Continued)
Listing of program CHAP30A.GPS.

The animation is given in Figure 30.2.
The animation is simple but does include the following:

1. The x–y plot uses green for the plotted data. By default, the plotted data is the same colour as the axes. It is also possible to plot multiple data on the same axis. The command would be something like

```
PLOT LOADS 50 100 51.5 101 color F2
```

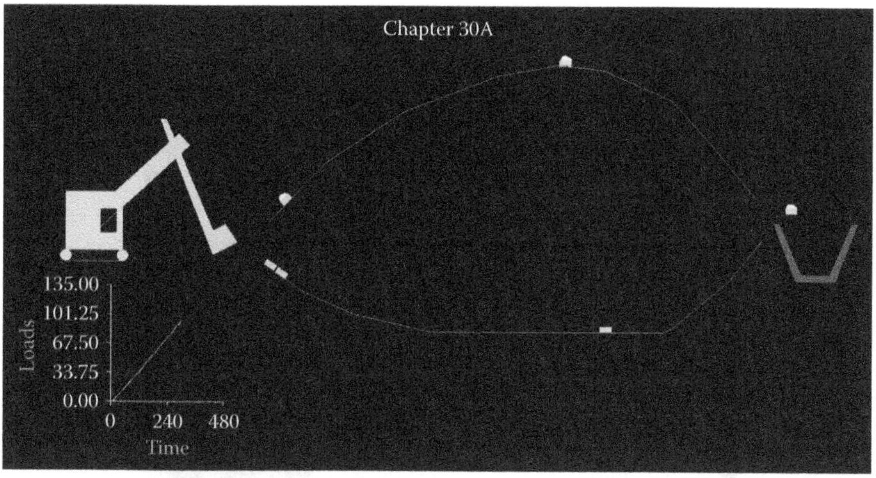

FIGURE 30.2
Screenshot of animation from program CHAP30A.GPS.

2. This would plot the data as given on the same axis as in the Example but with the colour RED (F2).
3. The loaded trucks are shown with exaggerated loads so that they show up better on the animation.
4. There are slight delays for trucks when a loaded truck replaces them. Otherwise, the animation would show the unloaded trucks immediately replacing the loaded truck at the shovel. The same slight delay is used at the crusher. In an actual mine, one might have short paths for the loaded truck to be placed on when it was done being loaded and another short path for the unloaded truck to spot at the shovel.

30.1.2 Bar Graphs

The icon to make a moving or *dancing* bar graph is as shown below:

When this is clicked, a pull-down menu appears as shown below.

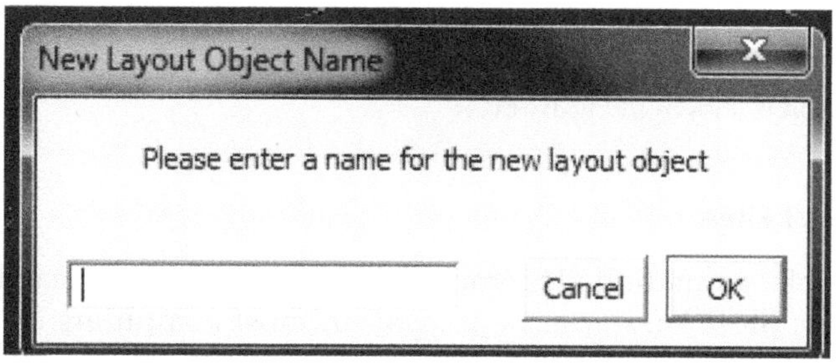

A name needs to be given to the bar graph; this is case sensitive. Once this is done, the following menu appears:

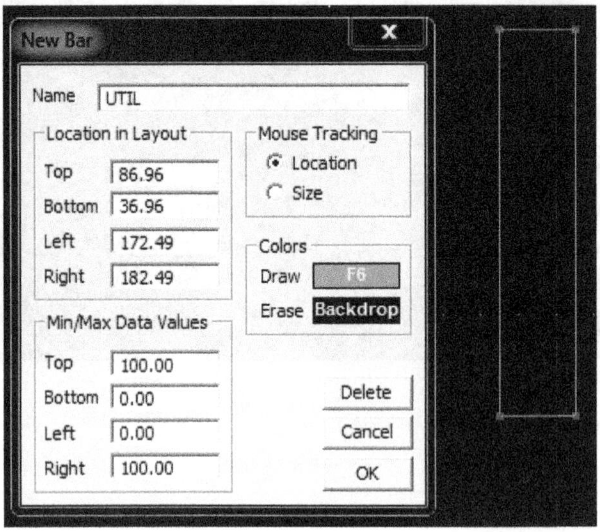

The menu gives the various options for the bar graph. The bar graph appears on the screen in red. This can easily be changed and the graph can easily be moved to any desired location. There are several options for the animation command for the bar graph but the one most commonly used is simply:

```
SET BAR (bar name) (value)
```

Thus, for Example 30.1, if one had the command

```
SET BAR UTIL 75.35
```

the bar would be filled to 75.35% of its size.

When the animation is saved, the bar graph disappears. It is only when the `SET BAR` command is executed that anything appears on the screen. Thus, in general, the animation will be blank when the animation begins. Hence, it is recommended that a frame be drawn for the bar graph. This is illustrated in the next program.

Example 30.2: The Miner Wants to Have a Bar Graph Added to the Animation

The program to have both a x–y plot and a bar graph for Example 30.1 is CHAP30B.GPS and is given in Figure 30.3.

The animation is shown in Figure 30.4.

30.1.3 Making .AVI Files

It is possible to make animation files from PROOF animation that can be exported and used as *stand-alone* files. These can be played on most computers by just clicking on

```
          SIMULATE
***********************************
*  PROGRAM CHAP30B.GPS            *
*  EXAMPLE OF A BAR GRAPH         *
***********************************
    ATF   FILEDEF    'CHAP30B.ATF'
          REAL       &X,&Y
          REAL       &XOLD,&YOLD,&YNEW
        GENERATE    1.5,,0,6      PUT TRUCKS IN THE MINE
          BLET       &X=1.0*FRN1+3.8
          BPUTPIC    FILE=ATF,LINES=4,AC1,XID1,XID1,XID1,&X
TIME *.****
CREATE TRUCK T*
PLACE T* ON P1
SET T* TRAVEL **.**
          ADVANCE    &X
 UPTOP    QUEUE      WAIT
          SEIZE      SHOVEL
          DEPART     WAIT
          BPUTPIC    FILE=ATF,LINES=2,AC1,XID1
TIME *.****
PLACE T* AT -18.5 -12.45
          ADVANCE    2.1,.75
          BLET       &Y=2.4*FRN1+5.4
          BPUTPIC    FILE=ATF,LINES=4,AC1,XID1,XID1,XID1,&Y
TIME *.****
SET T* CLASS TRUCKL
PLACE T* ON P2
SET T* TRAVEL **.**
          ADVANCE    1.      MOVE TRUCK AWAY FROM SHOVEL
          RELEASE    SHOVEL  FREE SHOVEL
          BPUTPIC    FILE=ATF,LINES=2,AC1,FR(SHOVEL)/10.
TIME *.****
SET BAR UTIL **.**
          ADVANCE    &Y-1
          QUEUE      CRUSH
```

FIGURE 30.3
Listing of program CHAP30B.GPS.

(Continued)

```
          SEIZE       CRUSHER
          DEPART      CRUSH
          BPUTPIC     FILE=ATF,LINES=2,AC1,XID1
TIME *.****
pLACE T* AT 78.15 -3.45
          ADVANCE     1.2,.5
          BLET        &X=1.0*FRN1+3.8
          BLET        &YNEW=&YNEW+1
          BPUTPIC     FILE=ATF,LINES=5,AC1,XID1,XID1,XID1,&X,&XOLD,&YOLD,AC1,&YNEW
TIME *.****
PLACE T* ON P1
set T* CLASS TRUCK
SET T* TRAVEL **.**
PLOT LOADS **.* **.*  **.*  **.*

          ADVANCE     1        LEAVE CRUSHER AREA
          RELEASE     CRUSHER
          ADVANCE     &X-1     TRAVEL BACK
          BLET        &XOLD=AC1
          BLET        &YOLD=&YNEW
          TRANSFER    ,UPTOP
          GENERATE    480
          TERMINATE   1
          START      1
          PUTPIC     FILE=ATF,LINES=2,AC1
TIME *.****
END
```

FIGURE 30.3 (Continued)
Listing of program CHAP30B.GPS.

them when they are loaded. These files have the extension .AVI. They can also easily be imported into a Powerpoint presentation. While easy to create, it should be noted that these files tend to be very large and some animations may not run as smoothly as under PROOF.

Example 30.3: Stand-Alone Animation File for Small Gold Miner's Problem

The program CHAP12D.GPS is the animation for a single shovel and multiple trucks (similar to Example 30.1 but a bit fancier). The steps to create an .AVI file are as follows:

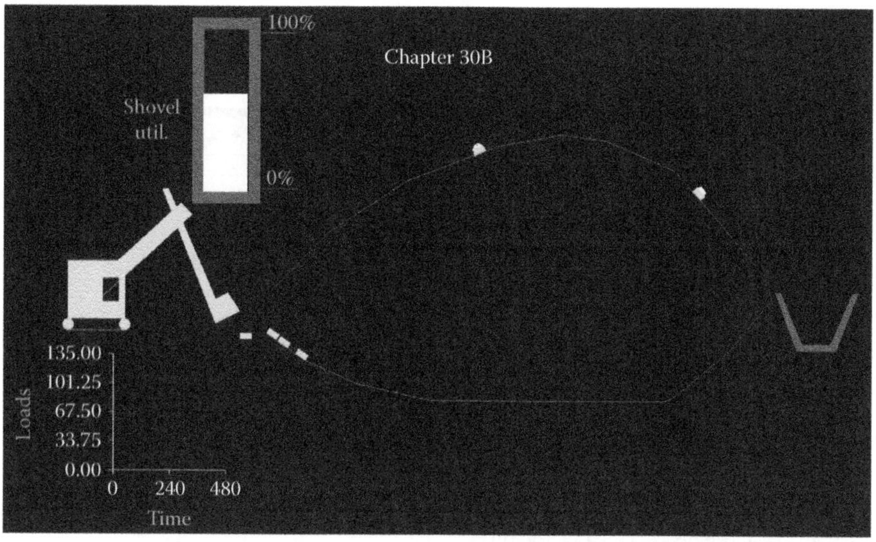

FIGURE 30.4
Screen grab for animation of program CHAP30B.GPS.

1. Have the animation up on PROOF.
2. Decide on the time period for the animation. For this example, take a starting time of 0 and create the animation until time 100.
3. Click on File → Create Special Files → Capture AVI.
4. A menu appears.

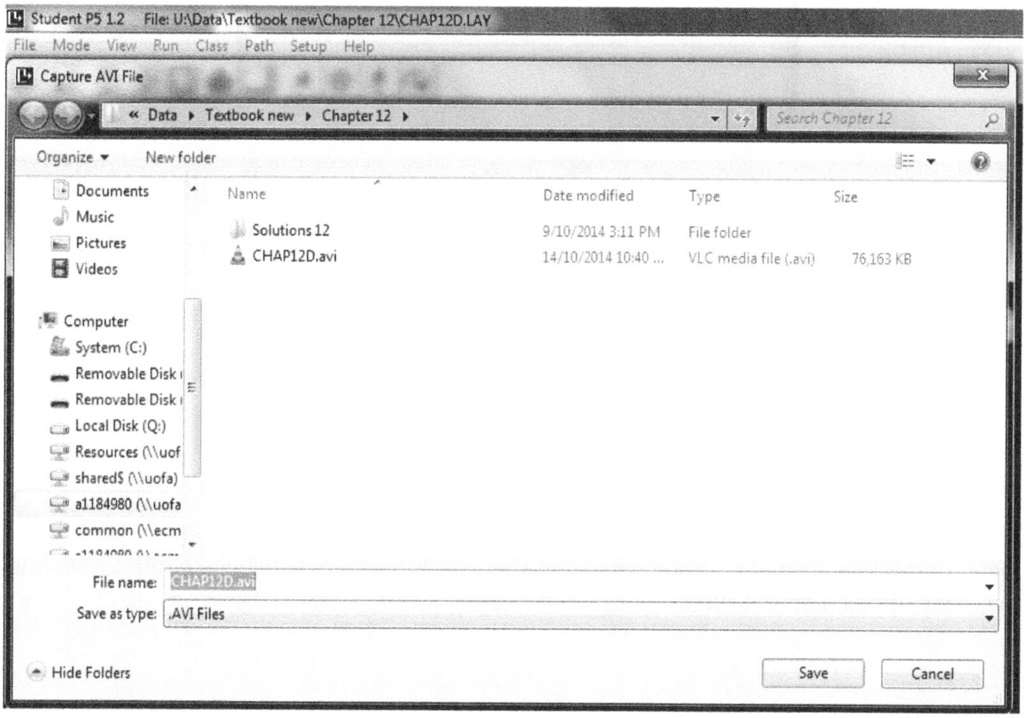

By default, the name of the .AVI is CHAP12D.AVI. This can be changed by keying in a different file name. When the <Save> button is clicked on, a new menu appears:

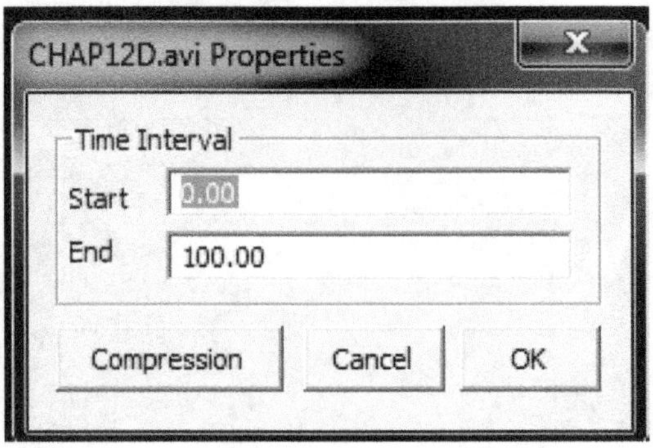

This is where the start and end of the animation is keyed in. In this case, the animation is to start at 0.00 and end at 100.00.

Clicking on Compression leads to another menu.

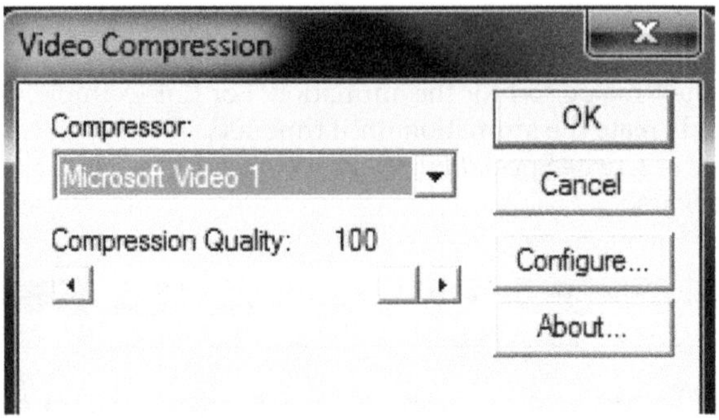

There are several options to select for the Compressor. Selecting the right one for a particular computer may take a bit of trial and error. In this case, Microsoft Video 1 has been selected. The important tab is the one for the Compression Quality. This should be high. In this case, it has been selected to be 100.

Clicking on the <OK> tab will start the creation of the .AVI file. The screen may show jerky motion while the file is being created. To view the animation file, CHAP12D.AVI, one need only to double click on it and the animation starts. As an example of using this in a Powerpoint presentation, the file Smallmine.pptx has been created. Both the animation file and this file are found on the sub-directory Animations.

30.1.4 Screen Grabs

The animation figures in this textbook have all been created using the PROOF pull-down menu under File. To make screen captures. The process is simple: FILE → Grab Screen and the menu appears on the screen:

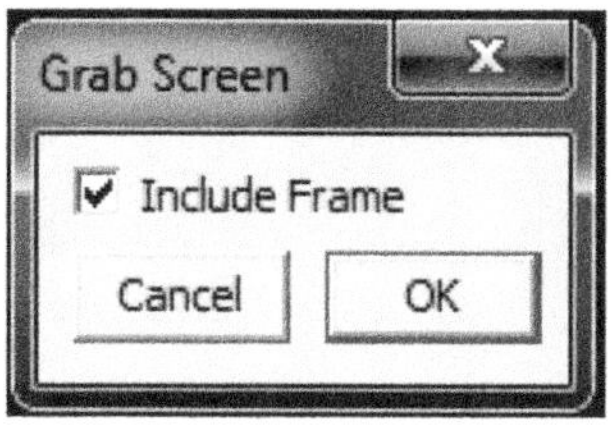

The frame can be included or not. Clicking on <OK> will make a bitmap image. The default name is CHAP12D.bmt, but this can be changed.

30.2 Exercises

30.1. The animation given for Example 30.1 is a bit simple. Add the code so that the boom moves when a truck is being loaded and have the actual loads delivered to the crusher show on the animation. Make the number of trucks in the mine a variable and also have this shown on the animation.

31

SAVEVALUES

Underground mine equipment—for seam-type deposits.

31.1 SAVEVALUES

At one time, GPSS had no equal sign that made doing arithmetic a bit awkward. There also were no ampervariables. GPSS/H was quick to correct these shortcomings. The way arithmetic was done was by means of Blocks known as SAVEVALUES. While these are rarely used today, there are a few examples where they can be used. One might see their use in older GPSS or even GPSS/H programs.

SAVEVALUES are user-defined SNAs. They are so named as they are *values* that are *saved*. These values are printed out in the output report, except for zero values. GPSS/H has four different types of these SAVEVALUES. They are as follows:

1. Full-word SAVEVALUES: These are integers, which range from –2,147,483,648 to +2,147,483,647. Their family name is XF. A full-word SAVEVALUE is referenced by XF (name) or simply X (name). Alternately, one can use a single dollar sign ($) such as X$name. This is an old way of referencing SAVEVALUES that still works in GPSS/H. A SAVEVALUE can be a number. In that case, it is referenced simply by XF_j or X_j.

2. Half-word SAVEVALUES: These are integers ranging from –32,768 to +32,767. Their family name is XH.
3. Bit-word SAVEVALUES: These are integers ranging from –128 to +127. Their family name is XB.
4. Floating-point SAVEVALUES: These are decimal values whose family name is XL.

Their size depends on the computer.

To reference a SAVEVALUE, it is necessary to do so using the family name with parentheses or a dollar sign.

Thus,

```
ADVANCE      XF(SPEED)
GENERATE     XL(AVER),XL(SPREAD)
QUEUE        XH(WAIT)
ASSIGN       1,XL(FIRST),PL
SEIZE        XB(MACH1)
GENERATE     XH$FAST)
```

are all examples of using SAVEVALUES as operands of other blocks.

If the second letter in the family name is omitted, it is assumed that the SAVEVALUE is a full-word SAVEVALUE. For example,

```
ASSIGN       1,XF(TEST),PH
```

and

```
ASSIGN       1,X(TEST),PH
```

are the same.

In both cases, the value of the SAVEVALUE TEST will be given to the transaction's first half-word parameter. In this text, the exact type of the SAVEVALUE will be fully specified. Thus, reference to SAVEVALUES as X(TEST) will not be done.

It is also possible to reference a SAVEVALUE using the dollar sign ($). XL$NUMB and XL(NUMB) are the same. This method of referencing SAVEVALUES will not be used here as it is a holdover from the original versions of GPSS. How to make changes to the values of SAVEVALUES is given next.

31.2 The SAVEVALUE Block

All SAVEVALUES *are initially zero.* How to give them initial values different from zero will be considered shortly. During the running of a program, it is often desired to change their values. This is done by the SAVEVALUE block. Its general form is:

```
SAVEVALUE    (name),value, type
```

where:

(name) is the name of the SAVEVALUE. These names are assigned according to the usual rules. This could be a number or any SNA

value is the value to be assigned to the SAVEVALUE. This could be an SNA

type is the family name and is either XF, XH, XB, or XL. If this specification is omitted, the SAVEVALUE is a full-word SAVEVALUE by default

Some examples of the SAVEVALUE block are as follows:

(a) SAVEVALUE JIM,2,XF
(b) SAVEVALUE TOMMY,-100,XH
(c) SAVEVALUE JOE,32,XF
(d) SAVEVALUE 1,4,XF
(e) SAVEVALUE FIRST,25.63,XL
(f) SAVEVALUE NEXT,25/2,XL
(g) SAVEVALUE OTHER,25.0/2,XL
(h) SAVEVALUE PH4,12,XH

In (a), the SAVEVALUE JIM is set equal to 2. It is a full-word SAVEVALUE. In (b), the SAVEVALUE TOMMY is set equal to –100 and it is a half-word SAVEVALUE. In (c), the SAVEVALUE JOE is set equal to 32. In (d), the SAVEVALUE 1 is set equal to 4. It, too, is a full-word SAVEVALUE. In (e), the SAVEVALUE FIRST is set equal to 25.63. In (f), the value of NEXT is set equal to 12.0000. It is *not* 12.5000 because division is integer division unless specified using floating point numbers. In (f), the division is floating point division because of the decimal in the 25.0 (GPSS/H converts the 2 to a floating point number). Here the value of OTHER is 12.5000. (The reason for the 4 decimals is because this is what GPSS/H prints out for the value of the SAVEVALUEs, not because this is the number of digits actually stored.) In (g), the SAVEVALUE NEXT is given the value of 25.0. In (h), it is not known what SAVEVALUE will have the value of 12. This will depend on the number stored in the transaction's fourth half-word parameter. If it is 5, then SAVEVALUE 5 will have the value 12.

The use of a number for the value of a variable may seem confusing as in other languages variables normally must be alphanumeric starting with a letter. For example, in Fortran, one might have statements defining variables such as JIM = 4, TOMMY = –100.0, JOE = 32, and X1 = 1, but *not* 2 = 1. In effect, these do the same thing as the first four examples of the SAVEVALUE block given above. Corresponding Fortran statements, for example, (e), (f), and (g) might be: FIRST = 25.63, NEXT = 25/2, and OTHER = 25.5.

The use of numbers for SAVEVALUEs will be avoided here if at all possible. There will be an example when this feature will come in very handy, especially when using a variable for the first operand of the SAVEVALUE. For example, consider the SAVEVALUE:

```
SAVEVALUE PH1,3
```

Here the SAVEVALUE to be given the value of 3 will depend on the transaction's first parameter. If it happened to be 4, then the SAVEVALUE 4 is given the value 3.

31.3 The FIX and FLT Mode Conversion

Suppose you want to have the value of SAVEVALUE FIRST to be equal to the sum of Q(ONE) plus Q(TWO) to be divided by Q(THREE) and you want an exact floating point result. It would *not* be correct to write the following:

```
SAVEVALUE      FIRST,(Q(ONE)+Q(TWO))/Q(THREE),XL
```

Since the queue lengths of queue are integers, the quotient will represent integer division. In order to have floating point division, one could do the following:

```
SAVEVALUE      JUNK1,Q(ONE)+Q(TWO),XL
SAVEVALUE      JUNK2,Q(THREE),XL
SAVEVALUE      FIRST,XL(JUNK1)/XL(JUNK2),XL
```

This is a bit awkward. GPSS/H offers two mode conversion operators. These are FIX and FLT to convert to fixed-point or floating-point mode. They work identically to the mode converters found in other languages such as Fortran, where the corresponding mode converters are IFIX and FLOAT. In GPSS/H, one might have the following:

```
FLT(Q(ONE))
FIX(XL(TEST))
FLT(FC(MACHA))
```

Thus, one could have written the original expression for the SAVEVALUE as:

```
SAVEVALUE      FIRST,(Q(ONE)+Q(TWO))/FLT(Q(THREE)),XL
```

Notice that it was not necessary to convert all three fixed-point queue values.

SAVEVALUES can be used in increment or decrement mode just as the ASSIGN block was used. Thus,

```
(a)  SAVEVALUE      LOAD+,25,XF
(b)  SAVEVALUE      COST-,XF(PRICE),XF
(c)  SAVEVALUE      PILE+,FN(TRUCK),XF
(d)  SAVEVALUE      PH1+,PH2,XF
(e)  SAVEVALUE      NEXT+,FN(LAST)+FN(FIRST),XL
(f)  SAVEVALUE      TOM-,Q(ONE)/3,XL
(g)  SAVEVALUE      TOM-,Q(ONE)/3.0,XL
```

In (a), the SAVEVALUE LOAD is incremented by 25. In (b), the SAVEVALUE COST is decremented by whatever the SAVEVALUE PRICE is. In (c), the SAVEVALUE PILE is incremented by reference to the function TRUCK. In (d), the SAVEVALUE specified by the transaction's first half-word parameter is incremented by the value in its second half-word parameter. In (e), the value of the SAVEVALUE is given by reference to the functions LAST and FIRST. This is added to the current value of the SAVEVALUE NEXT. In (f), the value of the SAVEVALUE TOM is decremented by the length of the queue ONE divided by 3 (integer division!). In (g), the SAVEVALUE TOM is decremented by the length of the queue ONE divided by 3.0 (floating point division). In the first four cases, the SAVEVALUES are full-word SAVEVALUES. In the other three, the SAVEVALUES are floating point. In a language such as Fortran, corresponding statements might be as follows:

```
LOAD = LOAD + 25
COST = COST - PRICE
PILE = PILE + F(TRUCK)
X(1) = X(1) + X(2)
NEXT = NEXT + F(LAST) + F(FIRST)
TOM = TOM - QONE/3
TOM = TOM - QONE/3.0.
```

A common error in programming is to omit the family name XF, XH, XL, or XB with the SAVEVALUE in parentheses when referencing a SAVEVALUE. If this is omitted, the value of the SAVEVALUE is taken to be zero no matter what it actually is. *This type of error is most insidious as it can be extremely hard to detect and no run time error takes place.* Thus, if you had intended to write

```
TEST EX        F(VALUE),1,AWAY
```

but instead wrote

```
TEST E         VALUE,1,AWAY
```

The test will *always be false.*

31.4 The INITIAL Statement

As indicated, the value of all SAVEVALUEs are set equal to zero when a program begins. This is done by the processor. Often, a programmer will want SAVEVALUES to be initially set to nonzero values. This is done with the INITIAL statement. The general form of it is as follows:

INITIAL $X_i(SV_j)$,value/$X_i(SV_k)$/....

where X_i is the family name (either XF, XH, XF, XB, or XL). SV_j, SV_k, and so on are the names of the SAVEVALUES. In case the SAVEVALUE name is a number, the parentheses are optional.

Some examples of this are as follows:

```
INITIAL        XF(FIRST),100/XH(TEST),-340/X1,10000
```

An alternate way to write the above is with dollar sign ($):

```
INITIAL        X$FIRST,100/XH$TEST,-340/X1,10000
```

This is a holdover from the early days of GPSS.

A shorthand form can be used for multiple initializations as follows:

```
INITIAL       XH1-XH10,3/XH(PLACE),125/XL(TOWN),1234.5432
```

This sets the half-word SAVEVALUE 1 through 10 equal to 3. It then sets the SAVEVALUES PLACE and TOWN equal to the value of the SAVEVALUE CITY + 3.

For students who have studied Fortran, the INITIAL statement in GPSS is analogous to the DATA statement.

31.5 Effect of RESET and CLEAR on SAVEVALUES

A RESET statement does *not affect* the values of the SAVEVALUES but the CLEAR statement sets all SAVEVALUES to zero. If the program is to be run again with the original initialized values, there are two things that can be done:

1. Reinitialize all the values with a new INITIAL statement (or statements).
2. Use a form of the CLEAR statement called the selective CLEAR.

The selective CLEAR is simply the CLEAR statement followed by a list of SAVEVALUES that are *not* to be set equal to zero.

Thus,

```
CLEAR  XH(TOM),XF(JOHN),XH(PLACE),XF7
```

will clear all the SAVEVALUES in the program *except* for TOM, JOHN, PLACE, and 7.

Example 31.1: Trucks Arrive for Inspection

This example will illustrate a case where the use of SAVEVALUES results in a shorter program than if ampervariables were used.

Four different trucks in a mine arrive at an inspection station where they are both inspected and weighed. They first travel on a short road from the mine to the inspection station that takes 8 ± 4 minutes. 25% of the trucks are type 1, 30% type 2, 30% type 3, and 15% type 4. Inspection depends on which type of truck it is. The times are all normally distributed with means and standard deviations as given in Table 31.1. Only one truck can be inspected at a time. When the trucks are done being inspected, they return to

TABLE 31.1

Inspection Time for Trucks

	Inspection Time	
Truck Type	Mean	Standard Deviation
1	8	1
2	13	2
3	10	1.5
4	9	1.2

TABLE 31.2

Material in Each Truck

	Load	
Truck Type	**Mean**	**Standard Deviation**
1		9010
2	95	5
3		10,010
4		12,013

the mine along a path that takes 7 ± 2 minutes. The amount of material in each truck is given in Table 31.2.

The simulation is to be for one shift of 400 minutes of work.

Solution

The program to do the simulation is CHAP31A.GPS. Its listing is given in Figure 31.1.

```
              SIMULATE
*****************************
*  PROGRAM CHAP31A.GPS       *
*  USE OF SAVEVALUES         *
*****************************
 TYPE      FUNCTION    RN1,D4
.25,1/.55,2/.85,3/1,4
 DUMP      FUNCTION    PH1,M4
1,RVNORM(1,8,1)/2,RVNORM(1,13,2)/3,RVNORM(1,10,1.5)/4,RVNORM(1,9,1.2)
 HOWMUCH   FUNCTION    PH1,M4
1,RVNORM(1,90,10)/2,RVNORM(1,95,5)/3,RVNORM(1,100,10)/4,RVNORM(1,120,13)
           GENERATE    11,4      TRUCKS ARRIVE
           ASSIGN      1,FN(TYPE),PH    WHAT TYPE TRUCK?
           ADVANCE     8,4       TRAVEL TO CRUSHER
           QUEUE       CRUSHER
           SEIZE       CRUSHER
           DEPART      CRUSHER
           ADVANCE     FN(DUMP)
           RELEASE     CRUSHER
           SAVEVALUE   PH1+,FN(HOWMUCH),XL  MATERIAL IN TRUCK
           SAVEVALUE   PH1+,1,XH            COUNT TRUCKS
           ADVANCE     7,2
           TERMINATE
           GENERATE    400
           TERMINATE   1
           START       1
           PUTPIC
LINES=7,1,XH1,XL1,2,XH2,XL2,3,XH3,XL3,4,XH4,XL4,FR(CRUSHER)/10.
             RESULTS OF SIMULATION
      TRUCK NO.    LOADS     AMT.
           *         **       ***.**
           *         **       ***.**
           *         **       ***.**
           *         **       ***.**
    UTIL. OF CRUSHER           **.**%
           END
```

FIGURE 31.1

Listing of program CHAP31A.GPS.

Notice the use of SAVEVALUES. One is used to determine how much material is in the truck. This is a floating point SAVEVALUE. The count of the trucks to arrive at the inspection station is also given by a half-word SAVEVALUE.

The output from the program is show in Figure 31.2.

The animation is given by program CHAP31B.GPS. In order to have the trucks that are finished being serviced leave the server before any waiting truck replaces it, a slight delay time of .5 is introduced. Thus, when a truck is done being serviced, it is placed on the path to leave the system and allowed to travel for .5 time units before a waiting truck replaces it. This leads to a slightly different result from the original program, CHAP31A.GPS. The listing of program CHAP31B.GPS is given in Figure 31.3. The animation is shown in Figure 31.4.

```
                  RESULTS OF SIMULATION

          TRUCK NO.     LOADS        AMT.

              1            8        790.70

              2           10        922.15

              3            8        794.15

              4            7        797.17

      UTIL. OF CRUSHER               86.43%
```

FIGURE 31.2
Output from program CHAP31A.GPS.

```
               SIMULATE

*****************************

*  PROGRAM CHAP31B.GPS       *

*  ANIMATION OF CHAP31A.GPS  *

*  USE OF SAVEVALUES         *

*****************************

 TYPE       FUNCTION    RN1,D4

.25,1/.55,2/.85,3/1,4

 DUMP       FUNCTION    PH1,M4

1,RVNORM(1,8,1)/2,RVNORM(1,13,2)/3,RVNORM(1,10,1.5)/4,RVNORM(1,9,1.2)

 HOWMUCH    FUNCTION    PH1,M4

1,RVNORM(1,90,10)/2,RVNORM(1,95,5)/3,RVNORM(1,100,10)/4,RVNORM(1,120,1
3)

  ATF       FILEDEF     'CHAP31B.ATF'

            REAL        &X,&Y

            GENERATE    11,4        TRUCKS ARRIVE
```

FIGURE 31.3
Listing of program CHAP31B.GPS.

(Continued)

```
          ASSIGN       1,FN(TYPE),PH    WHAT TYPE TRUCK?
          BLET         &X=8.*FRN1+4.
          BPUTPIC      FILE=ATF,LINES=4,AC1,XID1,XID1,XID1,&X
TIME *.****
CREATE TRUCKL T*
PLACE T* ON P1
SET T* TRAVEL **.**
          ADVANCE      &X         TRAVEL TO INSPECTION
          QUEUE        INSP
          SEIZE        INSP
          BPUTPIC      FILE=ATF,LINES=2,AC1,XID1
TIME *.****
PLACE T* AT -9 -1
          DEPART       INSP
          ADVANCE      FN(DUMP)
          BLET         &Y=4.*FRN1+5.
          BPUTPIC      FILE=ATF,LINES=4,AC1,XID1,XID1,XID1,&Y
TIME *.****
SET T* CLASS TRUCKU
PLACE T* ON P2
SET T* TRAVEL **.**
          SAVEVALUE    PH1+,FN(HOWMUCH),XL
          SAVEVALUE    PH1+,1,XH
          BPUTPIC
FILE=ATF,LINES=9,AC1,XH1,XL1,XH2,XL2,XH3,XL3,XH4,XL4
TIME *.****
WRITE M1 **
WRITE M5 ***.**
WRITE M2 **
WRITE M6 ***.**
WRITE M3 **
WRITE M7 ***.**
```

FIGURE 31.3 (Continued)
Listing of program CHAP31B.GPS.

(Continued)

```
WRITE M4 **
WRITE M8 ***.**
          ADVANCE     .5
          RELEASE     INSP
          BPUTPIC      FILE=ATF,LINES=2,AC1,FR(INSP)/10.
TIME *.****
WRITE M9 **.**%
          ADVANCE     &Y-1
          BPUTPIC     FILE=ATF,LINES=2,AC1,XID1
TIME *.****
DESTROY T*
          TERMINATE
          GENERATE    400
          TERMINATE   1
          START       1
          PUTPIC
LINES=7,1,XH1,XL1,2,XH2,XL2,3,XH3,XL3,4,XH4,XL4,FR(INSP)/10.
             RESULTS OF SIMULATION
      TRUCK NO.    LOADS      AMT.
          *          **        ***.**
          *          **        ***.**
          *          **        ***.**
          *          **        ***.**
    UTIL. OF CRUSHER             **.**%
          PUTPIC     FILE=ATF,LINES=2,AC1
TIME *.****
END
          END
```

FIGURE 31.3 (Continued)
Listing of program CHAP31B.GPS.

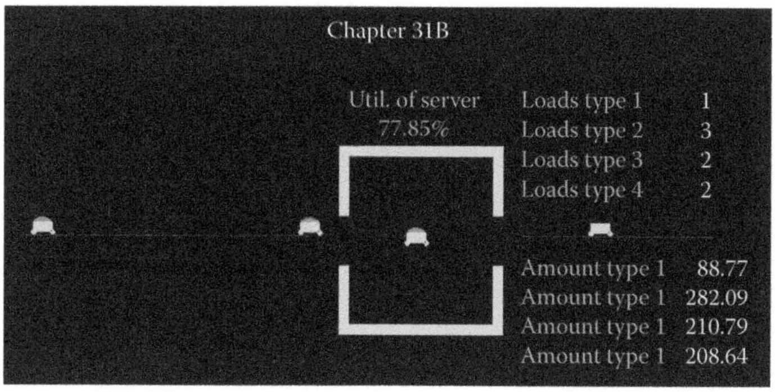

FIGURE 31.4
Screenshot of animation from program CHAP31B.GPS.

Index

Note: Locators "*f*" and "*t*" denote figures and tables in the text

U

V

W

X

Z

Printed and bound by CPI Group (UK) Ltd, Croydon, CR0 4YY
22/10/2024
01777611-0010